Periphyton of Freshwater Ecosystems

Developments in Hydrobiology 17

Series editor
H. J. Dumont

Periphyton of Freshwater Ecosystems

Proceedings of the First International Workshop on Periphyton of Freshwater Ecosystems held in Växjö, Sweden, 14–17 September 1982

Edited by
Robert G. Wetzel

1983 **Dr W. JUNK PUBLISHERS**
A MEMBER OF THE KLUWER ACADEMIC PUBLISHERS GROUP
THE HAGUE / BOSTON / LANCASTER

Distributors

for the United States and Canada: Kluwer Boston, Inc.,
190 Old Derby Street, Hingham, MA 02043, USA
for all other countries: Kluwer Academic Publishers Group,
Distribution Center, P.O. Box 322, 3300 AH Dordrecht,
The Netherlands

Library of Congress Cataloging in Publication Data

International Workshop on Periphyton of Freshwater
Ecosystems (1st : 1982 : Växjö, Sweden)
Periphyton of freshwater ecosystems.

(Developments in hydrobiology ; 17)
1. Periphyton--Congresses. 2. Freshwater ecology--
Congresses. I. Wetzel, Robert G. II. Title.
III. Series.
QH96.8.P43I57 1982 574.5'2632 83-11358
ISBN 90-6193-768-X

ISBN 90-6193-768-X (this volume)
ISBN 90-6193-751-5 (series)

Cover design: Max Velthuijs

PRINTED IN THE NETHERLANDS

Contents

II. PARAMETERS INFLUENCING GROWTH OF PERIPHYTON

III. PRODUCTIVITY AND UTILIZATION OF PERIPHYTON

IV. PERIPHYTON/SUBSTRATA INTERACTIONS

Participants of the First International Workshop on Periphyton of Freshwater Ecosystems held in Växjö, Sweden, 14–17 September 1982.

1. Max L. Bothwell, Canada
2. Lena Tullberg, Sweden
3. Gertrud Cronberg, Sweden
4. Jan Barica, Canada
5. Sven Björk, Sweden
6. Amelie Fritzon, Sweden
7. Robert G. Wetzel, USA
8. P.J. Roos, The Netherlands
9. Ryszard Bohr, Poland
10. K.H. Rogers, South Africa
11. Akira Shimizu, Japan
12. Elina Leskinen, Finland
13. Helga Kots, Austria
14. Alena Sládečková, Czechoslovakia
15. Arno Rosemarin, Canada
16. David M. Gordon, Australia
17. Timo Kairesalo, Finland
18. Christer Müller, Sweden
19. Pauli Snoeijs, Sweden
20. Andrzej Oleksowicz, Poland
21. Catarina Johansson, Sweden
22. Martin T. Auer, USA
23. Pertti Eloranta, Finland
24. Søren Wium-Andersen, Denmark
25. J.F.M. Geelen, The Netherlands
26. Hamish C. Duthie, Canada
27. S.P. Klapwijk, The Netherlands
28. Curt Forsberg, Sweden
29. Jytte Højmark, Denmark
30. Stanford L. Loeb, USA
31. Gunnar St. Jonsson, Denmark
32. Helene Godmaire, Canada
33. Marco A. Rodriguez, Canada
34. Herman J. Gons, The Netherlands
35. Jens Borum, Denmark
36. K.H. Nicholls, Canada
37. Roberto Antonietti, Italy
38. Bente Forberg, Norway
39. Robert J. Naiman, USA
40. Michael Turner, Canada
41. Christian Otto, Sweden
42. Eli-Anne Lindström, Norway
43. Charel Bruwer, South Africa
44. Gordon G.C. Robinson, Canada
45. G. van der Velde, The Netherlands
46. Pavel Punčochář, Czechoslovakia
47. John E. Reuter, USA
48. John T. Meulemans, The Netherlands
49. Kaj Sand-Jensen, Denmark
50. Peter G. Meier, USA
51. Eileen J. Cox, England
52. Edmond J.P. Delbecque, The Netherlands
53. Eugene B. Welch, USA
54. Herman Hillebrand, The Netherlands
55. Herman van Dam, The Netherlands
56. Boyan B. Boyanovsky, Bulgaria
57. Hans H. Riber, Denmark
58. Tor Traaen, Norway
59. Marta Luscinska, Poland

Participants not present when photo was taken:

Keld Andersen, Denmark
Bruno Berzins, Sweden
Ilppo Kettunen, Finland
Leif Kronborg, Sweden
Akira Kurata, Japan

Opening speech

S. BJÖRK

Mayor of the City of Växjö, Vice-Chancellor of Växjö College, Periphytologists:

Filled with enthusiasm from a successfully running SIL Workshop on Hypertrophic Ecosystems, Dr. Jan Barica said to me in a jubilant voice 'The next Växjö workshop ought to be on periphyton'. This was on the 11th of September, 1979, just outside the main entrance of Växjö College. By the 10th of October the suggestion of a periphyton workshop was forwarded to the Secretary General of Societas Internationalis Limnologiae, Professor Robert G. Wetzel. The proposal fell on good ground, and was fertilized by Professor Wetzel's special interest and own research in periphyton as well as his unique global overview concerning limnological activities. The idea rapidly developed into a draft for the program of today's workshop.

When Jan Barica first mentioned a workshop on periphyton, he talked about 'the next Växjö workshop'. There did not seem to be any doubts about the place, and in further planning no alternatives to Växjö were even mentioned. Thus, Mayor of the City of Växjö and Vice-Chancellor of Växjö College, in the international family of limnologists your city with its college was automatically chosen for the meeting. Your biotope is an attractive place to summarize present scientific knowledge and current research and to design the strategy to cover weak sections in the constantly moving front of research. We are most grateful for all the help we received from the City and College of Växjö.

I and several other generations of students who had Ruttner's Grundriss der Limnologie (later rebaptized to Fundamentals of Limnology) as the guidebook for our limnological philosophy, received our first impression of periphyton (Aufwuchs) from the classical drawing of a community on *Myriophyllum* in Lunzer Untersee (Ruttner 1952: p. 158, Fig. 43). For decades this textbook picture was, so to say, a rule for the general understanding of periphyton. Among the earlier drawings, a picture published in 1934 by Meschkat (p. 452, Fig. 2) gives a better idea of the often jungle-like appearance of periphyton communities. It is no wonder that many students have lost their way in these jungles as academic teachers have tried to encourage new recruits to periphyton research. The overwhelming number of taxonomically difficult algal and animal species and the variation in community composition and architecture from centimeter to centimeter with substratum changes pose, together with seasonal variation, thrilling difficulties in periphyton research.

The monumental methodological paper by Dr. Sládečková (1962) became a vertebra in the construction of a firm backbone of periphyton research. For many limnologists the excellent pictures published by Allanson (1973) gave new dimension and insight into connections concerning community architecture and species composition. As Allanson says, his findings 'amply confirm the reality and structural proximity of each periphyte component as required by the Wetzel and Allen model' (Allanson *op. cit.*, p. 539). This model illus-

Wetzel, R.G. (ed.), Periphyton of Freshwater Ecosystems.
 ISBN 90 6193 768 X.

trates interrelations within and through littoral periphyton communities (Wetzel & Allen 1972: p. 340, Fig. 3). One has the impression that the papers by Wetzel, Allen and Allanson acted as an injection for research, occasionally leading to too much enthusiasm and, as Professor Wetzel says in his summary paper for this workshop, to premature results. Let us regard these as evidence of how inspiring the papers by these authors were.

About ten years before the aforementioned milestone papers about intricate, internal interrelations in periphyton communities, Wetzel published the Borax Lake study (1964). This paper treated the total primary productivity in a shallow lake ecosystem and its shares in macrophytes, phytoplankton and *mirabile dictu* in periphyton. As a result, the limnological literature became enriched by facts showing students how important the so often overlooked periphyton could be in some types of lake ecosystems.

For this workshop it is an invaluable advantage that Professor Wetzel is the helmsman who has designed the scientific program. We are deeply indebted to Bob Wetzel for accepting, despite this innumerable duties, responsibility for publication of the workshop volume.

About 65 research workers from 15 countries are attending this first periphyton workshop. Distribution, species composition, and architecture of communities, functional relations within the communities and the role of periphyton in the ecosystems of inland water bodies, are, together with methodological problems, the topics of the workshop program. The workshop volume will certainly be frequently read and cited, though mainly by members of the already established family of research workers.

When planning for future growth of periphytological ideology, so important for true ecosystem-oriented limnology, it is necessary to secure new generations of periphytologists. We must confess that the majority of scientific papers – many in respected ecological journals – do not entice many students to become researchers. In the case of periphyton we know how difficult it is to enlist people. Periphyton research must be made as attractive as possible. It would, therefore, be to the credit of this workshop if somebody or groups of people decided – here in Växjö – to publish attractive and easily digestible articles and textbooks with seductive pictures to inspire young students to work with the fantastic world of periphyton. This world should not remain a hidden treasure; its existence and importance should be made part of the public biological consciousness. This ought to be the fundament for recruitment to make periphyton research a prosperously developing sector of limnology.

On behalf of the local organization committee I wish you all very welcome to Sweden and Växjö and to a successful and inspiring workshop.

References

Allanson, B.R., 1973. The fine structure of the periphyton of Chara sp. and Potamogeton natans from Wytham Pond, Oxford, and its significance to the macrophyte-periphyton metabolic model of R.G. Wetzel and H.L. Allen. Freshwat. Biol. 3:535–542.

Meschkat, A., 1934. Der Bewuchs in den Röhrichten des Plattensees. Arch. Hydrobiol 27:436–517.

Ruttner, F., 1952. Grundriss der Limnologie. 2. Auflage. Walter de Gruyter & Co., Berlin, 232 pp.

Sládečková, A., 1962. Limnological investigation methods for the periphyton ('Aufwuchs') community. Bot. Rev. 28:286–350.

Wetzel, R.G., 1964. A comparative study of the primary productivity of higher aquatic plants, periphyton, and phytoplankton in a large, shallow lake. Int. Revue ges. Hydrobiol. 49:1–61.

Wetzel, R.G. & Allen, H.L., 1972. Functions and interactions of dissolved organic matter and the littoral zone in lake metabolism and eutrophication. In: Kajak, Z. & Hillbricht-Ilkowska, A. (eds.), Productivity problems of freshwaters. PWN Polish Scientific Publishers, Warszawa-Krakow, pp.:333–347.

Author's address:
S. Björk
Institute of Limnology
University of Lund
S-220 03 Lund, Sweden

Opening remarks

ROBERT G. WETZEL

Periphytic communities

The word periphyton is something of a misnomer in that a literal translation is too restrictive. The history of terminology associated with microbial communities attached to substrata is long, circuitous, and is filled with inconsistencies and redundancies. Having spent considerable time evaluating the etymological development of various terms for the attached communities, I conclude that it is perhaps best simply to accept the word periphyton as it is widely understood: a complex community of microbiota (algae, bacteria, fungi, animals, inorganic and organic detritus) that is attached to substrata. The substrata are inorganic or organic, living or dead.

Periphyton as a component of freshwater ecosystems

Those who work with periphytic communities have a deep appreciation for both the complexity of the constituents of these communities, as well as the importance that periphyton can have for overall lake metabolism. Similar to the wetland and other littoral communities, periphyton is one of the most neglected aspects of freshwater investigations.

The importance of periphytic photosynthesis and secondary productivity is often greatly underestimated, since most lakes are shallow and have extensive littoral regions and so little is known from direct measurements. Research on periphyton is still mainly in descriptive phases with relatively few analyses of primary productivity at the community level. A very few recent investigations are demonstrating the very complex structural and intimate metabolic couplings between periphytic algal and bacterial populations, and the substrata on which they are growing, whether living or dead.

I sense that analyses of periphyton are in the doldrums. Investigation of periphytic communities, as we all know, is exceedingly difficult. The extreme heterogeneity that exists, not only within periphyton communities but also among habitats of lakes and streams, makes study very difficult and frustrating. All but the most perserverant workers commonly retreat into the more comfortable, homogeneous pelagial zone.

Two major informational voids exist in contemporary limnology. Foremost is the dearth of analyses of quantitative flux rates of carbon and energy by bacteria of all regions – pelagial, littoral, and profundal. Secondly, the wetlands and interfacing littoral zone of fresh waters are among the most productive of all ecosystems, yet least understood in regard to their regulation of inorganic and organic biogeochemical cycling. Periphyton is an important component of both of these voids and deserves intensified study.

One of the major impediments to our understanding of the functional relationships among components of periphytic communities centers on methodology. We need not only new techniques

Wetzel, R.G. (ed.), Periphyton of Freshwater Ecosystems.
 ISBN 90 6193 768 X.

but application and adaptation of existing methods from other fields, especially physiology and microbiology. We must also perform metabolic analyses at micro and species levels. Furthermore, it is essential that we determine not only what effects the dynamic substrata have on periphytic populations but also what specific effect periphyton communities have on the metabolism of the substratum.

It is my sincere hope that the exchange and results of the forthcoming week will represent a significant step in evaluating where we are in our understanding and, very importantly, what directions research on periphyton should, most profitably, take in the future. The meetings were designed to be a true workshop. Following today's summary lectures on topical reviews of our present status and projections of urgent problems requiring attention, the most important analyses will begin. We suggest dividing into four working groups:

1. Structure and Dynamics of Periphytic Populations and Communities (K. Sand-Jensen and R. Bohr, Chairpersons)
2. Periphyton/Substrata Interrelationships (R.G. Wetzel, Chairperson)
3. Effects and Fate of Periphyton Communities on Freshwater Ecosystems (J. Barica, A. Sládečková, and C. Otto, Chairpersons)
4. New Methodological Approaches (G.C.C. Robinson, Chairperson)

We request that new results from your research be interjected only as appropriate to develop discussion and evaluation of current and future research directions. By the afternoon the groups should be formulating collective opinions about voids in understanding of periphytic communities, the most important research needs, and the most fruitful approaches to be taken in the future. We have asked the chairpersons to summarize these conclusions in recommendations before the collective participants for discussion and modification. The recommendations will be published at the conclusion of the workshop proceedings. Time is short to perform these tasks. We must resist the temptations to simply report on our specific research results, to be published in the proceedings where they can be analyzed in detail.

In this workshop we have an opportunity to stress the importance of periphyton in freshwater ecosystems. Many limnologists and colleagues will read what we have said. We have a responsibility to set the pace and future directions. Let us begin.

Author's address:
Robert G. Wetzel
W.K. Kellogg Biological Station
Michigan State University
Hickory Corners, MI 49060, U.S.A.

CHAPTER 1

Dynamics of periphytic communities

P.J. ROOS

Abstract. The dynamics of periphyton communities are on a long term the result of maturation of the system or eutrophication. Yearly fluctuations are brought about by a number of factors, e.g. availability of nutrients, light, or substratum, by pelagic influence, or grazing. For a better understanding of these population dynamics, standard methods in the separation and culture of periphytic organisms are needed.

Introduction

In the following discussion, the word periphyton is used in the sense of euperiphyton (Sládečková & Sládeček 1977); that is, the attached framework of the community. This framework consists of organisms attached to the substratum by rhizoids, stalks, tubes or other luting mechanisms.

The productive base for the community can be found within the periphyton in a palette of red, gold, green, blue-green, and brown algae. The autotrophic base of the food-chain can also be found in the surrounding water, in which case, these algae are not necessarily present in the periphyton. The periphytic community may then consist of pure stands of e.g. peritrichous ciliates, hydrozoans or bryozoans. This macrofauna may also bear its own periphyton, consisting mostly of diatoms or specific ciliates. In all these cases, there is a structural coherence or an architectural continuity (Roos *et al* 1981), in which specific relations may occur.

Most algae are overgrown by diatoms; some groups however, such as the Zygnemales, are devoid of them. Oedogoniales bear many diatoms, except for the young expanding cells. *Oedogonium,* with its intercalar growth, may develop in this way an alternation of densely covered and nude regions. A succession, or merely an aging of algae, may therefore implicate a parallel succession in the secondary periphyton. This succession is one of the sources of periphyton dynamics and structural changes.

Most macrophytes bear periphyton, but some, or parts of some, more than others. Finer dissected parts of the plant bear more than coarser parts. The bulrush (*Scirpus*) is mostly devoid of periphyton, except for the older stems. Reed (*Phragmites*) and cattail (*Typha*) are always overgrown, the older stems often more than younger ones. This aging, with both algae and macrophytes, implicates a steady change, at least in biomass of the periphyton, but at first sight not necessarily in the structure or architecture of the attached community. One may observe, however, that when the algal epigrowth becomes thicker and more densely packed, the longer stalked species become more abundant. This structure is particularly clear in the case of *Cymbella* and *Gomphonema* during the winter and spring maxima. One species may develop longer stalks, or may be replaced by other species with longer stalks, perhaps a consequence of crowding. Probably the same situation is true for

Wetzel, R.G. (ed.), Periphyton of Freshwater Ecosystems.
 ISBN 90 6193 768 X.

other gelatinous secretions like *Cymbella* tubes or the matrix of bacterial masses. In this way, dense algal mats are progressively formed.

Assuming that crowding results in competition for light, nutrients, or space, then in all three cases longer stalked organisms or those filling their territory with mucous or gelatinous masses benefit more than others, not to mention species with specific, more physiological, adaptive advantages. Growth implicates succession.

Succession may lead to a climax situation, in a plant-sociological sense. In northern Poland two periphytic associations can be distinguished, which reach their climax situation every year (Bohr 1981). This situation is characterized by a number of stable species, and does not coincide with maximum development of the algal assemblage. The spring peak of the association consists of species typical for the association together with other algae, which disappear during the growing season. A calculation of the diversity, based on the stable species only, shows a successive increase from spring to autumn. Two consistent features become clear from this example. First, the maximum biomass is found in the early spring; second, there is a seasonal shift in species composition. Both phenomena are widespread and a number of these conditions are described in the literature.

Long-term changes are the result of maturation of the system, resulting in a more or less stable climax situation, in which a yearly succession may occur or may be caused by pollution. Kann (1982) reports an increase of chlorophytes and a reduction of certain cyanophytes as a result of increasing eutrophication in some Austrian lakes over a period of 40 years.

Factors influencing seasonal changes

Annual fluctuations are brought about by a number of possible causes, such as availability of food, nutrients, light or substratum, inhibition by planktonic algae, infection by fungi, and grazing by invertebrate animals, each of which will be discussed briefly.

Food

Availability of food plays a role only for periphytic invertebrates. Hatano & Watanabe (1981) showed a peak for the peritrich ciliate *Vorticella* in winter. This maximum was strongly correlated with leaf litter decomposition, which brought about reduction of dissolved oxygen and high bacterial density. Roos & Trueba (1977) found that the algae-feeding *Vorticella campanula* dominated in spring and summer, while the other bacterivorous vorticellids dominated in autumn.

Nutrients

Nutrient-dependency finds its expression in a number of ways. The onset of winter-spring maximum takes place just after lake turnover in November (Brown & Austin 1973), which suggests a nutrient limitation in the preceding period. For periphyton on cellulose acetate substrates, Hooper-Reid & Robinson (1978) showed an increasing standing crop and productivity in September and October, after a low summer growth. In July and August evidence for silica limitation was evident. During the increase of the standing crop, coinciding with higher nutrient levels, they noted the absence of deficiency symptoms. Jørgensen (1957) noted that in a Danish lake the maximum of periphytic algae appeared in a period of low silica concentrations in the water, and demonstrated that periphytic diatoms use silica from decaying reed-stems. Shortage of nutrients may lead to competitive exclusion. The planktonic diatom *Asterionella* consumes a great amount of phosphate (Van Donk & Ringelberg, in prep.), which may account for the suppression of the winter maximum in the periphyton when there is a bloom of *Asterionella* in early spring. There is some evidence for this interaction in Lake Maarsseveen (Netherlands). There *Diatoma, Fragilaria,* and *Tabellaria* are dominant diatoms in winter, while *Achnanthes, Gomphonema* and *Synedra* occur in about the same numbers during the whole year. The ratio (winter diatoms):(whole year diatoms) varied from 1.6 to 0.0 during spring 1981 and remained 0.2 during the spring of 1982. During the suppression of the peak of the winter species in early spring 1982, the amount of *Asterionella formosa* in the lake was extraordinarily high.

Light

The influence of the light intensity on periphyton is poorly understood. Only a few P:I relationships have been established under laboratory conditions (McIntyre & Phinney 1965). So there is no reliable base for statements about light inhibition or limitation based on percentage attenuation of light intensity. At the bottom in a dense reed stand in a mid-European fish pond, Ondok (1978) measured a light intensity of less than 1% of the incoming light. It would be expected that such a low intensity may have an impact on periphyton production. Light limitation is indicated by McIntyre (1973) in a model study which shows that low periphytic biomasses in small streams in western Oregon are partly the result of the effect of light reduction by a dense canopy of terrestrial vegetation. There is also the possibility that high light intensities in spring, which are generally at their highest level in May, have an inhibiting effect because of the lack of a canopy-cover in the littoral region.

Substrata

Extension of the available substratum brings about an increase in periphyton biomass. Growth of oedogonial algae is commonly accompanied by an increase in number of diatoms like *Achnanthes* or *Synedra*. In the same way, the occurrence of large *Cymbella* species with stalks or tubes favours the settlement of the same diatoms. A shift in composition of the chlorophytes can drastically change the diatom flora when algae like *Oedogonium* or *Bulbochaete* are replaced by non-colonised algae like *Zygnema* or *Spirogyra*. An increase of the Chlorophyta during summer, as described by Casterlin & Reynolds (1977), is responsible for the constancy in number of *Achnanthes*, *Gomphonema*, and *Synedra* during the year, and the presence of *Anomoeoneis* during summer (Roos *et al* 1981). Finer dissection of leaves also brings about higher biomass of epiphytes (Cattaneo & Kalff 1980). The same relationship occurs among sites of epidermal damage of macrophytes where underlying structures become available for colonization, as is shown in Fig. 1. Obviously this situation will give difficulties in counting or separating the diatoms. The presence of a stalked specimen of the otherwise planktonic diatom *Fragilaria crotonensis* is also noteworthy.

Pelagic influence

Jørgensen (1957) also noted that peaks in planktonic algae and epiphytic algae never occurred at the same time, and showed the inhibiting effect of *Asterionella* on the growth of the epiphytic diatom *Nitzschia*. The bloom of the planktonic algae thus ends the peak of the periphytic ones. We have already seen that *Asterionella* can compete for necessary nutrients, and that this competition is the primary competitive interaction.

The planktonic diatom *Asterionella formosa* may be a regulating factor in the winter maximum of periphytic algae. *Asterionella*, in turn, is sometimes, depending on winter temperature, controlled by fungal infection (Van Donk & Ringelberg, in prep.). Under such conditions, a competitive sequence occurs, and results from low winter temperatures and fungal infection, planktonic diatoms, and then dominance in diatoms of periphyton. We may find a similar infection in the periphyton as well. Scanning electron microscope photographs taken at the end of February revealed a heavy fungal infection exclusively on *Diatoma* in a dense matrix of this and other sessile diatoms (Fig. 2). In this way, specific infection in the plankton may influence the periphytic community as a whole, and infection within the periphyton may change species composition. Although this hypothesis is attractive, it is of course rather speculative.

Grazing

More substantial are explanations for both the reduction of the winter-spring maximum and the changes in species composition caused by grazing. McIntyre (1973) in his model for small streams in

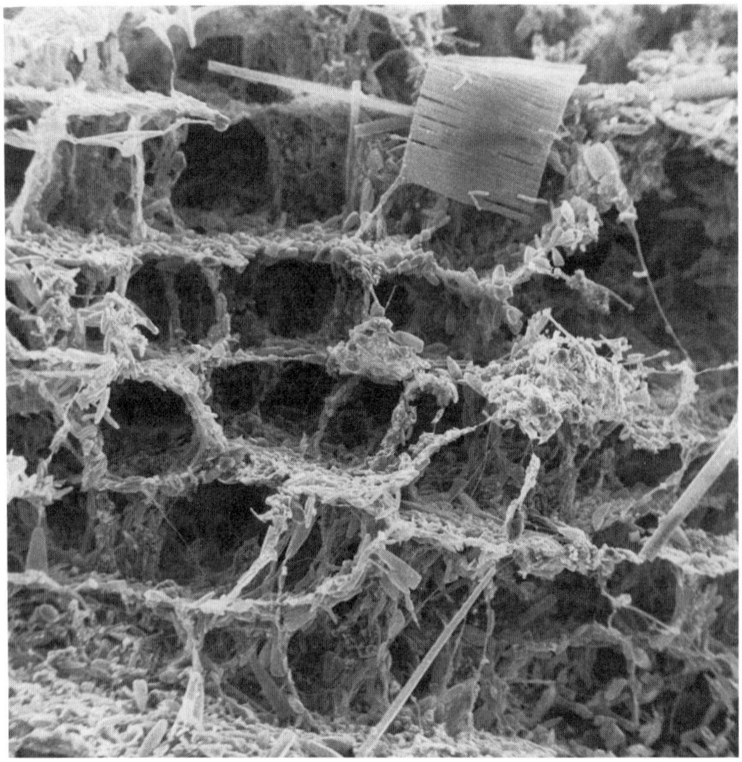

Fig. 1. Dense growth on subepidermal tissue of *Phragmites* after removal of the epidermis, predominantly consisting of *Achnanthes* and *Gomphonema*. *Cymbella*, *Synedra* and a stalked *Fragilaria crotonensis* are also visible.

western Oregon brings in the grazing activities of aquatic animals as a factor. For Lake Memphremagog, Cattaneo & Kalff (1980) suggest that the summer minimum is brought about by grazing. More specific grazers are mentioned by other investigators. Douglas (1958) found in the field that the caddis larva *Agapetus fuscipes* exhibited an inverse relation to the populations of the diatom *Achnanthes*. Mason & Bryant (1975) found a decline in standing crop that was largely due to the presence of chironomid larvae. Doremus & Harman (1977) found under laboratory conditions that the population density of two snails, *Physa heterostropha* and *Prometus exacuous*, was inversely correlated with periphyton standing crop. Influence of grazing on species composition was shown by Higashi *et al* (1981). The flat diatom *Cocconeis*, abundantly present in the lowest layer of the periphyton, remained uneaten by the grazing snail *Sinotaia historica*. Also Sumner & McIntyre (1982) showed that grazing by the snail *Juga plicifera* was associated with a reduction in the relative abundance of mostly the large, overstory species, which are most likely to be consumed or dislodged.

Conclusion

Study of dynamics and structure of periphytic communities offers many problems. Tackling these problems will fail until we are able to handle each problem without interference from other influencing factors. It is not easy to discriminate between influences of light and temperature, inhibition, and competitive exclusion. One controlling factor may have several consequences. Autumn turnover enriches the littoral as well as the pelagial with nutrients, but also results in a reduction of the available light.

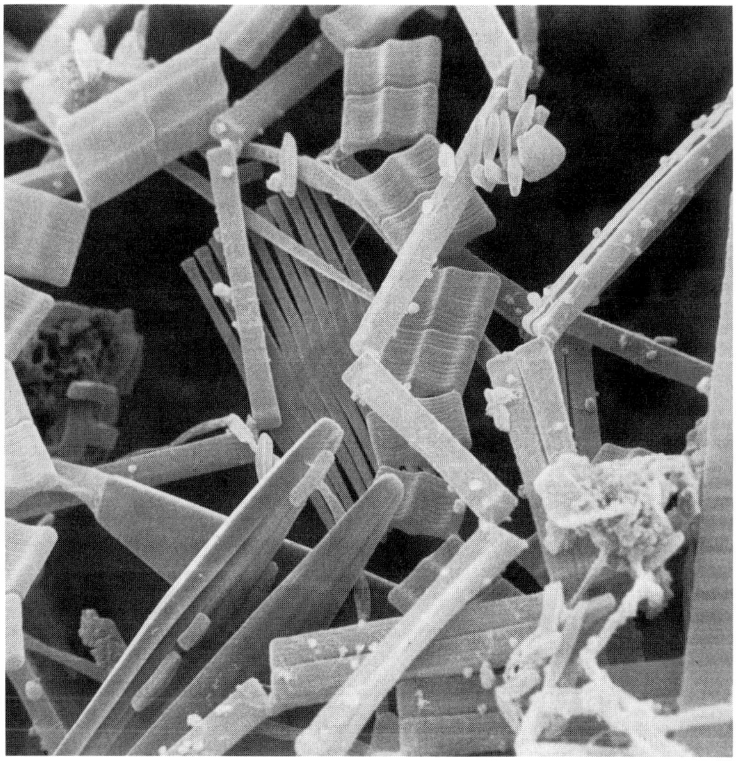

Fig. 2. Heavy fungal infection, exclusively on *Diatoma*. No sporangia were present on *Achnanthes*, *Cymbella*, *Gomphonema*, *Fragilaria capucina*, and *F. crotonensis*.

For quantification of the changes in the community, it is necessary to have good separation methods, by which it is possible to make a reliable comparison of algae on different substrates. Unfortunately, no generally acceptable method exists. Brushing, vibrating, sonification, or destruction of the substratum in the case of diatom studies or combinations of these methods have to be adapted to the kind of periphyton and/or substratum under study. Often the result is disappointing upon microscopic inspection, and comparable methods do not exist.

For a good understanding of the underlying processes it will be necessary to know more about the autecology of the periphytic algae and sessile animals, as well as the grazing animals. In order to make the truly periphytic organisms available for necessary laboratory studies, we must develop reliable methods for culturing of these sedentary organisms in ways that simulate field situations. Only then may we establish the relationships between light and productivity, study the roles of inhibition, stimulation, or competition, and evaluate in population studies of grazing animals their effects on periphytic communities.

References

Bohr, R., 1981. Do climax algal systems exist? Verh. int. Ver. Limnol. 21:1481–1483.

Brown, S.D. & Austin, A.P., 1973. Spatial and temporal variation in periphyton and physico-chemical conditions in the littoral of a lake. Arch. Hydrobiol. 71:183–232.

Casterlin, M.E. & Reynolds, W.W., 1977. Seasonal algal succession and cultural eutrophication in a North temperate lake. Hydrobiologia 54:99–108.

Cattaneo, A. & Kalff, J., 1980. The relative contribution of aquatic macrophytes and their epiphytes to the production of macrophyte beds. Limnol. Oceanogr. 25:280–289.

Doremus, C.M. & Harman, W.N., 1977. The effects of grazing by Physid and Planorbid freshwater snails on periphyton. Nautilus 91:92–96.
Douglas, B., 1958. The ecology of the attached diatoms and other algae in a small stony stream. J. Ecol. 46:295–322.
Hatano, H. & Watanabe, Y., 1981. Seasonal change of protozoa in a small pond with leaf litter supply. Hydrobiologia 8: 161–174.
Higashi, M., Miura, T., Tanimizu, K. & Iwasa, Y., 1981. Effect of the feeding activity of snails on the biomass and productivity of an algal community attached to a reed stem. Verh. int. Ver. Limnol. 21:590–595.
Hooper-Reid, N.M. & Robinson, G.G.C., 1978. Seasonal dynamics of epiphytic algal growth in a marsh pond: composition, metabolism, and nutrient availability. Can. J. Bot. 56:2441–2448.
Jørgensen, E.G., 1957. Diatom periodicity and silicon assimilation. Dansk bot. Ark. 18:1–54.
Kann, E., 1982. Qualitative Veranderungen der litoralen Algenbioconose österreichischer Seen (Lunzer Untersee, Traunsee, Attersee) im Laufe der lezten Jahrzehnte. Arch. Hydrobiol., Suppl. 62:440–490.
Mason, C.F. & Bryant, R.J., 1975. Periphyton production and grazing by chironomids in Alderfen Broad, Norfolk. Freshwat. Biol. 5:271–277.
McIntyre, C.D., 1973. Periphyton dynamics in laboratory streams: a simulation model and its implications. Ecol.Monogr. 43:399–420.
McIntyre, C.D. & Phinney, H.K., 1965. Laboratory studies of periphytic production and community metabolism in lotic environments. Ecol. Monogr. 35:237–258.
Ondok, J.P., 1978. Radiation climate in fishpond littoral plant communities. In: Dykyjová, D. & Květ, J. (eds.), Pond littoral ecosystems. Springer-Verlag, N.Y. & Berl.:113–125.
Roos, P.J., Post, A.F. & Revier, J.M., 1981. Dynamics and architecture of reed periphyton. Verh. int. Ver. Limnol. 21:948–953.
Roos, P.J. & Trueba, F.J., 1977. Epiphytic protozoans on reed rootlets from Dutch waters. Hydrobiologia 54:241–245.
Sládečková, A. & Sládeček, V., 1977. Periphyton as indicator of water quality 2. Pseudoperiphyton. Arch. Hydrobiol. Beih. 9:171–191.
Sumner, W.T. & McIntyre, C.D., 1982. Grazer-periphyton interactions in laboratory streams. Arch. Hydrobiol. 93:135–157.

Author's address:
P.J. Roos
Department of Aquatic Ecology
University of Amsterdam
Kruislaan 320
1098 SM Amsterdam, The Netherlands

CHAPTER 2

Seasonal changes in the taxonomic composition of epiphytic algal communities in Lake Wingra, Wisconsin, U.S.A.

R.C. JONES and K.B. MAYER

Introduction

As mixed assemblages of organisms, epiphytic algal communities exhibit seasonal patterns in species abundance reflecting the biotic and abiotic factors impinging on the community. Certain groups of algae are characteristic of certain seasons of the year. Numerous studies have suggested a relationship between the species composition of algal communities and nutrients (Lund 1950; Schelske & Stoermer 1971; Moss 1976), temperature (Klarer & Hickman 1975), irradiance (Lund 1949; Pechlander 1970), or grazing (Porter 1973). Depth variations in species composition may reflect differential adaptation to light.

The species composition of the algal community can, in turn, affect the rates of metabolic processes, such as primary productivity. Rates of photosynthesis at low light and at light saturation have been shown to vary widely between species (Parsons *et al* 1977), implying that differences in algal productivity could result at least partially from differences in species composition.

Seasonal patterns in species composition of epiphytic algal communities have been followed by a number of workers. Moss (1976) compared the seasonal succession of epiphytic diatoms in experimentally enriched and control ponds. Klarer & Hickman (1975) examined the relationship of thermal enrichment to epiphytic algal succession. Numerous other studies have followed seasonal succession on both natural (Cattaneo & Kalff 1978) and artificial (Brown & Austin 1973; Stockner & Armstrong 1971) substrata.

In the present work, the species composition of algal communities epiphytic on the submersed macrophyte *Myriophyllum spicatum* L. were examined seasonally and at different depths. In conjunction with concurrent photosynthesis studies the observed patterns of relative abundance suggested the following hypotheses:

1) The epiphytic algal community follows an orderly progression through the seasons.
2) Photosynthetic rate parameters are related to the seasonal succession of algal species composition.
3) Depth variations exist in algal communities possibly resulting from different adaptations to light intensity.

Methods

This study was conducted in Lake Wingra, Wisconsin, U.S.A., a small, shallow lake with a relatively large littoral zone. The lake is about 2 km long and 1 km wide with a mean depth of 2.4 m and a maximum depth of 4 m. The submersed freshwater macrophyte *Myriophyllum spicatum* L. occurs in nearly monospecific stands in the west end of the lake. Studies of the Lake Wingra macrophyte communities have been recently summarized by Adams & Prentki (1982).

Wetzel, R.G. (ed.), Periphyton of Freshwater Ecosystems.
 ISBN 90 6193 768 X.

Twenty-centimeter shoot sections of *M. spicatum* were collected from a standard depth at a station in Lake Wingra's west end. Water depth at this station was 1.6 m. Sampling depths were 60 and 125 cm in July 1978; 50 cm in June, August, and September of 1979; 110 cm in May and October of 1979; and 40 and 105 cm in July of 1979. Shoot sections were always at least 15 cm from growing tips to avoid immediate colonization effects. The shoot sections were incubated with ^{14}C for productivity determination and removed by a shaking technique as reported by Jones & Adams (1982). Aliquots of the resulting epiphyte suspensions were preserved with Lugol's iodine for later taxonomic analysis.

All six samples from both upper (depth – 0.6 m) and lower (depth – 1.24 m) stem sections were utilized in the analysis of 1978 depth variations. Since diatoms constituted the dominant portion of the algal community, only diatoms were counted in 1978. A known aliquot of each sample was cleaned by oxidation with hydrogen peroxide and potassium dichromate. After digestion cells were rinsed with distilled water several times in 600-ml beakers. Overnight settling of diatom frustules was allowed between rinses. After 4–5 rinses, the concentrated frustule suspension was stored in glass jars. To mount diatoms, twenty drops of water were placed on a cover slip together with one drop of Tween surfactant solution (a drop Tween per 40 ml water) and 0.4 ml of the suspension of cleaned diatoms was added. The coverslip was dried on a hot plate and then mounted in Hyrax. This technique resulted in a relatively uniform distribution of frustules on the slide. The algae were counted by transects until 300 frustules had been reached.

Samples from 1979 were counted on an inverted microscope to allow enumeration of all algae. Two samples were chosen randomly from each set of 12 samples available from the May, June, August, September, and October photosynthesis-light curves. For July, two samples were randomly selected from each of the two depths. Samples were mixed well and 1.0 ml of the epiphyte suspension was added to the 10-ml counting chamber. Then 9 ml of water were added and the glass cover was placed on the chamber. The algae were allowed to settle overnight and were counted the following day. Random fields were counted until 300 identified cells were enumerated. Two replicate slides were counted for each of the two samples. Some algae were evaluated as area (*Microcystis*) or length (*Fragilaria, Melosira*) and later converted to cell number.

Trends in species composition were examined using the relative abundances of each species on each sample date. Relative abundance of species A in a sample was calculated as the number of cells of species A divided by total cells of all species in that sample. In addition, principal components analysis (PCA) was used to integrate information on all species into a minimum number of dimensions. The computer program CANCOV written by R.R. Kowal was used for PCA. Since proportions were used as the multivariate observations, an arcsine square root transformation was required to approximate normality (Sokal & Rohlf 1969). PCAs conducted with and without this transformation demonstrated their value in producing more readily interpretable results.

Results

The major genera found and symbols used for each in the figures (Table 1) showed 19 dominant diatom genera present in the 1978 samples. In terms of relative abundance (as cell numbers), *Achnanthes* was the dominant diatom genus in both upper and lower sections (Fig. 1). Significantly greater relative abundance of *Achnanthes* and *Cocconeis* were found on the upper sections while *Navicula* and an unidentified araphate diatom (referred to as Unknown ‡2) were significantly more abundant on lower sections (paired-t test, $p<0.05$).

Principal components analysis (PCA) allows graphical depiction in a few dimensions of most of the variation in a series of multivariate observations. It provides a convenient summary for relationships among algal communities. Using PCA, lower sections were shown to be distinctly different than upper sections in July 1978 (Fig. 2). Little or no relationship was found between communities on the same shoot at different depths.

Table 1. Symbols for algal genera used in figures.

Symbol	Genus
ACH	*Achnanthes*
AMP	*Amphora*
ANA	*Anabaena*
ANO	*Anomoeoneis*
COC	*Cocconeis*
CYM	*Cymbella*
DES	*Desmidium*
FRA	*Fragilaria*
GOM	*Gomphonema*
LYN	*Lyngbya*
MEL	*Melosira*
MER	*Merismopedia*
MIC	*Microcystis*
NAV	*Navicula*
O	Others
SCE	*Scenedesmus*
STI	*Stigeoclonium*
SYN	*Synedra*
TRI	*Tribonema*
UK2	Unidentified araphate diatom

By examining the eigenvectors of each component axis, one can determine which species are associated with that axis. Axis 1, the first principal component, is associated with the contrast between *Navicula* and Unknown ‡2, on the one hand, and *Achnanthes* on the other. That is, most of the variation along Axis 1 was associated with differences in the relative abundance of these genera. In particular, *Achnanthes* was more abundant on upper sections while *Navicula* and Unknown ‡2 were more abundant on lower sections, confirming the results of the t-tests. Axis 1 explained only 21% of the total variation, but appeared to capture much of the between-depth variation. Axis 2, which explained 17% of the total variation, emphasized variation between samples at a given depth.

The relative abundances of algal genera for each monthly sample in 1979 are shown in Fig. 3. *Fragilaria* was the dominant alga in May and June, while *Achnanthes, Microcystis, Cocconeis,* and *Fragilaria* co-dominated the epiphytic algal flora of both upper and lower stem sections in July (Figs. 3 & 4). *Achnanthes* was more abundant on upper than lower stem sections, as in 1978. *Microcystis* was the

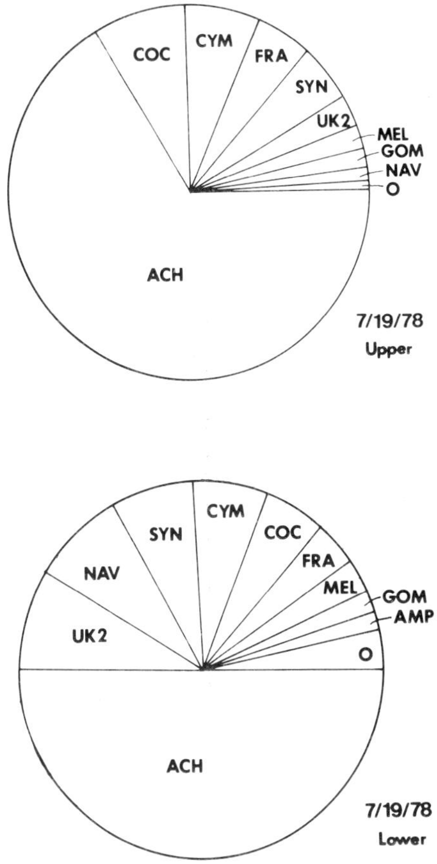

Fig. 1. Relative abundance of epihytic diatom species, July 1978.

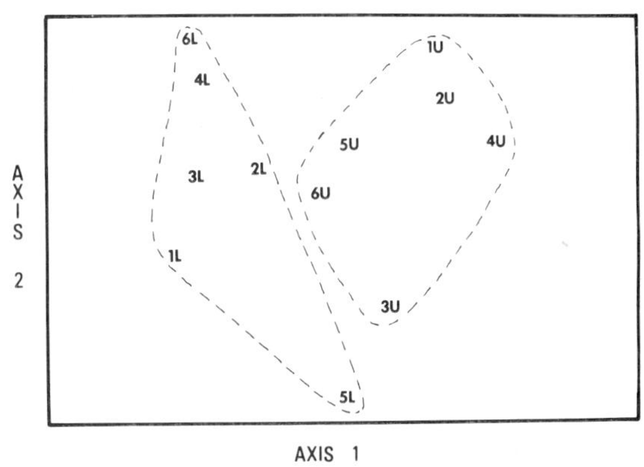

Fig. 2. Principal components analysis of epiphytic diatom communities, July 1978. U denotes communities from upper portion of the water column; L, those from lower shoots. Sections having the same number originate from the same shoot.

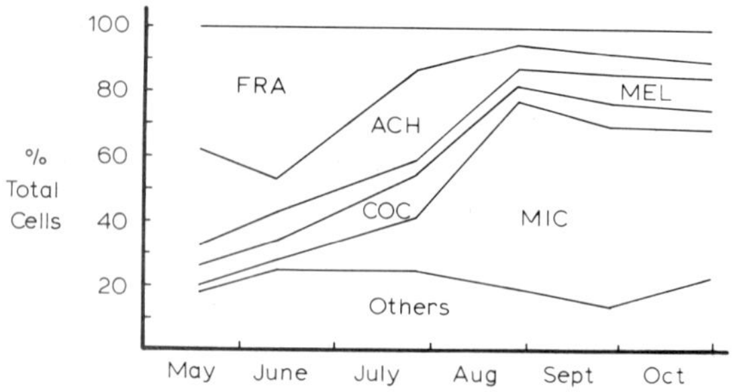

Fig. 3. Relative abundance of epiphytic algal species, 1979.

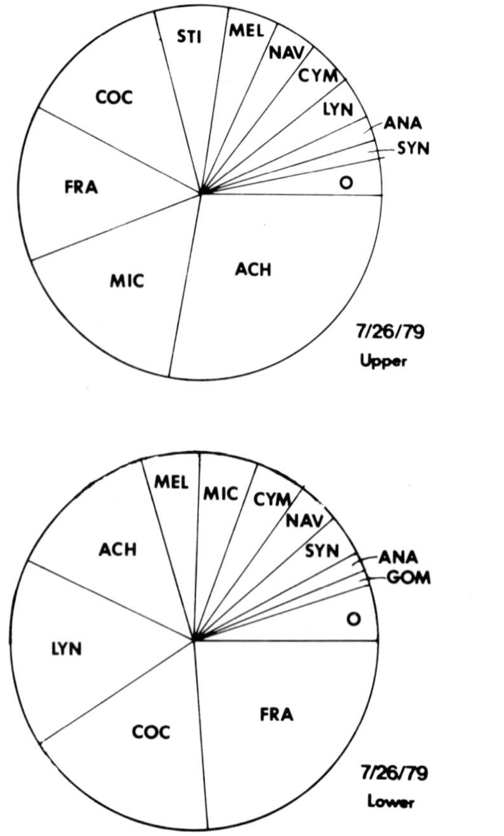

Fig. 4. Relative abundance of epiphytic algal species, July 1979.

overwhelming dominant during August, September, and October.

Principal component analysis (Fig. 5) clearly demonstrated the seasonal progression of epiphytic algal communities through the growing season. Each month is represented by four points: two counts of each of the two sections. The first principal compenent accounted for 25.5% of the total variance; the second component, 19.3%; and the third component, 12.3%. The difference between replicate counts of the same sample was often as large as the difference between samples. The communities of upper and lower stem sections in July were overlapping when only Axes 1 and 2 were plotted, but were well separated on the third principal component.

The first principal component was a contrast between *Fragilaria* and *Achnanthes*, the spring and early summer dominants, and *Microcystis*, the late summer and fall dominant. *Epithemia,* a relatively rare diatom present only in late summer and fall, was also associated with the first axis. Thus, the first principal component was a seasonal contrast: spring and early summer vs. late summer and fall. Axes 2 and 3, based on more subtle contrasts, provided distinct separation of all months except September and October. Algal communities in these two months were so closely related that samples overlapped until the fourth principle component was examined.

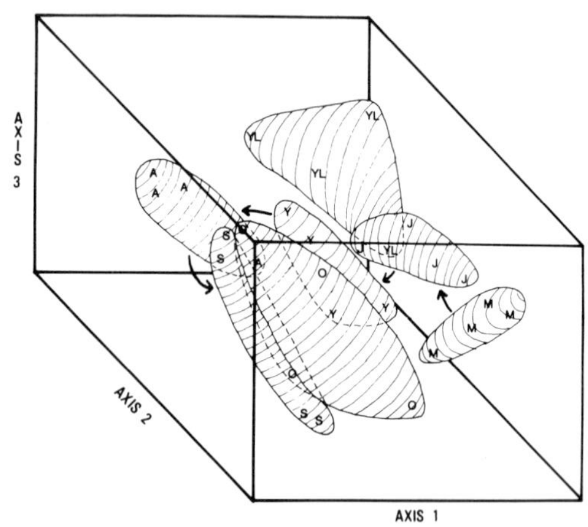

Fig. 5. Principal Components Analysis of epiphytic algal communities, 1979.

M = May
J = June
Y = July, upper
YL = July, lower
A = August
S = September
O = October

Discussion and conclusions

The epiphytic algal community followed an orderly seasonal progression. Except for September and October, the variation in species composition between monthly samples was substantially greater than between subsamples of the monthly sample. Distinct depth variations in algal community structure also existed during both 1978 and 1979, although the patterns were somewhat different.

Achnanthes was a dominant mid-summer genus during both 1978 and 1979. This period corresponded with low epiphyte density in Lake Wingra. Other studies (Klarer & Hickman 1975; Cattaneo & Kalff 1978) have shown *Achnanthes* to dominate epiphyte communities during the mid-summer minimum in other lakes. The spring maximum in *Fragilaria* observed in 1979 was also observed in 1978 (Prunty 1980) and has been reported by Klarer & Hickman (1975). The late summer and early fall dominance of *Microcystis* in the epiphyte community corresponds with the presence of massive numbers of this genus in Lake Wingra's phytoplankton.

Obvious differences in species composition with depth were found in both 1978 and 1979. In both years *Achnanthes*, a dominant, was more abundant on upper than lower shoot sections. Possible reasons for this depth variation include differing light regimes and differing colonization times (lower shoots being older).

Comparison of species composition to rates of photosynthesis at low light and light saturation must be made with care. Relative abundance as used here is based on cell number, not biomass. Small genera such as *Achnanthes* and *Microcystis* are over-represented. Quantitative conversion of cell numbers to cell volume was not possible due to the wide size ranges observed for some genera. Nonetheless, a few comparisons are appropriate. During August, September, and October of 1979 the relative abundances of algae in the epiphyte community were relatively constant with *Microcystis* as the overwhelming dominant. However, the photosynthetic parameters, α and P^B_{max}, varied widely (Jones & Adams 1982). Apparently, the trends observed in α and P^B_{max} over this period were not related to changes in species composition, but perhaps to changes in photosynthetic characteristics of species already present due to variations in environmental factors. In particular, temperature decreased markedly over this period.

Another comparison suggests that compositional changes can be important in explaining variations in photosynthetic parameters. P^B_{max} increased and α decreased between the May and June 1979 experiments (Jones & Adams 1982). At the time *Achnanthes* increased appreciably in relative abundance while *Fragilaria* decreased. Differences in temperature appear insufficient to explain the observed differences in photosynthetic parameters. Thus, changes in species composition may have some role in altering P^B_{max} and α for the epiphyte community in this case.

Acknowledgements

I acknowledge R.R. Kowal for his assistance in the execution and interpretation of principal component analysis, and M.S. Adams for his encouragement. Support for this work was provided by the National Science Foundation through grants DEB-19777, ENV T1-10188, and DEB 78 10640.

References

Adams, M.S. & Pretki, R.T., 1982. Biology, metabolism and functions of littoral submersed weedbeds of Lake Wingra, Wisconsin, USA: a summary and review. Arch. Hydrobiol. Suppl. 62:333–409.

Brown, S.D. & Austin, A.P., 1973. Spatial and temporal variation in periphyton and physicochemical conditions in the littoral of a lake. Arch. Hydrobiol. 71:183–232.

Cattaneo, A. & Kalff J., 1979. Primary production of algae growing on natural and artificial macrophytes in Lake Memphremagog (Que. & Vt.). Hydrobiologia 60:135–144.

Jones, R.C. & Adams, M.S., 1982. Seasonal variations in photosynthetic response of algae epiphytic on Myriophyllum spicatum L. Aquat. Bot. 13:317–330.

Klarer, D.M. & Hickman, M., 1975. The effect of thermal effluent upon the standing crop of an epiphytic algal community. Int. Revue ges. Hydrobiol. 60:17–62.

Lund, J.W.G., 1949. Studies on Asterionella formosa Hass. 1. The origin and nature of cells producing seasonal maxima. J. Ecol. 37:389–419.

Lund, J.W.G., 1950. Studies on Asterionella formosa Hass.2. Nutrient depletion and the spring maximum. J. Ecol. 38:1–35.
Moss, B., 1976. The effects of fertilization and fish on community structure and biomass of aquatic macrophytes and epiphytic algal populations: an ecosystem experiment. J. Ecol. 64:313–342.
Parsons, T.R., M. Takahashi & Hargrave, B., 1977. Biological oceanographic processes. Pergammon Press, N.Y., 332 pp.
Pechlaner, R., 1970. The phytoplankton spring outburst and its conditions in Lake Erken (Sweden). Limnol. Oceanogr. 15: 113–130.
Porter, K.G., 1973. Selective grazing and differential digestion of algae by zooplankton. Nature 244:179–180.
Prunty, J.A., 1980. Structural dynamics of a diatom community epiphytic on Myriophyllum spicatum. M.S. Thesis, Univ. Wis. Madison.
Schelske, C.L. & Stoermer, E.F., 1971. Eutrophication, silica depletion, and predicted changes in algal quality in Lake Michigan. Science 173:423–424.
Sokal, R.R. & Rohlf, F.J., 1969. Biometry. Freeman, San Francisco, 776 pp.
Stockner, J.G. & Armstrong, F.A.J., 1971. Periphyton of the Experimental Lakes Area, northwestern Ontario. J. Fish. Res. Bd Can. 28:215–229.

Authors' address:
R.C. Jones
Department of Biology
George Mason University
Fairfax, VA 22030, U.S.A.
K.B. Mayer
Department of Botany
University of Wisconsin-Madison
Madison, WI 53706, U.S.A.

CHAPTER 3

A comparison of the periphytic communities in two sections of the Danube River

J.N. NOSEK and N. OERTEL

Abstract. The macroinvertebrates of periphytic communities were studied in several sections of the River Danube (from 1848 to 1634 river km). Samples were collected from floating landing stages.

Periphyton mainly consisted of *Cladophora glomerata* (L). Kütz. The development of the *Cladophora* web depends on the effects of shipping (e.g. frequency of landings, cleaning and re-painting of the landing stages). The number of species and individuals of macroinvertebrates were greater in places where the flow velocity was relatively slower (e.g. river sections with low gradient, banks farther from the stream center, sides of landing stages near to bank).

The periphytic communities consisted of two groups: 1) Species found permanently both in habitat and time, which constituted the greater part of the total number of individuals. 2) The 'coloring' species found without pattern in a small percentage of the total number. On the basis of the latter group, it may be possible to delineate characteristics between river habitats and seasonal variations.

Introduction

Among the communities of running water, the periphyton forming on various substrata play an important role. In streams their habitat is mainly littoral. River control structures and revetments established for navigation and water regulation have greatly increased the habitat suitable for periphyton colonization, particularly in such cases where such structures resemble natural substrata in composition and texture (Weber 1967; Dvihally 1977).

In addition to phytoplankton, periphyton phytomass also contributes to river primary production, and in some cases may even attain greater importance than phytoplankton (Bayly & Williams 1973; Dvihally 1977). In running waters, the relationships between the periphyton of the littoral region and plankton, nekton, and benthos can have much greater significance than in certain lakes. The littoral communities provide a habitat for feeding, reproduction, as well as refuges under unfavourable circumstances (Bothár 1979).

Studies performed thus far on periphyton communities in the Danube River were principally of faunistic or floristic nature, limited to individual groups of organisms (Andrássy 1960; Berczik 1965, 1966b, 1969; Bothár 1966; Soós 1967; Richnovszky 1970; Unger 1916; Boros 1961; Szemes 1961, 1967; Tamás 1966).

Aims of the study

The composition of animal communities in the periphyton was investigated qualitatively and quantitatively in the section of the Danube River between Rajka and Budapest. The compositional changes among seasons and years and space (vari-

Wetzel, R.G. (ed.), Periphyton of Freshwater Ecosystems.
 ISBN 90 6193 768 X.

ous sections of the Danube, right and left shore, both sides of the landing-stages) were analyzed. The main objective of these studies was to obtain a general survey, since often harmful anthropogenic effects can be abated or at least reduced only with the thorough knowledge of changes from the original or nearly original conditions (Fekete *et al* 1979). In this paper, we have omitted detailed and space-demanding faunistic data, and present the general conclusions of our investigations.

Materials and methods

The Hungarian section of the Danube consists of two hydrodynamically different parts (Berczik 1971). The first section is between Rajka and Gönyü (from 1848 to 1794 river km). Here the average slope of the streambed is 0.35%, with a flow velocity of 2.0–3.2 m/s^{-1} depending on the water level. In the second part (between Gönyü and Mohács, from 1794 to 1443 river km) the slope is 0.07%, with a flow velocity of 1.0–2.5 m/s^{-1}. It should also be mentioned here, that the whole Hungarian section of the Danube River has been strongly regulated with revetments, rockfill dams, spurs, etc. These hydraulic structures can affect the flow-velocity in different ways, depending on the water level, and alter the position of the stream line (Berczik 1969).

During 1975 and 1976, samples were taken from the periphyton of the landing stations (Fig. 1). These were (numbers in parentheses indicating river km marks): Rajka (1848.5), the bridge at Medve (1806.4), Gönyü (1791.3), Szob (1707.0), Nagymaros/Visegrád (1694.5), and Budapest/Ujpest (1658.0). At the first three stations (R-M-G section in the following discussion) samples were taken three times a year, and at the other three stations (Sz-N-B section) monthly from March to December. In 1975, only the sides of the landing-stages that faced the stream line were studied; dur-

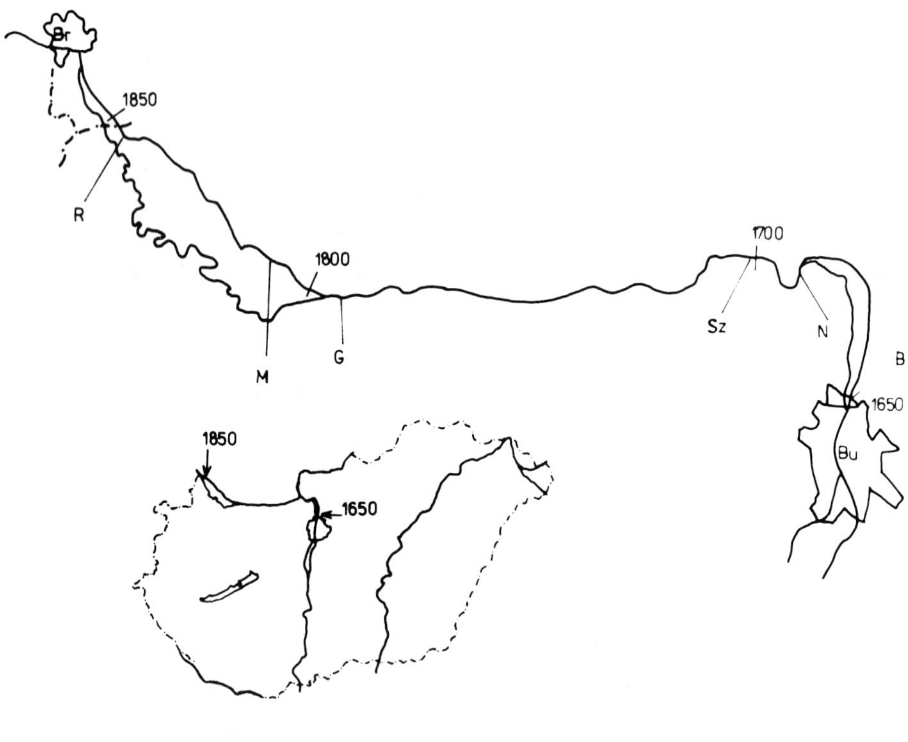

Fig. 1. The investigated section of the Danube River. Br – Bratislava, Bu – Budapest. Sampling stations: R – Rajka, M – bridge at Medve, G – Gönyü, Sz – Szob, N – Nagymaros/Visegrád, and B – Budapest/Ujpest.

ing 1976 both sides of the landing-stages (the side facing the stream line and that facing the shore) were sampled.

For zoological analyses, the samples were taken with a net of 0.5-mm mesh size, supplied with a drawing blade, from a surface area of 3x0.1 m^2 on each occasion. The samples were immediately fixed with 4% formalin. The data were not considered adequate for the determination of absolute abundance, but were suitable for making relative comparisons for temporal and spatial differences.

Animals were separated from plant material with a sugar solution of 1.12 specific weight (Anderson 1959; Biró 1972). The method developed for sediments and detritus worked well also with the live or preserved plant material making up the greatest proportion of periphyton. To make separations more complete, after three subsequent flotations, the plant material was analyzed under the microscope. By these methods nearly 100% of 'smooth-bodied' organisms with few fulcrums (e.g. chironomid larvae, larvae and nymphs of other dipterans) could be separated, and 40–70% of other organisms having several fulcrums (e.g. Oligochaeta, Amphipoda, Isopoda). The mollusks remained in the plant material. The material was preserved in 70% alcohol.

From 150 samples collected during the two years, members of 16 taxonomic categories could be identified. 24 taxa were identified by species belonging to the following systematic categories: Gastropoda, Lamellibranchiata, Cladocera, Copepoda, Isopoda, and Amphipoda.

Results

The periphyton was contained in a web of *Cladophora glomera* (L.) Kütz. During the vegetative period, *Cladophora* exhibited differences in the time of appearance and development between individual sampling stations. These changes resulted from both natural effects and stress effects by navigation (e.g. cleaning, repainting during winter maintenance, the frequency of landing), which resulted in data unsuitable for the determination of absolute abundance values.

The number of animals present in the periphyton during March increased rapidly in April. This increase continued until the end of May or the beginning of June, and then decreased. At landing stages affected by frequent navigation this process occurred 1–1.5 months later. In autumn a second increase in individual numbers was observed, but was generally less than the increase in spring. The individual maxima occurred in the same month on both sides of each landing-stage, but exhibited, however, higher values on the side towards the shore. The number of taxa also increased until the end of May or the beginning of June, and later remained at approximately the same level.

The organisms found during these investigations can be divided into two main categories on the basis of their occurrences: 1. Organisms regularly occurring in time and space, and dominating total individual numbers: Nematoidea, Oligochaeta, Bryozoa, Chironomidae and other Diptera. On the basis of the occurrences of these organisms, it was not possible to distinguish between the single sites or sections (Table 1). 2. Organisms irregularly occurring in space and time, and making up only a small proportion (1–3%) of total individuals. Among these, the members of Cladocera, Copepoda, and Amphipoda were the most important. In addition, other taxa of sporadic occurrence were also found: Ephemeroptera, Trichoptera, Gastropoda, Lamellibranchiata, Coleoptera, Hydridae, Hirudinoidea (in decreasing order of occurrence). The Amphipoda, Ephemeroptera,

Table 1. Percentage values for the yearly total individual numbers of the four dominant taxonomical categories in the R-M-G and Sz-N-B sections.

	1975		1976	
	R-M-G	Sz-N-B	R-M-G	Sz-N-B
Chironomidae	53.1	59.1	50.1	45.1
Oligochaeta	10.2	19.4	27.2	37.9
Nematoidea	13.1	10.1	13.7	11.2
Other diptera	4.4	6.9	2.4	3.9
	80.8	95.5	93.4	98.1

Hirudinoidea occurred predominantly, the Trichoptera exclusively in the Sz-N-B section, and the Gastropoda mainly in the R-M-G section. In connection with Cladocera and Copepoda, the following should be mentioned. Each species found possesses a wide ecological valency (euryök), and often exhibit a preference for detritus-rich sediment or periphyton (e.g. *Pleuroxus trigonellus, Alona affinis, Paracyclops fimbriatus*). In the periphyton these organisms principally occurred on the side of landing-stages towards the shore. The collected specimens were almost exclusively females, and among them nearly 40% were gravid. Part of the non-sessile organisms of the periphyton were benthic organisms (e.g. Gastropoda, Lamellibranchiata, Isopoda, Amphipoda), which in running water find better living conditions in the periphyton (cf. Richnovszky 1970).

With the exception of oligochaetes, the percentage contribution of the various taxonomic groups to total individual number was very similar in the two years (Table 2). Concerning the four dominant groups, the same can be said in connection with the whole sampling section, both the R-M-G and the Sz-N-B sections (Table 1).

Table 3 shows the average yearly percentage distribution of organisms collected at sampling stations at the two river shores. Except among the samples taken at Nagymaros in 1975, the numbers of organisms were always greater on the left shore. At each of these sampling stations, the stream line is nearer the right shore, and therefore on the left shore the flow conditions were reduced and more favorable. During 1975, the water level was also higher on the average, and the maximum of the flood stage occurred in mid-summer. At Nagymaros/Visegrád, where the streambed narrows and curves, flow conditions changed to a greater extent relative to other sampling sites, which could be the cause of the separation of values

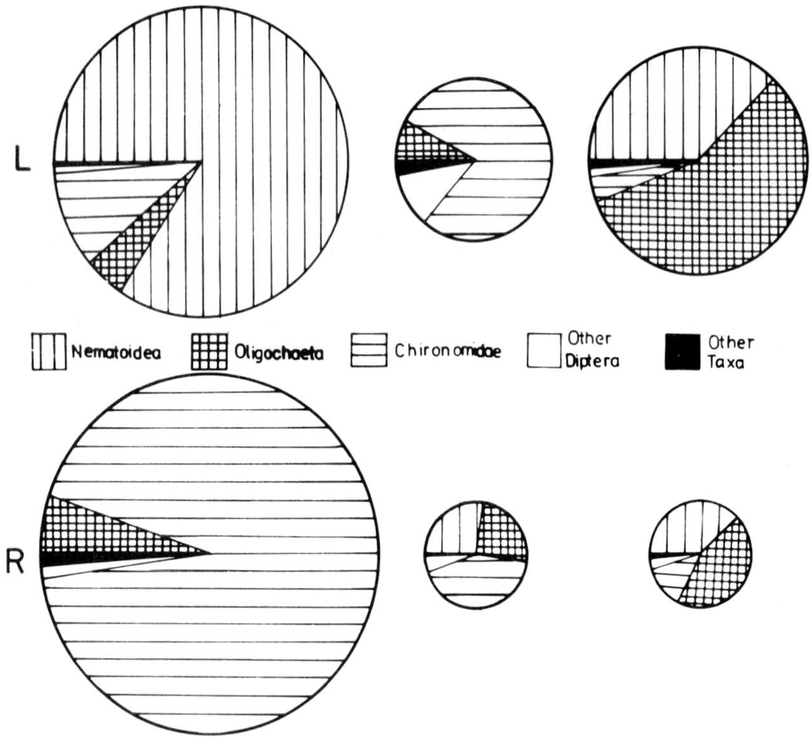

Fig. 2. Individual numbers of the four dominant taxonomical categories at Szob in 1976. The sections of the circles are proportional to the total individual numbers of organisms found (R – right shore, L – left shore; circles from left to right correspond to the seasons spring, summer, and autumn).

Table 2. Percentage distribution of yearly total individual numbers in the whole section of sampling (from Rajka to Budapest).

	1975	1976
Chironomidae	58.7	45.5
Oligochaeta	18.7	37.0
Nematoidea	10.3	11.3
Other diptera	6.6	3.8
Amphipoda	2.7	0.6
Cladocera, Copepoda	0.4	0.5
Ephemeroptera	0.8	0.3
Other taxa	1.8	1.0

Table 3. Yearly average percentage distribution of the individual numbers of organisms on those sides of the landing stages which face the stream line (average monthly values) in the Sz-N-B section.

	1975		1976	
	left shore	right shore	left shore	right shore
Szob	57.3	42.7	52.0	48.0
Nagymaros	44.0	56.0	69.4	30.6
Budapest	83.1	16.9	89.2	10.8

Table 4. Yearly percentage distribution of the individual numbers of organisms found on the two sides of one landing stage (average monthly values) in 1976.

	Left shore		Right shore	
	stream line	shore side	stream line	shore side
Medve	—		21	79
Gönyü	—		13	87
Szob	43	57	36	64
Nagymaros	63	37	45	55
Budapest	61	39	—	

at Nagymaros. At Budapest, the difference in the development of periphyton (see earlier discussion) must be responsible for the great differences in fauna found between the two shores.

Table 4 illustrates the yearly average percentual distribution of organisms found on both sides of one landing stage. Downstream from Medve, the contribution of the side facing the stream line increased.

Discussion

On the basis of these investigations, a general picture of the fauna of the periphyton emerged. Individual numbers exhibited similar yearly changes with a spring-early summer maximum, and occasionally an autumn-early winter maximum. The temporal occurrence of these maxima was correlated principally directly with temperature. The decrease of individual numbers during summer was the result of the emergence of hemihydrobionts.

The composition of fauna and the percentage distribution of dominant taxonomic categories was very similar along the whole river section investigated. Differences between the R-M-G and Sz-N-B sections appeared only in taxa occurring irregularly in small numbers. Most periphytic animals were generally benthic, but many benthic organisms probably found better conditions of living here than in the sediments. It also seems that the periphyton has an important role in the reproduction of plankton, as suggested by occurrence of the gravid females of cladocerans and copepods.

The effects of discharge fluctuations manifests itself partly in water-level fluctuations, partly in flow velocity changes. The water-level fluctuations did not effect the biota of the periphyton growing on floating substrates, in our case the landing stages. Changes related to flow velocity alterations, however, can be critical. With increasing flow velocity (when passing upstream), the organisms exhibited a tendency to concentrate on the shore-side parts of the landing stages. The relative difference between the two sides increased directly with the absolute flow velocity of the stream. On the side facing the shore, the situation was more 'lake-like', with slightly higher temperatures, greater detritus, and less abrasion by sediment and drifted material, which suggests that this side is more protected in every respect than the side facing the stream line. The effect of flow velocity is evidenced also by the fact that at the same sampling station more organisms were found at the shore lying farthest from the stream line.

The sampling station at Nagymaros/Visegrád deserves particular mention, because the results obtained here differed in many cases from the other places (cf. Tables 3 and 4). Here, in a short section, the river morphometry is similar to an upper reach, and differs in many respects from the preceding and following reaches of lowland character.

Acknowledgement

We wish to express our thanks to Dr. Anna Bothár for the identification of species of Cladocera and Copepoda, and to Professor Robert G. Wetzel for his revisions and for improving the English of our manuscript.

References

Anderson, R.O., 1959. A modified flotation technique for sorting bottom fauna species. Limnol. Oceanogr. 4:223.

Andrássy, I., 1960. Nematoden aus dem Periphyton der Landungsmolen der Donau zwischen Budapest und Mohács. Ann. Univ. Sci., Sect. Biol. 3:3–21.

Bayly, I.A.E. & Williams, W.D., 1973. Inland waters and their ecology. Longman, Victoria, Australia, 314 pp.

Berczik, Á., 1965. Die Chironomiden-Larven aus dem Periphyton der Landungsmolen im Donauabschnitt zwischen Budapest und Mohács. (Danubialia Hungarica 83). Acta Zool. Hung. 11:227–236.

Berczik, Á., 1966a. Über den Einfluss einiger hydrologischen Faktoren auf die Besiedlungsmöglichkeiten der Fauna der Mittleren Donau. (Danubialia Hungarica 30). Ann. Univ. Sci. Bpest, Sect. Biol. 8:25–32.

Berczik, Á., 1966b. Chironomidenlarven aus dem Aufwuchs der Schwimmkörper im Donauabschnitt zwischen Rajka und Budapest. (Danubialia Hungarica 39). Acta Zool. Hung. 12:41–45.

Berczik, Á., 1969. Die Chironomiden der Uferregion des ungarischen Donauabschnittes. (Danubialia Hungarica 50). Opusc. Zool. Bpest 9:249–254.

Berczik, Á., 1971. Die Chironomiden und ihre Lebensstatten auf dem ungarischen Donauabschnitt. Limnologica 8:61–71.

Berczik, Á., 1973. Benennung der zwei ökologischen Gruppen Wasserbewohnender Wirbellosen. Opusc. Zool. Bpest 12:33–41.

Biró, K., 1972. Állatok elkülönitése üledékmintákból cukoroldat segitségével. Vituki Vizmin. Kut. Eredm. 2:83–88.

Boros, Á., 1961. A Duna vizimohái. Hidrol. Tájékoztató, Bpest 47 pp.

Bothár, A., 1966. Beiträge zur Kenntniss der Weichtierfauna der ungarischen Donau. (Danubialia Hungarica 36). Opusc. Zool. Bpest 6:93–107.

Bothár, A., 1979. A közép-Duna Crustaceainak vizsgálata. Kandidátusi Értekezés. Bpest 149 pp.

Dudich, E., 1967. Systematische Verzeichnis der Tierwelt der Donau mit einer zusammenfassenden Erläuterung. In: Liepolt, R. (ed.), Limnologie der Donau. Stuttgart: 4–69.

Dvihally, Z., 1977. A Duna. In: Tihany, Elsödleges termelés 1. 321 pp.

Fekete, G., Précsényi, I., Molnár, E,.N. & Nosek, J.N. 1979. Szerkezet és müködés egy természetes növénytársulásban. Eredmények, problémák és perspektivák a Tece-homokpusztagyep kutatásban (Tece Studies 18). MTA Biol. Oszt. Közl. 22:311–322.

Richnovszky, A., 1970. A magyarországi Duna-szakasz puhatestü faunájának ökológiai viszonyai. Állatt. Közl. 57:125–130.

Sládečková, A., 1962. Limnological investigation methods for periphyton (Aufwuchs) community. Bot. Rev. 28:286–350.

Soós, Á., 1967. On the leech fauna of the hungarian reach of the Danube. (Danubialia Hungarica 44). Opusc. Zool. Bpest 7:241–257.

Szemes, G., 1961. Die Algen des Periphytons der Donaupontons (Quantitative Analyse der Bacillariophyceen). (Danubialia Hungarica 11). Ann. Univ. Sci. Bpest, Sect. Biol. 4:179–215.

Szemes, G., 1967. Das Phytobenthos der Donau. In: Liepolt, R. (ed.), Limnologie der Donau. Stuttgart: 225–241.

Tamás, G., 1966. Mikroflora aus dem Periphyton der Landungsmolen der Donau zwischen Budapest und Mohács. (Danubialia Hungarica 37). Ann. Univ. Sci. Bpest, Sect. Biol. 8:345–356.

Tóth, J., 1972. Az áramlástan egyes kérdéseinek ökológiai vonatkozásai. Hidrol. Közl. 42:76–79.

Unger, E., 1916. Adatok a Duna faunájának és oekológiájának ismeretéhez. Állat. Közl. 15:262–281.

Weber, A., 1967. Stauregion. In: Liepolt, R. (ed.), Limnologie der Donau. Stuttgart:272–283.

Authors' address:
J.N. Nosek/N. Oertel
Hungarian Danube Research Station
of the Hungarian Academy of Sciences
Jávorka S. u. 14.
Göd, Hungary

CHAPTER 4

Phytosociological associations of algal periphyton[1]

RYSZARD BOHR, MARTA LUŚCIŃSKA and ANDRZEJ S. OLEKSOWICZ

Abstract. In the search for a method to distinguish ecological units of periphytic algal communities, the phytosociological method was applied. On the basis of numerous samples collected from 49 lakes, it was found that there occur two associations in the lakes of Northern Poland: Oedogonio-Epithemietum litoralae in eutrophic lakes, and Hapalosiphono-Tabellarietum calciphobum in dystrophic and 'Lobelia' lakes. These associations may be regarded as synecological units, which repeat their basic characteristics in definite ecological conditions of the habitat. On the basis of multiple observations and measurements, it is possible to construct a model of cenosis which is a reference point for the purpose of comparison of discrete sets.

Analysis of species contents and numerous measurements of diversity, production and biomass of many phytocenoses justify making synhetic statements concerning the development of the structure and the role of periphyton the lake ecosystem. The association Oedogonio-Epithemietum litoralae produces about 5% of the total organic matter, while the association Hapalosiphono-Tabellarietum calciphobum 40%.

Introduction

We can ordinate and systemize any set of material objects or notions in various ways, depending on our objective and the point of view of a given set. If the subject of our interest is the biosphere taken as a whole, then the natural order will be the sequence of units: specimen, a set of specimens of the same species, or a population, and a community consisting of multi-species populations – that is, supraorganismal levels. The distinction between and the precise definitions of synecological units of the biosphere, despite the many attempts which have been made and the various concepts which exist, continue to be a controversial problem. Man has been classifying nature in order to learn about the things and phenomena amongst which he lives, and, consequently, to make use of this knowledge for practical application according to his needs.

Acceptance of the concept, first expressed by Ramensky (1924) and later by Gleason (1926), of the continuum of the biosphere, does not prevent us from seeking ways to define particular elements – units of biosphere – that would permit operative knowledge of consecutive development stages that interest us at a specific site and time. Establishing limits to such units is a matter of convention, depending on adopted concepts and principles of division. There exists objectively, regardless of a scholar's viewpoint, a material set of elements in nature, the characteristics of which can be given in many different ways, from a simple quantitative external

[1] This work is part of Problem R-III-15 entitled 'The development, the productivity and the functional correlations of typical ecosystems in selected areas of Tuchola Forests, of the Lake Districts Ilawa and Brodnica and of the farming district of Kujawy'.

Wetzel, R.G. (ed.), Periphyton of Freshwater Ecosystems.
 ISBN 90 6193 768 X.

description to a presentation of functional relations between elements of the set. Ecology seeks ways and possibilities to define model units of biosphere, such as in the present use: ecosystem, biogeocenosis, biocenosis, phytocenosis, association, sociation, etc.

Phytosociology as a branch of ecology has made many contributions to the field of classification of vegetation (cf. review of Aleksandrova 1969). Versatile and deep analysis and discussion can be found in the fifth part of Handbook of Vegetation Science entitled: Ordination and classification of communities (Whittaker 1973). Despite the differences in methods and concepts which exist between particular phytosociological schools, they have one fundamental link, namely, the approach to vegetation, which is contained in the very word 'sociology' and views vegetation as a set of interdependent elements of mutual interaction. What is plant association? As Tansley sees it, association or phytocenosis is a real and objectively existing set of plants growing together. To learn about phytocenosis by means of collecting information, let us say, about floristic combination, quantitative proportion of particular components, their spatial distribution, function, etc., makes it possible to classify a given phytocenosis into a unit defined on the basis of features selected by a scientist. In this sense it is an abstract notion.

After a long period of trials and experiences botanists and ecologists have progressed in the use of phytosociological methods. They successfully use the concept of 'plant association' as a basis unit of vegetation. Today, when ecologists all over the world focus much attention on productivity and other functions of ecosystems, it is of particular importance to have a reliable unit for making syntheses and comparisons.

Is it possible to adapt phytosociological methods used for macrophytes to the study of microphytes? In our opinion, phytosociological methods for periphyton investigation are only methods which can be used for describing a particular algal community. Further, more detailed investigation, like production measurement, population analyses etc., can be carried out by means of the 'plant association' concept, and then generalizations and syntheses can be made. We have stated that algae in the periphyton do form associations. Each of them is connected with a particular limnological type of lake.

Material and methods

In 1960–1981 a study on periphytic algae communities was carried out in 19 soft-water lakes and 29 hard-water lakes in the area of the Pomerania Lake District in Northern Poland. Two algal associations were found in examined lakes, depending on habitat: Hapalosiphono-Tabellarietum calciphobum in soft waters, not rich in calcium, as e.g. dystrophic and 'Lobelia'-type lakes; and the association Oedogonio-Epithemietum litoralae in lakes with semi-hard and hard waters, i.e. calcium-rich.

Samples of periphyton were collected from various substrata. Most often, however, they were macrophytes such as: *Phragmites communis, Nuphar luteum, Equisetum limosum, Typha latifolia, Typha angustifolia, Lobelia dortmanna, Stratiotes aloides, Schoenoplectus lacstris, Potamogeton* sp., *Carex* sp., *Heleocharis* sp., *Glyceria* sp., *Lysymachia* sp. and submergent rocks and dead tree branches. A total of 263 samples were collected from 49 lakes (164 samples from hard-water lakes and 99 from soft-water lakes). Some of the results were presented as phytosociological records in the works by Bohr (1962, 1967), Luścińska (1976, 1979) and Oleksowicz (1978). Detailed procedures of establishing minimum surface area of a sample, quantitative evaluation of individual species or cover degree, and spatial distribution of an association based on the biology of species participating in the association were described in detail by Bohr (1962, 1967, 1973).

Determination of productivity and biomass of an individual association of periphytic algae makes it possible to evaluate the function of periphyton in the lake ecosystem. Periodic measurements of primary production, biomass, and species diversity index with a large number of samples of a defined association enable some syntheses to be made. Results of many measurements of primary production and biomass were averaged for particular months in different years of investigation. Measurements of primary production were carried out for 11 lakes in two consecutive seasons and in 6 other lakes in one season. Primary production was measured by the light and dark bottle oxygen difference method, after introducing into them periphyton removed from a known area of substrate. Biomass was calculated from the individual volume of algae species and their numbers on the same area. Results of primary production and biomass measurements of periphyton were presented in the publications by Adamczak *et al* (1979), Bohr & Lúscińska (1975), Bohr & Luścińska (1975), Bohr & Zawiślak (1969), Bohr & Miotk (1979), Lúscińska (1976, 1979), Luścińska & Oleksowicz (1982), Oleksowicz (1974, 1980, 1982a, 1982b), and Oleksowicz & Luścińska (1982).

Assuming that the degree of species diversity may be regarded as the degree of stability and maturity of community, the species diversity of certain associations was calculated from Shannon's formula (1948) or, less frequently, the formula of Lloyd *et al* (1968).

Results

Characteristics of the Oedogonio-Epithemietum litoralae Bohr (1962) association

The collective phytosociological records (Table 1) of the association Oedogonio-Epithemietum litoralae includes about 100 taxa of algae permanently in association in a layered system in which they are separated spatially (Bohr 1.c). The filamentous algae are the core of the association, attached to the substratum layer 'A' (e.g. species of genus *Oedogonium* or *Cladophora* and filamentous algae not attached, layer 'A_1'), and reach the highest values of cover degree (Table 2). The filamentous algae from layer 'A_1', such as *Ulothrix* sp. div., *Mougeotia* spp., *Spirogyra* spp., may attach themselves to the substratum thus forming layer 'A', so it is difficult to differentiate them univocally. Crustose algae, attached to the substratum with an entire thallus surface, layer 'B' (e.g. species of genus *Coleochaete*), attain a smaller cover degree. Layers of second order epiphytes (on algal thalli, 'C'), and unicellular algae loosely connected with the substrate, 'D'), are relatively rich in species and quite numerous, and their cover degree is within the range of the average. Over 60% of species in the association Oedogonio-Epithemietum litoralae are diatoms occurring as single cells, e.g. *Epithemia, Gomphonema, Cymbella, Navicula,* or as filaments e.g. *Fragilaria, Melosira* etc. Filamentous and coccal green algae also form a large group. Other groups of algae, including blue-green algae, are represented in this association by few taxa.

High cover values of species indicate that other excluded taxa do not play any large role in the community. The high degrees of stability of these species in many samples show that a combination of species characteristic of the association emerge (Table 1). The characteristic combination of species exhibit remarkably consistent occurrence and species dominance may permit one to distinguish the associations as synecological units that occur repeatedly in lakes of high calcium concentrations.

The occurrence of specific periphytic algal associations among different types of lakes makes it possible to generalize on the ecological roles of the association, its biomass and productivity in the entire lake.

Mean values of biomass and production measurements were determined over several years in hard-water lakes and are summarized in Figure 1. The association develops in a season and reaches maturity in autumn. In spring, because of improved light, algae quickly overgrew the surface of newly growing stems of macrophytes. Springtime biomass increase was caused by quickly reproducing populations of mainly unicellular diatoms. In summer (June, July) the decrease of periphytic biomass was probably caused by reduced clarity of water and partial exploitation of the habitat by the abundance of developing planktonic algae and macrophytes. Renewed biomass increases from August until November were caused by biomass increases of filamentous algae. Filamentous algae of layer 'A' and 'A_1' were the core of the association and constituted 60% of the total (Table 2). In

Table 1. Collective phytosociological table of the periphytic algal association Oedogonio-Epithemietum litoralae.

Number of records	1–14		15–69		70–99		100–132		133–140		141–148		149–153		154–164	
List of taxa	S	CD	S	CD	S	CD	S	CD	S	CD	S	CD	S	CD	S	CD
Layer 'A'																
Cladophora glomerata	V	3540	V	4982	II	178	IV	204	IV	2317	IV	880	III	5250		
Oedogonium sp. div	V	3965	V	4160	V	656	V	473	V	17	V	3500	V	7158	V	8295
Bulbochaete sp. div.			II	721	I	34	I	1							IV	3703
Stigeoclonium farctum			II	42	I	2	V	176			II	10			I	1
Layer 'A_1'																
Mougeotia sp. div.	V	219	II	222	III	364	I	4	II	5	III	1533	IV	5204	V	5776
Spirogyra sp. div.			II	252	I	244	I	212			II	7	V	8750	IV	12
Tolypothrix sp. div.	III	1635	II	1100	II	430			II	283			II	104	III	3980
Lyngbya sp. div			IV	313			V	17			I	62	I	4	V	3351
Oscillatoria sp. div.							III	9			III	1315	IV	3980	I	1
Diatoma sp. div.	I	47	III	512	IV	203	V	53			I	2			I	5
Melosira sp. div.			II	182			IV	71			V	1171	IV	124	IV	14
Fragilaria sp. div.	IV	49	V	1077	III	84	V	509	V	1954	V	3364	V	256	V	181
Cymbella ventricosa	I	1	V	453	V	83	V	20	V	140	V	2377	III	12	V	1058
Binuclearia tectorum															IV	14
Layer 'B'																
Coleochaete sp. div.	II	174	IV	2501	I	39	IV	302	II	7	V	2553	II	12	IV	2409
Protoderma frequens							I	492	IV	12			I	4		
Layer 'C'																
Calothrix brevissima							IV	29							I	5
Aphanochaete repens	III	42	I	2	I	2			II	5	II	7	IV	16		
Epithemia sp. div.	V	4310	V	2040	V	1467	V	1428	V	1668	V	1919	IV	2536	V	2691
Cymbella sp. div.	IV	314	V	410	IV	432	V	377	V	276	IV	477	V	40	V	1166
Cocconeis sp. div.	V	1247	V	592	V	19	V	134	IV	15	V	84	V	1612	V	23
Gomphonema sp. div.	IV	337	V	803	V	79	V	157	IV	279	V	111	III	378	I	30
Rhoicosphaenia curvata	IV	6	IV	115	III	9	V	49	III	10	IV	15	II	8	I	1
Amphora ovalis	I	1	III	202	III	10	V	17					II	8	II	7
Layer 'D'																
Navicula sp. div.	II	3	V	68	IV	109	V	279	II	127	IV	29	IV	24	V	2128
Cyclotella comta							V	17			IV	15	II	8	V	238
Stephanodiscus astrea							IV	14			I	2			V	54
Nitzschia sp. div.			IV	66	II	7	V	20			III	25	IV	16	I	5
Gyrosigma sp. div.	I	3					IV	12					II	8		
Asterionella formosa							III	11			II	7	III	12	V	220
Synedra sp. div.	IV	46	V	499	V	1102	V	224	V	87	V	142	II	108	IV	26
Ceratium hirundinella			IV	26			I	1			I	2	III	12	I	1
Pediastrum sp. div.	V	121	V	28	V	18	V	50	V	54	V	29	V	40	II	166
Scenedesmus sp. div	IV	8	III	10	III	21	V	30	V	49	III	32	IV	20	V	361
Ankistrodesmus fussiformis							I	1	V	17	III	10	I	4		
Cosmarium sp. div.	II	23	IV	103	V	19	III	10	III	10	III	12	III	20	V	59
Staurastrum sp. div.			IV	26			I	1					II	8	II	10
Rhopalodia gibba	III	5			V	122	V	31	V	77	II	5	III	12	V	18
Cymatopleura solea							IV	14			I	2	II	8		
Closterium moniliferum	I	1											IV	16		
Coelestrum sphaericum											III	10				
Microcystis aeruginosa											III	12				
Other little diatoms		64				3310				2250		4750				4616

Explanations: S = class of constance species present in 1–20% of samples (I), 20–40% (II), 40–60% (III), 60–80% (IV), 80–100% (V);

CD = cover degree: $[CD = \frac{\text{the sum of mean cover percentage of a given species (all records in table) where this species occurred}}{\text{total number of records}} \cdot 100]$

Table 2. The sums of mean cover degree in various layers of the associations.

Layers	Associations			
	Oedogonio-Epithemietum litoralae		Hapalosiphono-Tabellarietum calciphobum	
	cover degree	%	cover degree	%
A	6283	31.0	3857	27.6
A_1	6675	32.8	8963	64.2
B	1063	5.2	2	0.0
C	3502	17.2	348	2.5
D	2810	13.8	790	5.7
Total	20333	100.0	13960	100.0

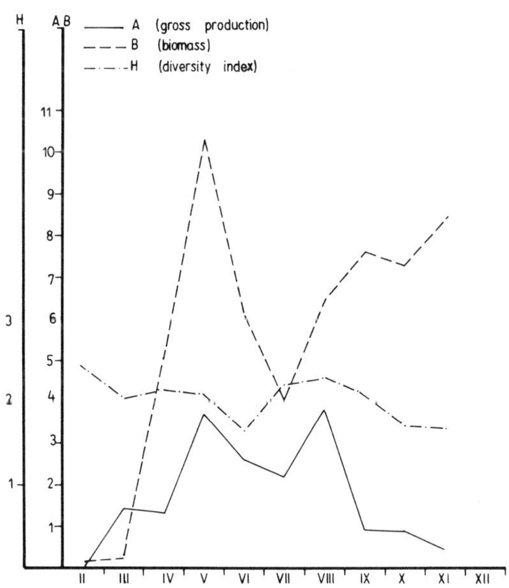

Fig. 1. Changes in gross production (kcal m^{-2} d^{-1}), biomass (kcal m^{-2}) and species diversity of periphytic algae association Oedogonio-Epithemietum litoralae.

autumn, after the period of biomass accumulation at lower production intensities, the algae decompose and the community disintegrates. Variability of species diversity index (Fig. 1) during the vegetative season was small, and more or less constant. This confirms the thesis that the association Oedogonio-Epithemietum litoralae takes its characteristic shape from the beginning of periphyton formation on the new substrate and continues throughout the season. Analysis of species occuring in spring, summer, and autumn showed that, qualitatively, the association changes somewhat, while quantitatively it charges greatly. Many species of early settlement (Connell & Slayter, 1977) that occupy the substrate in spring are replaced by the species that form the combined characteristics of the association (Bohr, 1973). Production and biomass correlate with organization of the species content. The periphyton production curve can be similar to the biomass curve. It is only during autumn that the community produces less despite high biomass, which is probably caused by lower temperature, less light, and gradual degradation of the association. This association in eutrophic lakes during vegetation season (±240 days) produced between 350 and 2800 kcal m^2 at average intensity (P/B about 15, turnover about 16 days), which constituted only about 5% of the total primary production of the lakes (Bohr & Luścińska, 1975; Adamczak *et al,* 1979; Oleksowicz, 1982a).

Characteristics of the Hapalosiphono-Tabellarietum calciphobum Bohr (1967) association

This association of periphytic algae included a varying number of species. The minimum number of taxa found in a single record was 77, and the maximum 231; the number of taxa usually oscillated about 100. The Desmidiales were predominant groups in periphyton phycoflora of soft-water lakes. Smaller numbers of species of other algal groups were found, with the most varied species composition shown by diatoms.

The characteristic combination of species, common occurrence, and prevalence showed that the algae characteristic for this association had a similar structure in all the tested lakes (Table 3). This relation is also illustrated by the sum of mean cover degree factors for particular layers (Table 2).

Filamentous algae forming layers 'A' and 'A_1' were dominated by the following species: *Mougeotia* spp., *Spirogyra* spp., *Zygnema* spp., *Tribonema* spp., *Oedogonium* spp., *Ulothrix* spp., *Tabellaria flocculosa, Hapalosiphon fontinalis* and *H. intricatus.* The species forming layers 'C' and

Table 3. Collective phytosociological table of the periphytic algal association Hapalosiphono-Tabellarietum calciphobum.

Number of records	1–6		7–27		28–33		34–39		40–45		46–51		52–57		58–79		80–89		90–99	
List of taxa	S	CD	S	CD	S	CD	S	CD	S	CD	S	CD	S	CD	S	CD	S	CD	S	CD
Layer 'A'																				
Hapalosiphon sp. div.	V	2531	IV	3357	V	8			V	877	II	3	V	95	I	4	IV	1471	I	4
Oedogonium sp. div.	IV	2004	V	4450	V	641	V	458	V	1167	V	3375	V	5500	V	20	V	3090	V	3931
Bulbochaete sp. div.			V	2548	II	3					V	260			V	15	I	3		
Microthamnion sp. div.									II	3	III	5					I	3	II	6
Batrachospermum sp.	III	2033													I	1				
Layer 'A_1'																				
Tabellaria flocculosa	V	63	V	517	V	10	III	5			III	5	V	92	IV	14	V	2176	V	1908
Mougeotia sp. div.	IV	809	IV	900	V	1218	V	6541	V	3218	V	5083	V	1250	V	3412	V	8750	V	8750
Ulothrix sp. div.					V	20	V	346	IV	174	V	4212	III	5	III	13	I	583	I	2
Binuclearia tectorum	III	199	I	71	II	3	III	87	II	3	V	90	II	3			V	2916	V	8750
Spirogyra sp. div.			II	595	II	9	II	10	II	85	V	3291	V	2000	III	154	V	7583	V	4554
Zygnema sp. div.			IV	3108	IV	88	II	167					II	3	I	82	I	3		
Tribonema sp. div.					III	5	V	385	III	8	IV	1470	II	3	I	2	V	3012	III	18
Lyngbya sp. div.							IV	7					IV	297			I	6	II	6
Spondylosium sp. div.							III	5	IV	7	II	3	V	13					IV	387
Fragilaria virescens	IV	58															I	3	I	6

'D' did not become significant quantitatively, but were decisive in the number of species of the association and would frequently constitute over 75% of all species. Of the most frequently and abundantly found species in layers 'C' and 'D', the following should be mentioned: *Netrium digitus, N. oblongum, Micrasterias truncata,* and richly represented species of *Cosmarium, Staurastrum, Closterium, Eunotia,* and *Frustulia.*

Lack of algae forming layer 'B' (the entire surface of their thallus attached to the substate) was a significant structural characteristic of the association. Species forming this layer occurred rarely and were not numerous and were without any large stability in the association.

This association varied greatly seasonally among different lakes and in particular years as a result of habitat differences. In 11 lakes the association developed its typical form, in 3 lakes it formed a variant with *Frustulia* et *Micrasterias truncata* (Bohr, 1967), in 4 lakes a variant with *Mougeotia* spp. et *Spirogyra* spp., and in 1 lake a variant with *Oedogonium* spp. et *Cosmarium pseudobroomei* (Oleksowicz, 1978). In a number of 'Lobelia' lakes, the association occurred in greatly impoverished form, and lacked a number of characteristic species (Luścińska, 1979).

The development rates of the Hapalosiphono-Tabellarietum calciphobum association during one season can be rapid. Biomass and production increase from the beginning of the vegetative season (Fig. 2). During this period, the association was characterized by the highest species diversity index ($\bar{H}$). The next stage of development, continuing until autumn, was the period of quantitative predominance (biomass and cover degree) of several species, mainly *Spirogyra, Mougeotia,* and declining species found in the initial period of colonization. The filamentous algae not attached to the substrate from layer 'A_1' reached 64.7% of the cover degree (Table 2). The periphyton biomass curve during the vegetation period was similar to the production curve. The relatively low production of periphyton compared with biomass deserves mention. It is believed that it was caused by relocation of algae from other parts of the lake, mainly from the bottom, to the association of pe-

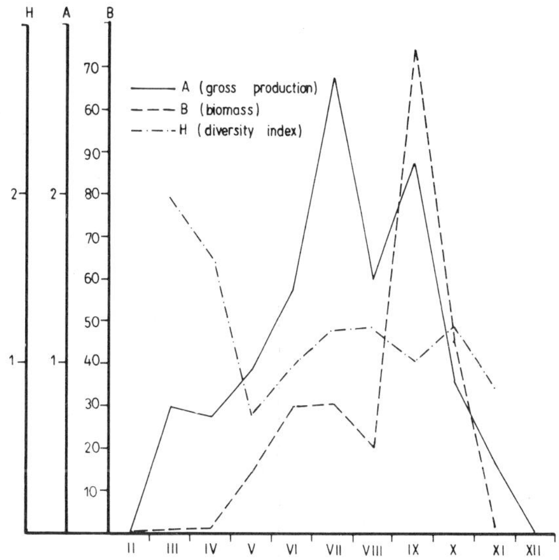

Fig. 2. Changes in gross production (kcal m^{-2} d^{-1}), biomass (kcal m^{-2}), and species diversity of periphytic algae association Hapalosiphono-Tabellarietum calciphobum.

riphyton (Oleksowicz, 1982a) and multiple accumulation of many new layers of physiologically more active algae. As a result, the outer layer, apparently more active physiologically, shaded cells of algae inside the periphyton concentrations around shoots of macrophytes. Moreover the presence of animal organisms and bacteria caused respiration processes to increase, thus lowering the results of measurement of photosynthesis by oxygen production. Degradation of the association at the end of the vegetation season proceeded rapidly and was indicated by rapid declines of biomass, production, and species diversity index. The process was accelerated by deterioration of submerged macrophyte substrata. The discussed association plays a remarkable role in dystrophic and oligotrophic lakes. As shown by Oleksowicz (1980), this association is responsible for about 40% of the total primary production of the lake.

Conclusions and discussion

The studies of periphytic algal communities in the lakes of Northern Poland justify the statement that by means of phytosociological criteria it is possible to distinguish the associations Hapalosiphono-Tabellarietum calciphobum in soft-water lakes and Oedogonio-Epithemietum litoralae in hard-water lakes.

The associations may be regarded as synecological units, with the same basic characteristics – species, spatial structure, numbers of particular components, biomass, and production – in certain habitat conditions.

On the basis of many observations and measurements it is possible to construct a model of cenosis of an association that would be the point of reference for comparative purposes of discrete sets. The differences between a real community and the model will result as the habitat is altered (e.g. by hypereutrophy or pollution).

The species diversity index can be an objective measure of association differences. Calculated for a model cenosis, it shows the optimal condition for development of an association, which can be regarded as a final stage, closest to maturity with stability. Lower values of $\bar{H}$-index are evidence of the development process during a season, or emergence of conditions limiting development of the association. Higher values may be caused by participation of species alien to the association from planktonic, benthic, or other periphytic communities, as well as a result of changing habitat conditions.

References

Adamczak, B., Luścińska, M., Oleksowicz, A.S. & Rejewski, M., 1979. The primary production of Lake Tynwald. AUNC-Prace Limnol. 47(11):3–9.

Aleksandrova, V.D., 1969. Classification of Vegetation. Izd. Nauka, Leningrad. 175 pp.

Bohr, R., 1962. Sociological research of the plant periphyton in the lake Mamry. Stud. Scien. Torun., Sec. D 6(1):1–44.

Bohr, R., 1967. Communities of periphyton algae in the lakes of North Poland. Zesz. Nauk. UMK w Toruniu. Nauk. mat.-przyr. 17:33–104.

Bohr, R., 1973. Phytosociology of periphyton. Pol. Arch. Hydrobiol. 20:185–188.

Bohr, R. & Luścińska, M., 1975. The primary production of the periphyton association Oedogonio-Epithemietum litoralae. Verh. Intern. Verein. Limnol. 19:1309–1312.

Bohr, R. & Miotk, G., 1979. The primary production of periphytonic algae in the south part of the Jeziorak Lake. AUNC 47(11):11–17.

Bohr, R. & Zawiślak, W., 1967. Production of periphyton in the Jeziorak Lake. Zesz. Nauk. UMK 18:3–10.

Connell, J.H. & Slayter, R.O., 1977. Mechanisms of succession in natural communities and their role in community stability and organization. Am. Nat. 111:1119–1144.

Gleason, H.A., 1926. The individualistic concept of plant association. Bull. Torrey Bot. Club. 53:7–26.

Lloyd, M., Zar, J.H. & Karr, J.R., 1968. On the calculation of information-theoretical measure of diversity. Am. Midl. Nat. 79:257–272.

Luścińska, M., 1976. Communities of epiphyton algae in the Lobelia-lakes of Tuchola Forest. Dr. Sc. Thesis. Copernicus Univ., Torun. 43 pp.

Luścińska, M., 1979. Structure of epiphytic algal communities of the Lobelia-lakes in the region of Bory Tucholskie (Poland). Arch. Hydrobiol. Suppl. 56 (Algological Studies 24):-324–338.

Luścińska, M. & Oleksowicz, A.S., 1982. Algal communities in five eutrophic lakes of the Tuchola Forest Area. AUNC Limnolog. Papers 14 (in press).

Oleksowicz, A.S., 1974. The production of periphytonic algae in Lake Tynwald. AUNC Limnol. Papers 40(10):41–51.

Oleksowicz, A.S., 1978. Epiphytonic algae of five dystrophic lakes in Bory Tuicholskie Forest, Northern Poland. AUNC Nauk. mat. przyr. 45:3–66.

Oleksowicz, A.S., 1980. Planktonic and epiphytic algal communities in three, limnologically different lakes in the region of the Tuchola Forest. Dr. Sc. Thesis, Copernicus Univ., Torun. 144 pp.

Oleksowicz, A.S., 1982a. Interactions among algal communities in three lakes of the Tuchola Forests area (Northern Poland). Arch. Hydrobiol. Suppl. 63 (Algological Studies 30):77–90.

Oleksowics, A.S., 1982b. The primary production of epiphyton association Hapalosiphono-Tabellarietum calciphobum Bohr, 1967. AUNC Limnol. Papers 14 (in press).

Oleksowicz, A.S. & Luścińska, M., 1982. Volumes of algae cells of the most frequent occurrence in 16 lakes in Pomerania Lake District. AUNC Limnol. Papers 14 (in press).

Ramensky, L.G., 1924. Die Grundgesetzmössigkeiten im Aufbau der Vegetationsdecke. Vest. opyt. Dela sred. chernoz. Obl., Voronezh, Abstr. Bot. Z.N.F. 7:453–455.

Shannon, E., 1948. A mathematical theory of communication. Bell. Syst. Tech. J. 27:379–426, 626–656.

Stugren, B., 1974. Grundlagen der allgemein Ökologie. VEB Fischer Verlag, Jena. 223 pp.

Whittaker, R.H. (ed.), 1973. Ordination and Classification of Communities. Dr W. Junk Publ., The Hague. 737 pp.

Authors' address:
R. Bohr/M. Luścińska/A.S. Oleksowicz
Copernicus University
Institute of Biology
Department of Taxonomy
Plant Ecology and Nature Protection
Torun, Poland

CHAPTER 5

Development and dynamics of floating clusters of filamentous algae

H. HILLEBRAND

Abstract. Most species of filamentous freshwater algae can occasionally be observed growing in floating clusters. The phenomenon seems to be characteristic for filamentous algae in shallow habitats. Dominant species may vary and show different periodicities. General characteristics can be observed in the development and dynamics of these floating clusters, called *flab* in the Dutch language. Occurring in shallow and often even ephemeral waterbodies, flab shows a pronounced stratification. The first development of algae is usually at the bottom. After a period of submerged growth, formation of oxygen bubbles leads to an ascent of the algal mass. During flotation, these algae are exposed to extreme conditions, especially of radiation, oxygen content, pH, and nutrient depletion. These conditions are greatly influenced by the weather and thus show rapid changes.

Introduction

Shallow eutrophic freshwater bodies in spring and summer are often covered with green or yellow masses of entangled algal filaments. These masses attain great densities and have a high primary production (Komárková & Marvan 1978; Velichko 1977). Eutrophication is indicated as the primary factor that enhances the development of floating algal masses (de Lange & van Zon 1973; Hillebrand & Klapwijk 1976; Hillebrand 1977; Komárková & Marvan 1978; Marvan *et al* 1978a, 1978b). Floating masses of filamentous algae have been used as fertilizers in the western part of the Netherlands. This application explains the existence of a specific word in the Dutch language for floating masses of filamentous algae: *flab*. For convenience this word is used in the present paper.

Most species of filamentous freshwater algae can be found in flab (Marvan *et al* 1978a, 1978b; Hillebrand & Klapwijk 1976, Hillebrand 1977). Well-known are many Zygnemataceae and species from the genera *Oedogonium* and *Cladophora*. Less often observed are species from the genera *Rhizoclonium, Microspora, Ulothrix, Tribonema,* the species *Enteromorpha intestinalis* (L.) Greville, *Vaucheria dichotoma* (L.) Martius, *V. sessilis* (Vauch.) De Candolle, *Hydrodictyon reticulatum* (L.) Lagerheim and various Cyanophyta. Several groups (*Zygnemataceae* and *Oedogonium*) draw attention because of their extraordinary high species numbers. These great numbers of potentially flab-forming species sustain the supposition that at least several of them will grow or reproduce optimally in flab. The species richness of these groups further suggests that favorable conditions exist in flab for speciation, probably caused by one or more of the following factors: isolation, mutagenic radiation (UV light), environmental diversification, and the ephemeral character of the water bodies. Flab develops in shallow water bodies, so the chance of desiccation is great. Many species have drought

Wetzel, R.G. (ed.), Periphyton of Freshwater Ecosystems.
 ISBN 90 6193 768 X.

resistant spores. A well-timed induction of spore formation is necessary for population survival, and it must occur in close connection with the changing environment. Artificial spore induction is difficult (Reichart 1967; Grote 1977; Yamashita & Sasaki 1979). This problem necessitates a better description of the life circumstances of the algae, especially in flab, where often exuberant conjugations of various Zygnemataceae can be observed. A systematic study on stratification or diurnal variations in flab is not yet available. Philip (1927) observed diurnal variations in flab of pH 7.35 to 9.6 in a shallow part of Crystal Lake (Minnesota, U.S.A.). An extensive literature exists on diurnal variations of respiration and photosynthesis in artificial ecosystems (e.g., Beyers 1965) or in open cultures of unicellular algae (Cohen & Church 1981). The present paper deals with a description of the flotation process and stratification phenomena of temperature, pH, and oxygen content as well as the diurnal aspects of these factors.

Sampling locations / Materials and methods

The observations were made at various sites in the area south of Amsterdam, the Netherlands, especially in the nature reserves Botshol near Abcoude and Naardermeer near Naarden and in a ditch in the polder Demmerik near the village of Vinkeveen. All waters are eutrophic and rich in bicarbonate, which is the common type of water in this area. The water quality was oligosaprobic, sometimes ranging to beta-mesosaprobic (Naardermeer and Botshol).

The succession in the Naardermeer area was observed in 1980 and 1981 in a shallow pond called Binnenzij, with a depth ranging from 0.40 to 0.80 m. Algal quantities were estimated by visual observations from a rowing boat; it was impossible to use a more exact method because of the undefined shapes and rather mobile character of the flab masses. The temperature stratification was studied on 21 June 1979 in a concrete experimental pond (depth 0.40 m) near the laboratory partly filled up with a dense perennial growth of *Chara vulgaris* L. Temperature stratification was further observed from 30 June to 1 July 1980, in Naardermeer (Binnenzij) in a part with a layer of *Oedogonium* spp. More extensive measurements of several environmental parameters were conducted on 13/14 July 1980, in the Demmerik ditch (depth 0.40 m) covered with a flab of *Spirogyra* spp., *Zygnema* spp., *Oedogonium* spp., and *Cladophora glomerata* (L.) Kuetzing.

Irradiation was measured with a Li-Cor 185A radiation meter with Quantum sensor. Light quantities were expressed in microeinsteins per m^{2} per second, a measure for the photon flux in the photosynthetic active light. Global radiation data were expressed in Joules per m^{2} per day (means of 10 days) and obtained from the weather report service of the KNMI (Koninklijk Nederlands Meteorologisch Instituut) in De Bilt, about 20 km from the sampling locations.

Temperature stratification was measured in three ways. A Grant G820 temperature recorder was used with 10 sensors mounted along a rod at 10-cm intervals. The need for more small-scaled observations led to the use of a glass embedded thermistor (2×15 mm) of suitable resistance mounted on top of a calibrated 4-mm diam. glass tube. The parts were sealed and connected with Parafilm. This sensor was very useful for measurements at very small intervals without disturbance of the stratification and it could easily be immersed in the algal mass. The resistance of the thermistor was calibrated and measured with a Keithly 160 digital multimeter. Positive experience with the thermistor led to the construction of a sensor with 14 thermistors. The sensor was made of a strip of double-sided epoxy print material on which the necessary wiring was etched. The small (1×3 mm) thermistors were soldered to the narrow side of the $200 \times 10 \times 1.5$-mm print, and the whole was isolated with a thin layer of epoxy enamel and connected to the multimeter by a multicoiled cable and a 20-position selector switch. A layout of the print is available from the author upon request. The thermistors on the sensor were calibrated in a waterbath.

Oxygen content of the water was measured with a Precision Scientific Galvanic Cell Oxygen Analyzer. Values of pH were obtained with a Knick 650 digital pH meter with an Ingold combined electrode. The time was expressed in Dutch Summer Time (GMT +2 h) with solar noon at 13.45 h.

Results

The ascent process

Observations at various locations showed that the updrift of flab proceeds with a general pattern (Fig. 1). First, algal mats develop from spores or vegetative material on the bottom of the water, and these mats can be observed already in February, as soon as the frost period is ended. These mats grow rapidly during March and April. During the last weeks of April, gas bubbles develop in the algal accumulations during periods of sunny weather and the process of updrift begins. On dark days the ascent is interrupted. Already by mid-May many waters are covered with a thick layer of floating filamentous algae, the flab. The layer becomes more compact during the summer and disappears during August or September but remaining fragments can be found until winter ice forms.

The ascent process was observed in detail in a

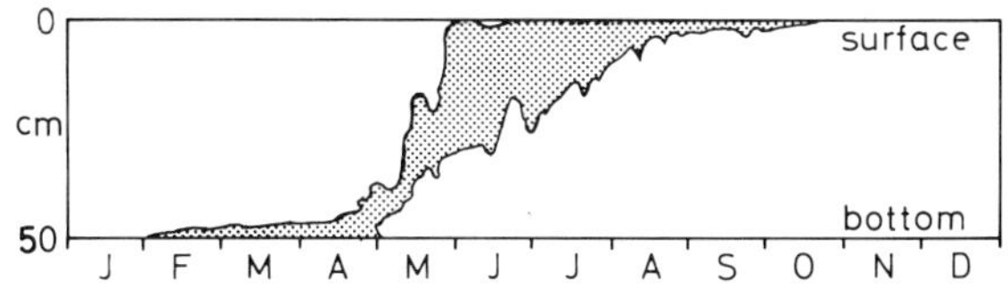

Fig. 1. General scheme of the ascent of flab in natural waters in the Netherlands.

glass aquarium filled with pond water at room temperature behind a window exposed to the West (Fig. 2). The species involved was *Cladophora glomerata* grown in the aquarium in the months preceding the observations. In early February gas bubbles developed during a sunny period in the algal mat on the bottom of the aquarium. The number and the size of the gas bubbles increased rapidly, and the algal mat was locally lifted up. This process proved to be irreversible, and during a number of dark days the updrift stopped but was not reversed. Within two weeks almost the whole mass floated at the water surface and it remained there during the entire summer until it was consumed by two introduced snails (*Lymnaea stagnalis* L.).

Individual algal species show variations on the theme given in Fig. 1. Dominance of species varied

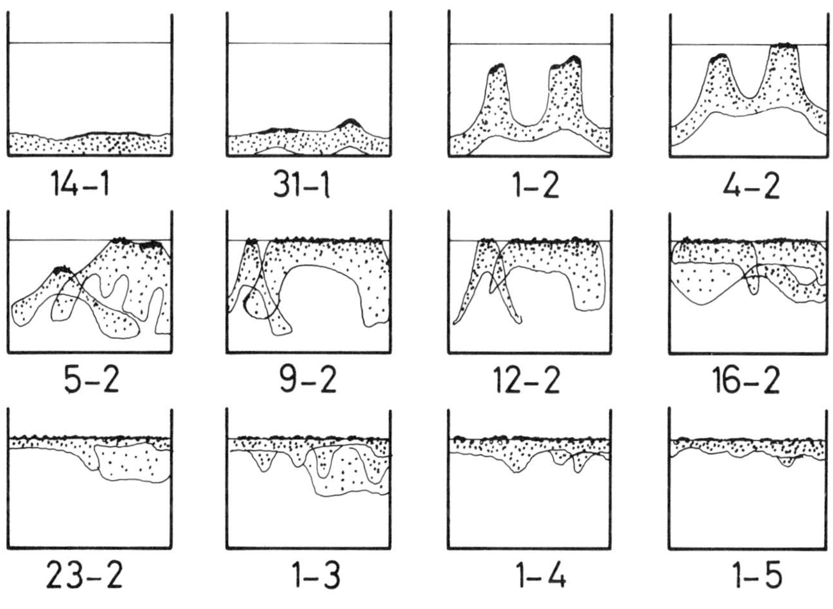

Fig. 2. Ascent of a flab of *Cladophora glomerata* (L.) Kuetzing in a laboratory aquarium in 1981.

during the growth season in Naardermeer in 1980 and 1981 (Fig. 3). Annual sequences of species exhibited differences between the years but it was evident that species show preferences for particular periods. Most species developed earlier in 1981. *Hydrodictyon reticulatum* and *Enteromorpha intestinalis* only attained mass development in 1981. Species of the genera *Tribonema* and *Microspora*, in addition to some *Spirogyra* species, were observed relatively early (April). Most *Spirogyra* species moved to the surface in May-June. *Mougeotia* species were mainly observed in July-August, but they were mostly preceded by *Cladophora glomerata* and *Oedogonium* spp. The two latter genera sometimes developed remarkably persistant flab that could be found until winter. Besides the temporal pattern, a clear spatial pattern was present, with patches of different age and species composition. The picture observed in the Naardermeer area seemed to be rather representative, and it was observed in numerous other waters during the observation period. The main period for flab development in the Netherlands was in the months of May and June. These months are characterized by relatively high levels of global radiation and rather low temperatures in comparison to those of the following summer months.

Stratification

Stratification in a floating thready mass-like flab can be expected during periods of intense radiation. Therefore measurements were conducted with the glass thermistor sensor at several locations around Amsterdam at various depths. All measurements were carried out during periods with fair weather in the afternoon. Table 1 shows the results of these observations. The temperature differences

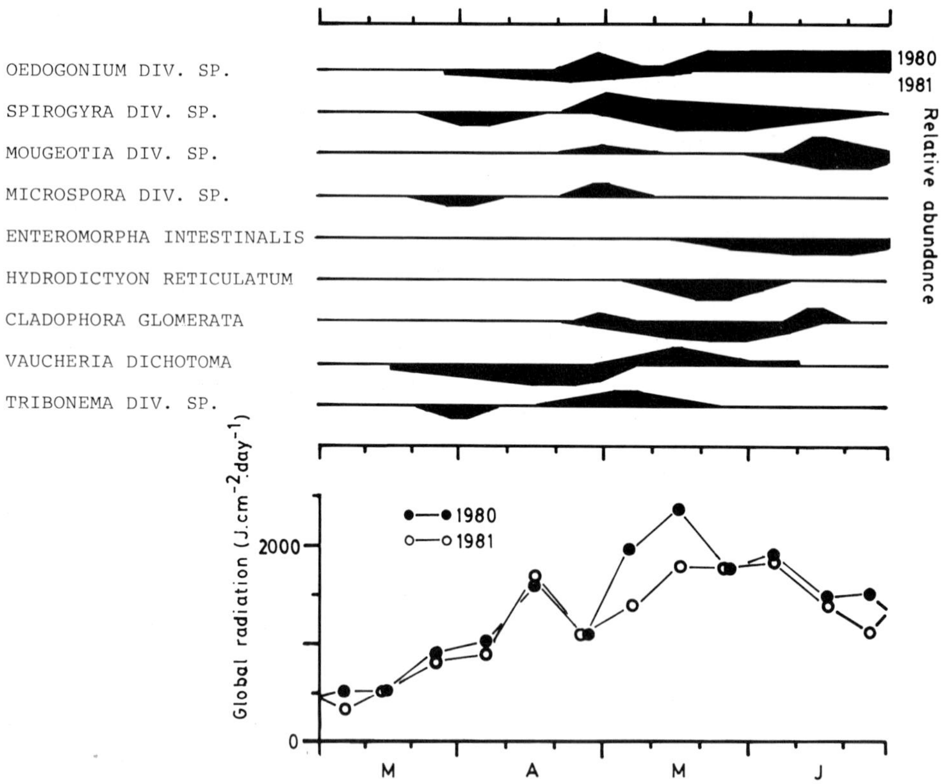

Fig. 3. Relative abundance in 1980 and 1981 of dominant flab algae in the Naardermeer area (Binnenzij) and global radiation in the same period.

Table 1. Temperature of the air, flab surface (s) and below the flab (b), and temperature differences between surface below the flab (s-b) in various locations and species.

Location	Date	Daytime (h)	Temperature (°C)			
			Air	Surface (s)	Below algae (b)	(s-b)
Experimental pond						
(Chara vulgaris)	13.6.77	1300	21	29	15	14
		1500	30	30	15	14
		1530	30	30.5	15	15.5
	20.6.78	1400	30	32	22	10
	21.6.78	1400	25.9	31.2	21.0	10.2
Botshol						
(Cladophora glomerata)	22.6.78	1225	18.2	26.8	21.0	5.8
Naardermeer						
(Oedogonium spp.)	2.9.81	1400	21.6	20.2	19.2	1.0
(Mougeotia spp.)	7.9.81	1400	24.9	23.8	20.6	2.2
Demmerik polder						
(several species)	14.7.80	1600	25.0	30.2	24.3	5.9

between the surface of the algal mass and the water below ranged from 1 to 15.5°C. Maximum temperature sometimes reached values about 30°C. A more detailed presentation is given of the stratification of temperature in the experimental pond with *Chara* (Fig. 4). Values in the algal mass are compared with those in a section of the pond without algae. The temperature in the algae even rose above air temperature.

Dynamics of stratification

The temperature stratification in shallow water will

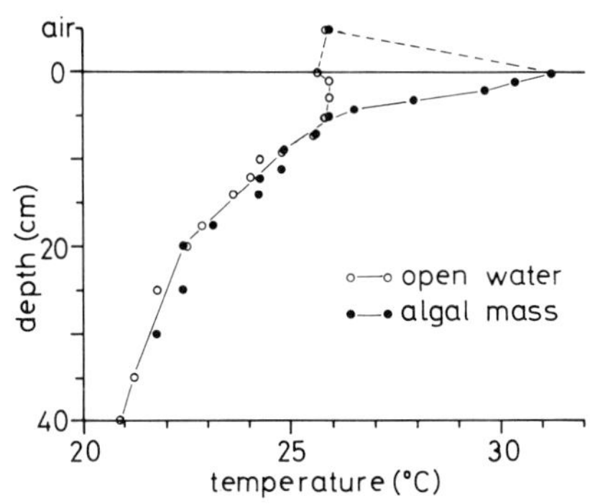

Fig. 4. Temperature stratification in *Chara vulgaris* L. in an experimental pond with a depth of 40 cm at 1400 h in the afternoon of 21 June 1979.

certainly be caused by radiation phenomena. Therefore, variations can be expected during a 24-h cycle. For open water this diurnal variation was already shown by Happey (1970). The diurnal variation in the Naardermeer in and below a cover of *Oedogonium* spp. is presented in Fig. 5. During rather cool and windy weather a distinct stratification was observed in the late afternoon. The inversion of stratification in the early morning was marked. More detailed observations were made with the epoxy sensor during 24 h in the Demmerik ditch. The sensor was mounted in a floating mass of various algal species and temperatures were observed at hourly intervals (Fig. 6). Simultaneously, light intensity and air temperature were measured. A distinct stratification developed, with a surface temperature of ca. 30°C at 1600 h and 25°C at –20 cm. The gradient was inverted in the early morning: at 0600 h the surface temperature was 16°C and the –20 cm temperature was then 20°C. The surface temperature showed a daily range of 9°C; at –20 cm it was 5°C. The gradient was clearly less steep than the one in Fig. 4. The temperature rose very rapidly, as is shown in Fig. 7, giving a series of observations obtained at 10-min intervals in the morning during the same 24-h experiment with the epoxy sensor. Between 1040 and 1140 h the temperature rose 4 degrees. At the same time a thin

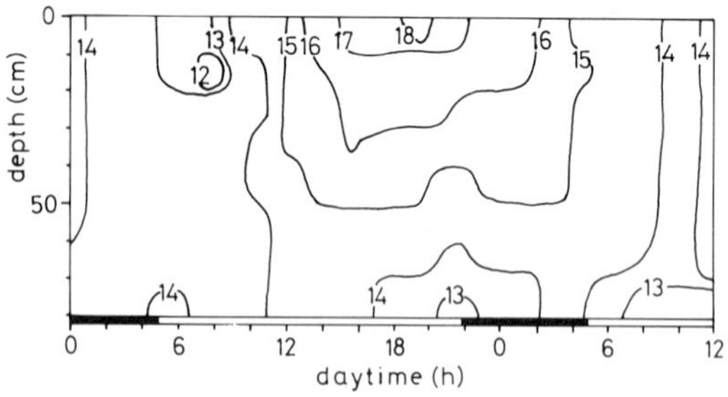

Fig. 5. Diurnal variation and stratification of temperature in the Naardermeer area (Binnenzij) on 30 June/1 July 1980, at a site with a flab of several *Oedogonium* species. Water depth 80 cm.

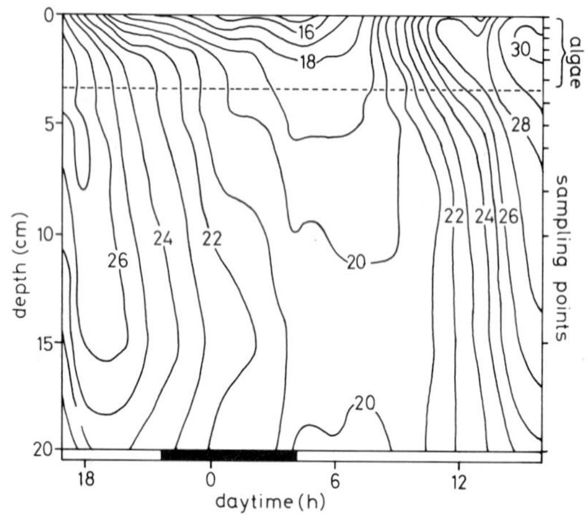

Fig. 6. Diurnal variation and stratification of temperature on 13/14 July 1980, in a ditch in Demmerik polder with a depth of 25 cm and a dense cover of several species of filamentous algae.

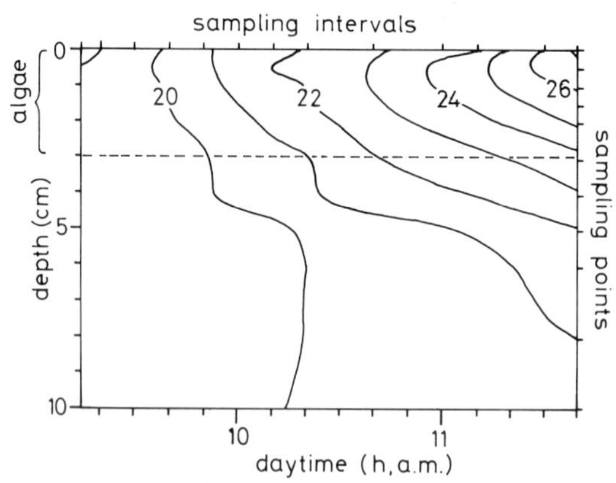

Fig. 7. Detailed presentation of rapid warming of the water in the Demmerik ditch (see also Fig. 6) on 13/14 July 1980.

surface film of a few millimeters was somewhat cooler than in the directly underlaying water. The difference between −0 and −5 mm varied from 0.3 to 1.0° C.

Stratification of pH and oxygen content

Because of the robustness of the electrodes, oxygen and pH could only be measured at rather large intervals. Measurements were carried out at some distance from the temperature sensor in order to avoid interferences. Further, pH and oxygen were measured each time at another site to avoid the effect of disturbance by the sampling procedure. The results are presented in Fig. 8, together with air temperature and irradiation. It was clear that pH and oxygen showed a distinct stratification between the depths of the measurements: at the surface of the algae (–0 cm), in the algae (–2 cm), and below the algae (–5 cm). The pH attained highest values (pH 9.9) at the surface about 1200 h. Lowest values occurred at dawn (pH 7.5 at –2 cm). The surface pH was lowest at 00 00 h with a value of 8.0. Great variations were observed in the oxygen content of the flab. At the surface oxygen varied between 6 and 16 mg $^{-1}$ (at 05 00 and 17 00 h, respectively). Below the flab values of 5 to 11 mg $^{-1}$ occurred (at 07 00 and 12 00–18 00 h, respectively).

Interactions between pH and oxygen content

It is well known that pH and oxygen content are mainly regulated by photosynthesis and respiration in freshwater systems (e.g. Welch 1980). Because the animal component was rather small in the systems investigated, the observed variations can be regarded as being caused mainly by the algae in the flab. The pH reflected almost directly the changes in the bicarbonate buffer system of the hard waters. The relation between pH and oxygen content at –2cm in flab and in open water (at a few meters distance in the same ditch) is presented in Fig. 9. The course of the observed daily variation was marked. As soon as it became dark after 20 00 h, pH and oxygen decreased until 00 00 h. Then oxygen continued to decline, but pH remained at a level of about pH 8. At dawn pH and oxygen

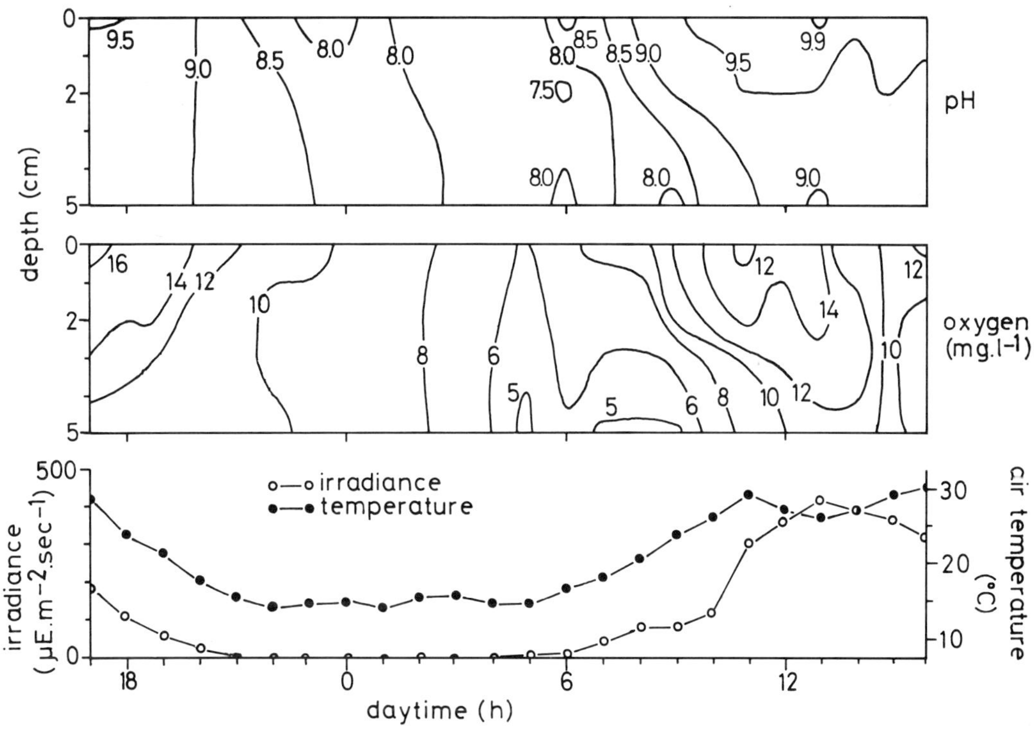

Fig. 8. Diurnal variation and stratification of pH and oxygen content in Demmerik ditch on 13/14 July 1980, and of air temperature and irradiance.

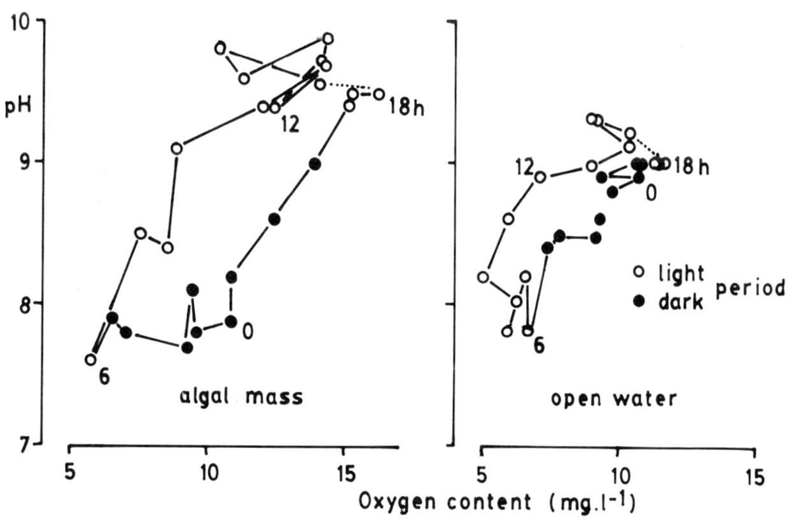

Fig. 9. The relation between pH and oxygen content during a 24 h cycle on 13/14 July 1980, in Demmerik ditch, in the algal mass (– 2 cm) as well as the open water.

increased immediately until 10 00 h. Then the pH remained at a level of about 9.5 until the end of the day. Oxygen, however, decreased after a few hours of stagnation, from 14 00 h to 16 00 h, and afterwards it rose again, in the so-called afternoon depression. During the morning pH was relatively high in comparison with oxygen, and in the evening it was relatively low. The morning and evening trajectories were more or less parallel but at a different level. Similar phenomena occurred in the open water of the Demmerik ditch, but the amplitude of the variations was less pronounced and the upper values were lower. The afternoon depression was almost absent.

Discussion

Marvan *et al* (1978b) mention that flab develops in the littoral region of ponds by means of outgrowth of algal filaments from reed stems and other substrates. Only *Vaucheria dichotoma* rises from the bottom upwards. Which process will prevail in the building up of flab, horizontal outgrowth or ascent from the bottom, will most probably depend on the morphometry of the habitat. We also observed such horizontal outgrowths in the littoral areas of the ponds investigated, but at some distance from the borders the vertical ascent prevailed. Of course, vertical ascent is only possible when a sufficient amount of light reaches the bottom. Thus in deeper or polluted waters flab development might be restricted to horizontal outgrowth from the littoral area. The aquarium observations show the importance of temperature. The ascent of *Cladophora glomerata* in the field occurred mainly in the second half of May or even later when water temperatures had increased to 12–15 °C. In the aquarium the process occurred after 1 February, during the first sunny period of the year.

Periodicity and variations in species composition of flab in the Naardermeer area strongly resembled the observations of Hillebrand (1977) and Marvan *et al* (1978). Thus there is evidence that the present observations showed a general pattern occurring in the temperate region of Europe. Observations from other parts of the world are not systematically available, but the findings from Northern America (e.g. Transeau 1916 and Philip 1927) fit fairly well into the useful scheme generated by Marvan *et al* (1978b).

Stratification of temperature in flab has not been described at all until now. The results, however, agreed with the expected stratification in water when turbulence is restricted by means of thin filaments, acting as an isolation blanket. The effect of this warming most probably can be explained completely by physical processes like radiation, isolation, and reduction of turbulence. The extreme temperature values support the supposition that, at least in the uppermost layer, lethal values for the algae are reached. This result might explain why so often after sunny periods the flab is covered with a yellow layer of dead algae. It is also possible that ultraviolet radiation plays a role in the senescence of algae in the upper layers. The rapid and strong variations of surface temperature are most important for the algae and the organisms living among them. Many invertebrates (*Trichoptera, Mollusca*) can be found living among the algae during the occurrence of high temperature values. These organisms must be able to survive changes of 15 °C or more during the diurnal cycle. Besides the diurnal variations, a distinct vertical gradient can be observed: differences of 10 °C across a few decimeters are common.

The observed variations of pH were similar but larger than the observations of Philip (1927). He found in Crystal Lake (Minnesota, U.S.A.) diurnal variations in flab between pH 7.35 and 9.8, using a colorimetric method. It can be concluded that the pH variations in the present paper were not exceptional: they most probably are normal for flab in the temperate climate region. Stratification of pH or oxygen in flab was not described before. The high supersaturation values in the second half of the light period explain the development of oxygen bubbles, causing flotation of the flab. At dawn low values were found just below the algal layer (5 mg l^{-1}). These low values are unfavorable for animal life and could interfere with the faunistic composition of shallow waters. Low values were especially found at sites where the flab covered almost the entire small water body.

Some remarks can be made on the afternoon depression. The phenomenon is defined as a decrease of oxygen production during the first half of the afternoon and is most probably caused by depletion of bicarbonate by photosynthetic processes (Welch 1980; King 1972). Values of pH remained at a constant and high level during the depression, the level being determined by the buffering capacity of the water and the dominating primary producers. The depression has been observed in phytoplankton blooms of high density, but it was not described before from flab. Further observations are necessary for a better understanding of the ecological implications of the phenomenon. The occurrence of the afternoon depression means that at least three nutritional factors can be limiting during the algal growth period: phosphorus, nitrogen, and carbon, the latter with a diurnal rhythm.

Flab is shown to be a community in which extreme values of temperature, pH, and oxygen occur in a diurnal rhythm with a distinct vertical stratification. This variation justifies further investigations in this remarkable community and the organisms that are adapted to grow and reproduce in it.

Acknowledgments

The author wishes to express his gratitude to the students J.C.A. van Alphen (1980), H.J. de Wit and J. van der Oost (1981) for substantial assistance in the Naardermeer observations and to R.W.J.R. Dekker and J.M. Stap for the execution of the 24-h experiment in Demmerik. Prof. Dr. M. Vroman, Dr. J. Simons, Drs. J. van der Heide, Drs. H.J. Sluiman and Drs. P.J.R. de Vries have contributed to the text by means of fruitful discussions or aid in writing the manuscript. Important technical assistance was obtained from L.J. Valkier and K. Wattel with the development and construction of the temperature sensor. The manuscript was typewritten by Ms. D.H.M. Hoonhout.

References

Beyers, R.J., 1965. The pattern of photosynthesis and respiration in laboratory microecosystems. Mem. Ist. Ital. Idrobiol. 18 Suppl.:61–74.

Cohen, R.R.H. & Church, M.R., 1981. A method for studying primary productivity and photosynthesis in unenclosed batch cultures and field samples. Arch. Hydrobiol. 91:231–241.

Grote, M., 1977. Studies in the copulation process of the green alga *Spirogyra majuscula*. Protoplasma 91:71–82.

Happey, C.M., 1970. Some physico-chemical investigations of stratification in Abbot's pool, Somerset: the temperature cycle. J. Ecol. 58:419–434.

Hillebrand, H., 1977. Periodicity and distribution of multicellular green algae in two lake areas in The Netherlands. Thesis, Free University, Amsterdam, 178 pp.

Hillebrand, H. & Klapwijk, S.P., 1976. Distribution of multicellular benthic algae in an eutrophic ditch. Hydrobiol. Bull. 10:48–58.

King, D.L., 1972. Carbon limitation in sewage lagoons. In: Nutrients and eutrophication, Special symposium, Vol. 1, Amer. Soc. Limnol. and Oceanogr. pp. 98–110.

Komárková, J. & Marvan, P., 1978. Primary production and functioning of algae in the fishpond littoral. In: Dykyjová, D. & Květ, J., (eds.) Pond littoral ecosystems. Ecological Studies, Vol. 28, Springer Verlag, Berlin. pp. 321–334.

Lange, L. de & Zon, J.C.J. van., 1973. Proposal for a numerical description of the development of aquatic macrophytic vegetation as an aid for the asessment of water quality. Wasser- und Abwasserforsch. 6:125–128.

Marvan, P., Komarék, J., Ettl, H. & Komárková, J., 1978a. Structural elements. Principal populations of algae. Spatial distribution. In: Dykyjová, D. & Květ, J., (eds.) Pond littoral ecosystems. Ecological Studies, Vol. 28, Springer Verlag, Berlin, pp. 295–313.

Marvan, P., Komarék, J., Ettl, H. & Komárková, J., 1978b. Dynamics of algal communities. Ibid., pp. 314–420.

Philip, C.B., 1927. Diurnal fluctuations in the hydrogen ion activity of a Minnesota lake. Ecology 8:73–89.

Reichart, G., 1967. Die Synchronkultur von Spirogyra. Ber. Deutsch. Bot. Gesellsch. 80:177–186.

Transeau, E.N., 1916. The periodicity of freshwater algae. Am. J. Bot. 3:121–133.

Velichko, I.M., 1977. Production of some green filamentous algae in natural associations (In Russian). Gidrobiol. Zh. 13:23–27.

Welch, E.B., 1980. Ecological effects of waste water. Cambridge University Press. U.K., 337 pp.

Yamashita, T. & Sasaki, K., 1979. Conditions for the induction of the mating process and changes in contents of carbohydrates and nitrogen compounds during the mating process of Spirogyra. J. Fac. Sci. Hokkaido Univ., Ser. V 11:279–287.

Author's address:
H. Hillebrand
Department of Biology
Free University
P.O. Box 7161
1007 MC Amsterdam, The Netherlands

CHAPTER 6

A comparison of the periphyton of *Nuphar lutea* and *Nymphaea alba*

The distribution of diatoms on the undersides of floating leaves

EDMOND J.P. DELBECQUE

Abstract. The periphyton of *Nuphar lutea* (L.) Sm. and *Nymphaea alba* L., the dominating species of the vegetation of an old branch of the River Waal near Nijmegen, was studied from May until October 1981. The study evaluated the influence of the architecture of nymphaeid plants on the periphyton-composition of different plant parts. In particular the composition of the epiphytic diatom-flora was studied. The first part of this study consisted of analyses of the diatoms present on the underside of healthy looking floating leaves of both species. For this purpose, leaves with a distinct visible horizontal zonation were collected. Although light could be expected to be an important factor affecting the distribution of diatoms on the underside of floating leaves, the results of the analyses indicated that the period, during which the particular leaf-area becomes exposed to be surrounding water, is most important. The spatial differences in the studied area were not determined by their position but they reflect the differences in time of colonization, which are sufficient to explain the differences in the various diatom populations. Differences in the diatom flora between the floating leaves of the two nymphaeid species could not be detected.

Introduction

As part of a comprehensive research on nymphaeid-dominated systems (see e.g. Van der Velde 1980, 1981), the influence of the architecture of nymphaeid plants on the structure and composition of their periphyton was the subject of this study. This paper only deals with the diatom community present on the undersides of healthy floating leafblades of *Nuphar lutea* (L.) Sm. and *Nymphaea alba* L.

Due to the gradual development of floating leaves of these species, the periphyton shows a distinct zonation (Fig.2B). Originally the new leaf is tightly rolled and only a narrow, arrowlike zone on both sides of the midrib is exposed to the surrounding water (Fig. 2A). On growth ascent or at the surface the leaf progressively unrolls itself, during which process the colonization of the newly exposed parts begins. Due to differences in, for example, colour and texture of *Nuphar* and *Nymphaea* leaves, an effect on their periphyton is possible.

Study area

The study area was in a former branch of the River Waal near Nijmegen (The Netherlands), which is normally flooded by this river during the first three months of the year. During the growing season *Nuphar lutea* and *Nymphaea alba* cover extensive parts of the water surface. To avoid disturbance of the main area, which is designated as a nature reserve, the main sampling site (Fig. 1, I) was chosen in the adjacent area. With respect to the

Wetzel, R.G. (ed.), Periphyton of Freshwater Ecosystems.
 ISBN 90 6193 768 X.

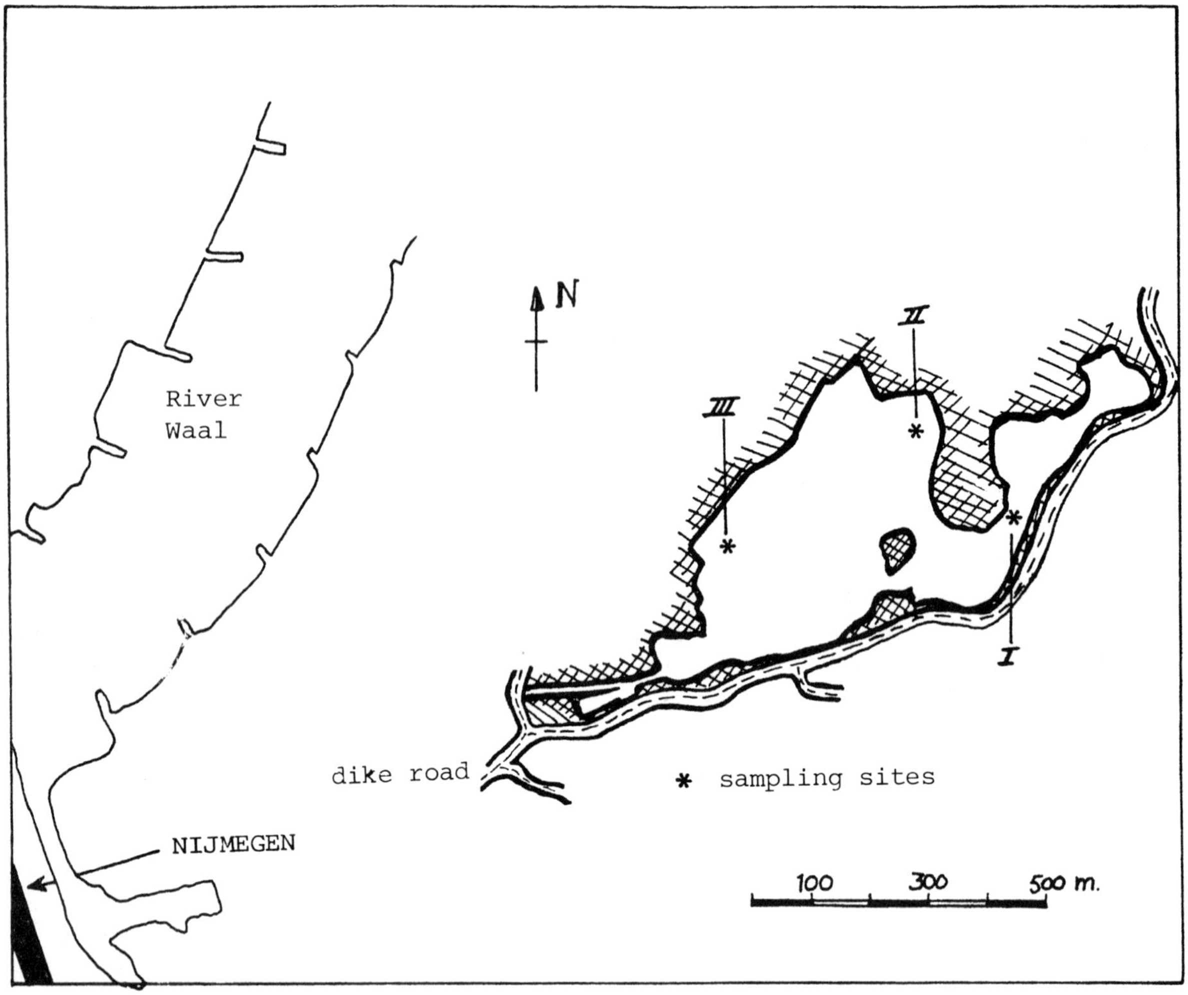

Fig. 1. Sampling locations at Oude Waal.

exact position, care is taken to eliminate the influence of the bottom and banks as much as possible. During the sampling period from May until October 1981, the water depth varied between 60 and 100 cm; the pH and alkalinity varied between 7.6 and 8.2, and 4.7 and 6.7 meq l^{-1}, respectively.

Material and methods

The samples taken consisted of undamaged leaves with an estimated age between 1 and 2 weeks (counted from when the leaf starts to unroll) and a distinct zonation of periphyton on their undersides. In order to determine how representative the chosen site I was on 12 May, samples were also taken from two other remote sites (Fig. 1, sites II and III).

Subsamples, consisting of two circular pieces bored out of the leaf with a diameter of 18 mm and a total surface of 510 mm^2, were taken in each zone along a line perpendicular to and at both sides of the midrib near the origin of the petiole (Fig. 2B).

To avoid loss of tightly adherent diatoms such as *Cocconeis placentula* Ehr., the pieces of leaf were oxidized together with the diatoms in a mixture of sulphuric and nitric acid. After washing, the cleaned diatom frustules were filtered onto a Sar-

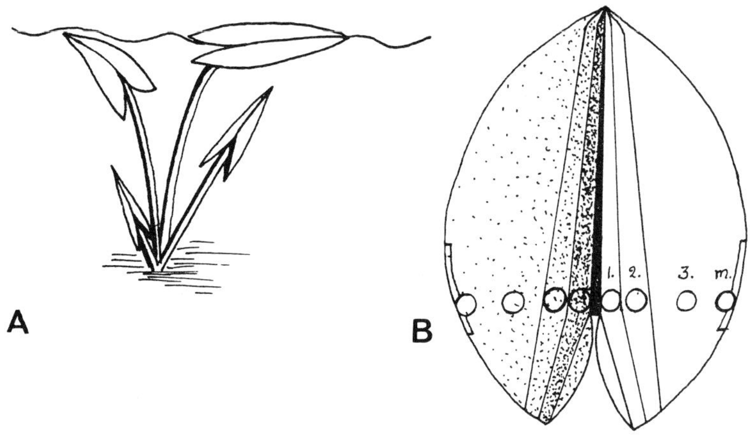

Fig. 2. Development of a floating leaf (A) and the resulting zonation on the underside of a floating leaf (B).
○: samples taken per zone;
□: sampling of the margin, during the summer.

torius membrane filter with a pore diameter of 3 μm. Subsequently, the filter was cleared with immersion-oil and at least 500 frustules were counted from each sample; especially when one species dominated, the counts were continued until 500 other frustules were encountered.

Results and discussion

The tendencies in diatom colonization and succession can be demonstrated with the aid of genera and species that were chosen because of their continued moderate to high relative abundance and their more or less clear description. According to their 'diatom life-form strategies', a term introduced by Kuhn *et al* (1981), the following divisions can be made.

1. Tychoplanktonic diatoms belonging to the genera *Melosira* (mainly *M. polymorpha* subsp. *distans* (Kütz.) Bethge var. *alpigena* (Grun.) Bethge and *M. varians* Ag.) and *Fragilaria* (mainly *F. intermedia* Grun.); both genera with chain-forming species.

2. Periphytic diatoms, that can be subdivided according to their mode of attachment: (a) diatoms belonging to the genus *Navicula* (mainly *N. gracilis* Ehr. and *N. cryptocephala* Kütz. sensu Germain), which glide along surfaces; (b) diatoms as *Cocconeis placentula* Ehr. (mainly var. *euglypta* (Ehr.) Cleve), that can cover the substrate as a carpet, and (c) the stalked diatoms *Gomphoneis olivacea* (Lyngb.) Dawson and *Gomphonema parvulum* (Kütz.) Grun., the latter of which includes possibly also individuals of *G. angustatum* (Kütz.) Rabenh.

Following the division of Sládečková (1962), groups 1 and 2a can be taken together under the name pseudo-periphyton, in contrast to the euperiphyton that consists of sessile, attached organisms, as 2b and 2c. According to their substrate, the last mentioned group can also be called epiphyton.

A comparison between sampling sites

The relative abundance of the selected diatoms is given per site, plant, and leaf-zone (Fig. 3). The absolute numbers of diatom cells per cm^2 are also plotted (Fig. 3, bottom). The relative abundances of the diatoms showed no consistent differences between the two macrophyte species at sites I and III. Site III, situated at the less exposed SW-corner of the study area, showed relatively somewhat fewer individuals of the *Navicula*-group, although these differences were not very striking.

The observation of Kuhn *et al* (1981), that the periphyton shows a succession from a dominance

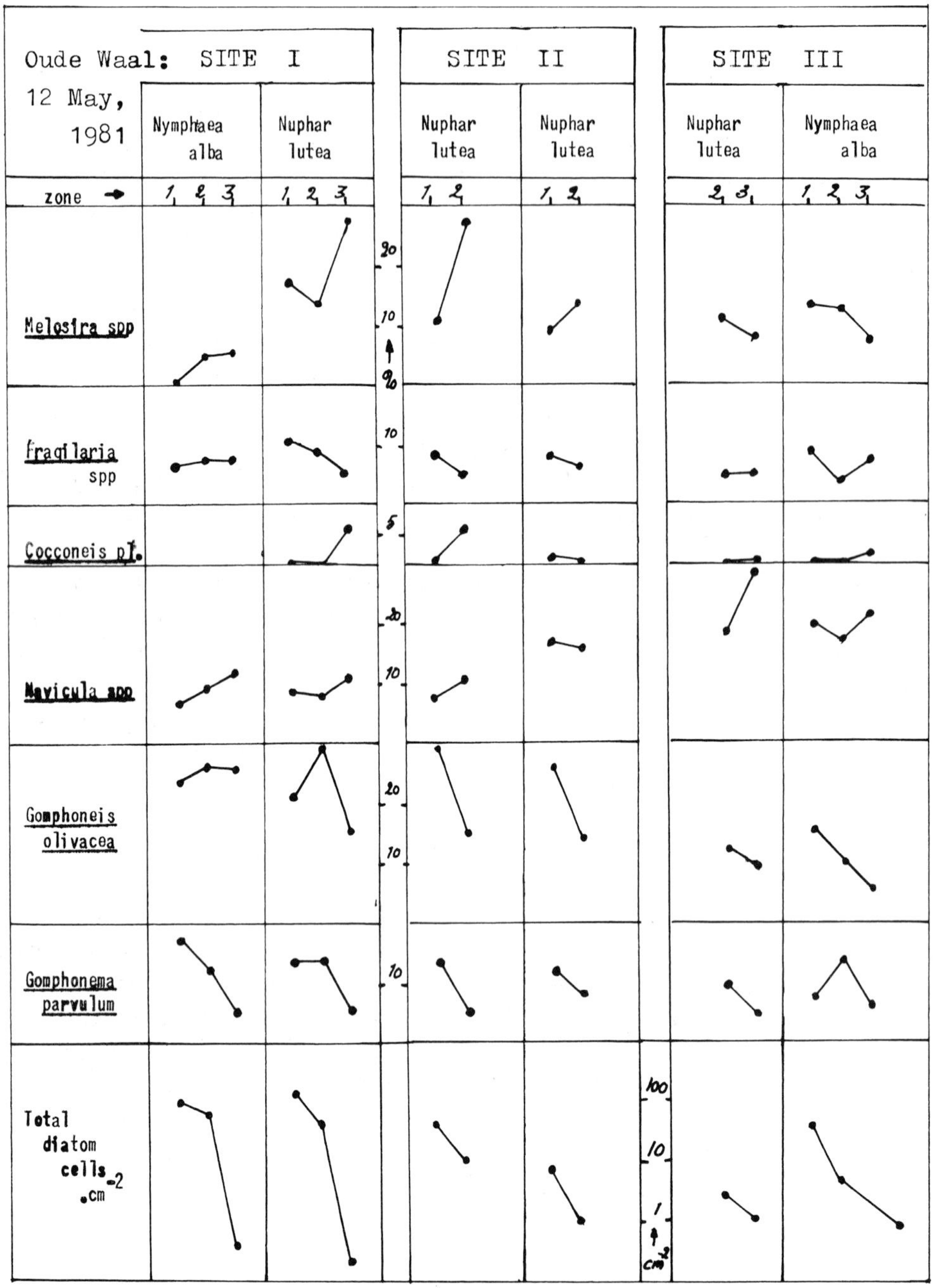

Fig. 3. The relative abundances (percentages of the total number of diatom cells) of some diatoms at different sampling sites.

of (tycho) planktonic species to a dominance of periphytic species, seemed to be confirmed by changes in relative abundance of *Gomphoneis* and *Gomphonema* from the most recent colonization in Zone 3 to the older settlement in Zone 1. One must, however, keep in mind that the analyses described by Kuhn *et al* (1981) began on one date, after which samples were taken at regular intervals, in contrast to conditions described in this study.

Periodicity

When only one or a few species dominate the periphyton, the relative abundance of the remaining species will be affected considerably. To avoid this dependency the counts are expressed as absolute numbers (Fig. 4).

The total number of diatoms on the undersides of the leaves exhibited a decline between 14 July and 11 August. Although the population density sequences of the different diatoms from high to low was normally correlated with the age of the zones, on 8 September the epiphytic species *Gomphoneis olivacea* and *Gomphonema parvulum* exhibited an inverse relation.

A plausible explanation for these changes is that the colonization of the different zones had taken place during conditions that became progressively more favourable for the propagation of species from the surrounding populations. Although the considered species did not show a comparable inverse relation on a very young, few-day-old leaf of *Nuphar,* collected on the same day, the moment of colonization by these highly dynamic, short-lived diatoms can play an important role in the composition of the periphyton at the different zones.

It is not probable that the mentioned effect was causal in the distribution of *Cocconeis placentula,* which exhibited a comparable distribution on *Nymphaea* on 11 October. Neither the distribution of green algae, e.g. *Oedogonium,* nor the presence of micro- and/or macrofaunal elements, could offer an explanation for the observed distributions.

The tychoplanktonic diatoms belonging to the genus *Melosira* exhibited a similar inversion, in contrast to *Fragilaria.* Because of their nature this distribution could also be derived from the distribution of the mentioned epiphytic species by colony entanglement.

The distribution of the mobile periphytic species belonging to *Navicula* resembled the distribution of 'total diatom cells', which indicated that the mobility of these diatoms did not lead to an even distribution over the surface.

The influence of the light intensity

Both because of the different depths at which the plants in differing zones became colonized and because the light intensity decreased at the undersides of the leaves from the margin to the midrib, light can be assumed to play an important role in the distribution of the diatoms. One must however remember that the time differs in the different zones in which the algae were first exposed to the surrounding water and the colonization started. The early colonizing algae received light more directly than algae after the macrophyte leaves unroll.

At the time of sampling of 1- to 2-week-old leaves, no influence of light intensity, e.g. a qualitatively differentiated diatom composition of the zones, could be detected. Continued sampling of marked leaves, which have a mean longivity of about 40 days (Van der Velde 1980), would be required to determine whether the light regime plays a more important role in the further development of the periphyton.

Acknowledgments

I thank Prof. Dr. C. den Hartog and Dr. J.F.M. Geelen for their critical remarks and Professor Robert G. Wetzel for improvements of my English. I am especially indebted to Els Geertman for her substantial contribution to the achievement of this paper. The study is carried out as a part of a project of the Netherlands Organization for the Advancement of Pure Research. Contribution No. 24 of the nymphaeid project.

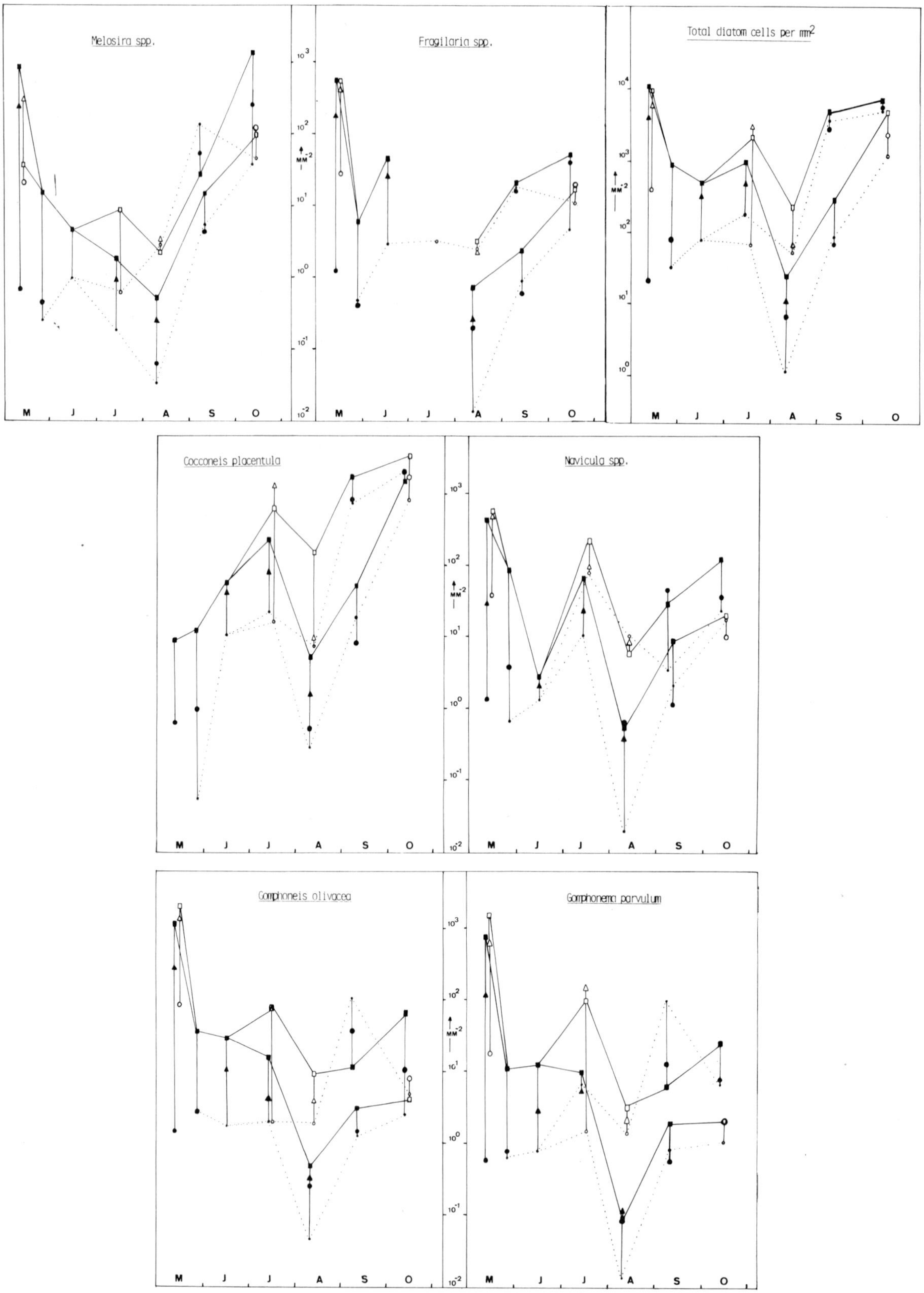

Melosira spp.
Fragilaria spp.
Total diatom cells per mm²
Cocconeis placentula
Navicula spp.
Gomphoneis olivacea
Gomphonema parvulum
MM⁻²
10⁴
10³
10²
10¹
10⁰
10⁻¹
10⁻²
M J J A S O

Fig. 4. Diatom numbers at different sampling dates per plant and per zone. *Nuphar lutea*: zone 1 (■), zone 2 (▲), zone 3 (●), margin (•). *Nymphaea alba*: zone 1 (□), zone 2 (△), zone 3 (○), margin (o). From two leaves sampled on the same day, the numbers of diatoms are plotted slightly shifted with respect to each other in order to avoid overlap. Results obtained from the first zones and from the margins are connected with a solid and a stippled line, respectively; the results obtained from one leaf are connected vertically.

References

Kuhn, D.L., Plafkin, J.L., Cairns J. & Lowe R.L., 1981. Qualitative characterization of aquatic environments using diatom life-form strategies. Trans. Am. Microsc. Soc. 100: 165–182.

Sládečková, A., 1962. Limnological investigation methods for the periphyton ('Aufwuchs') community. Bot. Rev. 28:286–350.

Velde, G. van der., 1980. Studies in nymphaeid-dominated systems with special emphasis on those dominated by Nymphoides peltata (Gmel.) O. Kuntze (Menyanthaceae). Thesis Nijmegen. Studentenpers Nijmegen. 163 pp.

Velde, G. van der., 1981. A project on nymphaeid-dominated systems. Hydrobiol. Bull. 15:185–189.

Author's address:
E. Delbecque
Laboratory of Aquatic Ecology
Catholic University
Toernooiveld
6525 ED Nijmegen, The Netherlands

CHAPTER 7

Structural investigations of periphytic protozoan communities in three layers of the Danube River

I. The question of stratification

M.Cs. BERECZKY, N. OERTEL and J.N. NOSEK

Abstract. Investigations were carried out on the Danube River at Göd (1669 river km) in Autumn 1979 and late Spring 1980. Samples were taken a three layers of the water column (surface, middle, and near-bottom) on the 2nd, 4th, 8th, 16th, and 32nd day after exposition. Glass slides (uncoated and coated by ovalbumin) placed into metal baskets were used as substrata. Simultaneously, water temperature, dissolved oxygen content, conductivity, pH, and oxidation-reduction potential were recorded using an *in situ* Water Quality Analyzer.

In the protozoan community two layers could be distinguished, near the surface and in a deeper stratum. Stratification was influenced foremost by the water-flow regime in the case of physico-chemical parameters. The near-surface stratum seemed to be a less stable environment than the others.

Introduction

Besides the systematic plankton research performed for almost 20 years, studies have been started to assess the role of protozoans in the formation of periphyton. From the beginning, these studies were concerned with the question of whether vertical stratification could form, and under what conditions, in streams.

Since Hentschel's classical work (1916), the process of the formation of periphytic communities and the role of periphyton have been investigated by several authors (Kralik 1957/58, Sládečková 1962a, 1962b, Sládečková & Sládeček 1963, Wilbert 1969, Nusch 1970, Pätsch 1974, Jones *et al* 1976, etc.) From these studies many valuable data have come to light, but the relationship between periphyton formation and vertical stratification in streams has been little explored.

Aims of the study

The main objective of this work was to investigate the protozoan community of periphyton in relation to depth. Parallel with that, the changes of certain physical and chemical parameters were also investigated. The results obtained are discussed here from the aspect of vertical stratification. The process of colonization is discussed in another part of this work.

Material and methods

The investigations were performed at Göd (1669 river km) 20 km upstream from Budapest in two seasons of different water flow regimes (from 23 October 1979 to 24 November 1979 and 12 May 1980 to 13 June 1980). The formation of periphyton was studied on artificial substrata (microscope slides). The slides fastened into plastic frames and placed

Wetzel, R.G. (ed.), Periphyton of Freshwater Ecosystems.
 ISBN 90 6193 768 X.

into metal baskets (Fig. 1) were exposed at three different depths: at 10 cm below surface, at the middle part of the water column, and near the bottom. The two latter depths were adjusted with variations in the water level. The positioning of the slides was parallel with the direction of flow in order to eliminate selective effects (cf. Kralik 1957/1958). 50% of the slides were coated with ovalbumine, the other half was left without treatment.

Following exposure, samples were removed on days 2, 4, 8, 16, and 32. Two treated and the two untreated slides, carefully lifted from the water, were placed into 'Schifferdecker' cuvettes. The periphyton was immediately analyzed and after fixation as well. Quantitative data obtained were based on counts determined directly on three 1 cm^2 surfaces on each slide. In the evaluation of results, the averages of these three replications were used. On day 32 in 1979 and on days 16 and 32 in 1980, the periphyton on the slides was so thick that direct counting was impossible, and therefore it was washed from the slides into 50 ml filtered Danube water and the data were converted to 1 cm^2 surface. During the evaluation, the sessile and mobile parts of the community were treated separately.

Fig. 1. Microscope glass slides fastened into plastic frames and held in a metal basket.

In the study period, the following physical and chemical parameters were measured: water temperature, conductivity, dissolved oxygen content, pH, and oxidation-reduction potential (ORP). Measurements were performed on every second day at the three depths of exposition. For these investigations a Model 6-D *in situ* Water Quality Analyzer (Surveyor) type field apparatus was used. Values of conductivity and pH were referred to 25° C, and the values for oxidation-reduction potential to 25° C and a normal hydrogen electrode. In the interpretation of results allowance was made also to water-level changes and water discharge.

The results were evaluated by means of multivariate analysis of variance (MANOVA). In addition to the physical and chemical data, four structural parameters pertaining to the protozoan community of the periphyton were also processed: species number (S), total individual number (N), species-individuals diversity (H), and the evenness (E). Diversity was calculated on the basis of the Shannon-Weaver function:

$$H = - \sum_{i=1}^{s} (n_i/N) \ln (n_i/N)$$

where $\sum_{i=1}^{s} n_i = N$ and the evenness by the formula

$$E = \frac{H_{max} - H}{H_{max} - H_{min}}.$$

Results

Analysis of physical and chemical conditions

In the two seasons differing essentially in water flow regime, the range (e.g. dissolved oxygen, pH, oxidation-reduction potential) and the direction (e.g. water temperature) of parameter changes also differed considerably.

Correlation coefficients between the water discharge and the parameters, as well as among the parameters, showed a close correlation between

Table 1. Correlation coefficients between the various physico-chemical parameters.

	WD	WT	DO	CD	pH	ORP
WD		−0.180 −0.186 −0.169	0.376 0.332 0.278	−0.899+++ −0.897+++ −0.892+++	−0.609+ −0.609+ −0.380	−0.335 −0.465 −0.662++
WT	−0.457 −0.447 −0.438		−0.752+++ −0.761+++ −0.771+++	0.162 0.171 0.186	−0.154 −0.150 −0.284	0.643++ 0.531+ 0.308
DO	−0.679++ −0.725++ −0.745+++	−0.055 −0.058 −0.035		−0.265 −0.235 −0.172	0.349 0.399 0.431	−0.686++ −0.592+ −0.512+
CD	0.416 −0.439 −0.428	0.298 −0.350 −0.387	0.711++ 0.819+++ 0.766+++		0.611+ 0.607+ 0.350	0.144 0.211 0.350
pH	−0.810+++ −0.834+++ −0.834+++	0.220 0.161 0.135	0.886+++ 0.916+++ 0.923+++	0.499+ 0.635++ 0.601+		0.082 0.204 0.350
ORP	0.786+++ 0.900+++ 0.843+++	−0.618+ −0.635++ −0.640++	−0.520+ −0.537+ −0.453	−0.129 −0.168 −0.057	−0.689++ −0.725++ −0.640++	

The right upper semimatrix contains the values of 1979, the left lower semimatrix contains the values of 1980. The three values below one another correspond to the layers: surface, middle, and near-bottom. Abbreviations: WD – water discharge, WT – water temperature, DO – dissolved oxygen, CD – conductivity, pH – pH, ORP – oxidation reduction potential. Levels of significance: + = 5%, ++ = 1%, +++ = 0.1%.

water discharge and conductivity, respectively, and between dissolved oxygen content and water temperature during autumn (Table 1). During spring a close correlation was found between water discharge and dissolved oxygen content, water discharge and pH, water discharge and oxidation-reduction potential, dissolved oxygen and pH, and pH and oxidation-reduction potential.

In autumn, 1979, on the basis of variance analysis, it was possible to establish the presence of two separate layers in the five parameters. Of the two layers, one contained the near-bottom sample, and the other the middle and the surface samples. In the case of dissolved oxygen, the difference was significant at the 10% probability level, and of the other parameters at the 5% level. During spring, such stratification was not evident.

Figure 2 shows the inverse stratification of oxidation-reduction potential during autumn 1979. In the figure the data of the surface and near-bottom samples are illustrated relative to the data of the sample taken in the middle. In spring, 1980, a similar phenomenon was not observed, but during the progressive decrease of water-level, which began in midsummer, a situation developed similar to the previous year (Fig. 3).

Structural parameters

The variance analyses considered the lengths of exposure periods, the depth, and the treatment (ovalbumine). The 5% probability level was chosen as the significance level. On the basis of interaction values, the single effects were not independent from one another.

Stratification

In 1979, stratification within the sessile communities was demonstrable in individual numbers, and

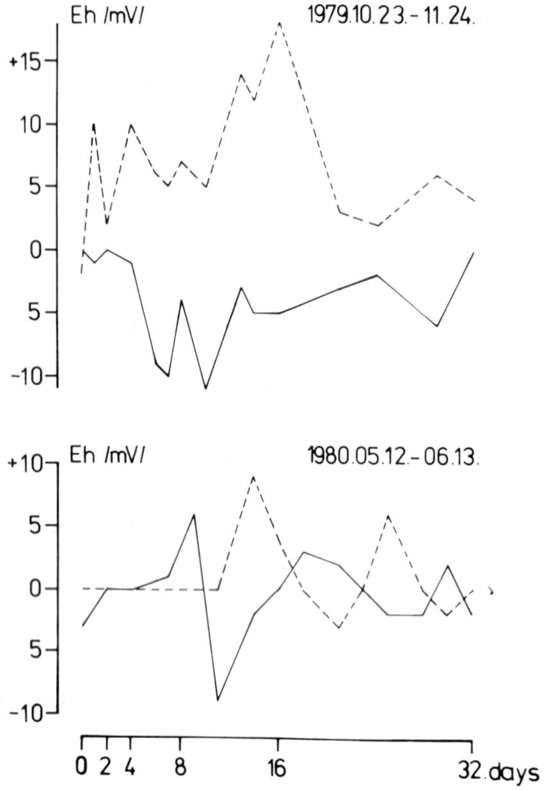

Fig. 2. Oxidation-reduction potentials in the surface and near-bottom layers relative to those in the middle part of the water column during the two investigation periods. —— surface, ----- near the bottom.

in 1980 in species number, individual numbers, and diversity. Concerning the mobile part of the community, stratification was found in both years and in each structural parameter.

When stratification occurred, the three depths formed two layers in most cases. The surface and the near-bottom layer separated in every case and the middle layer was linked in most instances to the near-bottom level, and less frequently to the surface stratum. In the sessile part of the community, the separation of layers took place on day 4 in 1979, and as early as day 2 in 1980. In the case of mobile organisms, the separation of layers became evident from day 4 in each year. From the point of view of stratification, there was no difference between the two kinds of slides.

In the sessile part of the community, apart from a few exceptions, the highest values of structural parameters always occurred in the near-bottom samples. In 1979, greatest species and individual numbers were found on the surface on day 32 both on the treated and the untreated slides (these total individual number values exceeded all experimental values). The greatest proportion of total individuals was made up of four species of *Vorticella* (Table 2).

During this experimental period, the water-level

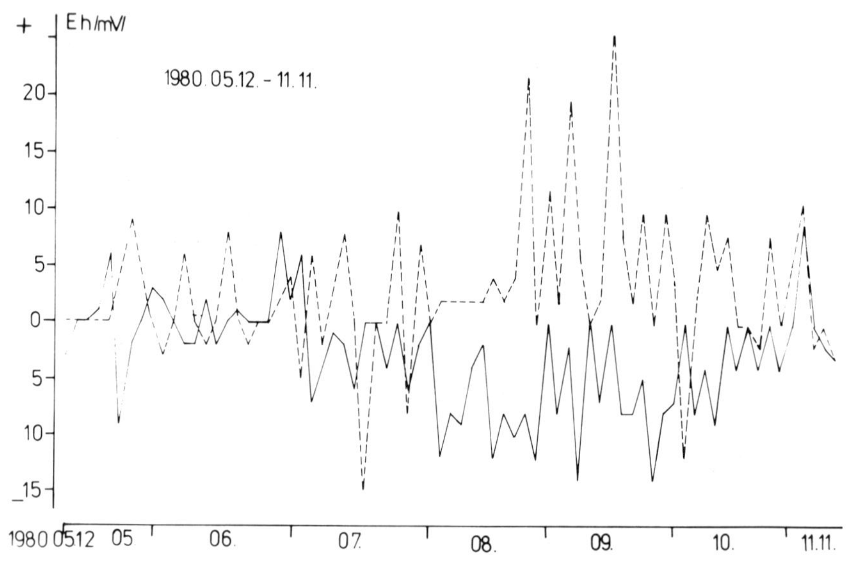

Fig. 3. Oxidation-reduction potentials in the surface and near-bottom layers relative to those in the middle part of the water column between 12 May and 11 November 1980. —— surface, ----- near the bottom.

increased rapidly from day 18, and as a result the formerly completely separated, lake-like side bay, about 450 m upstream from the sampling station, again became connected to the main reach. Individuals of the *Vorticella* population formed in the side bay drifted into the main arm and colonized the slides at the surface (cf. Wilbert 1969).

In the mobile part of the community in 1979, highest structural parameter values also occurred in the near-bottom samples. Here, too, day 32 was an exception but only in the case of untreated slides. In 1980, maximal values occurred predominantly in the samples from the middle stratum.

Treatment

In 1979, the effects of slide treatment manifested itself only in the individual numbers of the sessile community. In 1980, the sessile and mobile parts of the community behaved in the same manner in regard to species number, total individual number, and diversity.

The changes caused by treatment were not equal at the three different depths. In the case of sessile organisms, an appreciable effect was found in 1979 in the near-bottom and middle samples from day 4, at the surface from day 8 and in 1980 in the near-bottom samples on day 2, in the middle layer from day 4. In 1980, the effect of treatment was not evident on the surface. The effect of treatment manifested itself in the increase of the structural parameters.

With the mobile organisms, appreciable differences between treated and untreated slides developed only from day 8 in 1980. However, the corresponding structural parameters of the treated slides were found only on day 8 to be greater than the untreated slides. From that time, untreated slides exhibited greater values.

Table 2. Greatest proportion of total individuals of four species of *Vorticella*.

Species	Slide treatment	
	untreated (%)	treated (%)
Vorticella campanula	38.2	29.5
Vorticella convallaria	3.5	19.2
Vorticella margaritata	25.5.	26.2
Vorticella microstoma	20.1	15.8
	86.3	90.7

Summary

Under appropriate conditions, physical and chemical stratification of varying duration can occur transitorily in streams. From the point of view of stratification, the water-flow regime plays a primary role, since a sustained low water level assures favorable conditions for the formation of stratification. In the Danube River, such periods occur primarily in the autumn. Two layers can be distinguished, a 'thinner' one near the bottom, and a 'thicker' one ranging from the surface to about the lower third of the water column.

A similar stratification takes place in the protozoan communities of the periphyton, both in the sessile and the mobile parts of the communities but here the upper layer was 'thinner'. The formation of this stratification does not appear to depend as much on the discharge regime, since it was observed in both of the years. On the basis of structural parameter values and their temporal changes, the near-bottom layer is more stable than the upper one, and provides better living conditions for the sessile and mobile parts of the protozoan community of the periphyton.

The unpublished results of periphyton and plankton studies performed in the Danube River continuously since 1979–80 reinforce this general conclusion concerning the effects of stratification. Stratification of varying durations should be considered in studies of running waters. This phenomenon can be particularly important in the case of lower organisms.

Acknowledgement

We wish to express our thanks to Professor Robert G. Wetzel for his revisions and for improving the English of our manuscript.

References

Hentschel, E., 1916. Biologische Untersuchungen über den tierischen und pflanzlichen Bewuchs im Harburger Hafen. Mitt. Zool. Mus. Hamburg 33:1–172.

Jones, R.C., Cairns, J Jr. & Yongue, W.H. Jr., 1976. Vertical gradients in artificial substrate-associated Protozoan community structure in a stratified freshwater lake. J. E. Mitchell Sci. Soc. 92:1–8.

Kralik, U., 1957/58. Untersuchungen über den Bewuchs von peritrichen Ciliaten in einigen Fliessgewässern bei Leipzig. Wissensch. Zeitschr. d. Karl-Marx-Univ. Leipzig, Math.-Naturwiss. Reihe 7:309–328.

Märki, E. & Petrovic, G., 1971. Die physikalische und chemische Erforschung der Donau. Schweiz. Z. Hydrol. 33:269–286.

Nusch, E.A., 1970. Ökologische und systematische Untersuchungen der Peritricha (Protozoa, Ciliata) im Aufwuchs von Talsperren und Flusstauen mit verschiedenem Saprobitätsgrad (mit Modellversuchen). Arch. Hydrobiol. Suppl. 37:243–386.

Oertel, N., 1982. A Duna fö – és mellékágának kapcsolata Gödnél néhány vizkémiai komponens vizgálata alapján. (Relationship between the main and sidearm of the River Danube at Göd on the basis of some waterchemical parameters). Hidrol. Közl. 62:469–474.

Pätsch, B., 1974. Die Aufwuchsciliaten des Naturlehrparks Haus Wildenrath. Monographische Bearbeitung der Morphologie und Ökologie. Arb. Inst. Landwirtsch. Zool. Bienenkunde, Bonn, No. 1.

Sládečková, A., 1962a. Limnological investigation methods for the periphyton ('Aufwuchs') community. Bot. Rev. 28:286–350.

Sládečková, A., 1962b. Limnological study of the Reservoir Sedlice near Želiv. XX. Periphyton stratification during the second year-long period (August 1958–June 1959). Sci. Pap. Inst. Chem. Techn., Prague, Technol. Water 6:221–291.

Sládečková, A. & Sládeček, V., 1963. Periphyton as indicator of the reservoir water quality I. True-periphyton. Sci. Pap. Inst. Chem. Techn., Prague, Technol. Water 7:507–561.

Wilbert, N., 1969. Ökologische Untersuchungen der Aufwuchs- und Planktonciliaten eines eutrophen Weihers. Arch. Hydrobiol. Suppl. 35:411–518.

Authors' address:
M. Bereczky/N. Oertel/J.N. Nosek
Hungarian Danube Research Station
Hungarian Academy of Sciences
Jávorka S. u. 14.
Göd, Hungary

CHAPTER 8

Structural investigations of periphytic protozoan communities in three layers of the Danube River

II. The course of colonization

J.N. NOSEK and M.Cs. BERECZKY

Abstract. During two periods of investigations, 122 species of periphytic protozoans were found, of which 45 species occurred in both years. At the beginning of colonization in 1979 sessile Peritricha species dominated the community and in 1980 two Suctoria species (*Acineta tuberosa*, *Tokophrya lemnarum*) were dominant. During the first eight days of colonization of substrata, species that consume bacteria and microalgae were dominant; on the 16th day carnivores and omnivores also occurred. No species were characteristics within individual depth strata of the Danube River. The community reached its maximum development on the 8th day at the sessile, and on the 16th day at the mobile portions of the community. The rate of colonization was far less on the surface substrata than on substrata at lower depths.

Introduction

In this paper we deal with the process of protozoan colonization occurring at different depths. The objective of these investigations and the materials and methods used were presented in the first part of this work (Bereczky *et al* 1983).

Results

Quantitative and qualitative colonization of protozoan species

The distribution of the protozoan species of the periphyton was as is outlined in Table 1.

In the two investigation periods, 122 species were found, and 45 species occurred in both years.

In 1979, at the beginning of the experimental period, the sessile Peritrichida dominated [*Vorticella campanula* Ehr., *V. microstoma* Ehr., *Epistylis plicatilis* Ehr., and *Carchesium polypinum* L. appeared on the second day]. The rather balanced condition between the solitary and colony-forming Peritrichida lasted until day 8. The maximal individual numbers of *Carchesium* occurred on day 16. It is likely, however, that this value was essentially higher in the period between days 8 and 16, since the periphyton contained many stalks without zooids. Simultaneously, the numbers of the competitor *Stentor roeseli* (O.F. Müller) increased, particularly in the samples from the mid-

Table 1. Distribution of the protozoan species of the periphyton.

Species	1979	1980
Amoeba	7	6
Testacea	6	4
Heliozoa	–	1
Ciliata	61	71
Suctoria	5	6
Total	79	88

Wetzel, R.G. (ed.), Periphyton of Freshwater Ecosystems.
 ISBN 90 6193 768 X.

dle stratum. Several predatory species of *Litonotus* became an integral part of the periphyton, in addition to a few *Trachelius ovum* (Ehr.). Although in the first eight days principally species feeding on bacteria and microalgae dominated, on day 16 the predators and omnivores also appeared.

From day 16, both the species composition and the physiognomy of the periphyton changed. On both the glass and the ovalbumine-membrane coated substrata a thick layer formed, which consisted of detritus and hyphae. This layer was sufficiently thick to permit colonization by multicellular organisms such as nematodes, chironomids, and hydra.

In spring 1980, two species of Suctoria, *Acineta tuberosa* (Ehr.) *Tokophrya lemnarum* (Stokes) dominated from the beginning. Though the individual numbers of a few species *Vorticella* and *Carchesium* had also increased by day 8, by days 16 and 32 these species had completely disappeared from the periphyton, and were supplanted by Suctoria.

Investigating the different depths separately, it could be stated that no such species occurred which by their dominance might have characterized one or the other substrata depths. Species occurring only occasionally (e.g. *Enchelys simplex* (Kahl), *Nassula ornata* (Ehr.), etc.), however, were present in small numbers. Species considered as euperiphyton (e.g. *Cothurnia, Tintinnidium fluviatile* var. *emarginatum* Naskel) in other investigations (e.g. Sládečková & Sládeček 1963) also occurred, but in the Danube these species were never represented in great numbers in either of the seasons, perhaps because of some saprobity problem.

Course of colonization

The sampling times were significant in each case and for each structural parameter (most frequently at 0.1% probability level).

The sessile part of the community

The temporal changes of species number and individual numbers were similar on treated and untreated slides, both in 1979 and 1980. Numbers from the near-bottom and middle strata were simi-

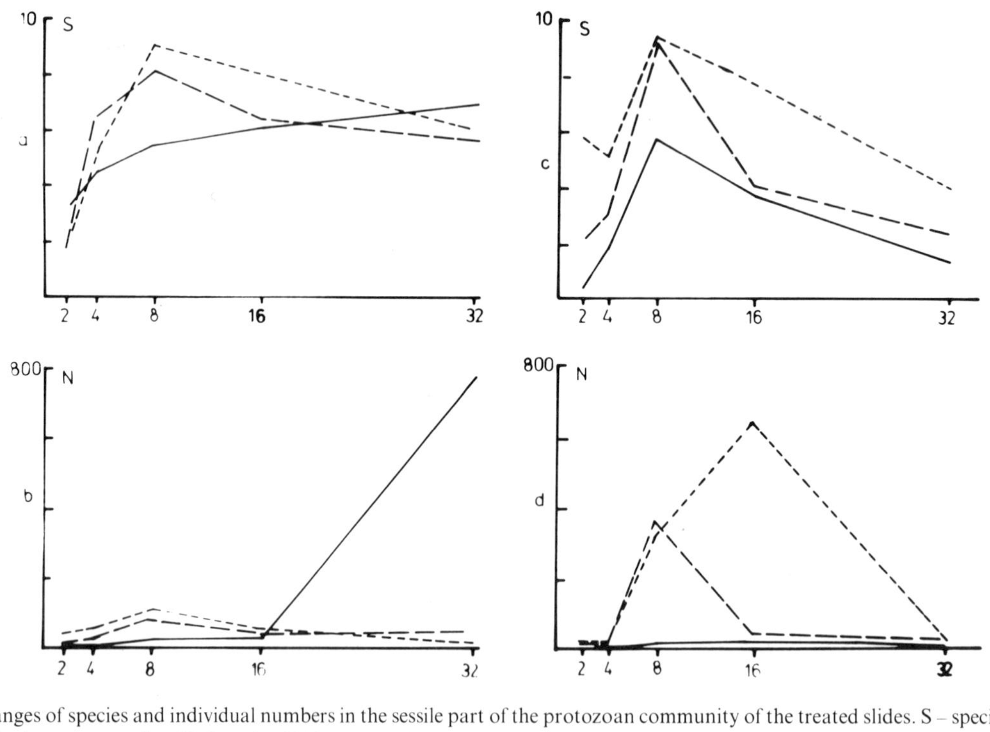

Fig. 1. Changes of species and individual numbers in the sessile part of the protozoan community of the treated slides. S – species number; N – individual numbers; a, b – 1979; c, d – 1980; —— surface, — — — middle, - - - - - near-bottom; exposure time in days on the x-axis.

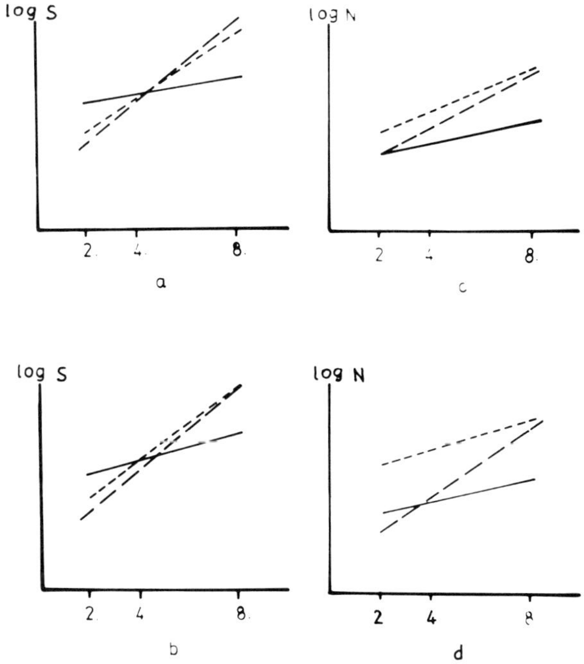

Fig. 2. Colonization changes on the exponential phase in 1979 (a, c – untreated slides; b, d – treated slides; other abbreviations and marks as in Fig. 1).

lar but differed from those of the surface stratum (Fig. 1). Colonization on the near-bottom and middle strata reached maximum values on day 8; the values diminished thereafter in most cases. Surface values were most variable. Colonization rate both in respect to species and individuals was much lower at the surface than at the other two layers during both years. This difference was also evident from the slope of curves fitted to the exponential phase of the data of 1979 (Fig. 2).

Diversity values reached their maxima on days 4 and 8 near the bottom and in the middle layer. On the surface, the maximum occurred later, on days 8 and 16. The maximum was followed by a decrease, after which the values remained more or less constant. The decrease was greater in 1980 than in 1979.

In 1979, evenness rapidly increased until day 4, remaining later at a high level with some fluctuations. On day 2, the values from the surface stratum were higher than at the two other depths on both types of slides. After day 8, there were no appreciable differences among the three layers. In 1980,

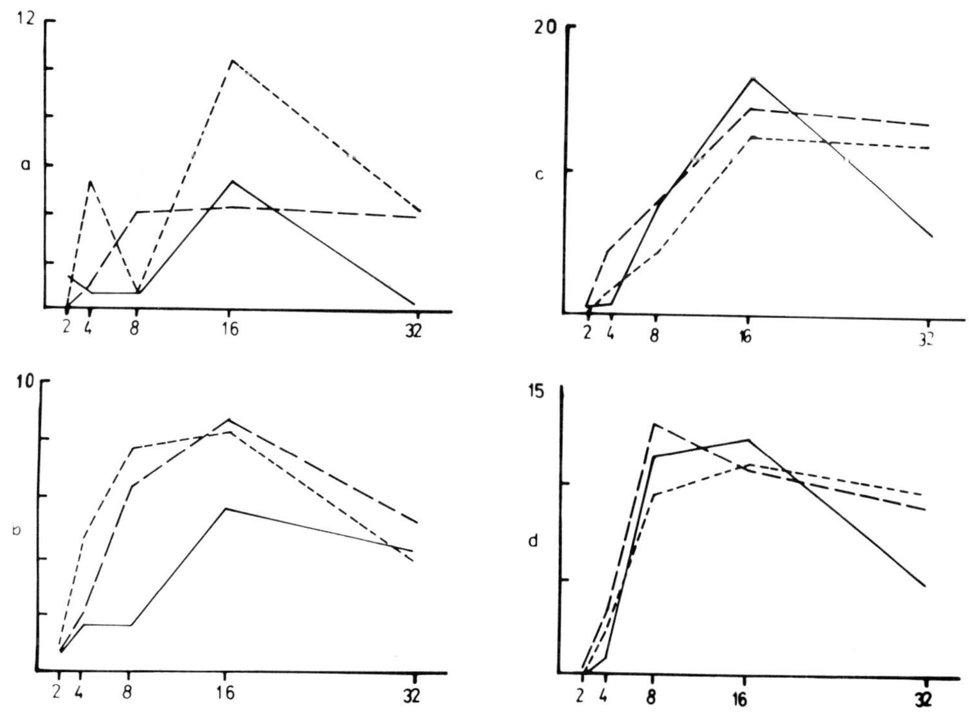

Fig. 3. Changes of species number of the mobile part of the protozoan community (a, b – 1979; c.d – 1980; a, c – untreated slides; b, d – treated slides; marks as in Fig. 1).

evenness increased until day 4 (near the bottom and in the middle layer) or day 8 (surface layer) on the untreated slides, and then decreased. On the treated slides evenness increased until day 8 on the surface, and then decreased. Evenness at the other two depths started from a higher level and behaved oppositely.

The mobile part of the community

In 1979, structural parameters reached their maxima on day 16 in each case, with the exception of the individual numbers on the treated slides exposed at the surface and the middle part of the water column. The changes observed on the two kinds of slides differed. In the case of treated slides, the increase was continuous and the decrease smaller. On untreated slides, individual numbers exhibited great fluctuations until the maximum was reached, and the decrease was greater. The changes on the surface samples, similarly to the sessile community, differed from those of the other two depths. The difference is more conspicuous in the case of treated slides.

The results indicate that colonization rate was lower on substrata in the surface stratum than at the other two depths both in respect of species and individual numbers. Colonization rates on the two kinds of slide did not differ in this respect.

The slopes of curves for 1980 and maxima were very similar in the case of treated and untreated slides, apart from a few exceptions occurring on day 16 (Fig. 3).

No important differences appeared in colonization rates between depths in regard to species and individual numbers. Of the two different types of slide, colonization rates were higher on the treated slides.

Summary

The temporal development of these protozoan communities agrees with the results of classical colonization studies (MacArthur & Wilson 1967). The initially rapid increase in species and individual numbers declined with the increase in competition for space and food. The appearance of non-protozoan predator species (Nematoda, Chironomida, Rotatoria) also played a role.

The sessile community reached its most developed state on day 8, the mobile on about day 16. The same phenomena, though less distinctly, could also be observed on substrata in the surface stratum. In both communities, colonization rates were higher at the near bottom and middle levels than at the surface.

The results of colonization studies, similarly to those of stratification, suggest that the near-surface water layer provides less favorable living conditions for the protozoan community of the periphyton. The results also suggest that studies of communities consisting of species of short generation times should be considered more critically, since their life cycles are generally commensurate with frequent short-term periodic loadings to river ecosystems.

Acknowledgement

We wish to express our thanks to Professor Robert G. Wetzel for his revisions and for improving the English of our manuscript.

References

Bereczky, M.Cs., Oertel, N. & Nosek, J.N., 1983. Structural investigations of periphytic protozoan communities in three layers of the Danube River. I. The question of stratification. In: Wetzel, R.G. (ed.), Periphyton of Freshwater Ecosystems. Developments in Hydrobiology 17. pp. 49–54.

Cairns, J. Jr., Plafkin, J.L., Yongue, W.H. Jr. & Kaesler, R.L., 1976. Colonization of artificial substrates by Protozoa: replicated samples. Arch. Protistenk. 118:259–267.

Cairns, J. Jr., Kuhn, D.L. & Plafkin, J.L., 1979. Protozoan colonization of artificial substrates. Special Technical Publication 690 ASTM. pp. 34–57.

MacArthur, R.H. & Wilson, E.O., 1967. The Theory of Island Biogeography. Princeton Univ. Press, Princeton. 203 pp.

Sládečková, A. & Sládeček, V., 1963. Periphyton as indicator of the reservoir water quality. I. True-periphyton. Sci. Pap. Inst. Chem. Technol., Prague, Technol. Water 7:507–561.

Authors' address:
J.N. Nosek/M. Bereczky
Hungarian Danube Research Station
Hungarian Academy of Sciences
Jávorka S. u. 14.
Göd, Hungary

CHAPTER 9

Seasonal changes in diatom-structure of periphyton from two localities in Lake Maarsseveen: community indices

P.J. ROOS

Abstract. The diatom composition of the diatom-dominated reed-periphyton from a Dutch lake varied in time and space. In February, March, April and May 1982, samples were taken from opposite sides of the lake from the lakeside, center, and bankside of the two reedstands. A seasonal shift in algae was demonstrated for the two reedstands as a whole as well as for the bankside and lakeside. Spatial variation in the center of the stands appeared to be greater than the seasonal variation. Diversity indices of the two sides did not differ, except in February when there was great variation in the number of *Tabellaria*.

Introduction

Lake Maarsseveen is a rectangular lake with a maximal depth of 37 m. The long axis runs NW SE and is 1700 m long; the lake is 400 m broad. Dense reed stands, 3–6 m broad, are found along the long sides. Periphyton from the submerged parts of this reed is described by Roos *et al* (1981). Samples from the reedstands show variation in algal composition, in space as well as in time.

In this study similarity indices were used to compare various algal samples from two stands, one from the NE and the other from the SW side of the lake, respectively. Objectives were: (1) to evaluate the use of these indices for determinations of variations in space or time, and (2) to discriminate between the general pattern of a seasonal shift in structural composition and local or temporal variations.

Materials and methods

Samples were taken on 9 February, 16 March, 13 April, and 11 May 1982 from the submerged parts of *Phragmites* stems from each of the two sides of the lake. Two stems of six were taken at the bankside of the reedstand, two stems came from the central part, and two were collected at the lakeside.

The periphyton was scraped and that from the two stems from one locality was carefully mixed by means of a whirlmixer. In this way six samples were obtained: bankside, central part, and lakeside sections from each side of the lake. Subsequently, these samples were divided into two equal parts and combined into five new samples, representing bankside, central part, and lakeside as well as NE stand and SW stand, respectively, consisting of two and three subsamples.

These five new samples were thoroughly mixed, and diatom preparations were made by chemical oxidation (Van der Werff & Huls 1957). In these preparations, the frustules of the following 13 genera of sessile diatoms were counted: *Melosira, Achnanthes, Amphora, Anomoeoneis, Cocconeis, Cymbella, Diatoma, Fragilaria* (non *crotonensis*), *Eunotia, Gomphonema, Rhoicosphaenia, Synedra* and *Tabellaria*. Between 625 and 890 frustules were

Wetzel, R.G. (ed.), Periphyton of Freshwater Ecosystems.
 ISBN 90 6193 768 X.

counted per preparation (average 757). Percentage contribution of the genera was calculated and similarities between the 20 samples were determined on the basis of this percentage.

Czekanowski's index of similarity between two samples was used (Clifford & Stephenson 1975):

$$S = 2w / (a + b)$$

in which w is the sum of the lowest contributions of one pair of corresponding genera in the two samples, a and b are the total contributions of the genera in the samples. In the case where the percentage contribution a and b are 100%, the percentage similarity can simply be expressed as

$$S = w$$

In order to compare the results of similarity indexing of the diversity indices from the NE and SW stands, the index of Brillouin was chosen, because this is the only appropriate index of diversity for a 'collection' (Pielou 1975), as the representability of the samples with regard to the 'indefinite' reedstands is unknown because of the combinations of the subsamples.

N.E.

	FEB	MAR	APR	MAY
FEB		74	72	67
MAR			92	81
APR				85
MAY				

S.W.

	FEB	MAR	APR	MAY
FEB		91	93	80
MAR			93	85
APR				84
MAY				

B.S.

	FEB	MAR	APR	MAY
FEB		91	80	74
MAR			85	79
APR				91
MAY				

C.P.

	FEB	MAR	APR	MAY
FEB		81	70	78
MAR			73	83
APR				80
MAY				

L.S.

	FEB	MAR	APR	MAY
FEB		93	84	79
MAR			85	79
APR				87
MAY				

Fig. 1. Similarity matrices. NE: north-eastern stand as a whole; SW: south-western stand as a whole; BS: bankside of combined stands; CP: central part of combined stands; LS: lakeside of combined stands.

Results

Similarities between the same localities at various sample dates are depicted in Fig. 1 by means of 5×5 matrices which are part of the complete 20×20 matrix. Generally, the highest similarities are arranged diagonally, which means that there was a greater similarity with neighbours in time than with samples taken longer ago or further in the future. This relationship did not hold for the central part of the reedstand. From the complete matrix, algal similarities of more than 90% were found between lakeside February and central part March, lakeside April and bankside April, lakeside April and central part May, and central part May and bankside May. High percentage similarity (88) was also found between lakeside March and central part April. All 19 similarities of the NE stand in February were low, between 59 and 79, with a mean of 69.

Percentage similarities between the NE stand and the SW stand as a whole were 67, 86, 88, 89, respectively, in February, March, April and May. Diversity indices were, in the same months, respectively, for the NE stand: 0.69, 0.53, 0.57, 0.63, and for the SW stand: 0.55, 0.56, 0.54, 0.63.

Discussion

From the matrices in Fig. 1, it is evident that a seasonal shift in generic composition was demonstrable in the two reedstands as a whole, at the lakeside and, even more clearly, at the bankside. The central part did not show any seasonal variation and has low similarities. It was therefore concluded that the central part of the reedstands shows local variation and is less dependent on weather conditions than the exposed sides of the stand facing the bank or lake. The strikingly high similarities of the central parts and the lakesides during the preceding months suggest that the seasonal shift started at the open side of the stand, under influence of light, wind, or wave action. These factors may account for differences observed between the two stands on opposing sides of the lake. From February until May, a progressively increasing similarity occurred between them. During this period wind and waves ran predominantly in the direction of the long axis of the lake, generally not battering one of the stands more than the other. Decrease of the directional effect of sunlight when the sun rises may explain the increase in similarity between the two stands during spring.

In order to quantify differences in generic composition between the stands, diversity indices were calculated. Differences declined from February until March, in accordance with increasing similarity. Throughout this period, the number of dominant genera was 12 (in April and May *Amphora* was replaced by *Anomoeoneis*); the difference contributed to the evenness component of the diversity only. In February this contribution was caused by an extraordinarily low presence (0.2%) of *Tabellaria* in the NE stand. In the other months, slight variations in relative contribution of various genera were responsible.

In this way, use of indices of similarity and diversity, based on relative countings only, reveal seasonal changes in periphyton structure and discriminates between long-term and temporal variations.

References

Clifford, H.T. & Stephenson W., 1975. An Introduction to Numerical Classification. Academic Press, New York. 229 pp.

Pielou, E.C., 1975. Ecological diversity. John Wiley & Sons, New York. 165 pp.

Roos, P.J., Post, A.F. & Revier, J.M., 1981. Dynamics and architecture of reed periphyton. Verh. Int. Ver. Limnol. 21:948–953.

Van der Werff, A. & Huls, H., 1957. Diatomeeenflora. Inleiding. Van der Werff. 26 pp.

Author's address:
P.J. Roos
Department of Aquatic Ecology
University of Amsterdam
Kruislaan 320
1098 SM Amsterdam, The Netherlands

CHAPTER 10

Physical and chemical parameters regulating growth of periphytic communities

KAJ SAND-JENSEN

Abstract. It is difficult to clarify when and how physico-chemical parameters regulate growth of periphytic algae. Many parameters are potentially involved, and these are usually measured in the external water. Growth rates of periphytic algae can be estimated during the early growth phase of the biomass, assuming that colonization and losses are of minor importance. During the later stationary or secondary phases it is necessary to quantify losses. During the early growth phase good correlations can be obtained between external parameters and the rate of biomass changes. In this phase the periphytic community is thin, and the rate of exchange with the external water, in proportion to internal processes, is relatively more important than in later phases, when the community becomes thicker and biologically more complex. Many submerged macrophytes continuously produce new tissue and slough off old tissue. The age of the tissue can sometimes be measured quite accurately. The rate of change in density of epiphytic algae with increasing age of tissue, or the mean density for all tissue ages, can be very sensitive to changes in inorganic nutrient loading. Carbon-14 and O_2 exchange methods should be applied to studies of periphytic processes, realizing that these processes occur within a boundary layer, and have great potential for analyzing effects of time and depth changes in chemical parameters.

By different approaches it is possible to obtain control of external parameters affecting natural periphytic communities, for example by creating a gradient of a single factor at one station, or by analyzing a natural gradient developed at several stations. Artificial systems can be suitable in certain functional analyses, but the results are often difficult to apply to natural systems.

Relatively simple measurements of structure and quantity of different autotrophic components may be a sufficient background for analyzing their internal balance, or for further measurements of their role in matter and energy transformation in the ecosystem.

Introduction

The problems involved in studies of growth of periphytic communities in relation to physico-chemical parameters are formidable. The periphytic community is an assemblage of highly dynamic groups of organisms: bacteria, fungi, algae, and animals associated with different types of substrates. For simplicity I have decided in this short review to concentrate mainly on the periphytic algae. Many aspects of the regulation of periphytic communities in artificial streams are thoroughly treated by McIntire (1968, 1973, 1975), McIntire & Phinney (1965), Phinney & McIntire (1965) and Eichenberger (1975). An up-to-date review of ideas and methods for studying algal growth in the plankton community is found in Morris (1980).

For several reasons it is not possible to describe

Wetzel, R.G. (ed.), Periphyton of Freshwater Ecosystems.
 ISBN 90 6193 768 X.

precisely when and how physico-chemical parameters regulate growth of periphytic algae:

1. There are potentially many physico-chemical parameters that affect growth. When we examine growth patterns in a fluctuating natural habitat, it is usually not possible to pinpoint a single regulating factor.
2. Measurements of physico-chemical parameters are usually confined to the free-water phase above the periphytic community. Conditions within this community, however, are often quite different.
3. Growth rates of periphytic algae are either difficult or impossible to measure directly.

At the end of this paper I shall discuss seasonal growth patterns of periphytic communities as related to external physico-chemical parameters and exemplify how we can select the effect of single parameters (point 1). I shall start by considering points 2 and 3.

Physico-chemical parameters in the periphytic community

The periphytic community is a microcosm with internal autotrophic and heterotrophic processes occurring within a boundary layer. Transport processes within this layer occur by molecular diffusion, in contrast to mixing or bulk movements in the free-water phase.

The internal processes change chemical conditions so that they can be quite different from those in the free-water phase. To what extent chemical conditions are modified in the periphytic community depends on the rate of exchange with the free-water phase in proportion to the rate of single processes, or the balance of opposing processes within the periphytic layer. Thick, dense, and active periphytic communities in standing waters, therefore, have a potential to show a large difference between internal and external chemistry.

Measurements of internal chemistry require the use of micro-techniques, e.g. micro-electrodes, and little work has been done along this line. Jørgensen *et al* (1979) examined diurnal changes in vertical patterns of light, O_2, H_2S and NH_4^+ in a cyanobacterial mat sediment in a hypersaline oligotrophic, desert lake in Israel (Solar Lake). The phytoplankton community had a low density and activity, but the cyanobacterial periphytic community on the sediment was dense and very active. Physico-chemical conditions were quite uniform and stable in the lake water, whereas extreme diurnal and depth variations occurred in the periphytic mat. The photic zone was confined to the upper 2.5. mm of the mat. During the day high photosynthetic activity developed an O_2 maximum of 0.4 to 0.5 mM at a depth of 1–2 mm in the mat (Fig. 1). The O_2 concentration in the lake water in equilibrium with the atmosphere was 0.11 mM, so the O_2 maximum represented almost 100% O_2 satu-

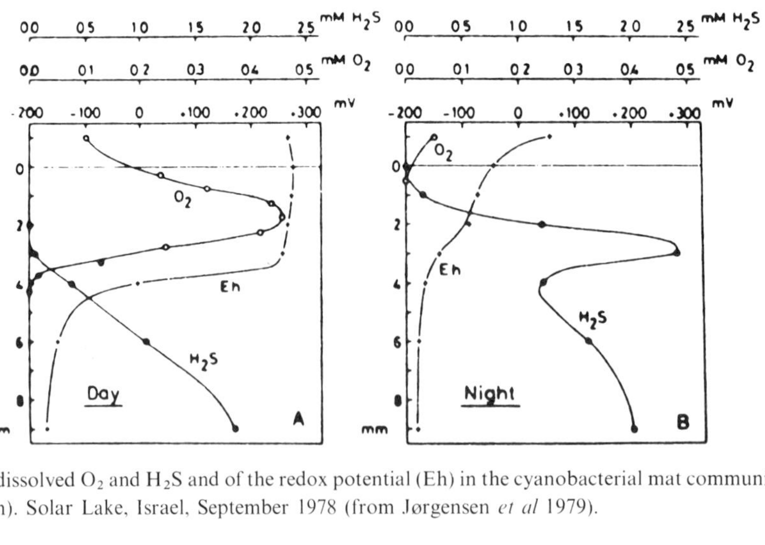

Fig. 1. Depth profiles of dissolved O_2 and H_2S and of the redox potential (Eh) in the cyanobacterial mat community during day (13 35 h) and during night (04 50 h). Solar Lake, Israel, September 1978 (from Jørgensen *et al* 1979).

ration. O_2 penetrated to about 4 mm of depth, whereas dissolved H_2S appeared at 2.5 mm, below which it rapidly increased. At night, the high heterotrophic activity had depleted the O_2 pool, and O_2 penetrated only 0.5 mm into the mat. A H_2S peak at 3 mm had replaced the former O_2 peak and H_2S reached to within 0.3 mm from the surface.

$NH_4^+$was not detectable in the lake water. A high peak of $NH_4^+$was built up in the upper 2 mm of the mat during the nigh and was consumed by photosynthetic organisms during the day (Fig. 2). This NH_4^+ peak was largely indiffusible and was thereby held in an internal recirculation within the mat. Further results also indicated that almost the entire production of photosynthates was continuously mineralized within the uppermost few mm of the mat.

This mat community was very efficient in recirculating and conserving essential inorganic nutrients. This function was of the utmost importance in the lake ecosystem, which itself was nutrient poor. A similar, closed nutrient cycling could not be attained in the lake water from which particle bound nutrients were lost by sedimentation. A prerequisite for the stability of the mat community was probably the seasonal stability in surface irradiance and temperature and lack of grazing.

The uppermost few mm of the mat is comparable with the entire photic zone in lake waters. To understand lake processes we are well aware that measurements are necessary both in the lake and in the effluent waters. Accordingly, refined micromethods, as used in the cited paper, are highly recommended in studies of internal dynamics of periphytic communities. Revsbech *et al* (1980 and 1981) have recently used O_2 micro-electrodes to study time and depth changes of O_2 concentrations and depth profiles of photosynthesis in marine microbenthic communities.

Growth rate of periphytic algae

Direct measurements of growth rates of algae in nature are difficult (e.g. Sakshaug 1981). Instead changes in numbers of either different algal species, algal biomass, or photosynthetic activity are usually measured.

Changes in algal biomass (ΔB) over a certain time interval is equal to growth or division (G) plus colonization (C) minus losses due to grazing (Gr) or mechanical detachment (M):

$$\Delta B = (G + C) - (Gr + M)$$

The biomass development of algae on permanent substrates (sediment, stone) in spring, following a low winter biomass, often has characteristically successive phases. Similar phases are observed on newly-exposed artificial substrates (McIntire 1975; Eichenberger 1975; Kevern *et al* 1966). The initial, early phase is characterized by a rapid exponential or sometimes linear increase in biomass. This phase in followed by a saturation or stationary phase where the biomass reaches a constant level, and later by a secondary phase where the biomass fluctuates widely (Fig. 3).

During the early growth phase, the periphytic community is relatively thin and the biological structure relatively simple. Algae or bacteria with high potential growth rates often dominate, and animals with larger generation times can be absent. Losses from grazing can be neglected, and losses by mechanical detachment are often small, when the community is thin. The growth rate will then approach the changes in biomass, if we assume that colonization is small in proportion to growth. Ex-

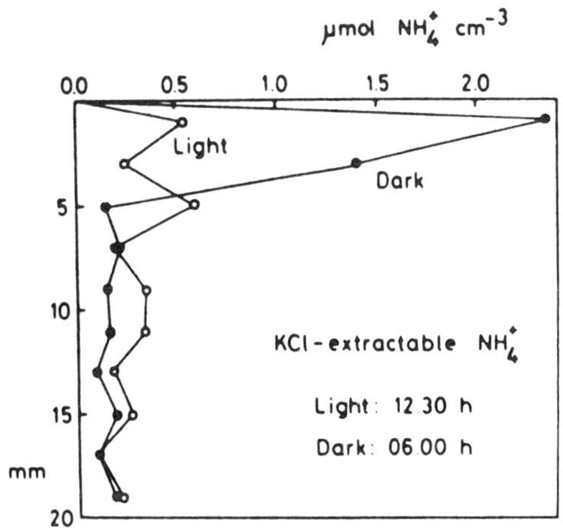

Fig. 2. Depth profiles of total KCl – extractable NH_4^+ in the cyanobacterial mat community during night and during day. September 1978 (from Jørgensen *et al* 1979).

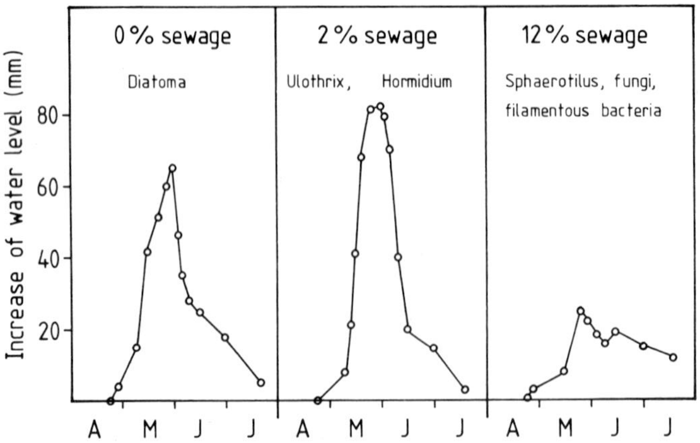

Fig. 3. The time course of development of periphytic communities in outdoor channels under different sewage concentrations. The ordinate axis shows the changes in water level in the channels at a constant discharge of 4 l sec^{-1}. For a given structure of the epiphytic community, changes in water level are approximately proportional to changes in biomass (redrawn from Eichenberger 1975).

changes with the external water phase are also more emportant than internal transformations at this early phase, since the community is thing, accumulation of senescent organisms and dead organic material is small, and the biological structure relatively simple. We can, therefore, obtain good correlations between water quality parameters and biomass changes. This relation is exemplified by the correlation between sewage concentration and the rate of biomass increase of autotrophic and heterotrophic organisms during the first 20 days in outdoor artificial channels (Eichenberger 1975; Fig. 4).

In the saturation and secondary phases, the complexity of the community increases, further ecological factors are involved (e.g. grazing), and the internal transformations become relatively more important than exchange processes with the external water-phase (viz. Figs. 1 and 2). We may, therefore, lose any correlation with external variables (e.g. Eichenberger 1975). To determine growth rates in the saturation or secondary phases it is necessary to follow changes in biomass and to quantify losses by grazing or mechanical detachment. This analysis has seldom been done, but is to be recommended whenever possible and relevant.

In a very fluctuating environment, for example a stream with frequent spates of water which tear away most of the periphytic community, the growth phases we observed during intervening, stable periods, may reflect either exponential or linear growth, and offer a possibility of obtaining good correlations between biomass changes and external variables.

The same growth is possible for certain epiphytic communities under more stable external conditions. Many submerged macrophytes continuously produce new tissue and slough off old tissue. For many species it is possible to define very precisely the age of different leaves by using a leaf marking

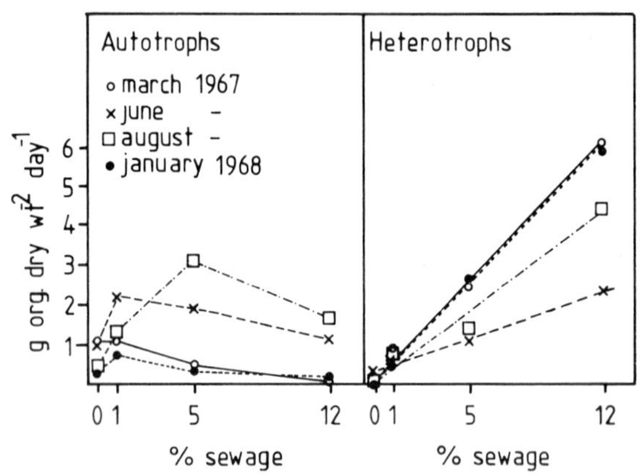

Fig. 4. The mean rate of changes in autotrophic and heterotrophic biomass during the early 20-day growth phase in outdoor channels under different sewage concentrations. Experiments in different seasons of the year are shown (redrawn from Eichenberger 1975).

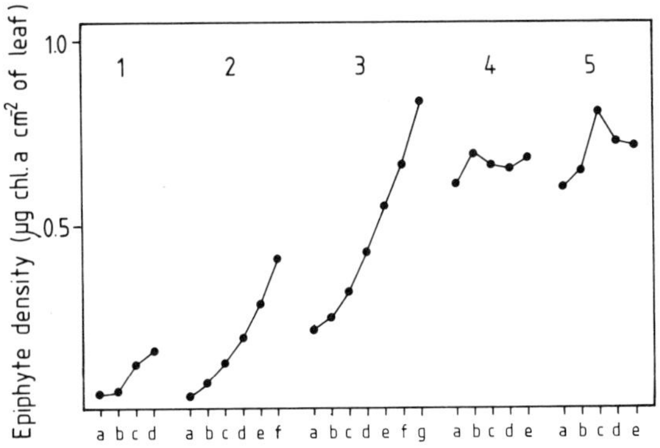

Fig. 5. The distribution of biomass of epiphytic algae within eelgrass turions (Roskilde Fjord, June 1981). The eelgrass turions were divided into leaves of increasing ages from 1 to 5, and each leaf was sectioned into 5 cm segments starting from the youngest part proximal to the sheath and ending at the tip (a to g) (from Borum, unpubl. data).

technique (e.g. Sand-Jensen 1975, Sand-Jensen & Søndergaard 1978, Nielsen & Nielsen 1982). The life span of eelgrass leaves in shallow Danish waters, for example, is slightly less than 2 months throughout the period of active growth from April to October. The density of the epiphytic algae increases with leaf age so that the three youngest leaves have a community in exponential growth and the two oldest leaves have a community in the stationary phase (Fig. 5). If the biomass of algae versus leaf age is plotted, the rate of biomass increase during the exponential growth phase can be calculated (Fig. 6, the doubling time of biomass in this example is 4 days).

Borum (unpubl. data) showed that the density of algal epiphytes on eelgrass along a natural gradient of inorganic N-concentrations in a coastal area was about 100-fold higher at the upper end compared to the lower end. This differential growth was a permanent situation throughout the period of rapid growth where leaf ages were always below 2 months. This difference was not present in early spring when many old, over-wintering leaves occurred, and when inorganic nutrient concentrations remaining from the previous winter were high at all stations.

The ^{14}C and O_2 light/dark techniques have frequently been used to study photosynthetic activity and community metabolism of periphytic communities. Problems of interpreting such data are thoroughly discussed for phytoplankton by Harris (1978). Rapid internal processes in the periphytic community in relation to exchange processes, as exemplified in Figs. 1 and 2, make both methods even more problematic for natural periphytic communities. I will not discuss results obtained by the two methods, but instead draw the attention to a few potential errors that should be considered in any future work. Considering a ^{14}C experiment, where ^{14}C is added to the water above the periphytic community, calculation of the ^{12}C assimilation of the periphytic algae implies that the ^{12}C concentrations and the ratio of $^{14}C/^{12}C$ at the site of fixation are known. ^{12}C concentrations and the ratio of $^{14}C/^{12}C$ in the bulk water are not necessarily representative. ^{14}C added to the bulk water does not necessarily reach diffusional equilibrium with the porewater in the periphytic community (e.g. Jørgensen *et al* 1979). The pool of ^{12}C may also change markedly with time and depth in the periphytic layer (e.g. Jørgensen *et al* 1979). A hetero-

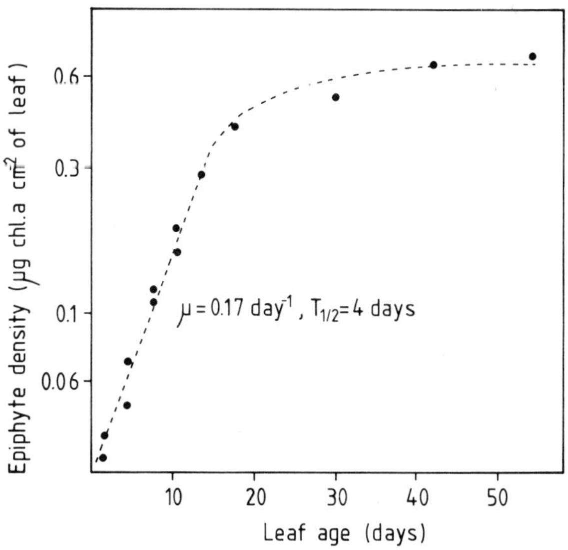

Fig. 6. The development of epiphyte density versus leaf age of eelgrass (Roskilde Fjord, June 1981). Calculated from the distribution of epiphyte density within eelgrass turions (Fig. 5) and from the rate of leaf growth and leaf renewal time in the preceding period. The exponential increase in epiphyte density during the first 14 days corresponds to a specific increase in biomass of 0.17 day^{-1} (μ) and a doubling time of biomass of 4 days ($T_{\frac{1}{2}}$) (from Borum, unpubl. data).

trophic respiration concurrent with algal photosynthesis in the light will also tend to decrease the $^{14}C/^{12}C$ ratio (Revsbech *et al* 1981).

The O_2 technique is usually applied to periphytic communities by measuring the rate of O_2 exchange between the periphytic layer and the stirred bulkwater in a closed chamber in light and dark. The O_2 exchange in the light is usually interpreted as the net community metabolism, i.e. the balance between algal net photosynthesis and O_2 consumption by other organisms or by chemical means. Referring the Fig. 1 and Revsbech *et al* (1980), it is evident that the O_2 pool within the periphytic layer can change markedly during incubation, which is not detected in the bulk water. Only by truly stready state experiments (i.e. flow-through chambers incubated under constant external conditions) is it possible to reach a steady-state between internal processes and rates of exchange. The O_2 consumption in the dark is often assumed equal to O_2 consumption in the light, which allows an estimate of algal gross photosynthesis. In periphytic communities with profound chemical changes in the periphytic layer from light to dark (e.g. Fig. 1), such an assumption is erroneous.

The problems discussed above are reduced in very thin and simple periphytic communities, e.g. communities dominated by algae. However, it is still not possible to make foolproof calculations of algal net photosynthesis either during the day, or day-plus-night to obtain an estimate of algal growth. Revsbech *et al* (1981) recently presented a technique to measure algal gross photosynthesis within a periphytic community by using O_2-microelectrodes. This technique deserves further attention.

Development of periphytic communities in relation to external factors

Under natural conditions, some parameters may show a characteristic seasonal trend (e.g. light and temperature), whereas other parameters vary more stochastically. To explain seasonal changes in the periphytic community of a single station, we often have to decide more or less subjectively, on the basis of correlations, whether one complex of factors is more important than another.

Microalgae in many periphytic communities, under temperate conditions, undergo rapid growth and increase in biomass during spring (e.g. Hickman & Round 1970; Hickman 1971; Marker 1976). This growth is illustrated by the development of biomass of microbenthic algae in the River Suså, Denmark (Fig. 7). This river is very rich in inorganic nutrients. It has a high winter discharge which removes all plants and a succession of autotrophic components occurs from April to September, when the discharge is several-fold lower. The biomass of microbenthic algae increases exponentially during spring, triggered by the increasing surface irradiance and declining discharge (Fig. 7). The declining discharge reduces both the mechanical stress and the light attenuation in the water

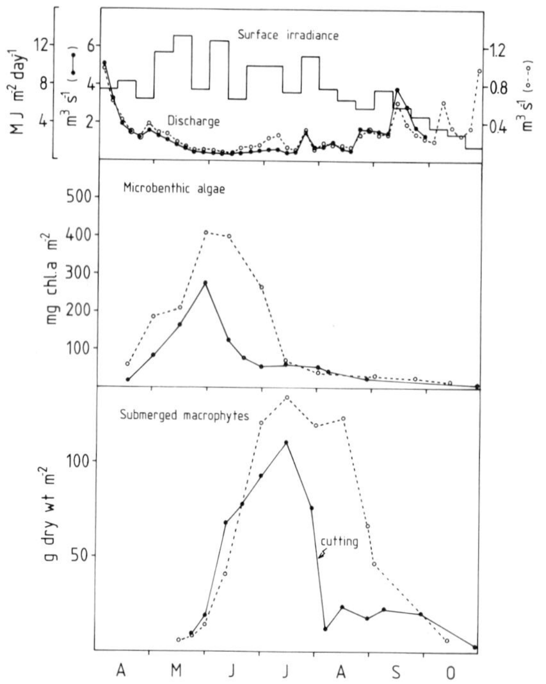

Fig. 7. Seasonal changes in surface irradiance, discharge, and biomass of microbenthic algae and submerged macrophytes at stretches in the upper part (○...○) and middle part (○—○) of the River Suså, Denmark, 1980. The surface irradiance and discharge are mean values for ten and five-day periods, respectively. The biomass of microbenthic algae as chl. *a* is corrected for the presence of degradation products (from the River Suså project, unpubl. data).

column through a reduction in water height and particle concentration. The exponential phase is followed by a saturation phase at about 400 mg chl a m^{-2}. Then the biomass of microbenthic algae decreases during the period of exponential growth of submerged macrophytes. The seasonal development of microbenthic algae is correlated with the irradiance reaching the sediment, taking into account the surface irradiance and the light attenuation by the water and by submerged macrophytes (Fig. 8).

The major importance of light for the development of microbenthic algae in the River Suså is further supported by two comparisons. The exponential spring development started earlier in years where there was an early reduction in discharge and a higher surface irradiance. The exponential increase started earlier, and the maximum biomass attained was also higher at an upper stretch than at a stretch in the middle part of the river. This difference correlated with a lower water height and lower flow velocity at the upper stretch. During the early phases, however, it is difficult to distinguish between the importance of increased light availability and sediment stability.

The major problem in these one-station analyses is often the great number of fluctuating variables that potentially can affect the periphytic community. The situation is simplified in the River Suså, because inorganic nutrients are available in such high concentrations that they are not likely to limit either the growth rate or the yield of periphytic algae. By comparison, the maximum microbenthic algal biomass was tenfold lower in the River Skærbæk, Denmark, than in the River Suså (Sode 1981). Inorganic nutrients were very low in the River Skærbæk (annual means of total P = 17 $\mu g\ l^{-1}$), whereas light conditions were better and the flow velocity fluctuated less. These conditions suggest the importance of P as a growth rate and yield-limiting factor for microbenthic algae in the River Skærbæk.

By further control of external parameters by three different approaches (outlined below) we can improve our understanding of growth regulating factors.

At a single station we can, for example, create a

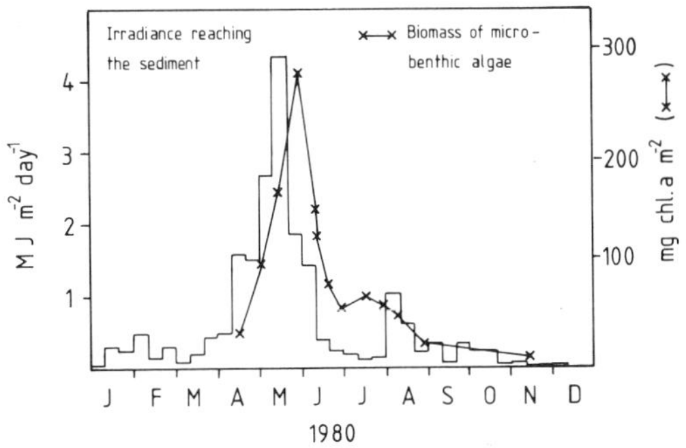

Fig. 8. Seasonal changes in biomass of microbenthic algae and the irradiance reaching the bottom at a stretch in the middle part of the River Suså, Denmark, 1980. The irradiance reaching the bottom was calculated from surface irradiance and attenuation in the water column and by submerged macrophytes (from the River Suså project, unpubl. data).

gradient of light conditions by artificial shading of certain spots, or create a gradient of inorganic nutrient loading. Harlin & Thorne-Miller (1981) used the latter technique in a uniform seagrass meadow with control areas and areas where they placed porous clay dispensers filled with different inorganic nutrients on the sediment. They then monitored the effect on composition and quantitative development of different autotrophic components. The main problem in this technique is to obtain a reliable description of the inorganic nutrient loading of the different areas.

A second approach is to select a gradient of, for example, inorganic nutrient loading by examining stations which have the same physical environment and chemistry of major ions, but differ in respect to inorganic nutrient loading. Cattaneo & Kalff (1980) showed a linear correlation between total P in water and ^{14}C fixation rate and chlorophyll content of epiphytes on *Myriophyllum spicatum* in bays of different trophic conditions in Lake Memphremagog during summer (Fig. 9).

The third approach is to determine the effect of specific physico-chemical conditions using artificial systems (laboratory channels, McIntire 1968; outdoor channels, Eichenberger 1975; ponds, Moss 1976). It is possible, by this approach, to achieve a general understanding of how these parameters can affect the structure and function of periphytic communities. One problem is that the internal biotic interrelations may become more important in such space-restricted systems and that, by chance, different organisms or communities dominate under the same external conditions (e.g. Eichenberger 1975). A further problem is that these artificial systems may represent ecological conditions quite remote from natural systems, so that the understanding we achieve is not necessarily applicable to, or completely relevant, for natural systems, about which we wish to learn.

The balance between different autotrophic components

The discussion up to now has illustrated the difficulties in determining growth rates of periphytic algae. Estimates of growth rates can be based on changes in biomass with time, during the early growth phase, whereas during later stationary or secondary phases it is necessary to include calculations of losses. Interpretation of rates of ^{14}C fixation and O_2 exchange during light and dark in relation to algal growth has the same inherent problems for periphytic algae as for phytoplankton. These techniques should also be applied with the knowledge that processes in the periphytic community occur within a boundary layer where chemical conditions are different and more dynamic than in the bulk water.

For many purposes, however, relatively simple measurements of composition and biomass of components in the periphytic community can be satisfactory. This condition is so, for instance, if we examine the balance in development of different autotrophic components and clarify the role of different physical (light, morphometry) and chemical (inorganic nutrients) factors for this balance. In this context, the biomass development of periphytic algae seems to be very sensitive to increased nutrient loading above an initially low level (Phillips *et al* 1978; Cattaneo & Kalff 1980; Goldman 1981; Sand-Jensen & Søndergaard 1981; Borum, unpubl. data). Localized inputs of nutrients in a lake or coastal waters are also reflected in the site-specific development of periphytic communities, whereas they are not as readily detectable in the phytoplankton (citations above). This relation points to the advantage of periphytic communities to assess the effect of inorganic pollution. Qualitative and quantitative descriptions of the autotrophic components usually also form the necessary background for measurements of processes such as the relative importance of different autotrophic components in the organic productivity or community metabolism in a specific ecosystem.

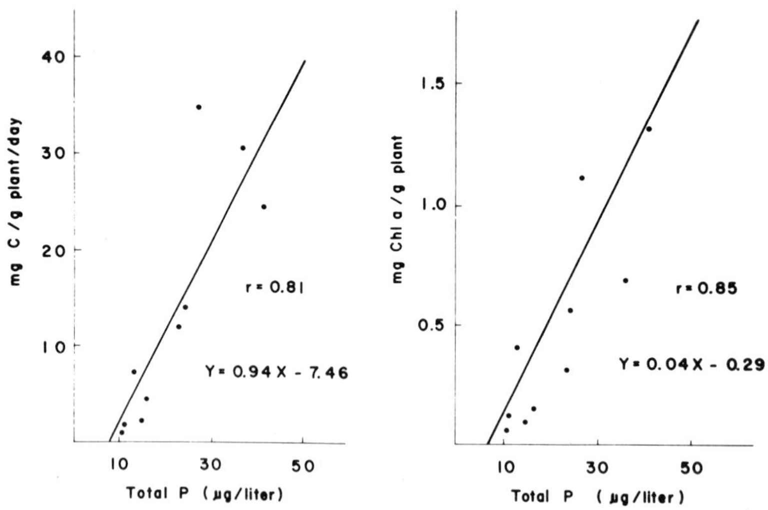

Fig. 9. Linear correlations between total P in water and ^{14}C fixation (left) and chlorophyll (right) of loose epiphytes attached to *Myriophyllum spicatum* in various bays in Lake Memphremagog during the summer of 1977 (from Cattaneo & Kalff 1980).

References

Cattaneo, A. & Kalff, J., 1980. The relative contribution of aquatic macrophytes and their epiphytes to the production of macrophyte beds. Limnol. Oceanogr. 25:280–289.

Eichenberger, E., 1975. On the quantitative effects of chemical factors on running water ecosystems. Schweiz. Z. Hydrol. 37:21–34.

Goldman, C.R., 1981. Lake Tahoe: two decades of change in a nitrogen deficient oligotrophic lake. Verh. Internat. Verein. Limnol. 21:45–70.

Harlin, M.M. & Thorne-Miller, B., 1981. Nutrient enrichment of seagrass beds in a Rhode Island coastal lagoon. Mar. Biol. 65:221–229.

Harris, G.P., 1978. Photosynthesis, productivity and growth. The physiological ecology of phytoplankton. Ergebnisse der Limnologie 10. 171 pp.

Hickman, M., 1971. The standing crop and primary productivity of the epiphyton attached to *Equisetum fluviatile* L. in Priddy pool, North Somerset. Br. phycol. J. 6:51–59.

Hickman, M. & Round, F.E., 1970. Primary production and standing crops of episammic and epipelic algae. Br. phycol. J. 5: 247–255.

Jørgensen, B.B., Revsbech, N.P., Blackburn, T.H. & Cohen, Y., 1979. Diurnal cycle of oxygen and sulfide microgradients and microbial photosynthesis in a cyano-bacterial mat sediment. Appl. Environ. Microbiol. 38:46–58.

Kevern, N.R., Wilhm., J.L. & Van Dyne, G.M., 1966. Use of artificial substrates to estimate the productivity of periphyton. Limnol. Oceanogr. 11:499–502.

Marker, A.F.H., 1976. The benthic algae of some streams in southern England. I. Biomass of the epilithon in some small streams. J. Ecol. 64:343–358.

McIntire, C.D., 1968. Structural characteristics of benthic algal communities in laboratory streams. Ecology 49:520–537.

McIntire, C.D., 1973. Periphyton dynamics in laboratory streams: a simulation model and its implications. Ecol. Monogr. 43:399–420.

McIntire, C.D., 1975. Periphyton assemblages in laboratory streams. In: Whitton, B.A. (ed.), River Ecology. Studies in Ecology II. Blackwell Scientific Publications, pp. 403–430.

McIntire, C.D. & Phinney, H.K., 1965. Laboratory studies of periphyton production and community metabolism in lotic environments. Ecol. Monogr. 35:237–258.

Morris, I. (ed.), 1980. The physiological ecology of phytoplanton. Studies in Ecology (VII). Blackwell Scientific Publications.

Moss, B., 1976. The effects of fertilization and fish on community structure and biomass of aquatic macrophytes and epiphytic algal populations: an ecosystem experiment. J. Ecol. 64:313–347.

Nielsen, K. & Nielsen, L., 1982. Grøden og dens chironomide fauna – biomasse, production og grødeskæringseffect. MS Thesis. Freshwater Biological Laboratory, Univ. of Copenhagen (in Danish).

Phillips, G.L., Eminson, D. & Moss, B., 1978. A mechanism to account for macrophyte decline in progressively eutrophicated freshwaters. Aquat. Bot. 4:103–126.

Phinney, H.K. & McIntire, C.D., 1965. Effect of temperatures on metabolism of periphyton communities developed in laboratory streams. Limnol. Oceanogr. 10:341–344.

Revsbech, N.P., Jørgensen, B.B. & Brix, O., 1981. Primary production of microalgae in sediments measured by oxygen microprofile, $H^{14}CO_3$-fixation and oxygen exchange methods. Limnol. Oceanogr. 26:717–730.

Revsbech, N.P., Sørensen, J., Blackburn, T.H. & Lomholt, J.P., 1980. Distribution of oxygen in marine sediments measured with microelectrodes. Limnol. Oceanogr. 25:403–511.

Sakshaug, E., 1980. Problems in the methodology of studying phytoplankton. In: Morris, I. (ed.). The Physiological Ecology of Phytoplankton. Studies in Ecology (VII). Blackwell Scientific Publication, pp. 57–91.

Sand-Jensen, K., 1975. Biomass, net production and growth dynamics in an eelgrass (Zostera marina L.) population in Vellerup Vig, Denmark. Ophelia 14:185–201.

Sand-Jensen, K. & Søndergaard, M., 1978. Growth and production of isoetids in oligotrophic Lake Kalgaard, Denmark. Verh. Internat. Verein. Limnol. 20:659–666.

Sand-Jensen, K. & Søndergaard, M., 1981. Phytoplankton and epiphyte development and their shading effect on submerged macrophytes in lakes of different nutrient status. Int. Revue ges. Hydrobiol. 66:529–552.

Sode, A., 1981. The effect of ferric hydroxide on periphytic algal production and biomass and total oxygen consumption of sediment in a small Danish stream. MS Thesis, Biological Institute, Univ. of Odense and Freshwater Biological Laboratory, Univ. of Copenhagen.

Author's address:
Kaj Sand-Jensen
Freshwater Biological Laboratory
Helsingørsgade 51
3400 Hillerød, Denmark

CHAPTER 11

Photosynthesis-irradiance relationships of epilithic algae measured in the laboratory and *in situ*

MICHAEL A. TURNER, DAVID W. SCHINDLER and ROGER W. GRAHAM

Abstract. Complementary laboratory and *in situ* techniques were developed to measure photosynthesis-irradiance relationships of epilithic algae in the littoral zone of an oligotrophic lake in the Experimental Lakes Area, northwestern Ontario, Canada. SCUBA divers placed colonized granite plates in acrylic chambers and returned them to a nearby laboratory. Changes in dissolved inorganic carbon (DIC) in water overlying the benthic sample were monitored in response to a gradient in intensity of artificial light. The field technique isolated natural communities on rock, and attenuated available light using diffusers. In both methods, dark chambers were monitored in parallel.

Both laboratory and *in situ* observations conformed well to Michaelis-Menten description of the dependence of photosynthesis upon irradiance ($r^2 > 0.94$) permitting estimation of the maximum rate of photosynthesis (C_{max}^{gross}) and the half-saturation constant ($I_{0.5}$). Carbon-14 uptake estimates of (C_{max}^{gross}) and ($I_{0.5}$) were 85% (±5%, SD) and 141% (±30%, SD) respectively, of corresponding DIC uptake estimates. In the light ^{14}C values were significantly different from and intermediate between net and gross DIC uptake ($P < 0.001$). The discrepancy between ^{14}C and net DIC uptake decreased loglinearly with increasing irradiance ($r = -0.76$, $P < 0.001$). Calculated depth profiles indicated substantial ^{14}C uptake below the DIC-predicted compensation depth.

Introduction

Haptobenthic[1] productivity has been measured by observing the rate of change in biomass per unit area of lake bottom (Castenholz 1960; Goldman & Amezaga 1975). Biomass accumulation is a net community process however (Wetzel 1964; Eichenberger & Wuhrmann 1975), so that its usefulness as a measure of benthic photosynthesis is limited (Sládečková 1962; Allen 1971). Measurement of change in concentration of dissolved inorganic carbon (DIC), a photosynthetic substrate, more directly represents photosynthesis. Haptobenthic studies have included *in situ* observations of DIC (Schindler *et al* 1973) and ^{14}C uptake (Wetzel 1964; Hunding 1971). However, the yield of predictive information from *in situ* observations of photosynthesis is relatively low. Variation in photosynthesis with depth is not directly causal, but is derived secondarily from the relationship which exists between photosynthesis and irradiance (Fee 1973a). Unfortunately irradiance at a particular depth fluc-

[1] 'Haptobenthos' is the community on solid and relatively inert surfaces as opposed to 'herpobenthos' which is the community on or within sediment. This terminology was originated by Warming (1923) and recently favoured by Hutchinson (1976). Haptobenthos does not distinguish among the different types of inert substrata within the littoral zone. It also emphasizes that benthic algae are part of a community, and not functionally separate from bacteria, fungi, viruses and zoobenthos.

Wetzel, R.G. (ed.), Periphyton of Freshwater Ecosystems.
 ISBN 90 6193 768 X.

tuates during an exposure period. Consequently a depth-photosynthesis profile is neither predictive nor particularly informative about the photosynthetic potential of the population sampled (Fee 1973a, 1979). Measurement of phytoplankton photosynthesis as a function of irradiance has progressed rapidly using laboratory incubator observations (early development reviewed by Strickland 1960; Fee 1973a, 1973b & 1980), while such studies are lacking for haptobenthic algae. Although single-irradiance observations have been made (Szczepanś 1968; Sheldon & Boylen 1975), an incubator incorporating a light gradient has been used to routinely study only laboratory-grown stream benthos (McIntire & Phinney 1965). This lack may be due to difficulty in obtaining reproducible benthic measurements, which is partly a result of large variations in surface area among samples.

The interrelationship of DIC and ^{14}C uptake is not well known. In phytoplankton studies, ^{14}C uptake is assumed generally to measure something between net and gross photosynthesis (Steemann Nielsen & Hansen 1959; Fogg 1974). Earlier tests of these assumptions using hapthobenthos (Hunding & Hargrave 1973; Schindler *et al* 1973) have been equivocal, partly because of uncontrolled variations in both haptobenthic surface area and exposure irradiance. Thus there is need for a more satisfactory test of the interrelationship of these methods.

To overcome the problems inherent in *in situ* observations of benthic photosynthesis, two incubator approaches were developed. In the laboratory technique, a gradient in intensity of artificial light and many of the advantageous features of phytoplankton incubator methods (Fee 1973a; Shearer 1976b) were incorporated. Smooth granite plates of known dimensions were used as substrates to eliminate both the difficulty of poorly known sample surface area, and the possible selectivity of artificial substrate types. Using enclosures similar to those of Loeb (1981), we developed an *in situ* variant which isolated natural communities on rock, and attenuated available light using diffusers. Furthermore, given the high photosynthetic rates characteristic of haptobenthos, in combination with the low epilimnetic DIC concentrations found

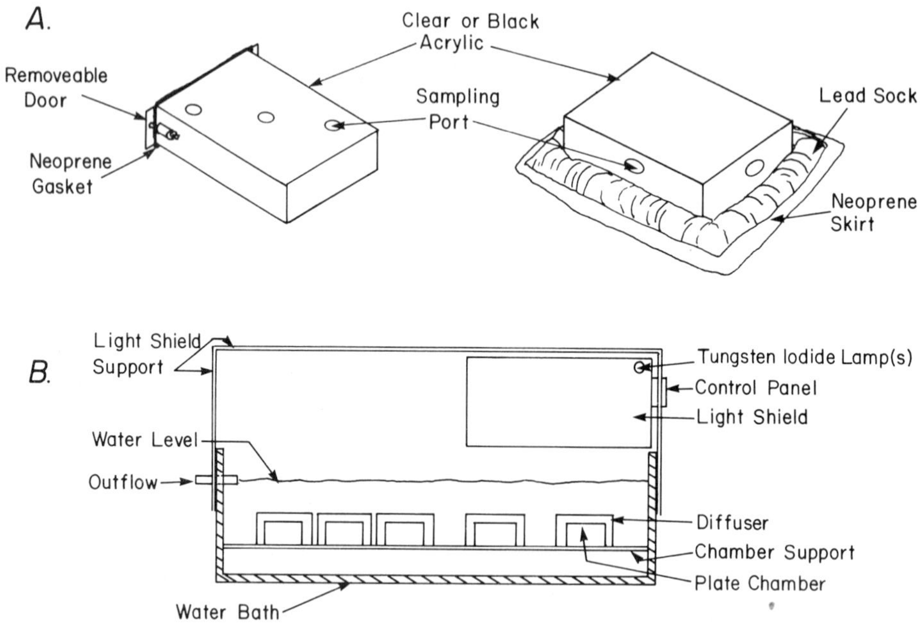

Fig. 1. A: Incubation chambers. The plate chamber (left) completely encloses a colonized plate, while the rock chamber (right) is sealed against a smooth rock surface.
B: Longitudinal section of laboratory incubator. Light shield, diffusers, chamber and water bath are shown schematically.

in many ELA lakes (Schindler *et al* 1973; Prokopowich 1979), we compared DIC and ^{14}C uptake by haptobenthos of known surface area under conditions of controlled irradiance.

Methods

Study area

Lake 239 is a small (56 ha, mean depth = 10.5 m) granite-basin lake located in the Experimental Lakes Area, a headwater region in the Precambrian Shield of northwestern Ontario, Canada (Brunskill & Schindler 1971). The oligotrophic waters have been characterized chemically by Armstrong and Schindler (1971) and Prokopowich (1979). Descriptions of the littoral zone can be found in Stockner and Armstrong (1971) and Schindler *et al* (1973).

Substrata and their retrieval

The substrata used were smooth plates cut from local granite (200 cm^2). Plates were overwintered on the bottom in the littoral zone in one to three meters of water among granite boulders and bedrock. During retrieval of the plates by a SCUBA diver, the communities attached to the sides and bottom of each plate were removed by careful scraping. Still working underwater, the diver placed each plate in an individual incubation chamber (Fig. 1A) filled previously with lake water. The chamber was then sealed immediately, and placed in a light-tight box which contained surrounding water for temperature control. One or two 20-liter samples of epilimnion water were also taken. Sampes were then returned to the field laboratory, and incubations were usually begun within two hours after completion of sampling.

Laboratory incubator technique

Incubation chambers (volume = 0.85 l) were made of clear acrylic with a removable door and neoprene gasket (Fig. 1A). In the upper surface, three holes were cut (Fig. 1A) to receive rubber serum stoppers. Dark chambers were either covered with aluminium foil and two layers of black PVC tape, or made entirely from black opaque acrylic.

The incubator (Fig. 1B) consisted primarily of a light source, light shield, diffusers, chamber tray, temperature bath and overflow basin. Seven 500-watt bulbs with tungsten iodide filaments were equally spaced along a line at one end of the light shield to approximate a line source. The light spectrum generated was dominated in the visual range by 500 to 700 nm wavelengths, similar to the spectra described by Fee (1973a). To achieve varying irradiances, chambers were located at different distances from the light source (Fig. 1B) and covered individually with diffusers made of neutral translucent acrylic, with fiberglass screening lining the interior. Irradiance varied among locations within a chamber by approximately 12% (coefficient of variation).

Tap water was circulated continuously throughout the incubator water bath to keep the chamber at epilimnion temperature during experiments. The inflow temperature and flow rate were adjusted so that the outflow temperature was within 1° C of the epilimnion temperature.

In seven ^{14}C versus DIC experiments approximately 1.5 l of $NaH^{14}CO_3$-labelled epilimnion water (≈10 μCi ^{14}C/l) was flushed through each chamber in about 10 min. Duplicate 50-mL (DIC) and 3-mL (^{14}C) samples were then withdrawn into polypropylene syringes from the middle port (Fig. 1A), such that water removed was replaced at the inlet. The order for initial samples was DIC, then ^{14}C.

Duplicate chambers were assigned randomly to each of four irradiances. After a 2- to 4-h exposure, chambers were removed and placed on a slight incline. Duplicate samples withdrawn from the lowest port were replaced with air at the opposite and highest port. Sampling order was ^{14}C and then DIC.

Field techniques

Chambers enclosed 200 cm^2 of natural rock community (Fig. 1A). Chambers were fabricated from clear or black acrylic (≈0.85 l internal volume) and ringed with an attached neoprene skirt. Lead 'socks' (Fig. 1A) were constructed using polyester or polyethylene fabric sewn into tubes which encircled the chambers. Each sock was filled with approximately 2 kg of lead shot.

To permit controlled attenuation of natural light, we made diffusers or boxes of double thicknesses of 1/8″ (3.2 mm) translucent acrylic, with or without fiberglass screening between the layers. Interior dimensions of the boxes permitted complete enclosure of the chambers. The two diffuser transmittances used routinely permitted 40% and 10% of naturally occurring light to reach the chambers.

The sample site selected contained bedrock and large boulders. Divers carefully placed chambers upon smooth surfaces, minimizing disturbance of the benthic community. Lead socks were then placed upon the neoprene skirt to seal the chamber to the rock. If a diffuser was required, it was placed immediately over the chamber. Start time and chamber depth were recorded. We usually monitored 10 chambers per run: three at each of 100% and 0% transmittance; and two at each of 40% and 10% of light at the rock surface.

Samples ($n \geq 6$) for initial DIC were collected either from the chambers directly or from the surrounding water. After approximately four hours of incubation, final DIC samples were collected from the chambers. To mix the water within each chamber 9 ml of lake water (or approximately 1% of chamber volume) was injected vigorously into each chamber using syringes. Five minutes later after suspended material had settled, triplicate 3 ml samples were collected using polypropylene syringes, one from each sampling port (Fig. 1A).

To quantify the exchange of water between the inside and outside of a rock chamber during an incubation, we injected a known quantity of ^{3}H-labelled water into several chambers. Chambers were syringe-mixed with approximately 1% of their volume, and initial samples taken in duplicate. After 4 h, final samples were obtained similarly. Aliquots (2.5 ml) of sample were added to 12 ml of PCS fluor and counted to <1% of coefficient of variation in a Beckman LS-8000 liquid scintillation counter. Exchange in the skirted chambers after four hours was calculated to be 8.1% ± 4.2% (± SD, n = 19). Given constant DIC in the water surrounding a chamber during an incubation, diffusion would reduce the change in DIC observed due to respiration or photosynthesis. Consequently, photosynthetic rates calculated by equation (1) were multiplied by an exchange factor of 1.04.

Chemical analyses

To obtain initial and final ^{14}C activity in the water, 2.00 ± 0.01 ml of each sample was immediately injected into a scintillation vial containing 15 ml of dioxane-based fluor (Schindler 1966) and 0.1 ml NCS (Amersham Corp.Ltd.). Using the channels ratio method and external quenched standards, samples were counted in a Picker Nuclear Liquimat 220 liquid scintillation counter to approximately 1% coefficient of variation.

After collection, DIC syringes were sealed and placed in ice and darkness until analysis, usually within 6 h. Samples in 1974 were analysed using the gas chromatographic method of Stainton (1973) which was calibrated against external standards and yielded a coefficient of variation of 2% (n = 20).

In 1982, samples were analysed using an infrared gas analyzer. One to 2.5 ml of sample were injected into a sparging column containing 0.5 ml of 0.1N H_2SO_4 purged previously of its CO_2. Carbon dioxide stripped from the sample was carried to the analyzer via CO_2-free nitrogen gas. Peak heights were calibrated against external gas standards. The coefficient of variation of the method was approximately 1% at ≈100 μmol C/l.

Light measurements

After each laboratory experiment, irradiance was measured at chamber locations in the incubator. A Li-Cor quantum meter with submersible photo-cell and flat cosine collector sensitive to light of 400 to 700 nm was used (Shearer 1976a). Units of measurement were $\mu E m^{-2} s^{-1}$

To determine the mean irradiance (I_c) received by a chamber during an *in situ* run, surface irradiance (I_o) was monitored continuously throughout each experimental period using a cosine-corrected, flat plate collector (Li-1905). Output was integrated over 10 minute intervals using a Campbell CR integrator, and the mean (I_o) determined for each chamber. The extinction coefficient (ε) for light in the enphotic zone was determined from light-attenuation profiles measured within one week of the experiment (Shearer and DeBruyn, unpublished data). In combination with the chamber depth (z) and diffuser transmittance (D), the mean irradiance within a chamber during an exposure period was calculated to be $I_c = D \times I_z = D \times I_o \times e^{-\varepsilon z}$.

Calculation of photosynthetic rates

An explanation of frequently used symbols is given in Table 1. Computer programs developed to perform work described here are presented in Turner (1981).

Observed net uptake of DIC (C^{net}) at a particular irradiance was calculated as the rate of change in DIC in water overlying the benthic sample of known surface area (A).

$$C^{net} = ([DIC]_i - [DIC]_f) \times V/(A \times t) \qquad (1)$$

'i' and 'f' subscripts refer to the initial and final samples, t is the duration of exposure and V refers to the volume of water in the incubation chamber. Units of C^{net} are μmol C m^{-2} h^{-1}.

For a chamber incubated in the dark, equation (1) yields a negative number which estimates community respiration in the dark (R_d). Although it is not strictly true that algal respiration is independent of light intensity (Rabinowitch & Govindjee 1969), for the sake of simplicity, and in the absence of an estimate of respiration at other irradiances, gross photosynthesis was estimated by equation (2).

$$C^{gross} = C^{net} + |R_d| \qquad (2)$$

where R_d is the mean of two or more dark chambers monitored in the corresponding experiment. Units of C^{gross} are μmol C m^{-2} h^{-1}-

Photorespiration is unaccounted for in equation (2). If photorespiration were important, the procedure described here would underestimate gross photosynthesis. Because only recently photosynthesized metabolites are respired (Tolbert 1974), ignoring photorespiration may not be of ecological significance to upper trophic levels.

Incorporation of DIC by phytoplankton during incubation would yield an overestimate of haptobenthic photosynthesis. From concurrent phytoplankton studies (Shearer 1976b), maximum rates of phytoplankton photosynthesis were calculated to be approximately 155 μmol C m^{-3} h^{-1}. Given a

Table 1. A list of frequently used symbols and their explanation.

Symbol	Explanation	Units
A	surface area of either plate or rock	m^2
C^{gross}	gross photosynthesis estimated by adding community dark respiration (R_d) to observed net DIC uptake (C^{net}). Defined by equation (2).	μmol C m^{-2} h^{-1}
C^{gross}_{max}	maximum rate of gross photosynthesis. Defined by equation (5).	μmol C m^{-2} h^{-1}
C^{net}	net photosynthesis estimated using observed DIC uptake measurements. Defined by equation (1).	μmol C m^{-2} h^{-1}
^{14}C	photosynthetic rate estimated using ^{14}C-labelled DIC uptake measurements. Defined by equation (3).	μmol C m^{-2} h^{-1}
DIC	dissolved inorganic carbon = $CO_2 + HCO_3^- + CO_3^= + H_2CO_3$	μmol C/L
I	light intensity	μeinsteins m^{-2} s^{-1}
$I_{0.5}$	irradiance constant used in the Michaelis-Menten-type mathematical formulation of photosynthesis (equation 4). It is numerically equivalent to that irradiance at which half the maximum rate of gross photosynthesis is achieved.	μeinsteins m^{-2} s^{-1}
I_0	light intensity observed at lake surface	μ einsteins m^{-2} s^{-1}
I_z	light intensity at depth z	μ einsteins m^{-2} s^{-1}
R_d	community dark respiration measured by DIC release	μmol C m^{-2} h^{-1}
t	time, on the order of hours. Frequently denotes length of an incubation.	h
z_m	maximum depth of interest. In this investigation, z_m usually means epilimnion depth.	m
ε	extinction coefficient	m^{-1}

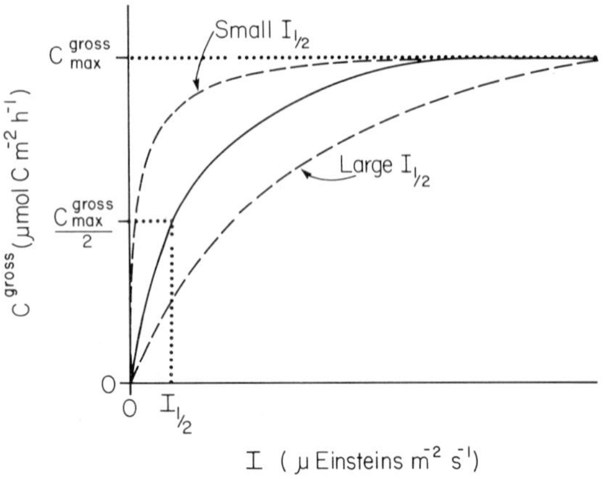

Fig. 2. Hypothetical presentation of C^{gross}_{max} and $I_{0.5}$ in relation to exposure irradiance (I) and gross photosynthesis (C^{gross}).

mean observed haptobenthic C^{gross} of approximately 465 μmol C m^{-2} h^{-1} (see also Table 4), an effective chamber volume of 0.61 l, and a sample surface area of 200 cm^2, the proportion of phytoplankton to haptobenthic photosynthesis was an insignificant 1.0%.

Uptake of carbon measured by the ^{14}C technique (^{14}C) is described by the following expression which is analogous to equation (1) except that changes in ^{14}C activity ([^{14}C]) in water are monitored rather than changes in DIC.

$$1.05 \times V \times [DIC]_i \times ([^{14}C]_i - [^{14}C]_f)/([^{14}C]_i \times A \times t). \qquad (3)$$

Units of the result are also μmol C m^{-2} h^{-1}. An isotope discrimination factor of 1.05 was included (Strickland 1960). Dark ^{14}C uptake was not substracted routinely from ^{14}C estimates in accordance with the recommendation of Morris *et al* (1971), and the suggestion of Foog (1974) that dark uptake may represent dark carboxylation reactions which do not occur in the light. Uptake of ^{14}C in the dark was observed to be only 5% of uptake at light-saturated irradiances.

Any dissolved or particulate organic ^{14}C released by the haptobenthos into the water in the incubation chamber would cause an underestimate when carbon uptake is calculated by the above expression. This organic ^{14}C was measured in four samples incubated for three hours at 168 to 217. micro-einsteins m^{-2} s^{-1}. Duplicate 2.00 ml water samples were injected into scintillation vials with 0.05 ml of 0.1 N HCl and bubbled with air for 20 minutes. Approximately 15 ml of dioxane fluor were added, and samples counted to approximately 2% coefficient of variation. The mean (± SD) percentage underestimation was 0.9% ± 0.2%, and was considered relatively unimportant.

Linearity of carbon uptake with time

Using the laboratory incubator method, the influence of incubation time upon carbon uptake was examined at irradiances of 168 to 218 micro-einsteins m^{-2} s^{-1} in early August of 1974. The ^{14}C technique was selected because of the ease of sample analysis, and small sample volumes required. Six haptobenthic samples were subsampled at 0.5, 1, 2, 3, 4, and 6 h of incubation (Fig. 3). The mean r^2 of the individual regressions on time was 0.961 ± 0.003 (± SD, n = 6); whereas, without the 6-h observations, the mean r^2 observed was 0.997 ± 0.001. Furthermore, the 6-h rate was 80.0% ± 0.7% (n = 6) of the 4-h rate. Therefore, the regression of carbon uptake on time was linear

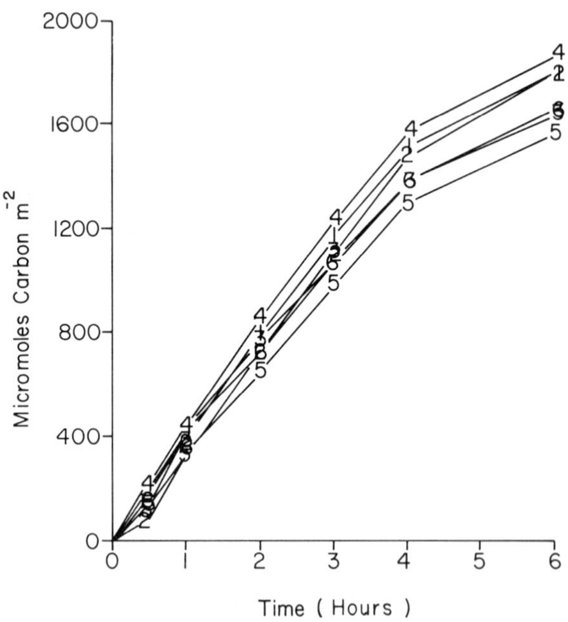

Fig. 3. Linearity of carbon uptake with time. Six colonized granite plates (1 to 6) were incubated at near saturation irradiances.

to 4 h; and the usual 2- to 4-h incubation periods appear satisfactory.

Variations in surface irradiance during the exposure period do not permit straightforward testing of linearity of carbon uptake during an *in situ* experiment. To overcome this difficulty, we constructed incubation chambers with a ratio of chamber volume: sample area which was double that of the usual chamber. If DIC or another component was limiting during an incubation, higher rates of photosynthesis in the oversize chamber would be expected. We tested this approach in August of 1982, monitoring depletion of DIC during a four-hour incubation period in three chambers of each type. Mean photosynthetic rates (μmol C m^{-2} h^{-1}) observed were 446 ± 13 and 423 ± 135 for the regular and oversize chambers, respectively. Because there was no difference between means, we concluded that carbon uptake was linear.

Estimation of other photosynthetic parameters

The relationship between gross DIC uptake and light intensity or substrate concentration (Fig. 2) can be described by the Michaelis-Menten-type equation

$$C^{gross} = C^{gross}_{max} \times I/(I + I_{0.5}) \qquad (4)$$

where C^{gross} is the rate of gross photosynthesis (μmol C m^{-2} h^{-1}) defined by equation (2), C^{gross}_{max} is the maximum rate of gross photosynthesis (μmol C m^{-2} h^{-1}), I is the exposure light intensity (μeinsteins m^{-2} s^{-1}); and $I_{0.5}$ (μeinsteins m^{-2} s^{-1}) is the Michaelis-Menten-type 'constant of convenience' (Riggs 1963). Numerically, $I_{0.5}$ is the intensity at which half the maximum photosynthetic rate is observed (Fig. 2).

To avoid the problems inherent in utilizing linear transformations to estimate C^{gross}_{max} and $I_{0.5}$ (Riggs 1963), the rate of change in discrepancy between observed (equation (2)) and predicted (equation (4)) results was examined with respect to the rate of change in C^{gross}_{max} and $I_{0.5}$ characterized by their respective partial differential equations (Turner 1981). The criterion for obtaining the overall best fit was the minimum obtainable variance.

Instantaneous and daily photosynthetic integrals

To quantify the significance of differences in C^{gross}_{max} and $I_{0.5}$ observed among the techniques, we calculated hypothetical integrals of daily photosynthesis with the epilimnion. The general form of the instantaneous depth integral for haptobenthic photosynthesis is $\int_z L_z \times C(I_z)dz$; where L_z is the length of the contour along the lake bottom at some depth z. The units of the solution are mass of carbon. If L_z is set equal to one meter, the integral represents photosynthesis of benthos colonized on an imaginary vertical surface of one-meter width, extending from the lake surface to the maximum depth of interest. Such an imaginary surface, with equal length at all depths, permits comparison of benthic activity among lakes, excluding differences due to morphometry or bottom type. The influence of these variables can be examined independently by varying L_z with depth. This approach is strictly analogous to the phytoplankton ecologist's column of water of one m^2 cross-sectional area, extending from the lake surface to lake bottom.

Light at depth z (I_z) can be approximated by the Beer-Lambert Law ($I_z = I_o \times e^{-\varepsilon z}$); where I_o is the surface irradiance, and ε is the extinction coefficient (m^{-1}). In combination with equation (4), the instantaneous depth integral for haptobenthic photosynthesis can then be solved for if L_z is held as an arbitrary constant. The indefinite solution for a layer which completely absorbs light is given by equation (5) with units of mass of carbon (Turner 1981).

$$\int_z L_z \times C(I_z)dz = L_z \times C^{gross}_{max} \times \{z - [\ln(I_o + I_{0.5} \times e^{+\varepsilon z})]/\varepsilon\} \qquad (5)$$

The daily depth integral is of the general form

$$\int_t \int L_2 \times C[I(z, t)]dzdt \qquad (6)$$

where t is time within a day. The computer solution for the iterative form of the daily integral for a depth interval which does not completely absorb light is presented in Turner (1981).

Daily integrals were calculated for haptobenthos on a vertical one-meter wide surface extending over a depth interval of zero to five meters. Two types of days were examined with maximum daily surface irradiances of approximately 2000 and 200 μeinsteins m^{-2} s^{-1}, corresponding roughly to a sunny midsummer (16 h) and a cloudy fall day (10 h), respectively. The variation of I_o with time of day was approximated by Vollerweider's (1965) cosine solution. Water column transparency was assumed to be constant and equal in value to the mean ε (0.78) observed in Lake 239 during the experimental period (calculated from Shearer 1976a).

Results

Precision of estimates

The mean coefficient of variation of C^{net} for duplicate chambers at the highest irradiance in the seven laboratory incubator experiments (Table 2) was 13.3% ± 6.0% (± SD). For duplicate dark chambers, corresponding values for R_d were 32.6% ± 31.6%. In four *in situ* analyses in June and July of 1982, the percent coefficient of variation of C^{net} for triplicate chambers with 100% transmittance was 8.1% ± 4.4%. Corresponding values for R_d were 20.5% ± 14.3%. Relative variation in estimates of R_d was larger than that of C^{net} because the absolute magnitude of C^{net} was greater than that of R_d (e.g. Table 2), although their sources of variance were similar. The reduced relative variation of field versus laboratory measurements for both C^{net} and R_d is due probably to the use of triplicate rather than duplicate sampling in the field technique.

Turbulence in enclosures

We were concerned that the use of enclosures would minimize the amount of turbulence to which benthic samples were exposed. If nutrients were depleted in the immediate vicinity of the algae, then photosynthesis might be underestimated. To test this possibility, we measured photosynthesis of colonized plates on two occasions in August of 1982. On the first occasion, one group of plates was shaken at 80 rpm in incubation chambers attached to a platform on a Junior Orbit Shaker. The photosynthetic rates observed were 300 ± 4 μmol C m^{-2} h^{-1} (± SD, n = 3) compared to 270 ± 28 μmol C

Table 2. Comparison of DIC with ^{14}C paired photosynthetic estimates. Photosynthetic measurements (micromoles C m^{-2} h^{-1}) were made of Lake 239 haptobenthic samples held at varying irradiances (I: microeinsteins m^{-2} h^{-1}). Negative DIC values indicate release of DIC. Negative values of ^{14}C observed in the dark are due presumably to counting error.

Experiment			Dark		Light											
No.	Date	Depth of samples (m)	DIC	^{14}C	I	DIC	^{14}C	I	DIC	^{14}C	I	DIC	^{14}C	I	DIC	^{14}C
1	5/7	1	−119	2	11.5	0	79	60	239	255	130	212	235	360	227	309
			−120	34	12.0	−22	95	65	278	285	135	330	338	370	334	354
2	15/7	1	−147	19	14.7	18	89	54	143	185	212	354	410	670	358	453
			−189	−7	14.9	65	145	58	180	240	227	285	366	670	263	315
3	21/6	2	−179	42	11.0	34	89	61	263	320	n.s.	n.s.	n.s.	483	305	413
			−125	25	11.5	86	174	68	209	293	n.s.	n.s.	n.s.	505	301	365
4	5/7	2	− 36	16	11.5	160	176	60	202	224	130	350	350	360	296	312
			−166	20	12.0	100	151	65	220	238	135	217	245	370	234	260
5	15/7	2	−142	9	14.7	−15	67	54	197	248	212	410	437	670	346	389
			−130	44	14.9	88	163	58	235	268	227	299	330	670	370	445
6	12/7	3	−178	19	14.0	102	149	44	227	250	166	377	396	620	283	345
			−114	−7	14.3	80	132	47	258	296	180	374	403	635	349	400
7	22/7	3	− 55	21	13.2	52	124	52	152	296	207	355	419	635	261	336
			−128	8	15.5	49	124	58	149	154	222	282	326	640	366	425

n.s. = not sampled

Table 3. The mean discrepancies between ^{14}C and C^{net}, or C^{gross} and ^{14}C, observed in experiments 1 to 7 of Table 2. Prior to estimation of the means, each discrepancy was normalized as a percentage of the corresponding experimental C^{gross}_{max}.

Irradiance (microeinsteins $m^{-2}s^{-1}$)	Number of observations	$\frac{(C^{gross} - {}^{14}C) \times 100}{C^{gross}_{max}}$ Mean %	($\pm$SD)	$\frac{({}^{14}C - C^{net}) \times 100}{C^{gross}_{max}}$ Mean %	($\pm$SD)
0	14	−3.0	(8.7)	30.5	(9.9)[a]
11 to 16	14	12.8	(7.7)[a]	14.6	(4.8)[a]
44 to 68	14	20.0	(5.4)[a]	9.2	(8.1)[a]
130 to 227	12	19.4	(7.1)[a]	7.2	(4.8)[a]
360 to 670	14	14.8	(8.0)[a]	12.6	(5.6)[a]

[a] Significance ($P < 0.001$) of a paired t-test of the differences observed. The remaining difference was not quite significant at $P = 0.05$.

$m^{-2}h^{-1}$ in the static condition. To ensure that we actually mixed the water overlying the benthic community, on the second occasion we increased the shaking to 160 rpm which suspended the loosely-attached component. Shaken and static rates were 324 ± 43 μmol C $m^{-2}h^{-1}$ (± SD, n = 3) and 300 ± 24 μmol C $m^{-2}h^{-1}$, respectively. The differences between static and shaken conditions were not significant ($P = 0.1$, factorial AOV), so we concluded that use of enclosures does not impart any bias due to reduction of turbulence.

Propriety of Michaelis-Menten formulation

To determine how well equation (4) corresponds with the photosynthesis irradiance observations, linear regressions were computed to describe the correspondence between measured and predicted values (Table 4). The small amount of variance left unexplained ($1-r^2 < 6\%$) indicates that there was good agreement between measured and equation (4) minus predicted values for all methods. Because the slopes and intercepts were not significantly different from one and zero, respectively, the fitting procedure appears unbiased for all methods. Consequently, differences in values of C^{gross}_{max} and $I_{0.5}$ (Table 5) are attributable to differences among the methods and not to the technique of fitting.

Table 4. Correlation between Michaelis–Menten predictions and actual photosynthetic measurements. Perfect agreement would yield a slope of 1, an intercept of 0, and an r^2 of 1.

Method of measurement	Number of observations	Slope	Intercept (mmol C m^{-2} h^{-1})	r^2
DIC, laboratory	88	0.957	0.011	0.949
^{14}C, laboratory	68	0.974	0.002	0.936
DIC, field	40	1.001	−0.003	0.981

Comparison between DIC and ^{14}C uptake

Excluding dark exposures, ^{14}C estimates were between net and gross DIC on 53 of 54 occasions (Table 2). Subtraction of dark ^{14}C uptake from ^{14}C estimates changed this to 47 of 54. Thus, ^{14}C uptake in the light represented something between net and gross DIC uptake (Fig. 4). But in the dark, there was usually a small amount of ^{14}C taken up (Table 2, Fig. 4) in contrast to the theoretical value of zero

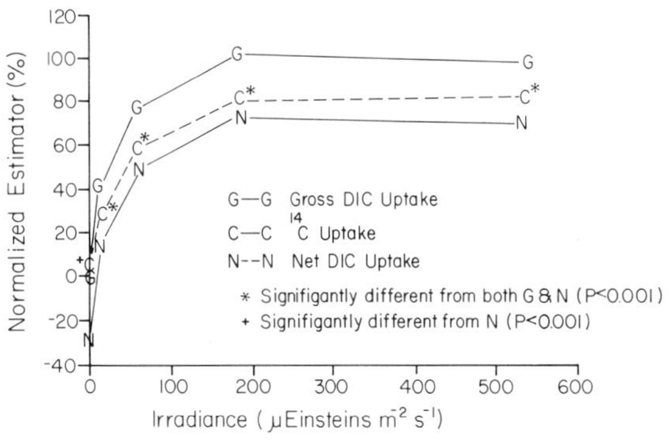

Fig. 4. Comparison of DIC and ^{14}C uptake as a function of light intensity. The mean ^{14}C uptake for each range of irradiances is compared with the corresponding mean of net and gross DIC uptake reported in Table 2.

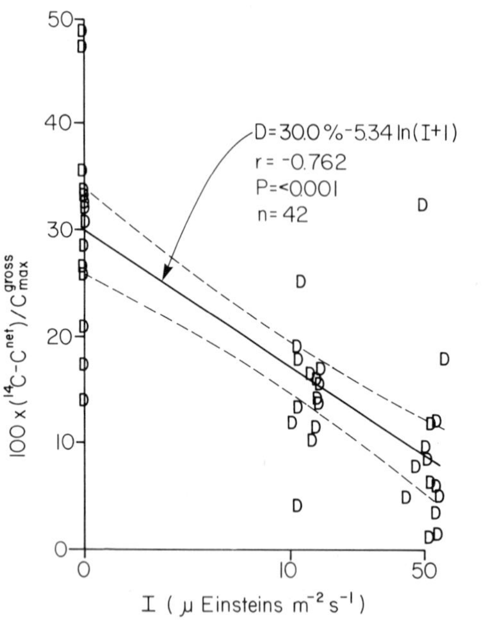

Fig. 5. The discrepancy (D) between ^{14}C and net DIC uptake expressed as a logarithmic function of light intensity (I). The regression line is shown with its attendant 95% confidence interval for the population mean. Discrepancies at irradiances >53 μeinsteins $m^{-2}s^{-1}$ were not different from those circa 50 μeinsteins m^{-2} s^{-1} (Table 3).

for gross DIC uptake. The magnitude of this dark ^{14}C uptake was unrelated to community dark respiration ($r = 0.01$, $n = 14$).

To examine whether any significant discrepancies existed between ^{14}C and C^{net} and C^{gross}, differences between paired measurements were normalized as a percentage of the C^{gross}_{max} estimated for that experiment (Table 5). Excluding the dark exposures, ^{14}C was always significantly less than C^{gross} (Table 3); but ^{14}C was always greater than C^{net} (Table 3). However, the discrepancy between ^{14}C and C^{net} decreased loglinearly from 0 to 68 microeinsteins $m^{-2}s^{-1}$ (Fig. 5). The magnitude of this discrepancy did not change appreciably at higher irradiances (Table 3).

Carbon-14 estimates of C^{gross}_{max} were observed to be consistently lower (85% ± 5%, $n = 7$) than corresponding estimates of gross DIC uptake (Table 5). Conversely, ^{14}C estimates of $I_{0.5}$ were always greater than corresponding DIC estimates (141% ± 30%, $n = 7$).

Daily photosynthetic integrals for a sunny midsummer day and a cloudy fall day were calculated. Estimates of R_d, C^{gross}_{max} and $I_{0.5}$ (Table 6) used for DIC integrals were the means of values obtained in experiments 1 to 7 (Table 5). Parameters for ^{14}C integrals were chosen similarly except that R_d was assumed equal to zero. To determine if subtraction of dark ^{14}C uptake from ^{14}C uptake in the light would alter our conclusions, we also set R_d equal to the mean rate of dark ^{14}C uptake observed (Table 2).

Carbon-14 integrals were less than those of gross DIC (Table 6), and the relative magnitude of this difference was a function of light intensity (Fig. 6) as one might expect from Figs. 4 and 5. Carbon-14 integrals were 80% and 73% of gross DIC on the sunny midsummer and cloudy fall days, respectively (Table 6). Therefore, the influence of the larger $I_{0.5}$ observed for ^{14}C (Table 5) became increasingly important with decreasing irradiance.

The relationship between ^{14}C and net DIC integrals was more complex. Carbon-14 integrals were always larger than those of DIC (Table 6), e.g., on the sunny day the ^{14}C estimate was thirty percent higher than the corresponding DIC value. Similarly, a compensation depth of 2.75 meters was observed in the DIC profile (Fig. 6), yet the ^{14}C

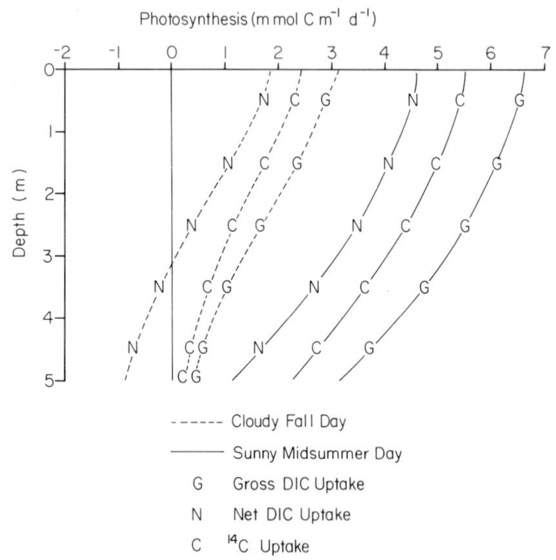

Fig. 6. Hypothetical depth profiles of ^{14}C, gross and net DIC uptake. Calculations were made for the daylight portions of a cloudless midsummer and a cloudy fall day.

technique continued to predict ^{14}C uptake. Correction of ^{14}C values for observed dark ^{14}C uptake did not change any of the above conclusions, reducing daily integrals by 7 and 15% on the cloudless and cloudy days, respectively (Table 6).

Discussion

Reliability of incubator observations

The control of light intensity afforded by the techniques described has made it possible to examine benthic photosynthesis with some degree of precision and to compare functional characteristics of communities. For example, in addition to observing community dark respiration (R_d), it is now possible to examine such characteristics as the maximum rate of gross photosynthesis C^{gross}_{max}), and a half-saturation light parameter ($I_{0.5}$) which is inversely proportional to the slope of the photosynthesis–irradiance curve at low irradiances.[2] Such information can be used to predict photosynthetic rates over a wide range of irradiances (e.g. Fig. 6), and at times other than the exposure period when the parameters continue to describe the community satisfactorily (Fee 1973a). This contrasts with other observations (e.g. Wetzel 1964; Hunding 1971; Loeb 1981) using the *in situ*-depth profile technique which only described photosynthesis that had occurred during the respective incubation periods. Additionally, the control over sample surface area afforded by the use of both the plates and the rock chambers eliminated sources of variation inherent in other investigations employing natural materials retrieved from the lake bottom (e.g. Schindler *et al* 1973).

Carbon uptake predicted from incubator measurements and *in situ* observations of haptobenthic photosynthesis has not been compared directly in this study. Such comparisons have been made for epilimnetic phytoplankton in the study area (Fee 1973b); and the agreement was very good (r = 0.96, n = 6). C^{gross}_{max} for *Chlorella pyrenoidosa* is independent of the wavelength of light, while $I_{0.5}$ is depen-

[2] When $I_{0.5}$ is used to compare photosynthetic efficiencies, care should be taken to ensure that the magnitudes of the corresponding C^{gross}_{max} are similar. Because $I_{0.5}$ is proportional to C^{gross}_{max}, erroneous conclusions may otherwise be drawn (Healey 1980). Thus $I_{0.5}$ is largely considered to be an indicator of the shape of the photosynthesis–irradiance curve.

Table 5. Best estimates of the photosynthetic parameters, maximum rate of gross photosynthesis (C^{gross}_{max}) and the half-saturation constant ($I_{0.5}$). Experiment numbers parallel those in Table 2.

Experiment	C^{gross}_{max} (μmol C m^{-2})		$I_{0.5}$ (μeinsteins m^{-2} s^{-1})	
	DIC	^{14}C	DIC	^{14}C
1	458	363	24.5	29.5
2	509	425	23.9	45.8
3	468	408	13.4	22.7
4	380	298	8.0	10.4
5	533	448	28.9	40.9
6	467	416	18.3	24.2
7	443	413	35.8	37.5

Table 6. Photosynthesis of haptobenthos colonized on a 1-m-wide × 5-m-deep surface for two different types of days. Hypothetical daily integrals were computed for each technique using estimates of parameters derived from Table 5. Details of the derivation are described in the text.

	Parameters			Photosynthetic integrals (mmol C m^{-1} d^{-1})	
	R_d mmol C / m^2 h	C^{gross}_{max} mmol C / m^2 h	$I_{0.5}$ μeinsteins / m^2 s	Sunny midsummer day	Cloudy fall day
Gross DIC	0.000	0.465	21.8	5.25	1.66
Net DIC	−0.127	0.465	21.8	3.21	0.39
^{14}C	0.000	0.396	30.7	4.18	1.21
^{14}C-dark ^{14}C	−0.018	0.396	30.7	3.89	1.03

dent (Pickett & Myers 1966). Thus, differences in light quality between *in situ* and laboratory conditions should not affect estimates of C^{gross}_{max}. Because photons with wavelengths near the chlorophyll *a* absorption maximum of 675 nm are used most efficiently (Rabinowitch & Govindjee 1969), efficiency would be overestimated in the 'reddish' light (500 to 700 nm) of our incubator. Consequently, an incubator-predicted half-saturation constant ($I_{0.5}$) may be too low; i.e. the community would be predicted to be too 'efficient' at low irradiances. Thus, close agreement between laboratory-predicted and *in situ* rates would be expected when the community was exposed to the relatively high irradiances characteristic of epilimnia during ice-free periods. Furthermore, comparison among estimates of $I_{0.5}$ by the ^{14}C and DIC techniques will be unaffected.

Estimation of photosynthetic parameters
Photosynthesis can be described as a function of irradiance by several equations (e.g. Jassby & Platt 1976). We initially chose Michaelis–Menten kinetics, or the rectangular hyperbola formulation (equation 4), to describe benthic photosynthesis because photons in the visible range can be considered to be the most fundamental substrate of photosynthesis and can be expected to saturate the pigment systems of benthic algae at high irradiances. Photo-inhibition – requiring at least one additional parameter – has not been observed in haptobenthic communites.[3] Lastly, the conceptual simplicity of the Michaelis–Menten formulation (equation 4) is exceeded only by that of the discontinuous linear approach of Blackman (1905). Because photosynthetic saturation is approached asymptotically (e.g. Fig. 4), the Blackman approach strongly overestimates photosynthesis at intermediate intensities (Platt *et al* 1975).

The Michael–Menten formulation was found to be a suitable descriptor of photosynthesis-irradiance observations (Table 4). This is contrary to observations for phytoplankton (Platt *et al* 1975; Jassby & Platt 1976) where good fits were not observed, and ^{14}C uptake was underestimated at low irradiances. However, Jassby and Platt used a two-stage fitting process in which $I_{0.5}$ was fitted to observations which were changing rapidly from gross to something intermediate between gross and net carbon uptake (Fig. 5, Table 3). Hence, the initial slope observed was too shallow, estimates of $I_{0.5}$ were too large, and subsequent fits of the equation to observed data were poor. Therefore, Jassby & Platt's (1976) conclusion that Michaelis–Menten kinetics are poor descriptors of photosynthesis does not apply to our study of benthic algae, and possibly in general for benthic studies.

Measurement of carbon uptake
At all light intensities the hypothesis that ^{14}C uptake is between net and gross DIC was confirmed. From darkness to approximately 68 microeinsteins $m^{-2} s^{-1}$, the discrepancy between ^{14}C and C^{net} decreased loglinearly (Fig. 5). Thereafter, the magnitude of this discrepancy did not change significantly (Table 3), and ^{14}C remained slightly larger than C^{net} (Fig. 4). However, this phenomenon may be an artifact of the ^{14}C technique. Metabolic DIC pools are unlabelled initially, and ^{14}C can only be taken up. Release of ^{14}C can occur only after some undetermined amount of photosynthesis. The ^{14}C technique will approximate net DIC uptake with subsequent respiration of photosynthetically-incorporated ^{14}C. However, ^{14}C estimates at higher photosynthetic rates will always be observed to be between net and gross DIC because of the initial bias imposed by the unlabelled DIC pools. Photosynthetic equilibration of populations with ^{14}C prior to beginning an incubation may be a solution.

Unfortunately the interrelationship among ^{14}C, net and gross DIC photosynthetic integrals was complex. As expected from Fig. 4, ^{14}C integrals were intermediate in value between net and gross DIC integrals (Table 6), but the correspondence between ^{14}C and net DIC became worse with decreasing light availability (Table 6, Fig. 6). Importantly, ^{14}C estimates always indicated net carbon fixation, although net DIC profiles indicated net respiration at low irradiances. This conclusion follows directly from the fact that the ^{14}C technique can demonstrate carbon uptake only, not release.

[3] Hunding (1971) demonstrated photoinhibition in epipsammic algae at 25–30 klux in late fall, but not in summer.

In contrast, a 16-fold decrease in total insolation induced only a nine percent change in relationship between ^{14}C and gross DIC photosynthesis (Table 6). Thus, the relationship between ^{14}C and gross DIC integrals is more constant than that between ^{14}C and net DIC integrals. Probably this relationship reflects the relative importance of C^{gross}_{max} (see equation (5); Turner 1981) in conjunction with the relationship between ^{14}C and DIC-derived estimates of C^{gross}_{max} used in the solution of these integrals. Thus, if ^{14}C measurements are the only photosynthetic data available, conclusions about availability of photosynthetically-produced carbon to upper trophic levels should be made with caution. Presumably conclusions about hypolimnetic populations, which live at relatively lower irradiances, would be more subject to the error associated with low light than would conclusions about epilimnetic populations.

Comparison of oxygen and DIC methods

Past studies have noted a number of problems in measuring the photosynthesis of attached plants by changes in oxygen concentration. In waters with high chemical oxygen demand, the method will overestimate productivity, and carbon uptake methods are preferred (Chrost & Sikorska 1976). Wetzel (1964) has noted that oxygen bubbles often form on the surface of photosynthesizing macrophytes, causing production to be underestimated. We observed bubbles adhering to the surface of the haptobenthic community in Lake 239, and small bubbles were occasionally seen in incubation chambers after incubation at higher irradiances. However, in two comparisons of DIC and oxygen methods, Turner (1981) found respiratory and photosynthetic quotients of 0.71 ± 0.14 and 1.08 ± 0.05, respectively. These are consistent with values for phytoplankton (Strickland 1960; Fogg 1974). Furthermore, photosynthesis-irradiance curves, depth profiles and daily integrals obtained by the two methods were similar. Overall, the oxygen method appears to be suitable for making haptobenthic production measurements, although occasional calibration with carbon techniques would be required.

Applications to waters with high DIC

Measurement of change in DIC in water overlying enclosed benthic samples involves summing the variance of both initial and final measurements of DIC. This currently restricts application of the methodology to lakes with low DIC concentrations where the photosynthetic signal is not obscured by noise of DIC analyses. Given that carbonate precipitation is measured or avoided, in lakes with high DIC, the incubation period could be increased, although the possibility that uptake of carbon could become non-linear must be investigated (cf. Fig. 3). Changes in oxygen concentration could also be monitored, but as we have discussed, such an approach has several potential problems. Alternatively, accumulation of ^{14}C in the benthic mat could be measured (Wetzel 1964; Schindler *et al* 1973; Loeb 1981), although in addition to the constraints we have observed (e.g., Figs. 5 & 6), subsampling the benthic mat will increase the total variance of the method. Furthermore, an estimate of respiration would be required.

Acknowledgments

Many members of the Experimental Limnology Project provided logistical support throughout this study. In particular we wish to thank Ed DeBruyn for SCUBA support, Sandee Meron for technical assistance, John Shearer and Ray Hesslein for several fruitful discussions. Michael Stainton, Ray Hesslein and Ed DeBruyn refined the infrared gas analyzer technique as it was used. Ms. S. Elliott, J. Shearer and Drs. J. Eales, E.J. Fee, F.P. Healey, R. Hecky, K. Patalas, A.E. Rosemarin, F. Ward and Robert G. Wetzel provided valuable criticisms of the manuscript. Early research by M. Turner was supported by a National Research Council scholarship.

References

Allen, H.L., 1971. Primary productivity, chemo-organotrophy, and nutritional interactions of epiphytic algae and bacteria on macrophytes in the littoral of a lake. Ecol. Monogr. 41:97–127.

Armstrong, F.A.J. & Schindler, D.W., 1971. Preliminary chemical characterization of waters in the Experimental Lakes Area, northwestern Ontario. J. Fish. Res. Board Can. 28:171–187.

Blackman, F.A., 1905. Optima and limiting factors. Ann. Bot. 19:281–295.

Brunskill, G.J. & Schindler, D.W., 1971. Geography and bathymetry of selected lake basins, Experimental Lakes Area, northwestern Ontario. J. Fish. Res. Board Can. 28:139–155.

Castenholz, R.W., 1960. Seasonal changes in the attached algae of freshwater and saline lakes in the Lower Grand Coulee, Washington. Limnol. Oceanog. 5:1–28.

Chrost, R.J. & Sikorska, U., 1976. The effect of pollution on photosynthetic activity of algae and physiological activity of bacteria in lake(s). Polskie Arch. Hydrobiol. 23:357–364.

Eichenberger, E. & Wuhrmann, K., 1975. Growth and photosynthesis during the formation of a benthic algae community. Verh. Internat. Verein. Limnol. 19:2035–2042.

Fee, E.J., 1973a. A numerical model for determining integral primary production and its application to Lake Michigan. J. Fish. Res. Board Can. 30:1447–1468.

Fee, E.J., 1973b. Modelling primary production in waterbodies: a numerical approach that allows vertical inhomogeneities. J. Fish. Res. Board Can. 30:1469–1473.

Fee, E.J., 1979. A relation between lake morphometry and primary productivity and its use in interpreting whole-lake eutrophication experiments. Limnol. Oceanogr. 24: 401–416.

Fee, E.J., 1980. Important factors for estimating annual phytoplankton production in the Experimental Lakes Area. Can. J. Fish. Aquat. Sci. 37:513–522.

Fogg, G.E., 1974. Oxygen-versus ^{14}C-methodology, pp. 76–78. In: Vollenweider, R.A. (ed.). A Manual on Methods for Measuring Primary Production in Aquatic Environments. Blackwell Scientific Publ., Oxford.

Goldman, C.R. & de Amezaga, E., 1975. Primary productivity in the littoral zone of Lake Tahoe, California – Nevada. Symp. Biol. Hung. 15:49–62.

Healey, F.P., 1980. Slope of the Monod equation as an indicator of advantage in nutrient competition. Microb. Ecol. 5: 281–286.

Hickman, M., 1971. Standing crops and primary productivity of the epipelon of two small ponds in North Somerset, U.K. Oecologia (Berl.) 6:238–253.

Hunding, C., 1971. Production of benthic microalgae in the littoral zone of a eutrophic lake. Oikos 22:389–397, Copenhagen.

Hunding, C. & Hargrave, B.T., 1973. A comparison of benthic microalgal production measured by ^{14}C and oxygen methods. J. Fish. Res. Board Can. 30:309–312.

Hutchinson, G.E., 1976. A Treatise on Limnology. Vol. III. Limnological Botany. John Wiley and Sons, Toronto. 660 p.

Jassby, A.D. & Platt, T., 1976. Mathematical formulation of the relationship between photosynthesis and light for phytoplankton. Limnol. Oceanogr. 21:540–547.

Loeb, S.L., 1981. An in situ method for measuring the primary productivity and standing crop of the epilithic periphyton community in lentic systems. Limnol. Oceanogr. 26: 394–399.

McIntire, C.D. & Phinney, H.K., 1965. Laboratory studies of periphyton production and community metabolism in lotic environments. Ecol. Monogr. 35:237–258.

Morris, I., Yentsch, C.M., & Yentsch, C.S., 1971. Relationship between light carbon dioxide fixation and dark carbon fixation by marine algae. Limnol. Oceanogr. 16:854–858.

Pickett, J.M. & Myers, J., 1966. Monochromatic light saturation curves for photosynthesis in Chlorella. Plant Physiol. 41:90–98.

Platt, T., Denman, K.L. & Jassby, A.D., 1975. The mathematical prediction of phytoplankton productivity. Can. Fish. Mar. Serv. Tech. Rep. 523. 110 p.

Prokopowich, J., 1979. Chemical characterization of epilimnion waters of unenriched lakes in the Experimental Lakes Area, northwestern Ontario. Can. Fish. Mar. Serv. Tech. Rep. 873. iv + 41 p.

Rabinowitch, E. & Govindjee., 1969. Photosynthesis. John Wiley and Sons, Inc. 173 p.

Riggs, D.R., 1963. The Mathematical Approach to Physiological Problems. M.I.T. Press. Massachusetts, U.S.A. 445 p.

Schindler, D.W., 1966. A liquid scintillation method for measuring carbon-14 uptake in photosynthesis. Nature 211: 844–845.

Schindler, D.W., Frost, V.E. & Schmidt, R.V., 1973. Production of epilithiphyton in two lakes of the Experimental Lakes Area, northwestern Ontario. J. Fish. Res. Board Can. 30: 1511–1524.

Shearer, J.A., 1976a. Light extinction measurements in the Experimental Lakes Area – 1974 data. Can. Fish. Mar. Serv. Tech. Rep. 615.

Shearer, J.A., 1976b. Phytoplankton primary production in the Experimental Lakes Area using an incubator technique – 1974 data. Can. Fish. Mar. Serv. Tech. Rep. 616.

Sheldon, R.b. & Boylen, C.W., 1975. Factors affecting the contribution by epiphytic algae to the primary productivity of an oligotrophic freshwater lake. Appl. Microbiol. 30:657–667.

Sládečková, A., 1962. Limnological investigation methods for the periphyton ('Aufwuchs') community. Bot. Rev. 28: 286–350.

Stainton, M.P., 1973. A syringe gas-stripping procedure for gas-chromatographic determination of dissolved inorganic and organic carbon in fresh water and carbonates in sediments. J. Fish. Res. Board Can. 30:1441–1445.

Steemann Nielsen, E. & Hansen, V.K., 1959. Measurements with the carbon-14 technique of the respiration rates in natural populations of phytoplankton. Deep-Sea Res. 5: 222–233.

Stockner, J.G. & Armstrong, F.A.J., 1971. Periphyton of the Experimental Lakes Area, northwestern Ontario. J. Fish. Res. Board Can. 28:215–229.

Strickland, J.D.H., 1960. Measuring the production of marine

phytoplankton. J. Fish. Res. Board Can. Bull. 122. 172 p.
Szczepanská, W., 1968. Vertical distribution of periphyton in the Lake Mikolajskie. Pol. Arch. Hydrobiol. 15:177–182.
Tolbert, N.E., 1974. Photorespiration, p. 474–504. In: Stewart, W.D.P. (ed.). Algal Physiology and Biochemistry. University of California Press, Berkeley, California.
Turner, M.A., 1981. Haptobenthic photosynthesis. M.Sc. Thesis, University of Manitoba, Winnipeg, Manitoba. 116 p. + xli.
Vollenweider, R.A., 1965. Calculation models of photosynthesis-depth curves and some implications regarding day rate estimates in primary production measurements. Mem. Ist. Ital. Idrobiol. Suppl. 18:425–457.
Warming, E., 1923. Økologiens Grunformer udkas til en systematisk Ordning. K. Dan. Vidensk.-Selsk. Skr. Naturw. Mat. Afd. 8 Raekke IV:119–187.
Wetzel, R.G., 1964. A comparative study of the primary productivity of higher aquatic plants, periphyton, and phytoplankton in a large, shallow lake. Int. Rev. ges. Hydrobiol. 49:1–61.

Authors' address:
M.A. Turner/D.W. Schindler/R.W. Graham
Department of Fisheries and Oceans
Freshwater Institute
501 University Crescent
Winnipeg, Manitoba, Canada R3T 2N6

CHAPTER 12

Selective light attenuation by the periphyton complex

RICHARD F. LOSEE and ROBERT G. WETZEL

Abstract. Selective attenuation of photosynthetically active radiation (PAR) was characterized for components of periphyton. Components included: three algal groups (diatom, green and blue-green), acid-cleaned diatom frustules, hypochlorite-bleached *Scenedesums* cells (to decrease pigment content), calcium carbonate, and bacteria. PAR attenuation was dominated by algal photosynthetic pigments (chlorophyll *b* and *c*, fucoxanthin, phycocyanins, and particularly chlorophyll *a*). Non-pigmented algal matter, diatom frustules, calcium carbonate and bacteria had low and nonselective absorption.

Geometric configuration of periphyton components also affected PAR attenuation. Although algal pigment covering a surface was a major factor influencing light absorption, clear areas, lacking absorbing pigments and transecting the periphyton layer, resulted in lower extinction coefficients than occurred if the surface was completely covered with pigmented cells. These clear areas may include weakly pigmented algae (especially diatoms), calcium carbonate, or open space. Algal extinction coefficients increased with increases in density until clear areas were covered with pigmented cells. At greater cell densities, extinction coefficients decreased as density increased.

Introduction

The selective attenuation of photosynthetically active radiation (PAR) by natural waters often influences the distribution of photosynthetic organisms in aquatic systems (Spence 1982). Further, several authors have concluded that macrophyte demise, in progressively eutrophicated waters, was a result of phytoplankton shading (Olsen 1964, cited in Morgan 1970; Moss 1976; Jupp & Spence 1977). Decreases in growth and photosynthetic rates, as well as the abundance and occurrence of submersed macrophytes, have also been attributed to PAR attenuation by the periphyton complex (Sand-Jensen 1977; Phillips *et al* 1978; Eminson & Phillips 1978; Sand-Jensen & Søndergaard 1981). Irradiance incident on a photosynthetic organism, such as a microalga embedded within the periphyton or a macrophyte underlying periphyton, is the result of selective light attenuation by the water column and the components of periphyton. Selective attenuation by the molecular characteristics of water, suspended particulate matter, and dissolved organic matter of lake waters has been investigated in detail (e.g. review of Hutchinson 1957; Wetzel 1983); however, the selective attenuation characteristics of periphyton have not been studied.

Periphyton consists of a mucoid matrix of algae, calcium carbonate, particulate detritus, and bacteria, which forms a complex optical system. Geometry and heterogeneity of periphyton components, and in particular of absorbing material, such as photosynthetic pigments, result in myriad scattering and absorbing interactions. Within the pe-

Wetzel, R.G. (ed.), Periphyton of Freshwater Ecosystems.
 ISBN 90 6193 768 X.

riphyton complex, the major effects of scattering (reflection, refraction, and diffraction) are the 're-emission' of light out of the complex and an increase in pathlength, thereby increasing probability of absorption.

The absorption characteristics of a component were assessed by comparing absorbance spectra of increasing component densities. With a homogeneous distribution of absorbing material, the Lambert-Beer relationship predicts that there will be a constant increase in absorbance with increase in density. As density is increased, deviation of absorbance spectra from this Lambert-Beer relationship implies an alteration in the distribution of absorbing material and, by inference, may be used to characterize the nature of the component effects on attenuation in the periphyton.

In this study, we investigated the selective attenuation of PAR by components of periphyton. Components included: three algal groups (diatom, green and blue-green), acid-cleaned diatom frustules, green alga cells treated with hypochlorite to decrease pigment content, calcium carbonate, and bacteria.

Methods

Algae (*Navicula pelliculosa, Fragilaria crotonensis, Scenedesmus,* and *Anabaena flos-aquae*) were grown in axenic cultures and concentrated with a Sorvall continuous-flow centrifuge the day prior to absorbance measurements. Two systems were used for measuring absorbance. Aliquots of the concentrated algal suspensions, acid-cleaned *N. pelliculosa* frustules, and hypochlorite-treated *Scenedesmus* suspensions were dispensed to Plexiglas settling chambers and allowed to settle. Algal density on the base of the chamber was adjusted by varying aliquot volume. After settling, excess medium was removed by aspiration to a standard depth. The chambers were made of transparent Plexiglas (0.64 cm thickness), which had nonselective absorbance and 89% transmittance of the incident radiance between 400 and 750 nm. This system was used for data given in Figs. 3, 4 upper, and 5.

The second system employed polystyrene petri dishes which also had nonselective absorbance and transmitted 90% of the incident radiation between 400 and 750 nm. Aliquots of components were pipetted from a well mixed, concentrated suspension, dispensed to a volumetric cylinder and brought to standard volume with the centrifugation supernatant. This suspension was then poured into the petri dish and allowed to settle with the dish on the diffuser plate of the spectroradiometer sensor. The petri dish system was used in absorbance measurments for the algae, calcium carbonate, and bacteria (Figs. 1, 2, 4 lower, and 6).

Acid-cleaned *N. pelliculosa* frustules were prepared by boiling the frustules in a concentrated nitric acid-potassium dichromate solution until all organic matter was digested. Cleaned frustules were rinsed copiously with ultrapure (Millipore Q) distilled water. *Scenedesmus* cells were treated with approximately 2.5% sodium hypochlorite for five seconds and immediately rinsed with distilled water. These cells were found to be virtually free of pigments under microscopic examination and to have cell walls and ornamentation intact. The calcite samples were precipitated from saturated solutions of CO_2–$CaCO_3$ in distilled water, by bubbling with nitrogen. Crystals were precipitated and settled directly in the settling chambers. Bacteria were obtained from a nutrient broth inoculated with material from decaying submersed macrophytes. Bacteria were concentrated and rinsed by centrifugation.

A Techtum Instrument QSM-2500 spectroradiometer was used to measure irradiance passing through the experimental components, media, and Plexiglas or polystyrene chambers. Chambers were placed directly on the diffuser plate of the sensor. Chamber bases were masked so that only light passing through the component was detected. An opaque, flat-black ring was placed over the top edge of the petri dish wall to prevent light scattered by the side walls from entering the sensor. The chamber base and sensor were positioned approximately 35 cm below a neutral diffuser and the light sources (two 15-watt General Electric cool white and two 15-watt Luxor Vita Lite fluorescent tubes). The apparatus was enclosed in a light box. A

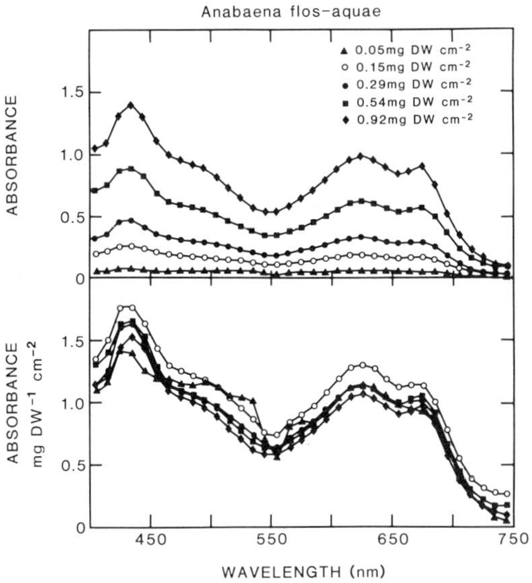

Fig. 1. Absorbance spectra for five densities of settled *Anabaena flos-aquae*. $A_{ij} = \log_{10}(I_i/I'_{ij})$ where I_i is mean intensity transmitted for the ith 10-nm waveband, by a GF/F filtered algal medium blank, and I'_{ij} is the mean transmitted intensity for the ith waveband and j^{th} density. Intensity measured as quanta $m^{-2}\ s^{-1}\ nm^{-1}$. Density was the oven-dried weight of the component settled on a 12.88 cm^2 area (mg DW cm^{-2}) (*upper*). Absorbance per unit density or extinction coefficient spectra (*lower*) for the five densities ($\varepsilon_{ij} = A_{ij}/\text{mg DW}_j\ cm^{-2}$).

Lambda voltage conditioner was used to stabilize the power supply to the fluorescent tubes. The surface area of settled material was larger than the diffuser plate of the sensor in both systems, which, combined with the diffuse light source, minimized the effects of scattering.

Transmitted irradiance as quanta $m^{-2}\ s^{-1}\ nm^{-1}$ was measured by scanning from 400 to 750 nm for each component and component density. Blanks were prepared by filtering media containing the component through a Whatman GF/F glass fiber filter (95% retention of 0.7-μm particle diameter). Absorbances (A) were calculated at 10-nm intervals (i) from 400 to 750 nm for a particular component and density (j), by taking the $\log_{10}$ of the quotient for mean blank intensity for a particular 10-nm waveband and component (I_{ij}), divided by the mean intensity of irradiance passing through the particular component for that waveband and density (I'_{ij}), as: $A_{ij} = \log_{10}(I_{ij}/I'_{ij})$. In order to make comparisons between components, component densities and mixtures of components, absorbance was normalized to mg dry weight (DW) resulting in extinction coefficients (ε_{ij}).

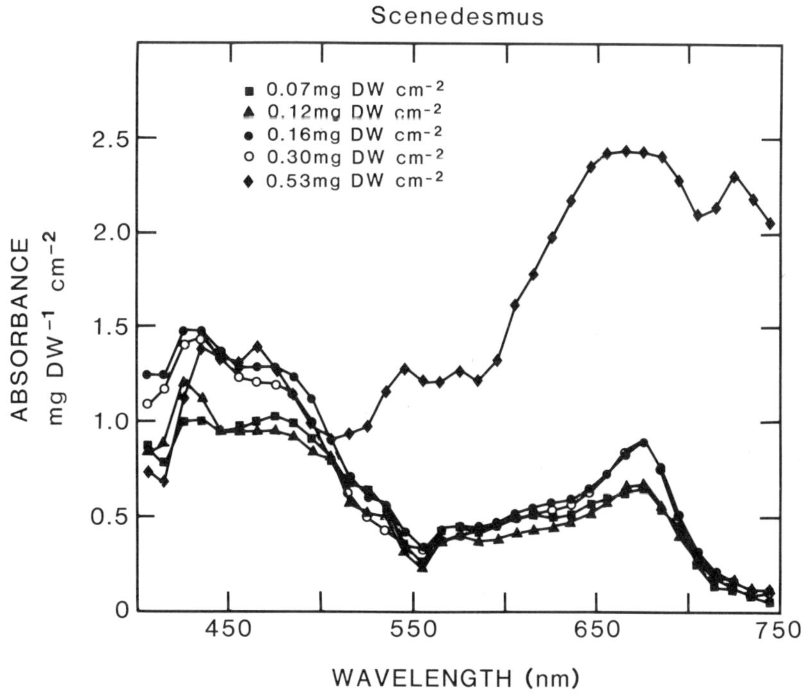

Fig. 2. Extinction coefficient (absorbance per unit density) spectra for five densities of settled *Scenedesmus*. See Fig. 1 for details.

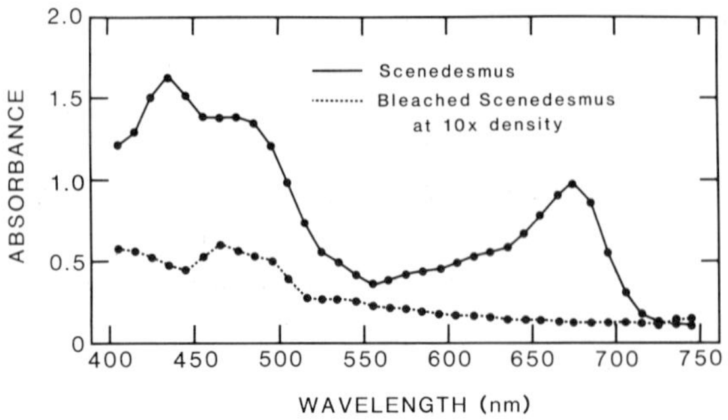

Fig. 3. Absorbance spectra for settled *Scenedesmus* (0.14 mg DW cm^{-2}) and hypochlorite-bleached *Scenedesmus* cells (1.5 mg DW cm^{-2}).

Densities of the components on the Plexiglas base were determined by measuring dry weight of five 0.71 cm^2 area subsamples. After making transmission measurements, subsamples were aspirated from the base of the chambers through a tube pressed against the chamber base and sealed with a basal 0-ring. The petri dish system was sampled with a much larger diameter tube (12.88 cm^2). Subsamples were filtered onto tared, precombusted (525° C), GF/F filters, and dried at 70° C for 48 h, cooled, and weighed.

Results

The absorbance spectra (Fig. 1, upper) and extinction coefficient or normalized absorbance spectra (Fig. 1, lower) for five densities of *Anabaena flos-aquae* had absorbance maxima in the wavebands of 430–440 nm, 670–680 nm, and 620–630 nm. These maxima correspond to *in vivo* absorbance peaks of chlorophyll *a* and phycocyanins, the main pigments of blue-green algae (Govindjee and Braun 1974). The extinction coefficient (Fig. 1, lower) is absorbance per unit density (density in this case is mg dry weight per cm^2). When absorbance measurements follow the Lambert-Beer relationship the extinction coefficients for a wavelength of radiation will be a constant for all densities. Within the range of densities measured (0.05 to 0.92 mg DW cm^{-2}), there was no significant reduction in absorbance per unit density as densities were increased.

The green alga *Scenedesmus* (Fig. 2) had absorbance maxima in the wavebands corresponding to *in vivo* chlorophyll *a* and absorbance shoulders in the range for *in vivo* chlorophyll *b* (470–480 nm and 650–670 nm) (Govindjee & Braun 1974). The extinction spectrum for the highest density of *Scenedesmus* (0.53 mg DW cm^{-2}) had the typical pattern from 400 to 500 nm; however, over 500 nm there was a marked increase in extinction coefficient and broadening of the longwave chlorophyll peak, over those at lower densities. At the lower four densities extinction spectra segregated into two groups with the lower density (0.07 and 0.12 mg DW cm^{-2}) extinction coefficients 10 to 15% lower than the higher density group (0.16 and 0.30 mg DW cm^{-2}).

The selectivity and degree of absorbance for hypochlorite-treated *Scenedesmus* cells (Fig. 3) were greatly reduced from that of pigmented cells as a result of this oxidative treatment. However, the organic matter of the cell had a small degree of selective absorbance at wavelengths less than 510 nm.

The absorbance peaks for the diatom *Navicula*

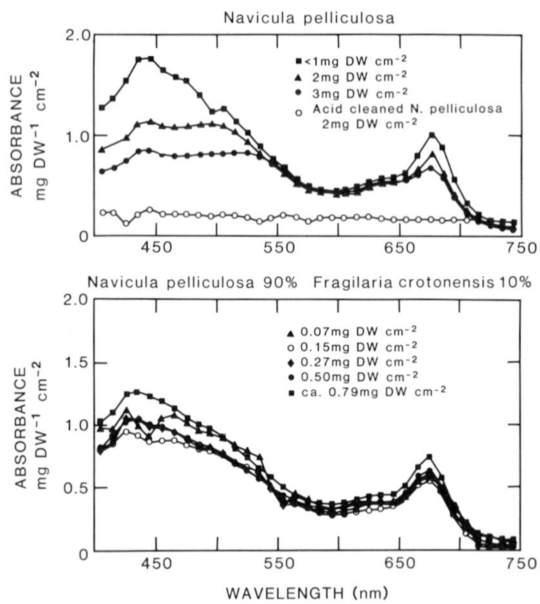

Fig. 4. Extinction coefficient spectra (absorbance per unit density) for four densities of the settled diatom *Navicula pelliculosa* and acid-cleaned *N. pelliculosa* frustules (*upper*), and for five densities of settled mixed diatoms (90% *N. pelliculosa* and 10% *Fragilaria crotonensis* by weight) (*lower*). See Fig. 1 for details.

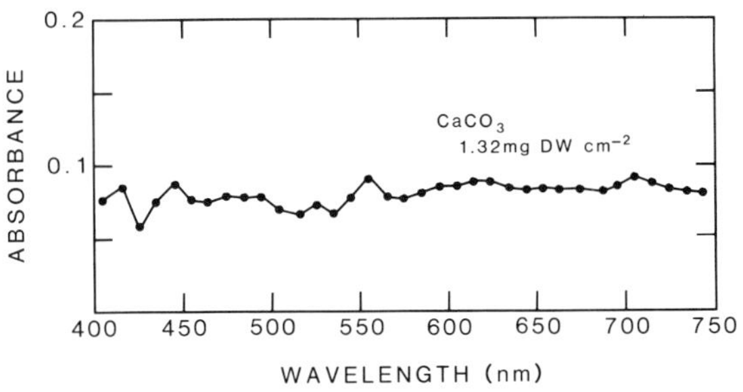

Fig. 5. Absorbance spectra for settled $CaCO_3$ (1.32 mg DW cm^{-2}).

pelliculosa and for *N. pelliculosa* combined with *Fragilaria crotonensis* (Fig. 4, upper and lower, respectively) were broad with maxima in the 420–440 nm and 670–680 nm wavebands. The broadness of the peaks resulted from the additional absorbance by fucoxanthin and chlorophyll *c* to the primary chlorophyll *a* absorbance. Acid-cleaned frustules of *N. pelliculosa* exhibited nonselective and very low absorbance (Fig. 4, upper). For *N. pelliculosa* the extinction coefficients decreased at densities of 2 and 3 mg DW cm^{-2} compared to extinction coefficents for densities less than 1 mg DW cm^{-2} (Fig. 4, upper). When *Navicula* and *Fragilaria* were combined 9:1 (Fig. 4, lower), extinction coefficients for densities less than 1 mg DW cm^{-2} were much lower than for *Navicula* alone. Also, the mixed diatom extinction coefficients exhibited a slight increase as densities increased.

Absorbance by calcium carbonate crystals was nonselective and very low (Fig. 5). A gradual decrease in extinction coefficients occurred with increasing density of calcite (data not shown). Bacteria (Fig. 6) also exhibited nonselective and low absorbance, and a decrease in extinction coefficients as settled bacteria densities were increased.

Discussion

Lambert-Beer's law states that successive increments in the number of identical absorbing molecules in the path of monochromatic radiation absorb equal fractions of radiant energy traversing them. In periphyton, where spatial inhomogeneity of absorbing substances (i.e. photosynthetic pigments) occurs, different parts of the incident radiation intersect different amounts of pigment. Amount of pigment traversed by a ray is a density probability function dependent of pathlength and pigment density. The Lambert-Beer relationship is valid for any one ray but is less so for the sum of rays in the area. The result for radiation traversing the periphyton complex is a sieve-effect, where any heterogeneity in pigment distribution can only diminish the absorption by a homogeneous distribution.

Periphyton component extinction coefficients increase with each increase in component density, until the dominant absorbing material of the component nearly covers the surface in a uniform layer. At this point absorbance per unit density is maximum, with nearly all rays traversing an equal amount of pigment. Eukaryotic algal photosynthetic pigments are in discrete chloroplast packets. The density at which these packets most nearly form, from the perspective of incident radiation, a pigment layer of uniform thickness, is dependent on cell geometry and the relative chloroplast to cell volume of the algae. When density is further increased, the sieve-effect again results in reduced extinction coefficients. Deviations from the Lambert-Beer relationship at these increased densities are not caused by some rays traversing the layer without crossing pigment, but rather by rays crossing unequal amounts of pigment.

In this study, absorbance is a measure of screen-

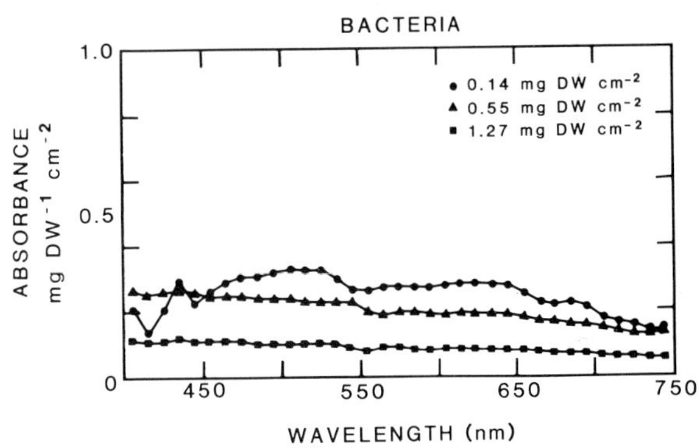

Fig. 6. Extinction coefficient spectra (absorbance per unit density) for three densities of settled bacteria. See Fig. 1 for details.

ing or attenuation of radiation by combined scattering and absorption. The apparatus used minimized the effects of scattering and, as a result, absorption measurements approached true measures of absorbance. The light source used in making measurements was diffuse as light is in the aquatic environment. The diffuse light source and juxtaposition of sensor and settled periphyton component combined to provide a light milieu at the sensor surface which was very similar to the light climate at a submersed macrophyte leaf surface. The lower density limit of resolution was demonstrated by the erratic nature of extinction coefficient spectra for the lowest densities of algae (in all cases less than 0.1 mg DW cm^{-2}).

Apparent absorbance of PAR by the nonselective and low light absorbing bacteria and calcium carbonate crystals was almost entirely the result of scattering. Particles in the size ranges of bacteria and carbonate crystals scatter light at low angles (Koch 1970). The light source used in the absorption measuring apparatus was strongly directionally oriented, upon entering the layer of material, despite its diffusivity. Part of the light scattered by a particle in the settled layer would be secondarily scattered by another particle and reoriented so that it fell upon the detector (Koch 1970). The frequency of this occurrence would increase as layer thickness increased and resulted in the decrease in extinction coefficient with increasing density as was observed for bacteria and calcium carbonate.

Extinction coefficient values for *Anabaena* were much greater than for calcium carbonate and bacteria, and were the result of photosynthetic pigment absorption to a much greater extent than from scattering. The relatively uniform distribution of pigment in *Anabaena* cells and the sheet-forming filamentous growth form provided a pigment layer of nearly uniform thickness at the lowest density and resulted in maximum extinction coefficients.

Scenedesmus showed an increase in extinction coefficients, with increasing density, over the range of densities measured. At the lower densities, *Scenedesmus* colonies may lie with their broad cell walls parallel to the chamber surface, and expose portions of the cells void of pigment to incoming radiation. At some density between 0.12 and 0.16 mg DW cm^{-2}, the cells may have been sufficiently dense to support colonies standing upright to some extent. This orientation could have resulted in a geometry which increased the absorption of radiation as it traversed a particular cell.

The marked increase in *Scenedesmus* absorbance and broadening of the long-wavelength peak at the highest density (0.53 mg DW cm^{-2}) is not understood at this time.

Extinction coefficients progressively increased as density was increased for the mixed *Navicula* and *Fragilaria*. Again, this increase was thought to be a result of the distribution of pigment, discretely packaged in the diatom frustules, becoming more nearly homogeneous with respect to incoming radiation as densities were increased.

Extinction coefficients for the mixed diatoms were lower than for the *Navicula* alone at densities less than 1 mg DW cm^{-2}. *F. crotonensis* cells were approximately four times longer than *N. pelliculosa* and form ribbon-like colonies with cells attached valve to valve. Additionally, the *Navicula* were much more densely pigmented than were *Fragilaria*. When settled, the *Fragilaria* cells traversed the layer of *Navicula* cells. The *Fragilaria* may have provided a pathway of lower absorption through the layer of *Navicula*, and thereby reduced the extinction coefficients.

In order for radiation to enter a macrophyte leaf it must penetrate the leaf surface/water interface. Reflection at this interface is a function of indicies of refraction of plant surface and water, and the angle which light strikes the leaf surface. Multiple reflections, which would occur at the periphyton surface/water interface and subsequent refractions and diffractions, may result in a lower light loss caused by reflection than would occur with a bare leaf surface.

Periphyton community structure may affect attenuation on two levels. On a gross level, clumping of colonial algae on a leaf surface would attenuate radiation to a lesser degree than a uniform distribution of cells with equivalent pigment content. On a finer scale, attenuation could be reduced as a result of a nonselective, low-absorbing material, such as calcium carbonate crystals, traversing the algal cell layer. Large senescent diatoms could function as excellent conduits of light through the periphyton complex, since chlorophyll destruction proceeds rapidly after cell death and lysis (Daley, 1973) leaving the relatively low, nonselectively absorbing cellular organic matter and frustule.

Calcium carbonate may attain densities of 11 mg DW cm^{-2} on *Potamogeton perfoliatus* (Kowalczewski, 1975). Using 0.056 as an average extinction coefficient (calculated from absorbance spectrum Fig. 6), this carbonate density could have an absorbance as high as 0.62 (log_{10} absorbance units). On these plants the non-carbonate density was 2.8 mg DW cm^{-2}. Presumably most of this non-carbonate dry weight was algal biomass. Using the lowest (most conservative) extinction coefficients found for algae for the red and blue chlorophyll absorbance bands (0.75 and 0.90, respectively), the above algal density results in calculated absorbances of 2.1 and 2.5, respectively. Even with these large carbonate depositions, algae have a much greater effect on absorption.

The most outstanding characteristic of periphyton component selective attenuation was absorption by algal photosynthetic pigments. Of these, chlorophyll *a* was dominant. Accessory pigments of the various algae, such as chlorophyll *b* and *c*, fucoxanthin, and phycocyanins, contributed to the absorption spectra by broadening the absorption bands around the red and blue chlorophyll *a* absorption peaks. Non-pigmented algal organic matter (e.g. hypochlorite-bleached *Scenedesmus* cells) had relatively low absorption and much less selectivity. Diatom frustules, bacteria, and calcium carbonate crystals also exhibited low absorption and nonselectivity.

The intense absorption by epiphytic algal chlorophyll is of particular importance because of the limited ability for chromatic adaptation of submersed angiosperms, which possess only chlorophyll *a* and *b*. In many instances, it is the submersed plant best able to utilize the limited light resource which is successful. Physiological adaptations and specific strategies have evolved which enable exploitation of these light-limited habitats.

Boardman (1977) has reviewed the physiological adaptations of shade plants. Shade tolerant plants generally have a reduced light compensation point which is largely a reflection of reduced dark respiration. Reduced dark respiration results from reduced enzyme levels. This reduction is apparent in the morphology of shade plant leaves which have relatively higher specific leaf area (area/weight), indicative of a reduction in mesophyllous tissue. As found by Van *et al* (1977) for *Hydrilla verticillata,* ratios of chlorophyll *a*:*b* may decrease with reduced light even though total chlorophyll ($a + b$) content does not change.

Submersed plants combine physiological adaptations with advantageous growth habit strategies in order to be successful in light-limited waters. The success of many plants is dependent on their ability to store energy in reproductive structures which will support rapid growth at the start of the subsequent growing season. *Hydrilla verticillata* is dependent on carbohydrate storage in turions, tubers, and stem fragments for rapid early growth, which enables the formation of canopies near the water surface (Van *et al* 1977). Some submergents, such as *Myriophyllum,* may grow at rates of 1 cm or more per day (Wetzel, unpublished), essentially out-growing epiphyte colonization.

Acknowledgments

We acknowledge with appreciation the technical assistance of Dr. Frederick Payne, the critical reviews of Mr. R.G. Carlton, Drs. R.E. Moeller and M.F. Coveney, and the support of the Department of Energy (EY-76-S-02-1599, C00-1599-221). Contribution no. 493, W.K. Kellogg Biological Station of Michigan State University.

References

Boardman, N.K., 1977. Comparative photosynthesis of sun and shade plants. Ann. Rev. Plant Physiol. 28:355–377.

Daley, R.J., 1973. Experimental characterization of lacustrine chlorophyll diagenesis. II. Bacterial, viral and herbivore grazing effects. Arch. Hydrobiol. 72:409–439.

Eminson, D. & Phillips G., 1978. A laboratory experiment to examine the effects of nutrient enrichment on macrophyte

and epiphyte growth. Verh. Internat. Verein. Limnol. 20:82–87.
Govindjee & Braun, B.Z., 1974. Light absorption, emission and photosynthesis. In: Stewart, W.D.P. (ed.). Algal Physiology and Biochemistry. University of California Press, Berkeley. pp. 346–390.
Hutchinson, G.E., 1957. A Treatise on Limnolgy. I. Georgraphy, Physics and Chemistry. Wiley, New York. 1015 pp.
Jupp, B.P. & Spence, D.H.N., 1977. Limitations on macrophytes in a eutrophic lake, Loch Leven. I. Effects of phytoplankton. J. Ecol. 65:175–186.
Koch, A.L. 1970. Turbidity measurements of bacterial cultures in some available commercial instruments. Anal. Biochem. 38:252–259.
Kowalczewski, A., 1975. Perphyton primary production in the zone of submerged vegetation of Mikolajaskie Lake. Ekol. Pol. 23:509–543.
Morgan, N.C., 1970. Changes in the fauna and flora of a nutrient enriched lake. Hydrobiol. 35:545–553.
Moss, B., 1976. The effects of fertilization and fish on community structure and biomass of aquatic macrophytes and epiphytic algal populations: An ecosystem experiment. J. Ecol. 64:313–342.
Mulligan, H.F. & Baranowski, A., 1969. Growth of phytoplankton and vascular aquatic plants at different nutrient levels. Verh. Internat. Verein. Limnol. 17:802–810.
Phillips, G.L., Eminson, D. & Moss, B., 1978. A mechanism to account for macrophyte decline in progressively eutrophicated freshwaters. Aquat. Bot. 4:103–126.
Sand-Jensen, K., 1977. Effect of epiphytes on eelgrass photosynthesis. Aquat. Bot. 3:55–63.
Sand-Jensen, K. & Søndergaard, M., 1981. Phytoplankton and epiphyte development and their shading effect on submerged macrophytes in lakes of different nutrient status. Int. Revue ges. Hydrobiol. 66:529–552.
Spence, D.H.N., 1982. The zonation of plants in freshwater lakes. In: A. Macfadyen and E.D. Ford (eds.) Advances in Ecological Research. Vol. 12. Academic Press, London. pp. 37–126.
Van, T.K., Haller, W.T., Bowes, G. & Garrad, L.A., 1977. Effects of light quality on growth and chlorophyll composition in Hydrilla verticillata, an aquatic weed in Florida. Hyacinth Control J. 15:29–31.
Wetzel, R.G., 1983. Limnology. 2nd ed., Philadelphia, W.B. Saunders Company. 860 pp.

Authors' address:
R.F. Losee/Robert G. Wetzel
W.K. Kellogg Biological Station
Michigan State University
Hickory Corners, MI 49060, U.S.A.

CHAPTER 13

Influence of current velocity on periphyton distribution

T.S. TRAAEN and E.-A. LINDSTRØM

Abstract. Current velocities were measured close to different periphyton elements in rivers. The results indicate that several algal species have a defined niche within the current velocity spectrum, while other species can inhabit a very broad range of current velocities. Many macroscopic visual algal colonies exhibit better development at velocities >20 cm/s, while only few thrive at velocities >100 cm/s.

Introduction

Among the factors that influence the habitat of attached algae in streams, current velocity is one of the most important. Current velocity is known to influence algal productivity, accumulation, and community structure (McIntire 1966; Horner & Welch 1981).

The aim of this study was to investigate the effect of current velocity on the patchy distribution of different algal colonies in streams.

Methods

In order to measure current velocity as close as possible to the algal colonies, we used a mini-propeller type current meter that can measure approximately 1 cm above solid surfaces (E. Schiltknecht ing.sia. type Micro-Mini-Water 642). Velocity measurements were taken above macroscopic recognizable algal colonies, and samples of the algal communities were gathered for further examination microscopically. Usually at least 30 measurements of velocity were taken of each colony type. The results are reported as frequency distribution at current velocity intervals of 20 cm/s.

Results and discussion

Examples of distribution of some dominating algal species at locations in three Norwegian rivers are given in Figs. 1, 2, and 3. At the River Alta station (Fig. 1), colonies of the green alga *Tetraspora cylindrica* were restricted to low current velocities. Nearly 90% of the colonies appeared at velocities <20 cm/s. The blue-green alga *Phormidium autumnale* was most frequent at velocities between 40 and 80 cm/s, while more than 90% of the *Hydrurus foetidus* colonies appeared at velocities >80 cm/s.

In the River Skjaersjøelva colonies dominated by the filamentous green algae *Hormidium rivulare* and *Zygnema* 'a' (israelson 1949) were most frequently observed between 20 and 60 cm/s. Examination of these green filaments microscopically indicated that *Zygnema* 'a' dominated the lower part of this velocity range, while *Hormidium rivulare* dominated at higher velocities. The blue-green alga *Stigonema mamillosum* exhibited a relatively even distribution from 20 to 100 cm/s. *Lemanea*

Wetzel, R.G. (ed.), Periphyton of Freshwater Ecosystems.
 ISBN 90 6193 768 X.

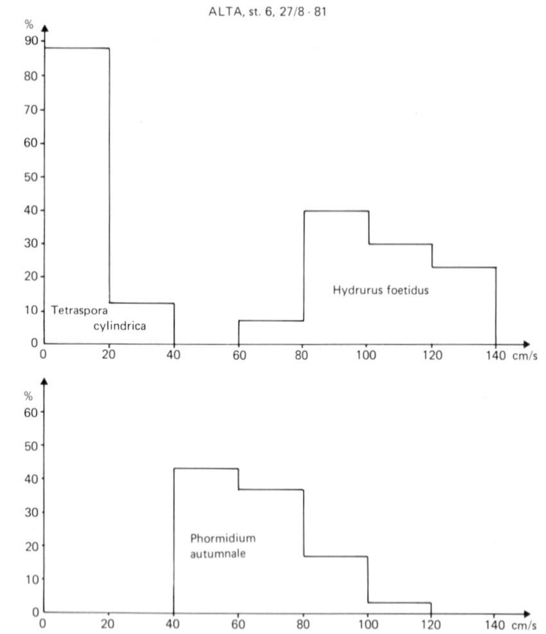

Fig. 1. Frequency distribution of *Tetraspora cylindrica*, *Hydrurus foetidus*, and *Phormidium autumnale* in relation to current velocity in the River Alta, 27 August 1981.

fluviatilis was restricted to high current velocities and was most frequently observed between 120 and 160 cm/s. The third example (Fig. 3), from the River Nitelva (St. SV1), exhibited a situation where stones in the lower velocity range (<40 cm/s) were colonized by *Batracospermum monoliforme* var. *densum*, while *Zygnema* 'b' (Israelson 1949) was most frequent at velocities >40 cm/s.

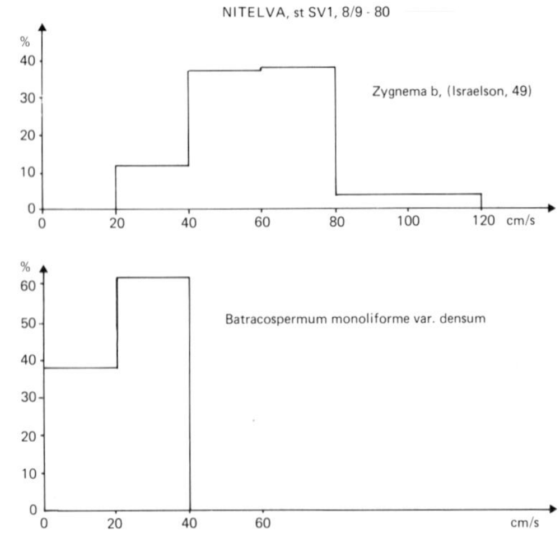

Fig. 3. Frequency distribution of *Zygnema* 'b' (Israelson 1949) and *Batracospermum monoliforme* var. *densum* in relation to current velocity in the River Nitelva, 8 September 1980.

At Station N1 in The River Nitelva, colonies of *Zygnema* 'b' were abundant. Between the algal-covered stones, stones were present that appeared suitable for algal growth but were not colonized by any visible colonies. Separate velocity measurements were taken just above these bare stones. As shown in Fig. 4, most of these stones had water velocities of <20 cm/s. These data, along with observations from other rivers, indicate that velocities <20 cm/s are not very suitable for development of many species of running water algae. On the other hand, it seems that only few species of

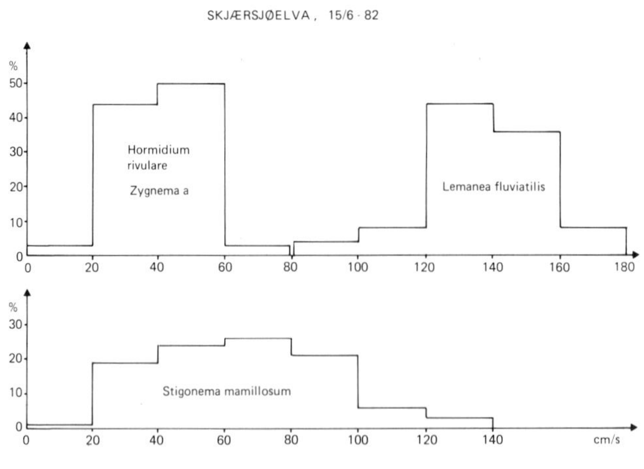

Fig. 2. Frequency distribution of *Hormidium rivulare*, *Zygnema* 'a', *Lemanea fluviatilis*, and *Stigonema mamillosum* in relation to current velocity in the River Skjaersjøelva, 15 June 1982.

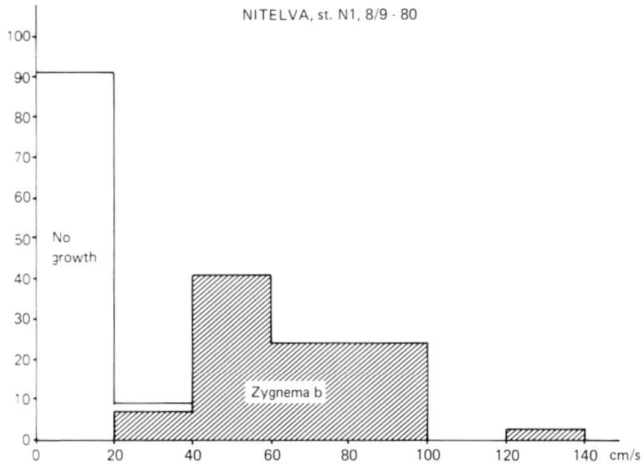

Fig. 4. Distribution of stones covered by *Zygnema* 'b' and stones with no visible colonies of algae. River Nitelva, 8 September 1980.

algae can develop high standing crops at velocities >100 cm/s.

The results indicate that different algal species have different current velocity niches which may help to explain the observed patchy distribution of algal colonies. How this distribution according to current velocities is modified by other ecological factors, e.g. substratum, nutrient concentrations, light, and flow variations, are suggested as topics for further research.

References

Horner, R.R. & Welch, E.B., 1981. Stream periphyton development in relation to current velocity and nutrients. Can. J. Fish. Aquat. Sci. 38:449–457.

Israelson, G., 1949. On some attached Zygnemales and their significance in classifying streams. Bot. Notiser 1949:313–358.

McIntire, C.D., 1966. Some effects of current velocity on periphyton communities in laboratory streams. hydrobiologia 27:559–570.

Authors' address:
T.S. Traaen/E.-A. Lindstrøm
Norwegian Institute for Water Research
P.O. Box 333
Blindern, Oslo 3, Norway

CHAPTER 14

Nitrogen fixation in periphyton of oligotrophic Lake Tahoe

JOHN E. REUTER, STANFORD L. LOEB and CHARLES R. GOLDMAN

Abstract. Heterocystous blue-green algae dominate the biomass of the upper sublittoral ($\sim$2–60 m) epilithic periphyton community in nitrogen deficient, oligotrophic Lake Tahoe. Representative genera are *Calothrix*, *Tolypothrix*, *Nostoc* and *Scytonema*. Samples obtained from three locations using SCUBA were assayed for N-fixation using the acetylene reduction technique. Seasonal rates ranged from 4–561 μgN m^{-2} h^{-1} with a distinct summer maximum and winter minimum. Laboratory experiments along with physico-chemical lake data suggest that temperature regulates the seasonal cycle of N-fixation. This community is perennial, actively fixes N throughout the entire year and is in sharp contrast to the ephemeral nature of planktonic N-fixers. The estimated loading rate of nitrogen resulting from periphyton N-fixation was 0.03 kg N ha^{-1} (of lake surface) yr^{-1}, which is very small ($<1\%$) in relation to other inputs of inorganic-N. The ability of this community to utilize atmospheric N_2 appears to be a successful adaptive strategy for survival in N-deficient environments.

Introduction

Nitrogen is an essential nutrient that can regulate algal production in a variety of aquatic ecosystems (Dugdale & Goering 1967; Axler *et al* 1982). ^{14}Carbon-determined nutrient enrichment bioassays conducted in oligotrophic Lake Tahoe have demonstrated that phytoplankton photosynthesis is often increased following additions of dissolved inorganic nitrogen (Arneson 1979; Goldman 1981). These observations suggest that the primary productivity of Lake Tahoe, along with subsequent transfers of energy to other trophic levels, is dependent on the availability of biologically useful forms of nitrogen. Biological N-fixation provides a source of new, combined nitrogen for incorporation into the aquatic N-cycle. N-fixing organisms not only supplement their intracellular pools, but they also influence the nitrogen regime in their surrounding environment.

Heterocystous blue-green algae are commonly associated with the phytoplankton of mesotrophic and eutrophic lakes and considerable effort has been directed towards characterizing patterns of N-fixation in these environments (Horne & Fogg 1970; Stewart *et al* 1971; Horne & Goldman 1972; Vanderhoef *et al* 1974). Since N-fixing blue-green algae are generally absent in the pelagic waters of oligotrophic lakes, measured rates of N-fixation in these systems were negligible (Rusness & Burris 1970; Flett *et al* 1980). In contrast, these organisms are common in periphyton communities over a wide range of trophic levels and are often very abundant in oligotrophic lakes (Fogg *et al* 1973; Loeb & Reuter 1981). Rates of periphytic N-fixation have been reported (Lundgren 1978; Moeller & Roskoski 1978; Lam *et al* 1979; Loeb & Reuter 1981); however, seasonal and spatial variations of

Wetzel, R.G. (ed.), Periphyton of Freshwater Ecosystems.
 ISBN 90 6193 768 X.

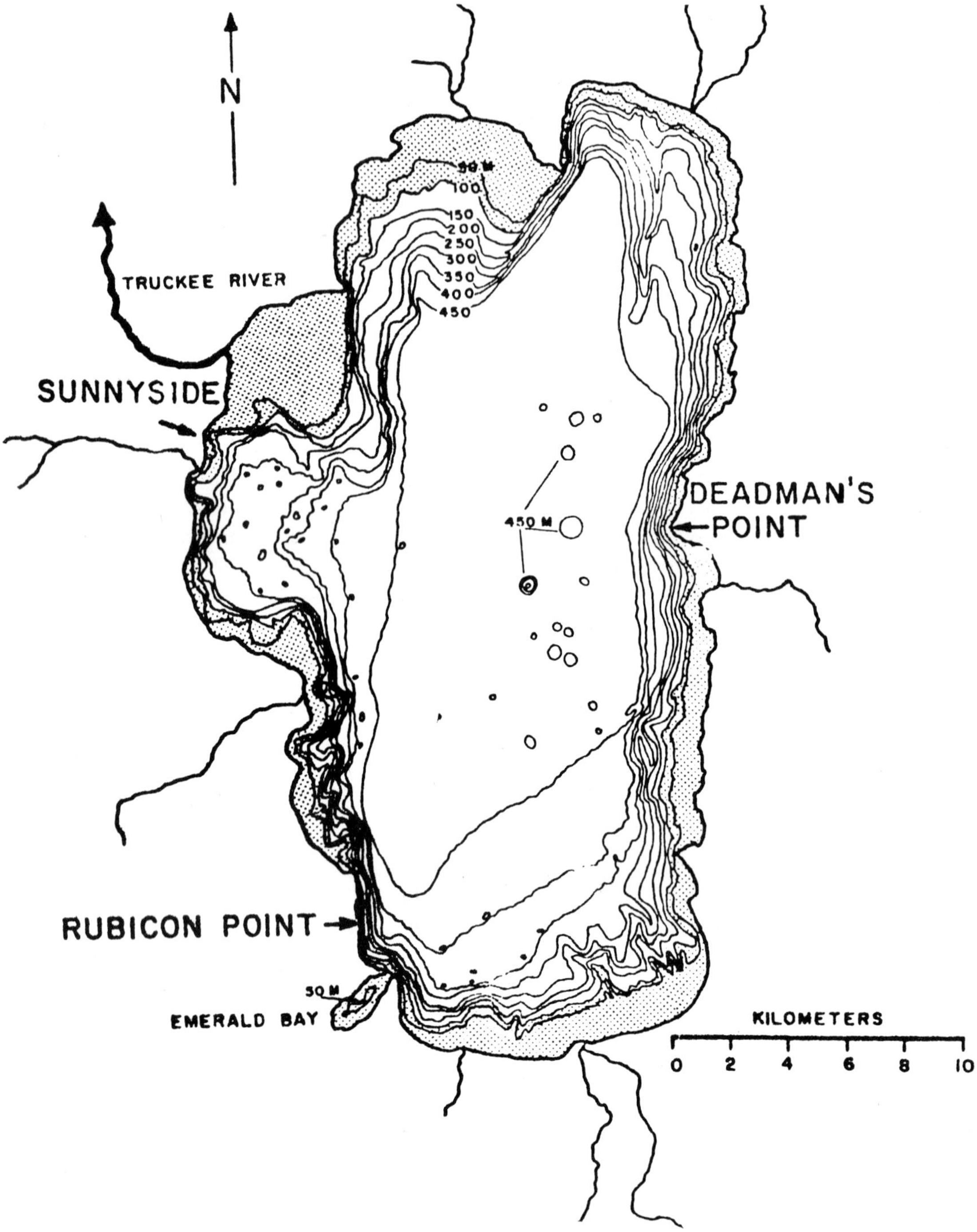

Fig. 1. Bathymetric map of Lake Tahoe. The shaded area denotes the littoral zone which extends down to 100 m. Sampling locations are indicated.

N-fixation rates in the benthic algal communities of freshwaters are not well documented.

Recently, we have shown that heterocystous blue-green algae dominate the biomass of the sublittoral epilithic periphyton in a number of subalpine, oligotrophic, nitrogen deficient lakes including Lake Tahoe and reported summer season rates of N-fixation (Loeb & Reuter 1981). The present study focuses on the annual variation of benthic algal N-fixation rates in Lake Tahoe and discusses the factors affecting these rates.

Study site

Lake Tahoe is located at an elevation of 1898 m in the Sierra Nevada mountain range of California-Nevada, USA, and lies in a graben fault. The lake basin is steep sided with a maximum depth of 501 m, a surface area of 499 km^2 and a volume of 156 km^3. The littoral zone (~1% surface light) represents only 19% of the lake surface and extends to a depth of approximately 100 m (Fig. 1). The watershed is small (800 km^2) in relation to the lake surface area (1.6:1) and is composed primarily of granitic and volcanic materials. The small area and nutrient poor soils of the drainage basin account for Lake Tahoe's oligotrophic condition. Lake Tahoe remains ice-free throughout the year.

Methods

Periphyton was collected from three locations (Fig. 1) on a monthly schedule. Sampling depths were 2 and 8 m at Deadman Point and Sunnyside, and 2, 8 and 16 m at Rubicon Point. The substrata at these areas ranged from cobbles (~0.25 m diam) to boulders (~20 m diam). Periphyton was removed from the natural rock substrate and returned to the laboratory for N- fixation assays and biomass determinations.

Rates of N-fixation were estimated using the acetylene reduction technique (Stewart *et al* 1967; Flett *et al* 1976). Periphyton (~0.2–0.6 g wet weight) was transfered to 35 ml serum vials containing 20 ml of lake water (final volume), capped with rubber septum stoppers and injected with 3 ml of C_2H_2. The resulting pC_2H_2 was ~0.2 atm and removal of N_2 from the gas phase was not necessary (Sorokin & Kadota 1972). Samples vials were shaken for 20 sec to equilibrate the C_2H_2 in the aqueous phase and incubated in environmental chambers under conditions approximating ambient light and temperature. Following a 3-h incubation the vials were again shaken and the gas phase was subsampled and stored in vacutainers (Monoject®; Schell & Alexander 1970). C_2H_2:C_2H_4 ratios for samples, blanks and standards were measured within a 2–3 week period using a flame ionization gas chromatograph (Varian 940 or 1440). $^{15}N_2$ (96 atom-%) was used to calibrate the acetylene reduction technique (Flett *et al* 1980). A C_2H_2:N_2 reduction ratio of 4.59 ± 1.29 (±SD; n = 6) was obtained for a 6-h incubation with Tahoe periphyton. This factor was used to convert moles of ethylene produced to moles of nitrogen fixed.

Samples for chlorophyll-*a* analysis were homogenized in 90% acetone with a tissue grinder and extracted for 24 h in the dark at 8° C. Chlorophyll-*a* concentrations were calculated using the trichromatic equations of Strickland and Parsons (1968). Ash-free dry weight was determined following combustion at 60° C and then at 475° C.

Samples for water chemistry analysis were collected at the substratum-lake interface. Nitrate-N ($NO_3^- - N + NO_2^- - N$) was determined using the hydrazine reduction method (Mullin & Riley 1955; Kamphake *et al* 1967). Analyses for ammonium-N were made using the indophenol method (Solórzano 1969). Ambient water temperature was measured using a hand-held mercury thermometer.

The influence of light intensity on N-fixation was determined using periphyton from the Sunnyside 8 m location. Incubation vials were enclosed in a series of neutral density screens which transmitted 1866, 1141, 706, 407, 274, 158, 71, 25 and 0 $\mu E\ m^{-2}\ sec^{-1}$ of natural sunlight as measured by a LI-COR submersible quantum sensor (LI-192S). Average summer-fall light readings range from 500–700, 300–350 and 180–200 $\mu E\ m^{-2}\ sec^{-1}$ for 2, 8 and 16 m, respectively. Sample vials were incubated in a water bath at 25° C for 3 h and assayed for nitrogenase activity.

The effect of temperature on N-fixation was also investigated. Samples were incubated at 5, 15 and 25° C and assayed for nitrogenase activity. Vials were pre-incubated for ~6 h at these temperatures prior to the introduction of C_2H_2. This temperature range corresponded to the annual range observed in Lake Tahoe.

Results and discussion

Heterocystous blue-green algae (*Calothrix, Tolypothrix, Nostoc* and *Scytonema*) dominated the biomass of the sublittoral periphyton community although the relative abundance of each genera varied with location. Non-heterocystous blue-green algae (*Chroococcus, Gloeocapsa*), filamentous green algae (*Zygnema*), stalked diatoms (*Gomphoneis*) and epiphytic diatoms (*Epithemia, Achnanthes, Fragilaria*) were also commonly found in this community.

Levels of dissolved inorganic nitrogen ($DIN = \sum NH_4^+ - N + NO_3^- - N + NO_2^- - N$) were extremely low throughout the entire year. Concentrations of nitrate-N never exceeded 3 μgN l^{-1} while concentrations of ammonium-N ranged from 0–11 μg N l^{-1} (Table 1). Concentrations of nitrite-N were always below the analytical limit of detection (<1 μgN l^{-1}). The seasonal range of temperature was 6.5–19.1° C within the water stratum from the surface to 16 m and was generally isothermal throughout the year (Table 1).

Rates of periphytic N-fixation, expressed in terms of horizontal rock surface area, ranged from 4–561 μgN m^{-2} h^{-1} (Fig. 2). Although a distinct seasonal trend in the rate of N-fixation was evident, the organisms retained their nitrogenase activity throughout the year. These results were consistent with the finding that the sublittoral periphyton community of Lake Tahoe is perennial and dominated by heterocystous blue-green algae (Loeb 1980).

N-fixation at Rubicon Point was greatest from June-September in both 1980 and 1981. Rates declined in the fall (October) and were low throughout the winter until spring (May) when rates began to increase reaching a maximum rate of

Table 1. Concentrations of NO_3^--N, NH_4^+-N and ambient water temperature over a 15-month period for each depth at the Rubicon Point sampling site. Values presented are representative of those measured at the Deadman Point and Sunnyside locations.

Date	NO_3^--N (μg l^{-1})			NH_4^+-N (μg l^{-1})			Temp. (°C)		
	2m	8m	16m	2m	8m	16m	2m	8m	16m
1980									
17 July	1	Bld	Bld	1	1	2	17.1	15.0	13.2
12 Aug.	3	2	2	4	2	1	18.8	17.7	15.0
12 Sept.	2	1	2	2	1	3	18.2	18.1	17.9
24 Oct.	2	2	2	3	2	7	13.9	13.9	13.9
17 Nov.	2	2	2	3	2	5	10.1	10.1	10.1
13 Dec.	3	2	2	3	2	7	8.3	8.3	8.3
1981									
10 Feb.	2	2	2	5	2	11	6.9	6.9	6.9
24 March	1	1	2	Bld	1	3	7.0	6.7	6.5
18 April	3	1	1	2	1	2	8.0	7.4	7.0
27 May	3	1	Bld	Bld	1	2	12.0	11.0	9.5
27 June	1	2	1	3	2	6	17.0	15.0	13.0
14 July	1	1	1	Bld	1	5	16.2	15.0	14.2
18 Aug.	1	1	1	Bld	1	1	19.0	19.0	16.5
17 Sept.	2	2	2	1	2	Bld	19.1	18.9	18.9
16 Oct.	–	–	–	–	–	–	13.0	13.0	13.0

Bld: below limit of detection or <1 μg N l^{-1}.

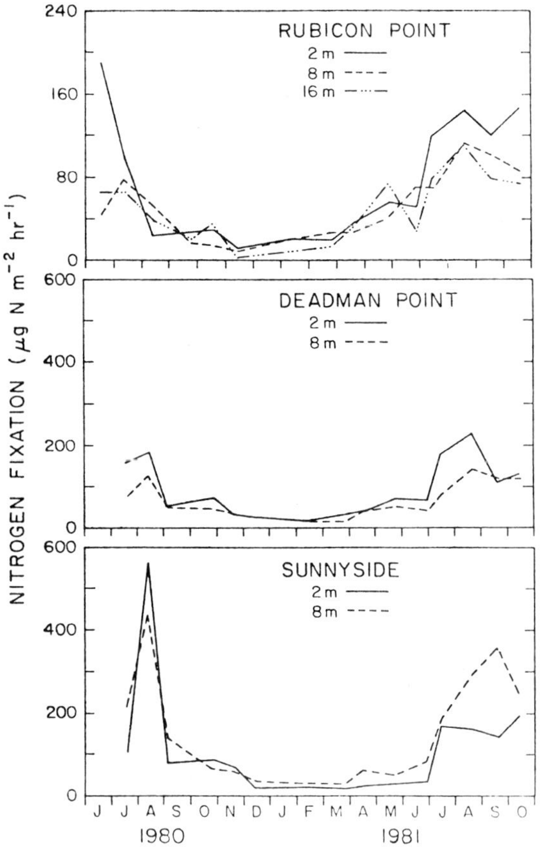

Fig. 2. Seasonal cycle of benthic N-fixation at three locations in Lake Tahoe. Note scale change at the Rubicon Point sampling site. n = 3 for each point.

145 μgN m^{-2} h^{-1} in August. Generally, N-fixation rates at all depths (2, 8 and 16 m) at Rubicon Point were similar. The seasonal patterns of N-fixation at the Deadman Point and Sunnyside sites were identical to that observed at Rubicon Point (Fig. 2). The ranges of measured rates at 2 and 8 m from Deadman Point and Sunnyside were similar to those measured at Rubicon Point for the period November-April (4–70 μgN m^{-2} h^{-1}). Differences in rates between 2 and 8 m at both these locations occurred during the summer when seasonal nitrogenase activity was maximal. Highest rates of periphyton N-fixation in Lake Tahoe were at Sunnyside and were 2–4 times those measured at the other sites. Loeb (1980) also observed that carbon fixation was highest at this location. Explanations for the observed differences in N-fixation include: 1) differences in the species composition of the periphyton community, 2) differential nutrient loading and, 3) differences in the amount of standing crop.

Although heterocystous blue-green algae have previously been reported in the epilithic, epipelic and epiphytic communities of lakes, few attempts have been made to quantify the rates of N-fixation in these benthic communities. Previous quantitative estimates of periphytic N-fixation range from 0–1313 μgN m^{-2} h^{-1} (Lundgren 1978; Moeller & Roskoski 1978; Lam *et al* 1979; Loeb & Reuter 1981). The rates reported in this present study are similar and represent a full annual cycle for periphyton N-fixation. The lakes from which the above observations were made were all oligotrophic and had no planktonic N-fixers.

Seasonal cycles of planktonic N-fixation illustrate the ephemeral nature of these N-fixers in lakes. Changes in the N-fixing activity of phytoplankton communities generally correlated well with the biomass of N-fixers which is typically very seasonal. In striking contrast, seasonal changes in periphyton nitrogenase activity appear to be due to a physiological response rather than changes in the biomass of the blue-green algae.

The average annual biomass of periphyton at the Rubicon Point study site was 14.0 ± 6.4 mg chl *a* m^{-2} (±SD) (range of 2.6–29.0 mg chl *a* m^{-2}) and generally increased with depth (2–16 m) (Fig. 3). The largest mean annual chl *a* concentrations were at Deadman Point 2 m (39.8 ± 20.2 mg chl *a* m^{-2}) and Sunnyside 8 m (34.4 ± 24.2 mg chl *a* m^{-2}). Summer chl *a* concentrations at these sites were greater in 1981 even though particulate carbon and nitrogen

Table 2. Summary of correlation and regression analyses performed on seasonal N-fixation and chl-*a* data.

Season	n	r^2	P	y-intercept	slope
Summer[a]	35	0.53	<0.01	66.5	2.78
Fall[b]	18	0.60	<0.01	9.70	2.86
Spring[c]	21	0.47	<0.01	29.9	0.91

[a] July, Aug., Sept.
[b] Oct., Nov.
[c] April, May, June.

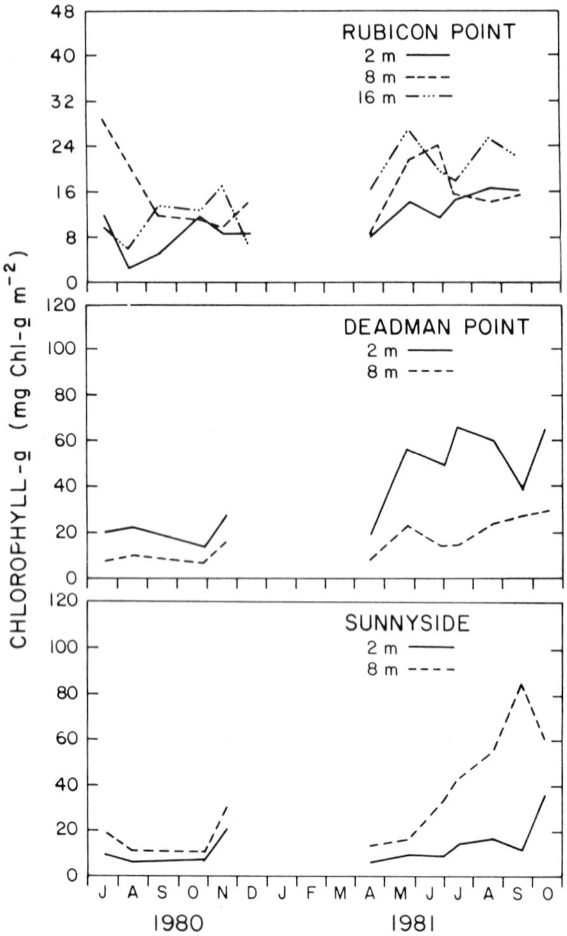

Fig. 3. Seasonal pattern of periphyton chlorophyll *a* concentration at three locations in Lake Tahoe. Chl *a* was not determined between December-March. Note scale change at Rubicon Point. n = 3 for each point.

were generally constant throughout the study (Reuter 1983). Although data are not available for the 1980 winter period, the sublittoral community of Lake Tahoe was been shown to remain structurally intact and physiologically viable during the winter months in other years (Loeb 1980).

The relationship between chl *a* and N-fixation was determined for the summer, fall and spring seasons using data combined from all locations and depths. These two parameters were significantly correlated ($p<0.01$) during each season. The slope of the regression equations was identical for the summer and fall seasons (Table 2) and was also similar to that reported by Loeb and Reuter (1981) for a five-lake survey during August-September 1979 which included Lake Tahoe. The correlation during the spring was also significant; however, the slope (N-fixation/chl *a*) was considerably less than observed for the summer and fall seasons. This finding suggests that: 1) N-fixation is not as important a pathway of nitrogen incorporation for the periphyton community in the spring as it is in the summer and fall, or 2) non-N-fixers may comprise a larger percentage of the community biomass in the spring. If chl *a* concentrations are elevated during the spring due to the presence of the more seasonal growth of diatoms and other non-N-fixing algae, a decreased slope (N-fixation/chl *a*) would result. As previously reported (Loeb & Reuter 1981), the strong positive correlation between these two parameters supports the microscopic observation that the standing crop of the sublittoral epilithic periphyton community is dominated by N-fixing blue-green algae.

Figure 4 illustrates the relationship between N-fixation and light intensity. When incubated in complete darkness, fixation was reduced to about 10% of the maximum rate. Exposure to even very low light intensities (25 $\mu E\ m^{-2}\ sec^{-1}$) resulted in a significant increase in nitrogenase activity which then remained essentially constant with increasing intensity. Full sunlight (1866 $\mu E\ m^{-2}\ sec^{-1}$) did not have an inhibitory effect on the rate of N-fixation. These data clearly indicate that nitrogenase activity was saturated at extremely low levels of irradiation

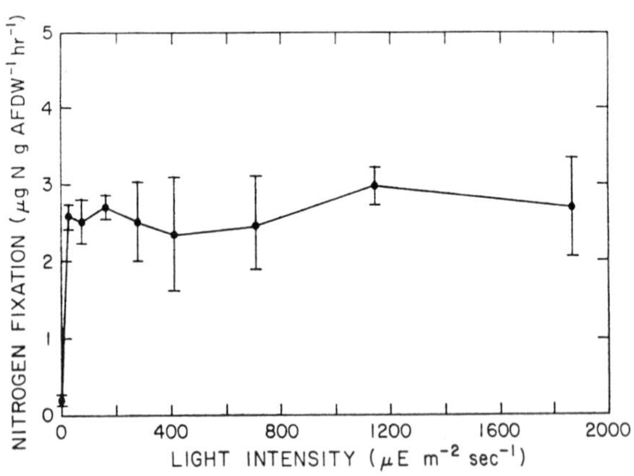

Fig. 4. N-fixation as a function of light intensity ($\bar{x} \pm 1$ sd). Samples were collected from Sunnyside (8 m) and incubations were in an outdoor water bath with natural sunlight. Water temperature was 20–25° C. n = 3 for each point.

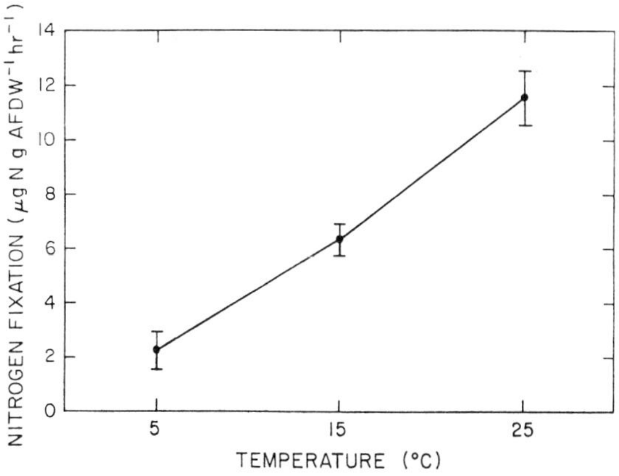

Fig. 5. The influence of temperature on N-fixation ($\bar{x} \pm$ 1ds). Samples were preincubated at 5, 15 and 25°C for 12 h prior to initiation of the nitrogenase assay. n = 3 for each point.

and suggest that at light levels >25 μE m^{-2} sec^{-1} (~5% mid-day surface light) the influence of solar radiation on periphyton N-fixation may be slight.

Studies by Mague *et al* (1974, 1977) and Flett *et al* (1980) have shown nitrogenase activity often increases with an increase in irradiance and inhibition is characteristically observed at high light levels. Goering & Neess (1964) also found that N-fixation varied with light; however, changes were not large (~30%). Dark N-fixation has frequently been observed with values ranging from 10–50% of light fixation rates (Goering & Neess 1964; Granhall & Lundgren 1971; Mague *et al* 1974). Significant rates of dark N-fixation may be attributed to non-photosynthetic N-fixers (i.e. bacteria) or may be the result of photosynthetic N-fixers ability to use organic compounds as sources to generate reducing power (Fay 1965). Dark N-fixation also appears to be dependent on the previous light history of the cell (FaY 1976). In the present study algal cells were pre-incubated in total darkness for 12 h prior to the initiation of the experiment; therefore, it is unlikely that the observed high rates of fixation at low light intensities were due to a physiological light memory effect.

The N-fixation versus light intensity experimental results suggest that the influence of light intensity on N-fixation is not important in Lake Tahoe above a depth of ~45 m (25 μE m^{-2} sec^{-1}). Loeb (1980) observed in Lake Tahoe that heterocystous blue-green algae no longer dominated the periphyton community at depths greater than 60 m; therefore, the majority of periphyton N-fixation may occur under light saturating conditions. Changes in the spectral quality of light with increasing water depth may influence N-fixation but this aspect was not investigated in our present study.

N-fixation rate and temperature were found to be positively correlated (Fig. 5). The rates of increase between 5–15°C and 15–25°C were 2.79 ± 0.89 (±SD) and 1.82 ± 0.23 respectively, however, these values were not significantly different. A similar linear increase in N-fixation with increasing temperature has also been reported for natural phytoplankton communities (Goering & Neess 1964) and a unialgal culture of *Anabaena cylindrica* (Fogg & Than-Tun 1960).

Many authors have shown a negative correlation between N-fixation and concentration of DIN (Stewart *et al* 1971; Fogg *et al* 1973) and Horne (1978) has stated that ammonium is a strong suppressor of both heterocyst formation and nitrogenase activity. Since the levels of DIN in Lake Tahoe are always low (<10 μgN l^{-1}), DIN availability is believed to have no regulating effect on the seasonal cycle of N-fixation in the sublittoral epilithic periphyton community. This conclusion is further supported by the observation that benthic N-fixation in Castle Lake, California was not suppressed following enrichments of ~80 μgN l^{-1} of

Table 3. Estimated annual N-fixation rates in various aquatic benthic systems. Values are expressed in m^2 of bottom substrate.

Habitat	gN m^{-2} yr^{-1}	Reference
Lake Tahoe – epilithic		
Rubicon Point	0.09–0.23	This study
Deadman Point	0.11–0.28	This study
Sunnyside	0.17–0.48	This study
Lake Windermere – epilithic	1[a]	Horne & Fogg 1970
Lake Taharoa – beach	0.15–0.62	Lam *et al* 1979
California stream	0.04–0.32	Horne & Carmiggelt 1975
Saltmarsh ecosystem	0.64	Carpenter *et al* 1978
Marine supra-littoral	2.5	Stewart 1967

[a] Indirect estimation.

NH_4NO_3 above the ambient DIN concentration of 0–10 μgN l^{-1} (Reuter *et al*, in press).

Solar radiation, temperature, and availability of nutrients are factors which most often influence the seasonal pattern of N-fixation in aquatic ecosystems. Benthic algal N-fixation in Lake Tahoe appears to be most influenced by changes in ambient water temperature. Further support of this conclusion that temperature is the primary factor regulating the annual cycle of N-fixation is the close similarity in the seasonal patterns of these two parameters (cf. Fig. 2 & Table 1). Other factors which have also been shown to influence N-fixation include levels of organic nitrogen, phosphorus and iron (Horne 1978) and their possible effects in Lake Tahoe cannot be dismissed.

The annual areal loading rate of inorganic nitrogen to Lake Tahoe contributed by periphytic N-fixers ranged from 0.09–0.48 μgN m^{-2} y^{-1}. These values are similar to those reported for a wide variety of benthic algal systems (Table 3). When applied to the whole lake, this equals a loading rate of 0.03 kgN ha^{-1} of lake surface area y^{-1} (range of 0.01–0.08). This loading rate is low in comparison to Lundgren's (1978) estimate of 0.1–0.5 kgN ha^{-1} of lake surface area y^{-1} for benthic N-fixation in a small, shallow, subarctic lake. The low areal loading rate in Lake Tahoe is also a function of the large lake surface area: littoral bottom area ratio. In relation to other input sources (Leonard *et al* 1979; Goldman *et al* 1982), loading of DIN to Lake Tahoe attributable to periphytic N-fixation is low ($<1\%$).

Conclusions

Depending on basin morphometry, the importance of benthic N-fixation as a source of new, combined nitrogen can be significant but has largely been overlooked in freshwater systems. Although N-fixation in oligotrophic lakes may be most important to the N-metabolism of the periphyton community itself, other organisms including bacteria, non-N-fixing periphyton and phytoplankton, and herbivores may benefit from this N-pool by 1) direct grazing of periphyton biomass, 2) uptake of nitrogenous compounds liberated following death and decomposition, or 3) uptake of both inorganic and organic nitrogenous extracellular products released from intact viable cells.

In N-deficient oligotrophic Lake Tahoe the ability of an algal species to fix nitrogen may provide a competitive advantage over other algae which cannot utilize this N source and which must depend on other forms and sources of DIN. To survive in this nutrient deficient environment, organisms must be able to scavenge nutrients for growth. The ability to fix nitrogen appears to be a successful strategy for benthic algae to overcome the N-deficiency in their environment.

Acknowledgements

We thank R.P. Axler for his critical review of this manuscript and his continuous support during this investigation. We also thank F.E. Broadbent for use of the mass spectrometer and J.C. Meeks for use of the gas chromatograph in Davis. D. Fellows assisted in various field and laboratory aspects of this work and E. deAmezaga and M. Smith aided in manuscript preparation. Research support was provided by NSF grants DEB78–23824 and DEB80–19918 to C.R. Goldman.

References

Arneson, P.A., 1979. Effects of nutrient enrichment on the natural phytoplankton of Lake Tahoe, California-Nevada. M.S. thesis, Univ. of California, Davis.

Axler, R.P., Gersberg, R.M. & Goldman, C.R., 1982. Inorganic nitrogen in a subalpine lake. Limnol. Oceanogr. 27:53–65.

Carpenter, E.J., Van Raalte, C.D. & Valiela, I., 1978. Nitrogen fixation by algae in a Massachusetts salt marsh. Limnol. Oceanogr. 23:318–327.

Dugdale, R.C. & Goering, J.J., 1967. Uptake of new and regenerated forms of nitrogen in primary productivity. Limnol. Oceanogr. 12:196–206.

Fay, P., 1965. Heterotrophy and nitrogen fixation in Chlorogloea fritschii. J. Gen. Microbiol. 39:11–20.

Fay, P., 1976. Factors influencing dark nitrogen fixation in a blue-green alga. Appl. Envir. Microbiol. 31:376–379.

Flett, R.J., Hamilton, R.D. & Campbell, N.E.R., 1976. Aquatic acetylene reduction techniques: solutions to several problems. Can. J. Microbiol. 22:43–51.

Flett, R.J., Schindler, D.W., Hamilton, R.D. & Campbell, N.E.R., 1980. Nitrogen fixation in Canadian precambrian shield lakes. Can. J. Fish. Aquat. Sci. 37:494–505.

Fogg, G.E. & Than-Tun, 1960. Interrelations of photosynthesis and assimilation of elementary nitrogen in a blue-green alga. Proc. R. Soc. B, 153:111–127.

Fogg, G.E., Stewart, W.D.P., Fay, P. & Walsby, A.E., 1973. The Blue-green Algae. Academic Press, London, 459 pp.

Goering, J.J. & Neess, J.C., 1964. Nitrogen fixation in two Wisconsin Lakes. Limnol. Oceanogr. 9:530–539.

Goldman, C.R., 1981. Lake Tahoe. Two decades of change in a nitrogen deficient oligotrophic lake. Verh. Internat. Verein. Limnol. 21:45–70.

Goldman, C.R., Leonard, R.L., Axler, R.P., Reuter, J.E. & Loeb, S.L., 1982: Interagency Tahoe Monitoring Program – 2nd Annual Report: Water Year 1981. Tahoe Research Group, Univ. of California, Davis. 201 pp.

Granhall, U. & Lundgren, A., 1971. Nitrogen fixation in Lake Erken. Limnol. Oceanogr. 16:711–719.

Horne, A.J., 1978. Nitrogen fixation in eutrophic lakes. In: Mitchell, R (ed.), Water Pollution Microbiology. Vol.2: 1–30.

Horne, A.J. & Fogg, G.E., 1970. Nitrogen fixation in some English lakes. Proc. Roy. Soc. Lond. B 175:351–366.

Horne, A.J. & Goldman, C.R., 1972. Nitrogen fixation in Clear Lake, California. I. Seasonal variation and the role of heterocysts. Limnol. Oceanogr. 17:679–692.

Horne, A.J. & Carmiggelt, J.W., 1975. Algal nitrogen fixation in Californian streams: Seasonal cycles. Freshwat. Biol. 5:461–470.

Kamphake, L.J., Hannah, S.A. & Cohen, J.M., 1967. Automated analysis for nitrate by hydrazine reduction. Water Res. 1:205–216.

Lam, C.W.Y., Vincent, W.F. & Silvester, W.B., 1979. Nitrogenase activity and estimates of nitrogen fixation by freshwater benthic blue-green algae. N. Z. J. Mar. and Freshwat. Res. 13:187–192.

Leonard, R.L., Kaplan, L.A., Elder, F.J., Coats, R.N. & Goldman, C.R., 1979. Nutrient transport in surface runoff from a subalpine watershed, Lake Tahoe basin, California. Ecol. Monogr. 49:281–310.

Loeb, S., 1980. The production of the epilithic periphyton community of Lake Tahoe, California-Nevada. Ph.D. thesis, Univ. of California, Davis.

Loeb, S.L., 1981. An in situ method for measuring the primary productivity and standing crop of the epilithic periphyton community in lentic systems. Limnol. Oceanogr. 26: 394–399.

Loeb, S.L. & Reuter, J.E., 1981. The epilithic periphyton community: a five-lake comparative study of community productivity, nitrogen metabolism and depth-distribution of standing crop. Verh. Internat. Verein. Limnol. 21:346–352.

Lundgren, A., 1978. Nitrogen fixation induced by phosphorus fertilization of a subarctic lake. Ecol. Bull. (Stockholm) 26:52–59.

Mague, T.H., Weare, N.M. & Holm-Hansen, O., 1974. Nitrogen fixation in the North Pacific Ocean. Mar. Biol. 24: 109–119.

Mague, T.H., Mague, F.C. & Holm-Hansen, O., 1977. Physiology and chemical composition of nitrogen-fixing phytoplankton in the Central North Pacific Ocean. Mar. Biol. 41:213–227.

Moeller, R.E. & Roskoski., 1978. Nitrogen-fixation in the littoral benthos of an oligotrophic lake. Hydrobiologia 60: 13–16.

Mullin, J.B. & Riley, J.P., 1955. The spectrophotometric determination of nitrate in natural waters, with particular reference to sea water. Anal. Chim. Acta 12:464–480.

Reuter, J.E., 1983. Studies of nitrogen assimilation in the periphyton communities of oligotrophic lakes. Ph.D. thesis, Univ. of California, Davis.

Reuter, J.E., Loeb, S.L., Axler, R.P., Carlton, R.G. & Goldman, C.R., 1983. Transformations of nitrogen following an epilimnetic nitrogen fertilization in Castle Lake, CA. 1. Epilithic periphyton responses. Verh. Internat. Verein. Limnol. 22nd Congress S.I.L. Lyon, France (in press).

Rusness, D. & Burris, R.H., 1970. Acetylene reduction (nitrogen fixation) in Wisconsin lakes. Limnol. Oceanogr. 15: 808–813.

Schell, D.M. & Alexander, V., 1970. Improved incubation and gas sampling techniques for nitrogen fixation studies. Limnol. Oceanogr. 15:961–962.

Solózano, L. 1969. Determination of ammonia in natural waters by the phenol hypochlorite method. Limnol. Oceanogr. 14: 799–801.

Sorokin, Y.I. & Kadota, H., 1972. Techniques for the Assessment of Microbial Production and Decomposition in Fresh Waters. IBP Handbook No. 23, Blackwell Scientific Publications, 112 p.

Stewart, W.D.P., 1967 Nitrogen turnover in marine and brackish habitats. II. Use of ^{15}N in measuring nitrogen fixation in the field. Ann. Bot 31:385–407.

Stewart, W.D.P., Fitzgerald, G.P. & Burris, R.H., 1967. In situ studies on N fixation using the acetylene reduction technique. Proc. Nat. Acad. Sci. U.S.A., 58:2071–2078.

Stewart, W.D.P., Mague, T., Fitzgerald, G.P. & Burris, R.H., 1971. Nitrogenase activity in Wisconsin lakes of differing degrees of eutrophication. New Phytol. 70:497–509.

Strickland, J.D.H. & Parsons, T.R., 1972. A Practical Handbook of Seawater Analysis. Bull. Fish. Res. Board Can. 167, Ottawa, Canada, 310 pp.

Vanderhoef, L.N., Huang, C.-Y., Musil, R. & Williams, J., 1974. Nitrogen fixation (acetylene reduction) by phytoplankton in Green Bay, Lake Michigan, in relation to nutrient concentration. Limnol. Oceanogr. 19:119–125.

Authors' address:
J.E. Reuter/S.L. Loeb/C.R. Goldman
Institute of Ecology and Division of Environmental Studies
University of California
Davis, CA 95616, U.S.A.

CHAPTER 15

Direct examination of growing filaments to determine phosphate growth kinetics in *Cladophora glomerata* (L.) Kütz and *Stigeoclonium tenue* (Agardh) Kütz

ARNO S. ROSEMARIN

Abstract. Semi-continuous (petri-plate) cultures were used for assessing growth kinetics in attached filaments of (Lake-Ontario derived) *Cladophora glomerata* and *Stigeoclonium tenue* relative to limiting levels of inorganic phosphate. Under enriched conditions, the filaments were allowed to produce zoospores which then attached to the base of the plastic petri-plates to produce germlings. Bottom attachment was achieved by exploiting positive phototaxis in providing bottom illumination. Six phosphate treatments, ranging from 0 to 2 μM PO_4, were run for 15 days with daily replacement of medium. Cell counts were carried out on the same filaments each time, every two days, using an inverted light microscope. Specific growth rate was calculated as the doubling rate (of the number of cells per filament) per day. The Monod kinetics model was used to determine the growth kinetic coefficients. The results reinforce previous findings, derived using the Droop and Kilham variable internal stores model, that *Stigeoclonium tenue* outgrows *Cladophora glomerata* under conditions of limiting phosphate. This finding provides further evidence that the dominance of *Cladophora glomerata* in the North American Great Lakes may not be related strictly to the response to phosphate.

Introduction

Although freshwater phytoplankton nutrient kinetic studies are now quite common (McCarthy 1981; Rhee 1980), those involving the larger filamentous algae are few. Recent work on *Pithophora* (Spencer & Lembi 1981), *Cladophora* (Auer & Canale 1982a, 1982b) and *Cladophora* and *Stigeoclonium* (Rosemarin 1982) has shown that nutrient-limited growth and uptake kinetics in these algae can be modelled using the enzyme-substrate kinetics model developed by Monod (1942). Nutrient uptake and growth studies involving benthic marine algae have shown a similar pattern (Chapman *et al* 1978; Hanisak & Harlin 1978; Haines & Wheeler 1978; Indergaard & Jensen 1981; Lapointe & Tenore 1981). Some work has been done on the nutrition of stream periphyton, and the bulk of this has been done on laboratory streams rather than in the field (Horner & Welch 1981).

The isolation and laboratory culture of epiphytic filamentous algae was described by Reynolds (1950) in a useful but surprisingly rarely cited paper. Other methods for macroscopic algae are described by Chapman (1973) and Venkataraman (1969). The culture of *Cladophora* using synthetic medium was only recently accomplished (Gerloff & Fitzgerald 1976). Up to this point, the specific nutritional requirements were rather poorly understood. That phosphorous alone controls the presence and spreading of *Cladophora* in the North American Great Lakes is not universally accepted (Mantai *et al* 1982; Rosemarin 1982). *Cladophora* has relatively low phosphorus and nitrogen requirements and may be limited by its relatively large requirements for the trace elements boron

Wetzel, R.G. (ed.), Periphyton of Freshwater Ecosystems.
 ISBN 90 6193 768 X.

and zinc and the vitamin (B_1) thiamine (Gerloff & Fitzgerald 1976). The culture of *Stigeoclonium* was described by Cox & Bold (1966) in a major taxonomic study and also by Francke & Cate (1980) and Trotter & Hendricks (1979) in nutritional studies.

The ecological stability of the *Cladophora* dominance in the Great Lakes was questioned previously (Rosemarin 1982). Evidence was provided, using data from uptake experiments and growth curves, that the dominance of *Cladophora* could not be explained on the basis of phosphate assimilation and it was hypothesized that other factors relating to life cycle and cell differentiation may be determining success. *Stigeoclonium* outgrew *Cladophora* in the laboratory (under phosphate-limited conditions) but failed to do so in the lake.

The objective of the present work is to provide further evidence, using other methods, that *Stigeoclonium tenue* should be able to outcompete *Cladophora glomerata* under conditions of limiting phosphate. Determination of the growth rate of such filamentous algae, by removing small subsamples from cultures over time and then counting the cells, is not practical because of enormous problems in heterogeneity. Direct observation of the same filaments from one day to the next was thought, however, to be a possible method for measuring growth with some accuracy. By directly observing filament growth under controlled conditions and over a range of low phosphate concentrations, *Stigeoclonium* could again be compared to *Cladophora* as a potential competitor.

Methods and materials

Algal isolation and culture medium

Stigeoclonium tenue and *Cladophora glomerata* were isolated from the Lake Ontario shore line at Kingston in the Canadian Province of Ontario. Unialgal isolates were made from epiphyte-free apical sections of the parent filaments in enrichment cultures using a modified Gorham No. 11 growth medium (Gerloff & Fitzgerald 1976). A description of this medium is also given in Graham *et al* (1982). Axenic clones of *Stigeoclonium* were obtained by exposing stationary-phase cells to antibiotics (penicillin, streptomycin and chloramphenicol) as in Guillard (1973), and plating out cells on growth medium plus peptone-agar. Isolates were stored on agar slants in darkness at 10° C. Only limited success was achieved in producing axenic cultures of *Cladophora*. Bacteria were eliminated from the cell surface using the antibiotics but seemed to survive inside empty zoosporangial cells. Zoospores and germlings did not survive the antibiotic levels required to eliminate bacteria.

Cultures and inocula for experiments

Stock cultures were maintained in standing Erlenmeyer or Fernbach flasks. Sub-cultures were made once per month if the flasks were kept in the light (about 50 $\mu E/m^2.sec$ from cool-white fluorescent lights) at 15° to 20° C. If kept in the dark or very low indirect light, sub-cultures were required only once every six to nine months to maintain healthy-appearing pigmented cells.

Growth experiments were carried out using standard-sized (9.0 cm diameter) plastic pre-sterilized petri-plates each containing 15 mL of medium. Experiments were initiated by transferring five loopfuls (5 mm diameter sterile platinum wire loop) of mature filaments from the standing cultures to each of the petri-plates (one plate per treatment for each species) containing full stength medium and 20 μM PO_4 (1 μM P equals 30.97 μg P/L) to nutritionally enhance zoosporogenesis. The top-side of the plates was covered with dark non-reflective paper to eliminate surface illumination. Zoospores, once released, responded phototactically and attached to the base of the bottom-illuminated petri-plates. A three day period for *Stigeoclonium* and a two- or three-week period for *Cladophora* were sufficient to provide adequate material for the growth experi-

Table 1. List of symbols used

Symbol	Explanation	Unit
S	external substrate concentration	μM
μ	specific growth rate	day^{-1}
μ_{max}	maximum μ	day^{-1}
K_s	half-saturation constant for growth	μM

ments. The start-up time for sporulation was phased, taking into account this time difference, so that the growth experiments could be carried out on both species (in separate petri-plates) simultaneously using the identical treatment medium.

Stage of development in *Stigeoclonium,* at the commencement of the experiments, was not variable since zoospore production was near-synchronous. Growth experiments were started, therefore, with an array of uniformly developed attached single cells. Since sporulation in *Cladophora* was sporadic in time and dependent on cell differentiation on the filament (apical cells produce the bulk of the zoospores but intercalary cells become zoosporangia as well), the resulting attached starting material varied in size from a few cells to 20 or 30 cells per filament. This variable was eliminated from the experiments by monitoring growth in each filament separately. The same filaments were returned to repeatedly for each count.

Physical conditions of incubation

A simple growth chamber was built to supply top and bottom source illumination and controlled temperature. A small portable fume hood was placed on a lab bench surface and an internal frame support was installed using standard laboratory aluminum bars (0.5-inch diameter) to hold the four 18″ (45 cm) fluorescent fixtures (one in each corner of the rectangular box frame) and the petri-plate support (a modified metal grill from a refrigerator shelf). Cooling was provided using on adjacent refrigeration unit equipped with a pump and exterior remote cooling coil (coolant used was 50% methanol). The coil was placed at the base of the growth chamber under the petri-plate support and a small fan equipped with a thermostat was used for air circulation. Warm air was vented passively through the upper port of the fume hood. Temperature was maintained at between 20° and 21° C and continuous light intensities varied from 100 to 150 $\mu E/m^2 \cdot sec$ depending on the spot location. Petri-plates were stirred and rotated daily to expose each plate equally over the 15-day experimental period. Sterile transfers were made using a laminar-flow sterile bench.

Phosphate treatments

Six treatments were used consisting of 5 concentrations of Na_2HPO_4 (2.0, 1.0, 0.5, 0.25 and 0.125 μM PO_4) and a blank, in full strength medium (see above for details on the medium). One petri-plate was used for each treatment for each alga, thus totalling twelve petri-plates for each experiment. Enough medium was made at time zero for the entire experiment so that no variation could exist within each treatment throughout the duration. The phosphate treatments were made up by serially diluting from a single standard. Medium was manually replaced every second day when the filaments were small (up to day seven) and daily as growth progressed, by decanting followed by immediate replacement.

Phosphate concentrations were checked especially near the termination of the experiments when cell density was highest, using the method of Murphy & Riley (1962). The low density of cells relative to the volume of medium in the petri-plates and the daily replacement of medium, ensured that the treatments were in effect continuous.

Growth was assumed to have 'equilibrated' to the treatments (the germlings were initially phosphorus-enriched) by the time growth was near zero in the blank treatment or when the growth rate became near-steady in the PO_4 treatments. The period required to reach near zero growth in the blank treatment was 3 or 4 days in the case of *Stigeoclonium* and 6 or 7 days in the case of *Cladophora.* The period required to reach near-steady growth in the PO_4 treatments was less than this starvation period. This 'period of equilibration' to each of the treatments was closely related to the initial size of the germling used in the experiment, the larger filaments taking a longer period.

At the termination of the experiment, in the case of *Stigeoclonium,* treatments 3 through 6 (i.e. 0.5, 0.25, 0.125 and 0.0 μM P) were shifted up 2 steps to treatments 1 through 4 (i.e. 2.0, 1.0, 0.5 and 0.25 μM) in order to test further the sensitivity of the methodology. The entire experiment for both species was then repeated from the beginning using fresh inocula from the same original stock culture and new growth medium.

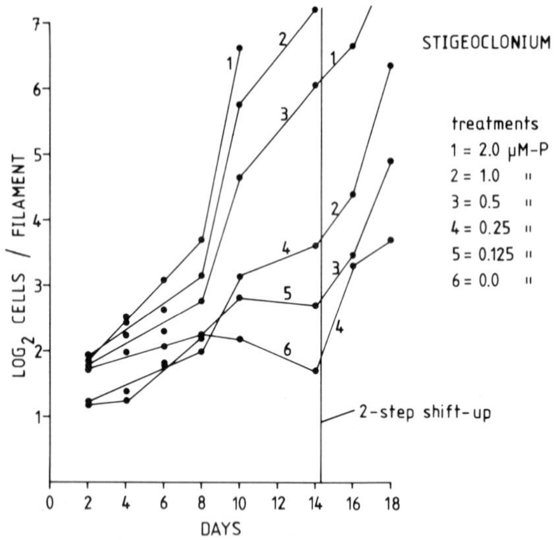

Fig. 1. Growth of *Stigeoclonium* ($\log_2$ cells per filament) versus time (days) for the six phosphate treatments. Treatments 3 to 6 were shifted up to treatments 1 to 4, respectively, at day 14.

Direct observation of cell growth

The petri-plates were examined using a NIKON inverted light microscope (Model M) equipped with a camera attachment. In the case of *Stigeoclonium,* a homogeneous 5-mm diameter field was chosen and marked on the outside bottom surface of the petri-plates so that the same cluster of attached filaments could be examined from one day to the next. The number of cells on fifteen to thirty filaments were counted each time. Growth was monitored every second day. In the case of *Cladophora,* the same filaments were examined each time (5 or 6 filaments per treatment). Each filament was given an identification number which was labelled to the underside of the petri-plate adjacent to the filament holdfast.

Specific growth rate was calculated as doublings per day as follows:

$$\mu = \log_{2\,(N_1/N_0)} / (t_1 - t_0)$$

where N_1 and N_0 represent the number of cells per filament at time t_1 and t_0, respectively. In this way, the specific growth rate was calculated from one sampling day to the next. The overall specific growth rate for each treatment can be determined by taking the mean of all growth rate estimates for each treatment. Alternatively, the log of the number of cells per filament can be plotted versus time and the average slope of this line also represents the overall specific growth rate. The growth rates used in the calculation of kinetic parameters were determined from straight sections of the log plots when growth was near-steady.

Since the *Stigeoclonium* cells were at the same one-cell stage of development at the commencement of the experiments, the filament cell-counts at each sampling for each treatment could be aggregated to provide a mean value of cells per filament. Specific growth rates were determined using these mean estimates. Since the *Cladophora* filaments were each at different stages of growth, each filament was regarded as a separate entity and the specific growth rate of each filament was determined separately from the change in cell number. Mean specific growth rates for each treatment were calculated by averaging the specific growth rates of the filaments in each treatment.

Results and discussion

Direct observation of filament growth provided estimates of cell doubling rates sufficiently accurate to model phosphate growth kinetics. Although the process of growth consisted of filament elongation followed by cell wall formation, growth was best determined by counting cell numbers and not measuring cell volumes. By counting enough filaments and monitoring the same filaments over short intervals, the variation in cell size was not a major source of error.

Stigeoclonium

Stigeoclonium mean cell numbers per filament were plotted as the logarithm to the base 2 ($\log_2$) versus time (Fig. 1). Cell numbers increased directly with the concentration of phosphate. Growth was at a relatively constant logarithmic rate (as determined by a straight line) in each treatment from day 2 to day 8. Growth increased substantially between days 8 and 10 and then returned to the previous levels between days 10 and 14. The filaments of treatments 1 and 2 became too dense to

Table 2. The specific growth rate (μ) determined from the mean of the cell numbers per filament for n filaments of *Stigeoclonium tenue* at different concentrations of phosphate.

PO_4 concentration (μM)	Specific growth rate for (n), replicate filaments (day^{-1})	Grand mean [($\bar{\mu}$) (day^{-1})]
2.0	0.53(14), 0.33(15), 0.30(19), 0.32(18), 1.28(22), 0.67(13), 0.78(15)	0.60
1.0	0.64(14), 0.26(15), 0.26(22), 1.14(17), 0.5(13), 0.39(21), 0.99(21)	0.60
0.5	0.49(15), 0.21(16), 0.20(29), 0.83(15), 0.49(17), 0.39(26), 0.71(25)	0.47
0.25	0.16(14), 0.23(16), 0.5(30), 0.35(37), 0.56(29), 0.19(32)	0.33
0.125	0.2(12), 0.29(18), 0.19(43), 0.26(34), 0.20(31)	0.23
0.0	0.18(15), 0.12(14), 0.04(24), 0.1(38), 0.0(37), 0.0(10)	0.07

count by day 14 and 16, respectively. The shift up of treatments 3 through 6 to treatments 1 through 4, respectively, resulted in data very closely resembling the original treatments. In all cases, the shift up to a higher PO_4 concentration resulted in a higher growth rate. The two lowest treatments (0 and 0.125 μM P) when exposed to 0.25 and 0.5 μM P, respectively, showed large immediate increases in growth when monitored two days after the shift (Fig. 1).

Specific growth rates (cell doublings per day) for each treatment were calculated from the overall slope of the log plots. The mean specific growth rates (Table 2) when related to the PO_4 treatments could be fit (Fig. 2) to the Monod model ($\mu = \mu_{max} \cdot S/K_s + S$)). Growth was maximal in the 1 and 2 μM P treatments and was near maximum at the 0.5 μM P level. The μ_{max} (maximum growth rate) as determined by the Lineweaver-Burke double reciprocal linear transformation of the model, was 0.71 per day (Fig. 2) and the K_s[1] value, 0.27 μM P. The experiment, upon repetition, resulted in a μ_{max} of 0.65 per day and a K_s of 0.13 μM P. The major source of variability contributing to the estimate of K_s is the growth rate at the lowest treatment levels (0.125 and 0.25 μM P). Small variations in these determinations of growth rate result in relatively large differences in the estimate of K_s. The high PO_4 μ_{max} values, on the other hand, do not appreciably affect the estimate of K_s.

[1] The K_s value represents that level of PO_4 where growth is half-maximal. For species with a similar μ_{max}, the K_s is a fair indicator of competitiveness for the substrate in question (eg. PO_4). The lower the K_s value, the more competitive the species because it grows faster at lower PO_4 levels. In the case where the two species have very different μ_{max} values, the ratio of K_s/μ_{max} is a better indication of competitiveness for the substrate in question (Healey 1980). The lower the K_s/μ_{max} ratio, the higher the growth rate at a given substrate level. The species exhibiting the lower K_s/μ_{max} ratio is therefore more competitive.

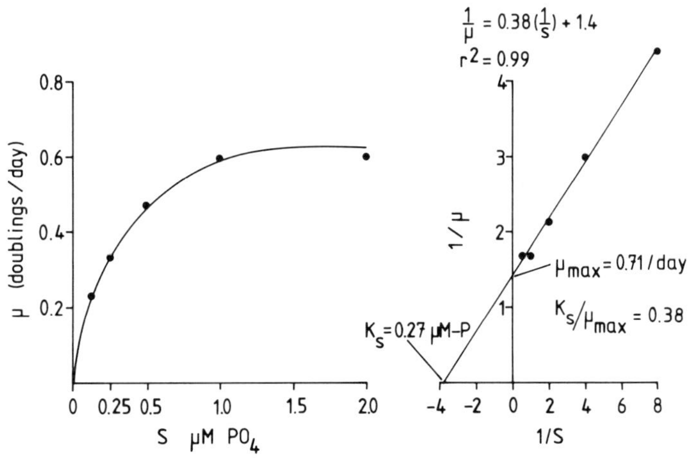

Fig. 2. Growth rate (μ) of *Stigeoclonium* at different concentrations of external phosphate (S).

Table 3. The specific growth rate (μ) determined from individual filaments of *Cladophora glomerata* at different concentrations of phosphate.

PO_4 concentrations (μM)	Specific growth rate (μ) (day^{-1})	Mean $\mu \pm$ standard deviation (day^{-1})
2.0	0.29, 0.29, 0.37, 0.25, 0.27, 0.16, 0.32, 0.22, 0.37, 0.42, 0.26, 0.25, 0.44, 0.19, 0.24, 0.37	0.29 ± 0.08 n = 16
1.0	0.19, 0.14, 0.24, 0.31, 0.17, 0.17, 0.2, 0.17, 0.16, 0.39,	0.21 ± 0.08 n = 10
0.5	0.21, 0.20, 0.15, 0.20, 0.13, 0.14, 0.21, 0.15, 0.26, 0.19,	0.18 ± 0.04 n = 10
0.25	0.17, 0.21, 0.13, 0.12, 0.16, 0.14, 0.11, 0.10	0.14 ± 0.04 n = 8
0.125	0.15, 0.11, 0.11, 0.04, 0.11, 0.11, 0.07, 0.03, 0.02	0.08 ± 0.04 n = 9
0.0	0, 0.06, 0.04, 0.09, 0, 0.05, 0, 0, 0, 0, 0, 0, 0, 0, 0, 0, 0, 0, 0,	0.01 ± 0.03 n = 19

Cladophora

In the case of *Cladophora,* each individual filament was monitored separately because the initial starting material varied in size and cell number (due to the nonsynchronous nature of zoosporogenesis in *Cladophora*). In order to graph the cell numbers per filament versus time would require a separate plot for each filament and treatment. For ease of presentation, the overall specific growth rate for each filament is listed in Table 3 along with the mean and standard deviation for each treatment. Growth appeared not to saturate but did reach a near assymptote at about 0.25 per day at about 1.5 μM P (Fig. 3). The μ_{max} determined from the Lineweaver-Burke double reciprocal linear transformation was 0.31 per day and the $K_s \cdot 0.33\,\mu$M P. In repeating the experiment, the values for μ_{max} and K_s were 0.24 per day and 0.15 μM P, respectively.

Table 4. Summary of phosphate growth kinetic parameters for *Stigeoclonium tenue* and *Cladophora glomerata.*

Parameter	*Stigeoclonium tenue*	*Cladophora glomerata*	Source
K_s (μM P)	0.27 } mean =	0.33 } mean =	Expt. 1
	0.13 } 0.20	0.15 } 0.24	Expt. 2
	0.18	0.61	Rosemarin (1982)[a]
μ_{max} (day^{-1})	0.71 } mean =	0.31 } mean =	Expt. 1
	0.65 } 0.68	0.24 } 0.28	Expt. 2
	2.0	0.8	Rosemarin (1982)[b]
K_s/μ_{max} (μM . day)	0.38 } mean =	1.06 } mean =	Expt. 1
	0.19 } 0.29	0.63 } 0.85	Expt. 2
	0.09	0.76	Rosemarin (1982)

[a] Kinetic coefficients were generated using PO_4 uptake and growth data and the variable internal stores model.
[b] Growth rates were determined using cell carbon analysis on cells grown in stir-aerated cultures (petri-plates in present study were not continually stirred).

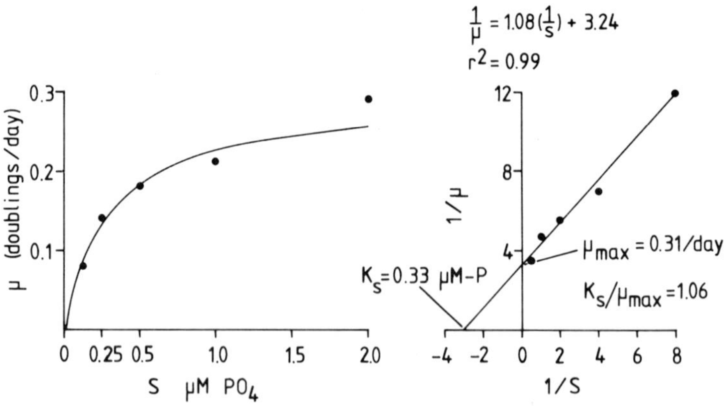

Fig. 3. Growth rate (μ) of *Cladophora* at different concentrations of external phosphate (S).

Comparing the growth kinetics of Stigeoclonium and Cladophora.

Under the conditions of the experiments, *Stigeoclonium* grew consistently faster than *Cladophora*. The mean μ_{max} from the two experiments was 0.68 per day for *Stigeoclonium* and 0.28 per day for *Cladophora* (Table 4). Comparing the values of μ for each treatment in a two-tailed t-test (Table 5) showed that *Stigeoclonium* grew significantly faster in all phosphate treatments (99% confidence level). This growth pattern is also clear from Fig. 4 which depicts the mean growth at each of the treatments for each species. The K_s values for the two species were more similar however. The means from the two experiments were 0.2 μM P for *Stigeoclonium* and 0.24 μM P for *Cladophora*. The respective growth rates at these levels of PO_4 were clearly divergent though in favour of *Stigeoclonium* (Table 5 and Figs. 2 and 3).

The K_s/μ_{max} ratios, which better express the overall competitive capability, as previously explained, were 0.29 for *Stigeoclonium* and 0.85 for *Cladophora* (Table 4). This difference agrees with previous work (Rosemarin 1982), indicating that *Stigeoclonium* is able to outcompete *Cladophora* under phosphate-limited conditions. The previous experiments (Rosemarin 1982) combined PO_4 uptake and growth data using the variable internal stores model (Droop 1968; Kilham 1978). Although the methods differed from those employed in the present case (eg. cultures were stir-aerated batch-type and biomass was measured as cellular carbon), the results were complimentary. Growth rates were much higher in the stir-aerated cultures in comparison to the still (once-daily stirred) petri-plate cultures (Table 4) but the K_s and K_s/μ_{max} ratios fell into the same range.

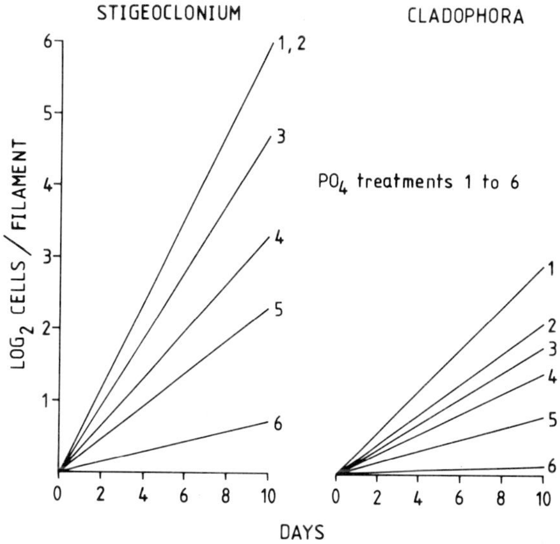

Fig. 4. Mean growth of *Stigeoclonium* and *Cladophora* ($\log_2$ cells per filament) over the time course of the experiment for the six phosphate treatments.

The methods described here, although somewhat labour intensive, are not impractical considering the question being tackled, and in addition are inexpensive. Materials used here can be found in most biological laboratories. Performing nutrient kinetic experiments with filamentous algae at

Table 5. Results of a two-tailed t-test comparing the specific growth rates of *Stigeoclonium* and *Cladophora* growing at different levels of phosphate.

PO_4 concentration (μM)	Mean specific growth rate (μ)				df	t-value	S	NS	C.L.
	Stigeoclonium	n	*Cladophora*	n					
2.0	0.60	7	0.29	16	21	3.35	X		99%
1.0	0.60	7	0.21	10	15	3.40	X		99%
0.5	0.47	7	0.18	10	15	3.86	X		99%
0.25	0.33	6	0.14	8	12	3.13	X		99%
0.125	0.23	5	0.08	9	12	5.79	X		99.9%

n = replicates.
df = degrees of freedom.
S = significant difference.
NS = no significant difference.
C.L. = % confidence level.

realistic low levels of limiting nutrients is therefore possible using these techniques. It is thus hoped that other phycologists and limnologists carry out such investigations involving these and other similar species in order to begin to fill this major gap in limnological research.

Acknowledgements

This work was supported by funds derived from the National Research Council of Canada. I thank Drs. S.R. Brown and D. Turpin of the Department of Biology at Queen's University, Kingston, Canada for advice in designing these experiments. I thank Drs. D.C. Mortimer and D. Miller of the National Research Council, Division of Biological Sciences in Ottawa for the use of laboratory facilities, assistance and invaluable expert opinions.

References

Auer, M.T. & Canale, R.P., 1982a. Ecological studies and mathematical modeling of Cladophora in Lake Huron: 2. Phosphorus uptake kinetics. J. Great Lakes Res. 8(1): 84–92.

Auer, M.T. & Canale, R.P., 1982b. Ecological studies and mathematical modeling of Cladophora in Lake Huron: 3. The dependence of growth rates on internal phosphorus pool size. J. Great Lakes Res. 8(1):93–99.

Chapman, A.R.O., 1973. Methods for macroscopic algae, pp. 87–104. In: Stein, J.R. (ed.) Handbook of Phycological Methods, Culture Methods and Growth Measurements. Cambridge Univ.

Chapman, A.R.O., Markham, J.W. & Lüning, K., 1978. Effects of nitrate concentration on the growth and physiology of Laminaria saccharina (Phaeophyta) in culture. J. Phycol. 14(2):195–198.

Cox, E.R. & Bold, H.C., 1966. Taxonomic investigations of Stigeoclonium. Phycol. Stud., 7:7–167.

Droop, M.R., 1968. Vitamin B_{12} and marine ecology. IV. The kinetics of uptake, growth and inhibition in Monochrysis lutheri. J. Mar. Biol. Assoc. U.K. 48:689–733.

Francke, J.A. & Cate, H.J. ten, 1980. Ecotypic differentation in response to nutritional factors in the algal genus Stigeoclonium Kütz. (Chlorophyceae). Br. Phycol. J. 15:343–355.

Gerloff, G.C. & Fitzgerald, G.P., 1976. The nutrition of Great Lakes Cladophora. EPA Report 600/3-76-044.

Graham, J.M., Auer, M.T., Canale, R.P. & Hoffmann, J.P., 1982. Ecological studies and mathematical modeling of Cladophora in Lake Huron: 4. Photosynthesis and respiration as functions of light and temperature. J. Great Lakes Res. 8(1):100–111.

Guillard, R.R.L., 1973. Methods for microflagellates and nannoplankton, pp. 69–85. In: Handbook of Phycological Methods, Culture Methods and Growth Measurements (J.R. Stein, ed.) Cambridge Univ.

Haines, K.C. & Wheeler, P.A., 1978. Ammonium and nitrate uptake by the marine macrophytes Hypnea musciformis (Rhodophyta) and Macrocystis pyrifera (Phaeophyta). J. Phycol. 14:319–324.

Hanisak, M.D. & Harlin, M.M., 1978. Uptake of inorganic nitrogen by Codium fragile subsp. tomentosoides (Chlorophyta). J. Phycol. 14:450–454.

Healey, F.P., 1980. Slope of the Monod equation as an indicator of advantage in nutrient competition. Microb. Ecol. 5: 281–286.

Horner, R.R. & Welch, E.B., 1981. Stream periphyton development in relation to current velocity and nutrients. Can. J. Fish. Aquat. Sci. 38:449–457.

Indegaard, M. & Jensen, A., 1981. Nitrate and phosphate uptake in small populations of Laminaria digitata (Phaeophyceae). pp. 411–417. In: Xth international Seaweed Symposium. Berlin: Walter de Gruyter & Co.

Kilham, S.S., 1978. Nutrient kinetics of freshwater planktonic algae using batch and semicontinuous methods. Mitt. Internat. Verein. Limnol. 21:147–157.

Lapointe, B.E. & Tenore, K.R., 1981. Experimental outdoor studies with Ulva fasciata Delile. 1. Interaction of light and nitrogen on nutrient uptake, growth, and biochemical composition. J. exp. mar. Biol. Ecol. 53:135–152.

Mantai, K.E., Garwood, P.E. & Peglowski, L.E., 1982. Environmental factors controlling physiological changes in Cladophora in Lake Erie. J. Great Lakes Res. 8(1):61–65.

McCarthy, J.J., 1981. The kinetics of nutrient utilization. pp. 211–233. In: Physiological bases of phytoplankton ecology. (Platt, T. ed.). Can. Bull. Fish. Aquat. Sci. 210.

Monod, J., 1942. Recherches sur la Croissance des Cultures bactèriennes. Paris: Hermann.

Murphy, J. & Riley, J., 1962. A modified single solution method for the determination of phosphate in natural waters. Anal. Chim. Acta 27:31–36.

Reynolds, N., 1950. Methods of culturing epiphytic algae. New Phytologist 49:155–162.

Rhee, G-Y., 1980. Continuous culture in phytoplankton ecology. pp. 151–203. In: Advances in Aquatic Microbiology Vol. 2 (Droop, M.R. and Jannasch, H.W. ed.). London: Academic Press.

Rosemarin, A.S., 1982. Phosphorus nutrition of two potentially competing filamentous algae, Cladophora glomerata (L.) Kütz and Stigeoclonium tenue (Agardh) Kütz. from Lake Ontario. J. Great Lakes Res. 8(1):66–72.

Spencer, D.F. & Lembi, C.A., 1981. Factors regulating the spatial distribution of the filamentous alga Pithophora oedogonium (Chlorophyceae) in an Indiana Lake. J. Phycol. 17:168–173.

Trotter, D.M. & Hendricks, A.C., 1979. Attached, filamentous algal communities. pp. 58–69. In: Methods and measurements of periphyton communities: a review (Weitzel, R.L. ed.) Philadelphia: American Society for Testing and Materials.

Venkataraman, G.S., 1969. The Cultivation of Algae. New Delhi: Indian Council of Agriculture Research.

Author's address:
A.S. Rosemarin
Department of Biology
Queens University
Kingston
Ontario, Canada, K7L 3N6
Present address:
Swedish Environmental Research Institute
IVL-Aneboda
S-360 30, Lammhult, Sweden

CHAPTER 16

Development of nuisance periphytic algae in laboratory streams in relation to enrichment and velocity

R.R. HORNER, E.B. WELCH and R.B. VEENSTRA

Abstract. Accrual of periphyton biomass in artificial channels responded more to increased SRP (soluble reactive phosphorus) concentration over a range of 2 to 75 $\mu g\ l^{-1}$ than to velocity ranging from 5 to 75 cm s^{-1}. Accrual increased in proportion to increased SRP up to about 25 $\mu g\ l^{-1}$; but further increase was not as pronounced above that concentration, presumably due to a saturation of uptake rates. Velocity enhanced accrual between 5 and 25 cm s^{-1}, apparently by maximizing the nutrient concentration gradient between surrounding water and the cell surface. This enhancement effect may prevail at velocities ≤15 cm s^{-1}, as had been suggested previously. Velocities up to 75 cm s^{-1} did not cause scouring and loss of biomass from surfaces, however, a result which was unexpected. Review of the literature suggested that a biomass of 100 to 150 mg chl *a* m^{-2} may represent a nuisance condition. That level was attained in these experiments at between 15 and 25 $\mu g\ l^{-1}$ SRP, which is suggested as a critical P concentration in running water. Although field observations are necessary to verify the validity of these levels, a mechanism-based model of periphyton accrual represented the observed pattern of development in these artificial channels.

This study was conducted to verify previous field observations on lotic periphyton accumulation in relation to current velocity and nutrients through more thorough and controlled study under laboratory conditions. The principal objective was to gather data that might better explain the observed nutrient-velocity interactions and to determine nutrient concentrations critical to the development of nuisance growths. Investigation of nuisance lotic attached algal growth and the conditions creating it has been much less extensive than the consideration given those issues in the case of lentic phytoplankton. Most previous work concerning the factors affecting stream periphyton has centered on the role of either nutrients, current velocity, light, or temperature (Whitford & Schumacher 1964; McIntire 1966, 1968; Sperling & Hale 1973; Rodgers & Harvey 1976), with few reports of simultaneous study of more than one of the variables. Furthermore, although it is recognized that filamentous algal growths produce negative effects on lotic ecosystems analogous to those in lakes, there has been little study of the conditions for the development of such nuisance communities in streams, or even what constitutes a nuisance. Therefore, the role of nutrient enrichment is insufficiently understood to develop guidelines on critical concentrations (or loadings) of nutrients to prevent nuisance levels of algal biomass in running water systems.

Through research based on six Western Washington streams, Horner & Welch (1981) demonstrated interactions between velocity and limiting nutrient in controlling chlorophyll *a* (chl *a*) accumulation on flattened rock substrates. Velocity increase up to approximately 50 cm s^{-1} enhanced

Wetzel, R.G. (ed.), Periphyton of Freshwater Ecosystems.
 ISBN 90 6193 768 X.

diatom accrual when soluble reactive phosphorus (SRP) concentration exceeded 35 μg l^{-1}. At lower P concentrations, accumulation was largely a function of phosphorus, with no positive velocity effect in evidence. Accrual declined with successive velocity increments above 50 cm s^{-1}. It was hypothesized that velocity increase created two opposing effects: improved turbulent diffusion of nutrients to algal cell surfaces, predominant at velocities under 50 cm s^{-1}; and increasing frictional shear stress, which becomes dominant at higher velocities.

Materials and methods

Apparatus

Laboratory streams were constructed to permit growing periphyton under controlled nutrient, velocity, temperature, and light conditions. The laboratory stream system consisted of nine channels, a water supply reservoir and distribution tubing, drainage piping, three nutrient supply reservoirs and associated valving and tubing, a compressed air manifold with valving and tubing, and overhead lights. Each channel, illustrated in plan view by Fig. 1, was constructed of 6.4 mm thick Plexiglas and had dimensions of 1 m length, 12 cm height, and 20 cm total width. A partial center wall subdivided the channel to allow for recirculation of water. Curved pieces of Plexiglas were installed in each end to avoid abrupt flow direction changes and excessive eddies. Placement of the drain pipe 8 cm above the channel bottom maintained water depth. For quantitative sampling of periphyton, small Plexiglas plates were installed on the base in one race of each channel. The plates were held in place between Plexiglas ridges glued to the base.

Water supply was piped to the laboratory from a concrete-lined campus pond under gravity flow. It discharged into a 100 l Plexiglas reservoir, from where it was fed by gravity to the nine channels through individual 9.5 mm O.D. Nalgene tubes with wall thicknesses of 1.6 mm. The tubing was wrapped with duct tape to prevent algal photosynthesis and clogging. Water supply to each channel was controlled with an adjustable pinch cock set according to timed volume collection to provide a desired channel water residence time of 16 minutes. During the course of experiments, flow rates were checked approximately every five days and rarely had to be reset. The pond water was low in nutrients (soluble reactive phosphorus: 2–5 μg l^{-1}; nitrate + nitrite-nitrogen: 100–150 μg l^{-1}; dissolved inorganic carbon: 4–6 mg l^{-1}) and, thus, was well-suited for use in fertilization experiments.

Nutrient solution was supplied by gravity under constant head from three 20 l polypropylene carboys. Near the base of each carboy, three shut-off valves were inserted and glued in place. Three of the channels were supplied from each of the three carboys via 6.4 mm O.D. Nalgene tubes, each fitted with intravenous valves for flow control. In

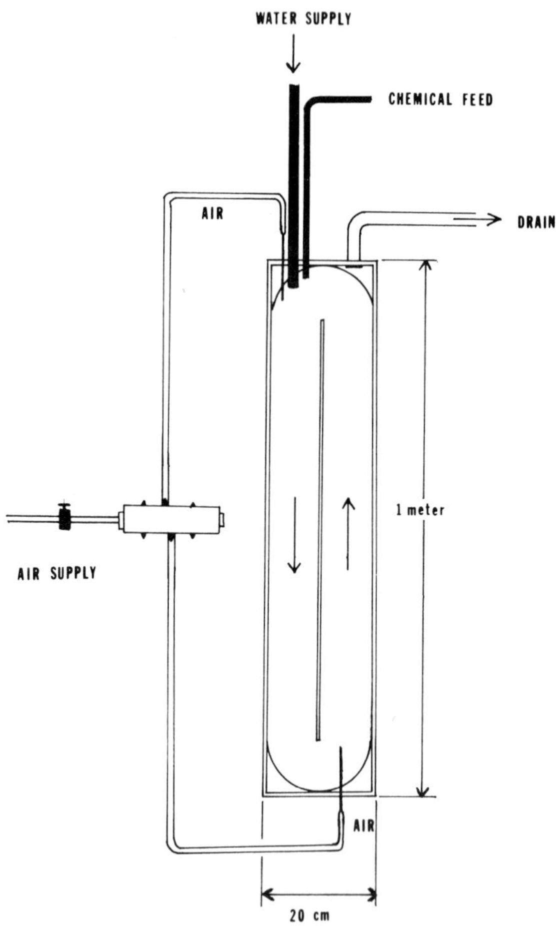

Fig. 1. Experimental stream design.

normal operation all valves were set at the same slow feed rate, according to timed drop tests, in order to conserve chemical solution. Nutrient solutions were made up with filtered pond water on the basis of mass balance calculations based on concentrations of each nutrient in pond water, target concentrations to the channels, and the set flow rates of feedwater and nutrient stock solution. The carboys were shrouded with black polyethylene sheeting, and the distribution tubing was wrapped with duct tape to prevent photosynthesis.

When the laboratory channels were first constructed, electric motor-driven Plexiglas paddlewheels, regulated by motor speed controls, were installed to create the current. After testing, the paddle-wheel system was rejected because of its inability to provide current velocity greater than 40–45 cm s^{-1} and frequent motor failures caused by attempts to attain and hold high velocities over an extended experimental period. Whereas larger motors, capable of higher speeds, theoretically could create higher velocities, the paddle-wheels caused excessive splash even at the velocities produced.

Velocities were achieved by a compressed air distribution system instead of the paddle-wheels. Air centrally supplied to the laboratory at 0.69 MPa pressure was fed through 19.1 mm O.D. rubber hose with 4.8 mm wall thickness to a manifold for each set of three channels. Manifolds were constructed of ABS plastic pipe (6 cm O.D. x 15 cm long), capped at one end and fitted with a hose connection at the other end. Air was distributed to each channel through two lengths of Nalgene tubing (6.4 mm O.D., 1.6 mm wall thickness). The two tubes were inserted at opposite ends of the channel, on opposite sides of the partial center wall, to create equal currents in both flow directions. The nutrient and air distribution systems permitted simultaneous experimentation involving nine treatments; three different nutrient concentrations were fed to each channel in groups of three, with each group flowing at a different velocity.

The current in each set of channels was controlled by flow control valves in the manifold supply lines. Velocity was set according to Gurley meter measurements taken in the portion of the channel where periphyton was sampled. Using a small Gurley meter, this technique proved adequate to measure the velocity actually affecting the algae, because the algal filaments cultivated under the experimental conditions extended into the flow from their surfaces of attachment. Once each air control valve was set, the air flow was monitored by checking a pressure gauge located in the main supply line.

The compressed air system was capable of generating current velocities from near zero to about 90 cm s^{-1}. Measurements demonstrated that velocities at a selected point in the channel could be set within five to ten percent of a target value and maintained at that accuracy over an extended period. Although greater variation occurred along the channel due to turbulence, the desired velocity could be set accurately at the point where periphyton growth was observed.

Light was provided by twelve 91.4 cm long General Electric Deluxe Warm White F 40/WWX fluorescent tubes, suspended over each set of three channels. This particular tube was selected because it provides relatively high intensity in the red wavelengths, which algae absorb preferentially (Wetzel 1975). The light fixtures were supported by a wooden frame at a height of 40 cm above the base of the channels. Mylar sheeting was mounted between the channels and the lighting to guard against splash. To maximize light availability, masonite sheets painted silver on the inside were placed on the four vertical sides of the lighting support frame to enclose each set of three channels. This lighting system provided an intensity of approximately 215 μ E m^{-2} s^{-1} at the channel base. No appreciable heating occurred within the enclosures.

Operating procedures

Six SRP and six velocity treatments were applied, as follows:

SRP – Ambient (~2), 8, 15, 25, 50, 75 μg l^{-1}

Velocity – 5, 25, 35, 45, 55, 75, cm s^{-1}

Experiments were conducted with all combinations of SRP and velocity, for a total of 36 dual treatments (four experiments of nine treatments each).

Nitrogen was added to the nutrient stock carboys in sufficient quantities to insure that P would be limiting in all experiments. Trace element algal bioassays, performed according to the standard algal assay procedure (Miller *et al* 1978), demonstrated no limitation by trace elements. There was no temperature control, but the four experiments were performed within an 80-day period in the summer and early autumn, such that mean temperatures for all experiments were within 5° C and temperatures always remained well above 15° C, in the optimum range for filamentous green algae (Bott *et al* 1973).

To seed the channels, rocks were gathered from a local stream. The rocks selected were of similar size and had a similar covering of periphyton, including filamentous green algae. One rock was placed in each channel for four days, during which time nutrient concentrations and velocities were the same in all channels. The experiments were then performed using the selected conditions for the next 17 days, a period shown to be adequate to attain maximum biomass in the 24-hour lighting but avoid extensive sloughing.

Sampling, analysis, and data reduction

During the 17-day experimental period, periphyton was sampled every two days from specific areas of the Plexiglas plates and channel side walls and from the circulating water (representing export). Periphyton collections were transported in the dark to the laboratory, suspended in a known water volume, homogenized, filtered (in duplicate), and frozen for later analysis of chl *a*. Chl *a* was determined according to the procedure outlined by Strickland and Parsons (1972), using a Turner model 110 fluorometer. Periphyton accumulation was expressed in terms of mg chl *a* m^{-2} of surface, while biomass in the water was expressed as μg chl *a* l^{-1} of water.

Specific areas of the Plexiglas plates were also sampled twice during each experiment for qualitative analysis of the taxonomic composition of the periphyton. These collections were preserved with Lugol's solution for later microscopic examination.

The pond feedwater and channel water were analyzed for SRP, total soluble phosphorus (TSP), total phosphorus (TP), nitrate + nitrite-nitrogen ($NO_3 + NO_2$-N), ammonia-nitrogen (NH_3-N), temperature, dissolved oxygen (DO), and pH, according to the following schedule:

Daily – temperature
Days 1, 3, 5, 9, 11, 13 – SRP
Days 5, 13 – TSP, TP, $NO_3 + NO_2$-N, NH_3-N
Day 3 – DO, pH

SRP was determined by the ascorbic acid method (American Public Health Association 1975). TP and TSP were analyzed by the same technique following persulfate digestion of unfiltered and filtered samples, respectively. $NO_3 + NO_2$-N analysis was performed using the automated cadmium-reduction method, while NH_3-N determinations were made according to the automated phenate method (APHA 1975). Both analyses were conducted on a Technicon AutoAnalyzer II.

Dissolved oxygen was determined by the Winkler method (azide modification) (APHA 1975), while pH values were obtained with an Instrumentation Laboratories pH meter. These quantities maintained similar values in all channels throughout all experiments, demonstrating that varying rates of compressed air introduction and levels of attached growth did not affect the quantity of oxygen available for uptake or the inorganic carbon equilibrium.

Light and velocity were measured throughout the experiments, the former with a Spectra Lumicon light meter and the latter with a Teledyne pygmy Gurley meter, each held approximately 1 cm above the Plexiglas plates.

The various treatments produced maximums in biomass after different periods of growth but always on day 10, 12, or 14. Because of the differing timing and fluctuations within the 10–14 day interval, maximum chl *a* attained was taken as the average of that measured on the three sampling dates. To evaluate the response of periphyton to the experimental treatments, maximum chl *a* accrual was plotted versus inflow SRP concentration and velocity. The average accumulation rate was computed as the quotient of maximum accrual and the time of growth.

It was noted that chl *a* concentrations in the circulating water (representing export) remained relatively uniform throughout the experiment for each treatment, despite a large increase in attached growth. From this observation it was concluded that export was largely a function of surface area covered, rather than the thickness of the covering (with the four-day seeding period, the channels were fully covered with thin algal coatings at the beginning of the experiments). Export rates (mg m^{-2} hr^{-1}) thus were estimated as functions of the measured chl *a* concentrations, the set water flow rates, and the area of coverage. With grazing absent, overall productivity was determined as the sum of accrual and export rates.

Results

Phosphorus utilization

In all cases a trend of decreasing SRP concentrations in the channels was observed over the first five days of the experiments. The algae apparently took up the available P during their exponential growth phase and depleted it in the channels until a point was reached in which release from decomposing cells balanced uptake or until uptake stabilized. Although SRP decreased within the channels, a consistent gradient in mean SRP between channels was maintained nevertheless. Fig. 2 illustrates the typical pattern, representing one of the four experiments.

Periphyton response, and hence effective P concentrations, were evaluated on the basis of the target inflow SRP concentrations, which were determined by means of phosphorus mass balance calculations and the set flow rates, as described and listed previously. This treatment implicitly assumes that the channels represent lengths of stream having 16-minute travel times which receive SRP concentrations equal to the set inflow concentrations at the upstream ends.

Biomass accumulation

The filamentous green alga *Mougeotia* dominated the periphyton in 30 of the 36 treatments. In the remainder of the cases, concentrated in the high nutrient and velocity ranges, a filamentous blue-green (*Phormidium* or *Lyngbya*) dominated. The results thus reflect the development of filamentous forms, which can represent a nuisance condition.

Figures 3 and 4 illustrate, respectively, the max-

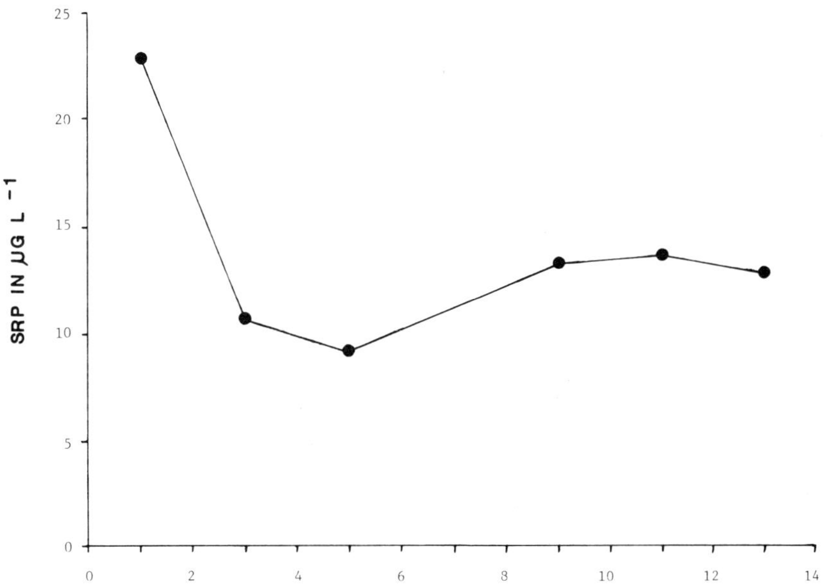

Fig. 2. Measured SRP concentration versus time for one treatment (velocity = 35 cm s^{-1}; inflow SRP = 25 $\mu g\ l^{-1}$).

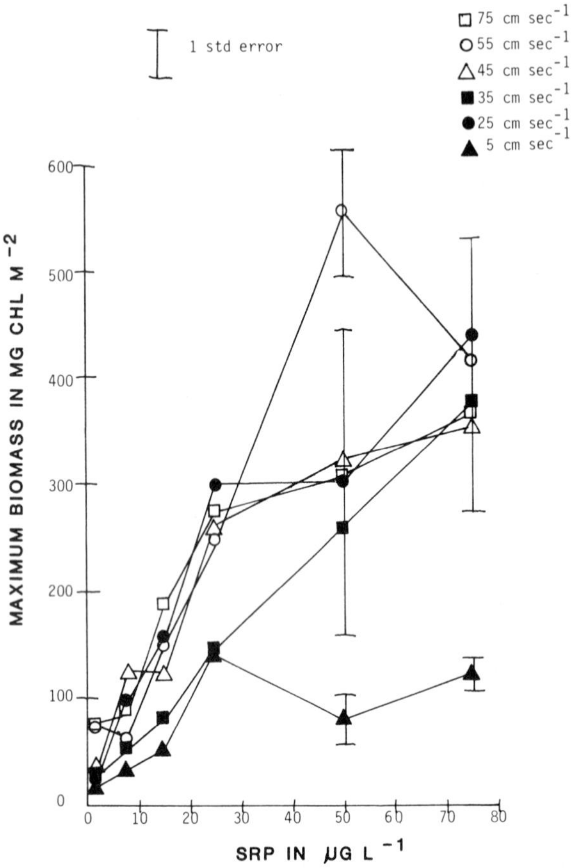

Fig. 3. Maximum biomass (as mg chl *a* m^{-2}) versus SRP concentration for six velocities.

imum biomass attained and average rate of accumulation versus inflow SRP for each velocity. Accumulation, and especially maximum biomass, increased with P at a relatively rapid rate below 25 μg l^{-1}. Increasing P above that level resulted in generally less rapid increases in both variables. In this region three distinct velocity groupings occurred, with maximum biomass and accumulation at 5 cm s^{-1} falling significantly below and that at 55 cm s^{-1} lying generally above the maximum chl *a* in the range 25–45 cm s^{-1}. At 75 cm s^{-1}, accumulation was comparable to that at 25–45 cm s^{-1}. These results indicate that SRP, in analogous experimental systems at least, must be reduced below 25 μg l^{-1} for substantial reductions in filamentous algal biomass accrual.

Maximum chl *a* accumulation is replotted in Fig. 5 relative to velocity, with SRP as a parameter. This graph points out that the most distinct increase in accrual occurred when velocity was raised from 5 to 25 cm s^{-1}. Beyond that velocity, a further trend in biomass accrual as a function of velocity was not apparent. Chl *a* increase, in general, was less pronounced with velocity increase than with P increase, which is considered a significant finding.

Export and productivity

Figure 6 presents the average loss rates over the full experimental periods as a function of SRP concentration for each velocity. Error bars ($\pm$1 standard error) have been included to illustrate the variation associated with each rate. It is apparent that export was not substantially influenced by either velocity or P differences and that it did not vary in any regular way with those factors. Variability was greatest at the lowest P concentrations ($\leq$25 μg SRP l^{-1}), where the range was 0.30–1.10 mg m^{-2} h^{-1}. At higher P concentrations the range was restricted

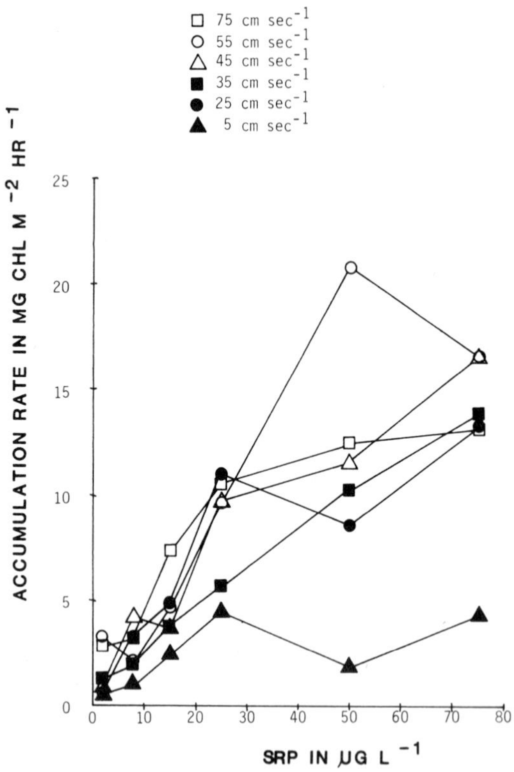

Fig. 4. Biomass accumulation rate (as mg chl *a* m^{-2} h^{-1}) versus SRP concentration for six velocities.

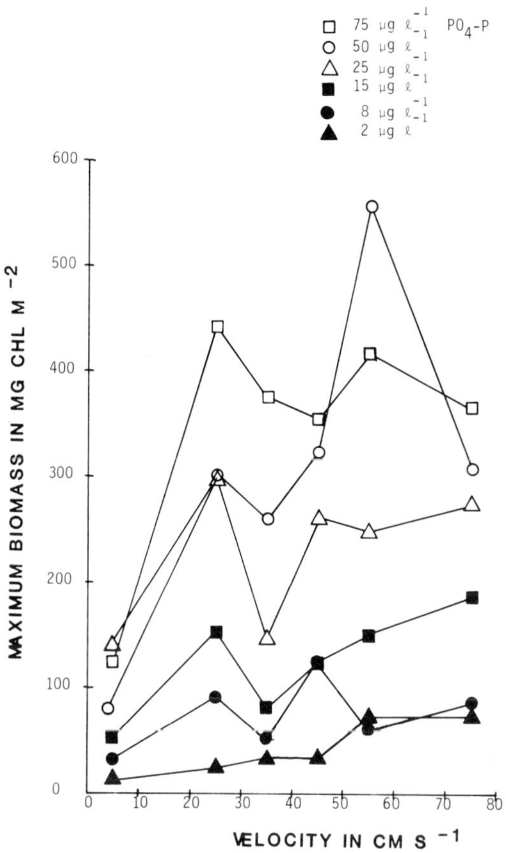

Fig. 5. Maximum biomass (as mg chl *a* m^{-2}) versus velocity for six SRP concentrations.

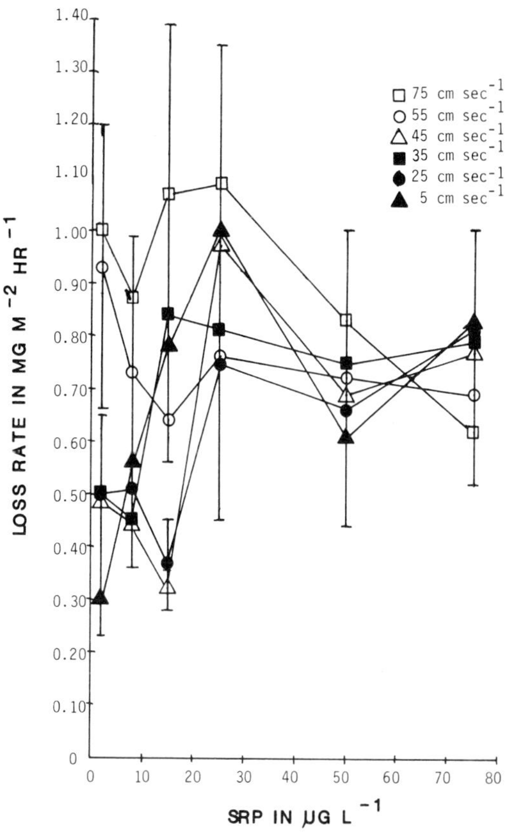

Fig. 6. Loss rate (as mg chl *a* m^{-2} h^{-1}) versus SRP concentration for six velocities.

to 0.60–0.85 mg m^{-2} h^{-1}, and no statistically significant differences occurred among velocities.

The overall productivity of the stream under each treatment was computed as the sum of chl *a* accumulation rate (Fig. 4) and export rate (Fig. 6). Since grazing did not occur in the laboratory channels, this aspect did not represent a loss. Fig. 7 presents plots of productivity versus inflow SRP concentration for each velocity tested. Since the export rates were small in comparison to accumulation rates and were fairly uniform among treatments, the plots are influenced more strongly by accrual than by export. With its similarity to the plot of accumulation rate versus SRP, Fig. 7 reinforces earlier observations concerning the response of periphytic algae to limiting nutrient and velocity increases and critical levels of those factors.

Because the expected strong influence of velocity on export was not evident in the results, an additional set of experiments was performed to verify the initial observations. This series employed each combination of 2, 25, and 75 μg SRP l^{-1} and velocities of 5, 35, and 75 cm s^{-1}, for a total of nine treatments. In addition to chl *a* analysis, particulate organic carbon (POC) content was determined (Beckman Model 865 infrared analyzer) on attached and exported periphyton samples. Otherwise, experimental procedures were identical to those in the preceding experiments.

Mean chl *a* export rates again failed to show a pattern relative to velocity or limiting nutrient variation. Chl *a* accumulation patterns were similar to those observed earlier. The same statements generally apply to POC, as well, except there was evidence of very high POC accumulation and productivity at low velocity (5 cm s^{-1}) and high SRP

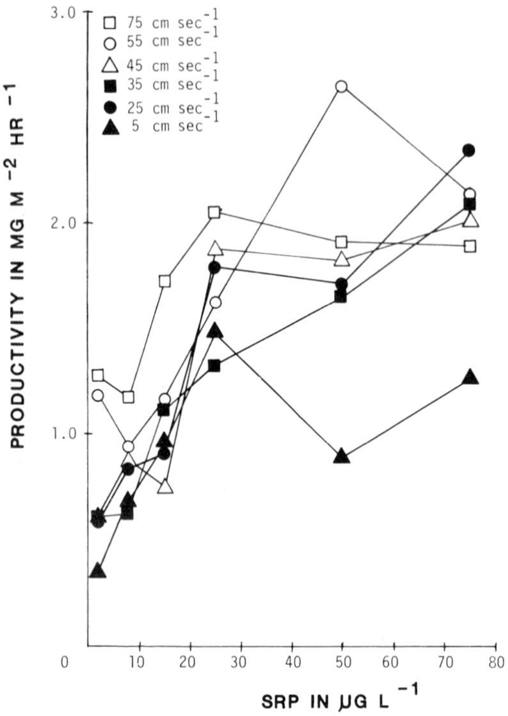

Fig. 7. Productivity (as mg chl *a* m^{-2} h^{-1}) versus SRP concentration for six velocities.

($75\,\mu g\ l^{-1}$). As a consequence, the mean chl *a*: carbon ratio at $75\,\mu g$ SRP l^{-1} declined from 0.083 at 75 cm s^{-1} to 0.046 at 5 cm s^{-1}, representing a statistically significant difference at $P < 0.001$. This result suggests accumulation of substantial heterotrophic or detrital material with the low velocity/high enrichment combination.

Periphyton accumulation model

The data were evaluated further by fitting with the following mechanism-based model of periphytic biomass accumulation rate proposed by Horner (1978):

$$\frac{dB}{dt} = K_1\, \mu L\, (k_f + k_{f_0})\, (B_{max} - B) - K_2 V^{\theta}$$

where: B is biomass (mg chl *a* m^{-2}), t is time (days); k_f is the mass transfer coefficient with turbulent diffusion (cm s^{-1}); k_{f_0} is mass transfer coefficient in a quiescent or laminar flow system (cm s^{-1}); μ is P uptake rate (day^{-1}); L is the light factor (dimensionless); B_{max} is the maximum sustainable biomass (mg chl *a* m^{-1}); V is velocity at the algal cell surface (cm s^{-1}); K_1, K_2, and θ are empirical factors derived from the data.

With initial condition B = 0 at t = 0, the integral form of the linear differential equation is:

$$B = [B_{max} - K_2 V^{\theta} / K_1\, \mu L\ (k_f + k_{f_0})]\ [1 - e^{-K_1 \mu L (k_f + K_{f_0}) t}]$$

Forms of k_f, k_{f_0}, μ, and L were selected from the literature. Table 1 summarizes the expressions or values selected by Horner (1978) for the various model components. Consideration of the conditions existing in this series of laboratory experiments yielded the following values for the respective components:

B_{max} = 560 mg chl *a* m^{-2}
μ_{max} = 1.92 day^{-1}
K_s = $7\,\mu g\ l^{-1}$
L = 0.755
k_{f_0} = 0.0094 cm s^{-1}

The data set was divided into a subset to recalibrate the model and a subset for verification. In calibration, constant values of 0.30 for K_2 and 0.45 for θ were found to be appropriate. However, K_1 varied with P concentration. At low SRP ($<15\,\mu g\ l^{-1}$), K_1 was essentially constant at 1.2. At concentrations of $25\,\mu g\ l^{-1}$ and above, linear regression was employed to establish a relationship $K_1 = 0.022\,P + 1.59$. The intermediate 15–25 μg SRP l^{-1} interval is a transition region between the constant and variable values.

Comparison of model predictions with actual biomass accrual during verification demonstrated reasonably good agreement, except at 5 cm s^{-1}, where predictions were uniformly high. Considering the hypothesized diffusion limitation at velocities <15 cm s^{-1} (Whitford 1960), it seemed inappropriate to use a turbulent mass transfer coefficient (k_f) for this range in the same form as employed at higher velocities. Close agreement with experimental data was achieved if turbulent diffusion was assumed negligible relative to thermal (non-turbulent) diffusion at low velocities, i.e. $k_f = 0$ and therefore D = 0 if V <15 cm s^{-1}. Fig. 8

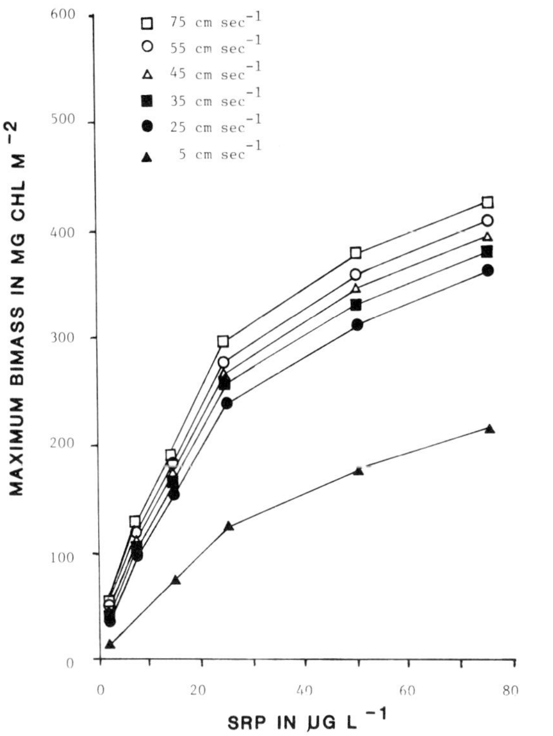

Fig 8. Periphyton accumulation model prediction of maximum biomass after 12 days of growth versus SRP concentration for six velocities.

presents predicted maximum chl *a* accumulation over a 12-day growth period versus SRP for the experimental velocity range, using the same scales as Fig. 3 to permit comparison.

Discussion

Biomass accumulation

These results show that, while the maximum biomass did increase with velocity, the response to increased nutrient levels was significantly more pronounced. The slopes of the lines in response to velocity increases were all rather consistent but relatively small throughout the range of velocities tested. On the other hand, biomass response to nutrient concentration increases was much more evident, both in magnitude and consistency. The ratio of the slopes of these lines (biomass versus nutrient: biomass versus velocity) was 7.7:1. It can be concluded that under these particular test conditions, nutrient concentration controlled the degree of biomass accumulation to a substantially greater degree than did velocity. This finding reinforces the conclusion of Sládečková (1964) that nutrients

Table 1. Summary of components of periphyton accumulation model (Horner 1978).

Symbol	Definition	Unit	Form or value[a]	Source
B_{max}	Maximum sustainable chl *a* biomass	mg m^{-2}	500	Data
K_1	Growth coefficient	—	1.1–2.7	Data
K_2	Scour coefficient	—	0.34–0.40	Data
θ	Velocity exponent	—	0.50–0.75	Data
μ	Nutrient uptake rate	day^{-1}	$\frac{\mu_{max} P}{K_s + P}$	Michaelis-Menten formulation
μ_{max}	Maximum uptake	day^{-1}	$0.22\ e^{T/10}$	Eppley (1972)
K_s	Phosphorus half-saturation constant	μg SRP l^{-1}	25	Data, with literature guidance
L	Light factor	—	$(R/Ro)e^{1-R/R_o}$	Steele (1962)
R_o	Optimum light intensity	μE m^{-2} s^{-1}	8	Data, with literature guidance
k_f	Mass transfer coefficient (turbulent diffusion)	cm s^{-1}	$(DV/\pi)^{0.5}$	Canale & Weber (1972)
D	Diffusion coefficient	cm^2 s^{-1}	$1.5 \cdot 10^{-5}$	Pasciak & Gavis (1974)
k_{f_o}	Mass transfer coefficient (non-turbulent condition)	cm s^{-1}	$0.009 \times (1.018)^{(T-20)}$	Canale & Weber (1972)

[a] T = temperature (... C); P = SRP concentration (μg l^{-1}); R = light intensity (μE m^{-2} s^{-1}); V = velocity (cm s^{-1}).

Table 2. Reported literature values of maximum periphyton biomass in natural and experimental streams.

Investigator	Stream type	Maximum biomass (mg m^{-2})	Reported as	Remarks
Bahls (1971)	Natural	104	Chl *a*	
		18,930	AFDW	
Beers & Newhold (1968)	Natural	960	Chl *a*	Downstream of sewage outfall
Brock (1967)	Thermal spring	687	Chl *a*	
Crawford (1981)	Growth chamber in natural stream	520	Chl *a*	With N enrichment 360μg l^{-1} above ambient
Cushing (1967)	Natural	22	Chl *a*	
		4,200	AFDW	
Dunn (1976)	Natural	335	Chl *a*	Productive
		67,710	AFDW	desert stream
Eichenberger (1975)	Experimental	40,000	AFDW	Sewage enrichment
		(200–400)[a]	(Chl *a*)	
Elwood & Nelson (1972)	Natural	658	AFDW	Mean
		(3.3–6.6)[a]	(Chl *a*)	5 μg TP l^{-1}
Elwood *et al* (1981)	Natural	810	AFDW	
		5.9	Chl *a*	
Erhlich & Slack (1969)	Experimental	10,400	AFDW	Inorganic enrichment
		(52–104)[a]	(Chl *a*)	
		13,900	AFDW	Organic enrichment
		(70–139)[a]	(Chl *a*)	
Ertl *et al* (1972)	Natural	330	Chl *a*	Danube River
		65,000	AFDW	
Flemer (1970)	Natural	111	Chl *a*	Maximum 110 μg SRP l^{-1}
Horner & Welch (1981)	Natural	420	Chl *a*	Mean 41 μg SRP l^{-1}
Kevern *et al* (1966)	Experimental	12,000	AFDW	
		(60–120)[a]	(Chl *a*)	
	Natural spring	5,200	AFDW	
		(26–52)[a]	(Chl *a*)	
King & Ball (1966)	Natural	3,800	AFDW	
		(19–38)[a]	(Chl *a*)	
Kobayasi (1961)	Natural	70	Chl *a*	
Lyford & Gregory (1975)		46	Chl *a*	
McConnell & Sigler (1959)	Natural	1,420	Chl *a*	Highly enriched
McIntire (1966)	Experimental	540	Chl *a*	38 cm s^{-1} current
		73,000	AFDW	
McIntire (1966)	Experimental	305	Chl *a*	35 cm s^{-1} current
		65,700	AFDW	700 ft-ca light
Nelson *et al* (1969)	Natural	3,000	AFDW	
		(15–30)[a]	(Chl *a*)	
Odum & Hoskins (1957)	Experimental	380	Chl *a*	Organic enrichment
Rodgers & Harvey (1976)	Experimental	66,710	AFDW	
Tezuka *et al* (1974)	Natural	680	Chl *a*	Organic enrichment
Veenstra (1982)	Experimental	560	Chl *a*	50 μg SRP l^{-1} 55 cm s^{-1} velocity
Yount (1957)	Natural spring	35	Chl *a*	

[a] Where only AFDW was given, chl *a* was estimated as 0.5–1 percent of the AFDW, based on ratios computed from data where both measures were given (Bahls 1971; Cushing 1967; Dunn 1976; Elwood *et al* 1981; Ertl *et al* 1972; McIntire 1966, 1968; Rodgers & Harvey 1976).

are most important among the factors influencing periphyton growth.

Velocity showed the greatest effect in the range of 5 to 25 cm s^{-1}. Increased velocity in this range noticeably increased biomass accrual and productivity when inflow SRP exceeded 15 μg l^{-1}. The velocity effect may exist at lower P concentrations but was not distinguishable from the more dominant nutrient influence. This result supports Whitford's (1960) contention that velocity increase above 15 cm s^{-1} enhances periphyton growth by increasing the nutrient concentration gradient and the consequent speed of diffusion from surrounding water to algal cells.

Some authors have noted increases in various measures of periphyton growth response in proportion to velocity increase up to approximately 40 cm s^{-1} (Whitford & Schumacher 1964; McIntire 1966, 1968; Sperling & Hale 1973; Rodgers & Harvey 1976). In this study accrual and productivity were similar at all velocities when SRP was <15–25 μg l^{-1} and similar in the velocity range 25–45 cm s^{-1} for higher P concentrations. Although accumulations generally were greater at 55 cm s^{-1} at the higher P concentrations, no proportional relationship between biomass accrual or productivity and velocity was evident. Biomass and productivity levels at 75 cm s^{-1} fell in the same range as those at 25–45 cm s^{-1}, which may indicate a saturating effect of nutrient at velocities exceeding 15 cm s^{-1}. Ball *et al* (1969) and Horner and Welch (1981) previously had noted a tendency for accrual rate to decline at high velocities under field conditions.

The fact that biomass accrual and productivity, in general, showed considerably less response to nutrient increases above 25 μg SRP l^{-1} than occurred at lower P concentrations indicates the phosphorus utilization rates of the organisms began to reach saturation at this point. From these results it appears that avoiding attached algal accumulations in nuisance quantities may require achieving inflow SRP concentrations below 25 μg l^{-1}, at least at velocities above 15 cm s^{-1}. This concentration was thus adopted as an initial approximation of a critical P level which may produce nuisance biomass levels in phosphorus-limited systems in the broad velocity range of 25–75 cm s^{-1}.

The identification of approximately 25 μg SRP l^{-1} as the saturation point conflicts with the results of Wuhrmann & Eichenberger (1975), who reasoned that lack of stimulation of periphyton accumulation in an artificial stream by fertilization with N and P was due to growth-saturation at the concentrations existing in the groundwater source employed (~8 μg SRP l^{-1} and 500–700 μg NO_3-N l^{-1}). More recently, Peterson *et al* (in press) documented enhancement of chl accumulation and CO_2 uptake by the addition of 10 μg l^{-1} SRP to an arctic tundra stream. They reasoned that Wuhrmann & Eichenberger's (1975) conclusions were accurate for their study but that P limitation would have become evident in a short distance if the same experiment were conducted in a natural stream. Another possibility is that a trace element was limiting in the water used by Wuhrmann & Eichenberger (1975).

General agreement exists on algal biomass levels that are associated with different trophic states in lakes, as well as the average nutrient concentrations and annual loadings which give rise to those states. Streams, however, have not been investigated with that objective; and nutrient criteria for trophic state classifications and nuisance algal levels have yet to be developed for running water. To address this question the literature was reviewed to determine the range in maximum periphytic biomass reported as chl *a* or ash-free dry weight (AFDW) in streams and, where given, the associated conditions (Table 2). Larger values typically were reported in experimental channels than in natural streams. This occurrence probably is the result of enriched conditions often created in experimental channels, the lack of periodic losses due to frictional scour, and more constant observation, which would insure that the maximums were recorded.

These results reveal that chl *a* >150 mg m^{-2} has been reported only in cases of enrichment above ordinary natural levels. Areal concentrations under 100 mg m^{-2} are rarely associated with enriched conditions, while those in the vicinity of 100–120 mg m^{-2} vary in their relation to enrichment. Maximum biomass levels <100 mg chl *a* m^{-2} probably would not pose a nuisance, while levels >150 mg chl *a* m^{-2}

most likely would. In light of these observations, a maximum periphyton biomass of 100–150 mg m^{-2} chl *a* may represent the onset of a potential nuisance situation. The validity of such a potentially nuisance biomass level in running water requires a thorough field verification.

In our laboratory channel experiments, a maximum accumulation of 150 mg chl *a* m^{-2} was attained in all currents at some SRP concentration. At some velocities it was reached at 15 μg SRP l^{-1}, whereas enrichment to 25 μg l^{-1} was required to achieve the hypothetical nuisance level at 5 cm s^{-1}. This observation suggests that the critical P concentration may have some velocity dependency but that it probably occurs in the fairly narrow range of 15–25 μg SRP l^{-1} regardless of velocity.

Export and productivity

It had been hypothesized that export rates would increase noticeably with velocity. This circumstance was observed by McIntire (1968) and Horner (1978). It is true that, unlike the other studies cited, the feedwater in these experiments was essentially free of suspended solids. Nevertheless, it was expected that velocity shear alone would cause measurable differences in export rate with velocity increase. Leopold *et al* (1964) have shown theoretically that shear stress increases with the square of velocity. In addition, there is a body of evidence in fluid mechanics that pipe wall shear stresses similarly increase with velocity (Vennard & Street 1975).

In the separate set of experiments conducted to reinvestigate export patterns, POC was analyzed along with chl *a* to determine if senescing algae were responsible for the surprisingly low loss rates of chl *a*. It was considered that exported material may not have been flushed from the channels at the same rate as water recirculated. Algae retained for any length of time may have undergone senescence and chl *a* deterioration. However, observations of POC generally confirmed the earlier findings: under the test conditions, periphyton export did not increase with velocity increase nor vary in any clear manner with phosphorus.

Two possible explanations exist for the link between export and velocity observed in natural systems and laboratory streams fed from a creek but not observed in the present work. First, the erosive action of suspended sediment may have a much greater influence on periphyton export than does velocity shear. Second, periphytic algae may adapt to the generally prevailing velocity (within limits). Losses observed in nature, then, would be more a function of an increase in velocity from the norm than the magnitude of the elevated velocity itself. Traaen & Lindstrøm (1983) have documented adaptations of certain periphytic algal species to high velocities (>80 cm s^{-1}) such that relatively large accumulations in patch distributions can result. Both phenomena no doubt play a role in periphyton export in natural communities. It is recognized (Ball *et al* 1969; Horner & Welch 1981) that more extensive erosion occurs during elevated storm flows than would be expected from consideration of velocity magnitude alone. Both abrasion and rather sudden velocity increase probably are responsible.

Periphyton accumulation model

The periphyton accumulation model presented here is not sufficiently developed for predictive use in management but has contributed to improved understanding by reinforcing certain conclusions arrived at through direct evaluation of the data. This reinforcement was provided by the ability of the mechanism-based formulation to reproduce the essential pattern of the observed biomass accrual.

As a case in point, the model represented the rapid increase in maximum biomass with P increase from 2 to 25 μg SRP l^{-1} at all velocities in a fashion very similar to the actual observations. The simulated patterns exhibited a discontinuity at 25 μg SRP l^{-1}, followed by substantially smaller response to further P increase, just as observed experimentally. The ability to reproduce this form with the mechanism-based model adds support to the tentative conclusion that SRP concentration of approximately 25 μg l^{-1} is a critical value for the development of nuisance periphytic algal biomass.

The notably lower simulated accrual at 5 cm s^{-1}

than at higher velocities also follows the experimentally observed pattern in form, although not exactly in magnitude. To achieve this agreement it was necessary to assume negligible turbulent diffusion at the low velocity. Success in approximating the observed pattern through manipulating the diffusion mechanism lends further credence to Whitford's (1960) assertion that diffusion limitation exists at velocities under 15 cm s^{-1}.

Acknowledgement

The research discussed was supported by National Science Foundation Award Number (CME) 79-18514.

References

American Public Health Association, 1975. Standard methods for the analysis of water and wastewater, 14th ed., Washington, American Public Health Association, Inc., 1193 pp.

Bahls, L.L., 1971. Ecology of the diatom community of the upper East Gallatin River, Montana, with in situ experiments on effect of current velocity on features of the Aufwuchs. Ph.D. dissertation, Montana State University, Bozeman, 145 pp.

Ball, R.C., Kevern, N.R. & Linton, K.J., 1969. The Red Cedar River report. II. Bioecology. Publications of the Museum of Michigan State University, Biology Series 4:107–157.

Beers, G.D. & Neuhold, J.M., 1968. Measurement of stream periphyton on paraffin-coated substrates. Limnology and Oceanography 13:559–562.

Bott, T.L., Patrick, R. & Vannote, R.L., 1973. The effects of natural temperature variations on riverine communities. *In* Effects and methods of control of thermal discharges. Report to Senate Committee on Public Works by U.S. Environmental Protection Agency, Serial No. 93–14, November 1973.

Brock, T.D., 1976. Relationship between standing crop and primary productivity along a hot spring thermal gradient. Ecology 48:566–571.

Canale, R.P. & Weber, Jr., W.J., 1982. Aeration and gas transfer. In: W.J. Weber, Jr. (ed.) Physicochemical processes for water quality control. New York, Wiley-Interscience, pp. 503–531.

Crawford, J.K., 1981. Limiting nutrients in streams: in situ trient enrichment experiments. Presented at American Society of Limnology and Oceanography Annual Meeting, Milwaukee, June 16, 1981.

Cushing, C.E., 1967. Periphyton productivity and radionuclide accumulation in the Columbia River, Washington, U.S.A. Hydrobiologia 29:125–139.

Dunn, R.W., 1976. Seasonal variations in periphyton, chlorophyll *a*, algal biomass and primary production in a desert stream. M.S. thesis, Department of Biology, Idaho State University, Pocatello, 312 pp.

Eichenberger, E., 1975. On the quantitative assessment of the effects of chemical factors on running water ecosystems. Schweizerische Zeitschrift für Hydrologie 37:21–34.

Elwood, J.W. & Nelson, D.J., 1972. Periphyton production and grazing rates in a stream measured with a ^{32}p material balance method. Oikos 22:295–303.

Elwood, J.W., Newbold, J.D., Trimble, A.F. & Stark, R.W., 1981. The limiting role of phosphorus in a woodland stream ecosystem: effects of P enrichment on leaf decomposition and primary producers. Ecology 62:146–158.

Eppley, R.W., 1972. Temperature and phytoplankton growth in the sea. Fishery Bulletin 70:1063–1085.

Ehrlich, G.G. & Slack, K.V., 1969. Uptake and assimilation of nitrogen in microbiological systems. ASTM STP 488. American Society for Testing and Materials, pp. 11–23.

Ertl, M., Juris, S. & Tomajka, J., 1972. Vorlaufige Angaben uber jahreszeitliche Veranderungen und die vertikale Verteilung des Periphytons im mittleren Abschnitt der Donau. Archiv für Hydrobiologie (Supplementband) 44:34–48.

Flemer, D.A., 1970. Primary productivity of the North Branch of the Raritan River, New Jersey. Hydrobiologia 35:273–296.

Horner, R.R., 1978. Stream periphyton development in relation to current velocity and nutrients. Ph.D. dissertation, Department of Civil Engineering, University of Washington, Seattle, 273 pp.

Horner, R.R. & Welch, E.B., 1981. Stream periphyton development in relation to current velocity and nutrients. Canadian Journal of Fisheries and Aquatic Sciences 38:449–457.

Kevern, N.R., Wilhm, J.L. & Van Dyne, G.M., 1966. Use of artificial substrata to estimate the productivity of periphyton. Limnology and Oceanography 11:499–502.

King, D.L. & Ball, R.C., 1966. A quantitative measure of Aufwuchs production. Transactions of the American Microscopical Society 85:232–240.

Kobayashi, H., 1961. Chlorophyll content in sessile algal communities of a Japanese mountain river. The Botanical Magazine (Tokyo) 74:228–235.

Leopold, L.B., Wolman, M.G. & Miller, J.P., 1964. Fluvial processes in geomorphology. San Francisco, W.H. Freeman, 522 pp.

Lyford, Jr. J.H. & Gregory, S.V., 1975. The dynamics and structure of periphyton communities in three Cascade Mountain streams. Verhandlungen Internationale Vereinigung Limnologie 19:1610–1616.

McConnell, W.J. & Sigler, W.F., 1959. Chlorophyll and productivity in a mountain river. Limnology and Oceanography 4:335–351.

McIntire, C.D., 1966. Some effects of current velocity on periphyton communities in laboratory streams. Hydrobiologia 27:559–570.

McIntire, C.D., 1968. Structural characteristics of benthic algal communities in laboratory streams. Ecology 49:520–537.

Miller, W.E., Greene, J.C., Merwin, E.A. & Shiroyama, T., 1978. Algal bioassay techniques for pollution evaluation. *In* Toxic materials in the aquatic environment. Seminar, Oregon State University, Water Resources Research Institute, Spring, 1978, Corvallis, SEMIN WR 027-78, pp. 9–16.

Nelson, D.J., Kevern, N.R., Wilhm, J.L. & Griffith, N.A., 1969. Estimates of periphyton mass and stream bottom area using phosphorus-32. Water Research 3:367–373.

Odum, H.T. & Hoskin, C.M., 1958. Comparative studies on the metabolism of marine waters. Publications of the Institute of Marine Science, University of Texas 5:15–46.

Pasciak, W.J. & Gavis, J., 1974. Transport limitation of nutrient uptake in phytoplankton. Limnology and Oceanography 19:881–888.

Peterson, B.J., Hobbie, J.E., Corliss, T.L. & Kriet, K. 1983. A continuous-flow periphyton bioassay: tests of nutrient limitation in a tundra stream. Limnology and Oceanography 28:583–591.

Rodgers, Jr. J.H. & Harvey, R.S., 1976. The effect of current on periphytic productivity as determined using carbon-14. Water Resources Bulletin 12:1109–1118.

Sládečková, A., 1964. Factors affecting the occurrence and stratification of sessile protozoans in artificial reservoirs. Technology of Water 8:483–490.

Sperling, J.A. & Hale, G.M., 1973. Patterns of radiocarbon uptake by a thermophilic blue-green alga under varying conditions of incubation. Limnology and Oceanography 18:658–662.

Steele, J.H., 1962. Environmental control of photosynthesis in the sea. Limnology and Oceanography 7:137–150.

Strickland, J.D.H. & Parsons, T.R., 1972. A practical handbook on seawater analysis. Bulletin of the Fisheries Research Board of Canada 167.

Tezuka, Y., Watanabe, Y., Hayashi, H., Fukunoga, S. & Aizaki, M., 1974. Changes in the standing crop of sessile microbes caused by organic pollution of the Tomagawa River. Japanese Journal of Ecology 24:43–49.

Traaen, T.S. & Lindstrøm, E.A., 1983. Influence of current velocity on periphyton distribution. In: R.G. Wetzel (ed.). Periphyton of Freshwater Ecosystems. Developments in Hydrobiology 17:97–99.

Veenstra, R.B., 1982. Effects of nutrients and velocity upon periphytic algal populations in laboratory streams. M.S. thesis, Department of Civil Engineering, University of Washington, Seattle, 99 pp.

Vennard, J.K. & Street, R.L., 1975. Elementary fluid mechanics, 5th ed., New York, John Wiley and Sons, Inc., p. 381.

Wetzel, R.G., 1975. Limnology. Philadelphia, W.B. Saunders Company, 743 pp.

Whitford, L.A., 1960. The current effect and growth of freshwater algae. Transactions of the American Microscopical Society 79:302–309.

Whitford, L.A. & Schumacher, G.J., 1964. Effect of a current on respiration and mineral uptake in Spirogyra and Oedogonium. Ecology 45:168–170.

Wuhrmann, K. & Eichenberger, E., 1975. Experiments on the effects of inorganic enrichment of rivers on periphyton primary production. Verhandlungen Internationale Vereinigung Limnologie 19:2028–2034.

Yount, J.L., 1956. Factors that control species numbers in Silver Springs, Florida. Limnology and Oceanography 1:286–295.

Authors' address:
R.R. Horner/E.B. Welch/R.B. Veenstra
Department of Civil Engineering
University of Washington
Seattle, WA 98195, U.S.A.

CHAPTER 17

Factors regulating the spatial and temporal distribution of *Cladophora* and *Ulothrix* in the Laurentian Great Lakes

MARTIN T. AUER, JAMES M. GRAHAM, LINDA E. GRAHAM and JAMES A. KRANZFELDER

Abstract. The attached filamentous green algae *Ulothrix zonata* and *Cladophora glomerata* are important members of the periphyton community in the rocky littoral zone of the Laurentian Great Lakes. When these algae occur together, *Ulothrix* occupies the splash zone and *Cladophora* grows in deeper water. The two algae show definite seasonal patterns of abundance as well as distinct geographic distributions within the Great Lakes basin. We examined the role of light, temperature, and phosphorus supply in regulating the geographic, seasonal, and spatial distribution of these two filamentous algae. Phosphorus availability plays a major role in determining the geographic distribution of attached algae in the Great Lakes. Light intensity, acting through the balance between photosynthesis and respiration, appears to have a significant effect upon the pattern of vertical zonation. Optimum photosynthesis in *Ulothrix* occurs at high light levels (1200 $\mu E/m^2 \cdot sec$), while the optimum for *Cladophora* lies at 300 $\mu E/m^2 \cdot sec$. Seasonal patterns of abundance in *Ulothrix* and *Cladophora* are consistent with their respective temperature optima for photosynthesis. It is suggested, however, that the timing and extent of reproduction in *Ulothrix* are also of importance in determining the shift in dominance of the periphyton community from *Ulothrix* to *Cladophora*.

Introduction

In the rocky littoral zone of the Laurentian Great Lakes there are two major species of attached filamentous green algae. The unbranched, green alga *Ulothrix zonata* occurs in cold water on wave-swept rocks at the air-water interface. *Cladophora glomerata* is a branched, green alga which can form massive growths as a result of phosphorus enrichment of nearshore waters. When the two algae occur together, *Ulothrix* occupies the splash zone and *Cladophora* grows in deeper water. The distribution of these two algae can be examined from several aspects: 1) geographical distribution at sites throughout the Great Lakes basin, 2) seasonal distribution at a specific site where the algae vary in their abundance, and 3) vertical zonation patterns at a specific site and time. Vertical zonation patterns appear to be comparable to algal zonation in marine intertidal zones. Our investigations have been directed toward the physical and chemical factors which regulate the distribution of *Ulothrix* and *Cladophora*.

Although *Ulothrix zonata* is one of the most common filamentous algae in the Great Lakes, published reports on the genus are sparse. Observations on the seasonal and spatial distribution of *U. zonata* have been made during floristic surveys and studies of *Cladophora*. *Ulothrix zonata* was the most common filamentous green alga reported in a survey of the north shore of Lake Superior (Parker 1979). Herbst (1969) described a seasonal growth pattern in Milwaukee Harbor on Lake Michigan in which *Ulothrix* was dominant along breakwalls un-

Wetzel, R.G. (ed.), Periphyton of Freshwater Ecosystems.
 ISBN 90 6193 768 X.

til the end of June when increasing water temperatures permitted replacement by *Cladophora*; *Ulothrix* was again dominant by late September. In western Lake Erie, Taft & Kishler (1973) observed that cold-water dominant *Ulothrix* was replaced by *Cladophora* when water temperatures reached 10° C in spring; *U. zonata* resumed growth in shallow water in the fall. A similar seasonal pattern was reported by Moore (1978) and by Kirby & Dunford (1981) at sites close to a power generating station on the northeastern shore of Lake Erie, where *Ulothrix* persisted together with *Cladophora* at water temperatures up to 15 °C. At Harbor Beach, Michigan on Lake Huron, Auer *et al* (1982) reported that *Ulothrix* first occurred on March 25 near the discharge point of a wastewater treatment plant. *Cladophora* did not appear at this site until 1 May when *Ulothrix* growth was peaking. After 1 May water temperatures rose above 10 °C, and *Cladophora* expanded its range and displaced *Ulothrix* from deeper waters. By late May *Ulothrix* was confined to the splash zone of rocks or close to the shore. These field observations suggest that *Ulothrix* grows well at low temperatures and in shallow waters where light intensity is high, but no quantitative studies of growth or photosynthesis as functions of light intensity and temperature have been reported.

Field observations may be interpreted to suggest that *Ulothrix* is relatively efficient at acquiring nutrients such as phosphorus because this alga occurs in the oligotrophic waters of Lake Superior (Parker 1979), Lake Winnipeg (Evans & Stockner 1972) and Lake Tahoe (Smith & Ludwig 1968) as well as more eutrophic waters. In the cold, phosphorus-poor waters of Lake Superior, Parker (1979) observed that although *Cladophora* was a common component of the periphyton community in streams and near stream mouths, it was conspiciously absent from the littoral zone. Parker & Drown (1982) found that phosphorus enrichment stimulated *Ulothrix* growth in Lake Superior and observed an abundant growth of *Ulothrix* near the discharge of the Grand Marais, Minnesota, wastewater treatment plant, where high levels of soluble orthophosphate were detected. It is interesting to note that *Ulothrix,* rather than *Cladophora*, is the dominant littoral filamentous alga in Lake Superior, even at sites receiving significant phosphorus enrichment. In the other Great Lakes, however, when *Cladophora* is observed at a site, *Ulothrix* will also be observed as a seasonal component of the littoral community. *Cladophora* displaces *Ulothrix* from deeper waters when temperatures rise above 10°–15° C, and such displacement produces seasonal patterns of abundance and vertical zonation. Auer & Canale (1980, 1982a) have shown that phosphorus limits the growth of *Cladophora* in Lake Huron. Where phosphorus is limiting, species differences in nutrient kinetics are widely believed to influence algal competition (cf. Tilman & Kilham 1976). The role of nutrients in regulating the relative abundance of *Cladophora* and *Ulothrix* in Lake Huron (and the other upper Great Lakes) can be examined through an evaluation of phosphorus uptake kinetics.

The relationship between dissolved phosphorus concentration and phosphorus uptake rate is generally described by a mathematical expression developed by Michaelis & Menten (1913) for enzyme-substrate phenomena:

$$\varrho = \hat{\varrho} \cdot (S/(K_m + S)) \qquad (1)$$

where:

ϱ = phosphorus uptake rate (10^{-8}gP/mgDW · h),
$\hat{\varrho}$ = maximum phosphorus uptake rate (10^{-8} gP/mgDW · h),
S = external phosphorus concentration (μgP/L),
K_m = half-saturation constant for uptake as a function of external phosphorus concentration (μgP/L).

This relationship describes the substrate driven portion of the uptake process. The potential for interspecific nutrient competition may be understood through a comparison of the values of the kinetic coefficients $\hat{\varrho}$ and K_m. If one alga has a lower value of K_m than another, the first alga will reach a higher rate of uptake at a given level of external phosphorus even if their $\hat{\varrho}$ values are identical. Similarly, if two algae have the same value of K_m, the alga with the higher value of $\hat{\varrho}$ would reach a higher uptake rate at any given external phosphorus level. Values for these kinetic coefficients vary widely among species, and even change with the physiological state of the organism. Tilman & Kilham (1976) have examined the ability of two species of planktonic algae (*Asterionella formosa* and *Cyclotella meneghiniana*) to acquire and utilize nutrients. Predictions of competitive dominance of each organism were based on the results of an analysis of nutrient kinetics under conditions of phosphorus and silicate limitation. Phosphorus uptake kinetics have recently been studied in *Cladophora* (Auer & Canale 1982a, b), but there are no previous reports of phosphorus kinetics in *Ulothrix* for comparison.

Cladophora glomerata is the most extensively studied filamentous green alga in the Great Lakes. Many important aspects of *Cladophora* ecology have been reviewed by Bellis & McLarty (1967) and Whitton (1970). Nutrient requirements have been investigated by Gerloff & Fitzgerald (1976), Lin (1977), Mantai (1978) and Auer & Canale (1980, 1982a, b). The role of light and temperature in *Cladophora* growth has been studied by Bellis (1968), Storr & Sweeney (1971), and Gerloff & Fitzgerald (1976). Photosynthesis has been examined by Jackson (1966), Adams & Stone (1973), Lester *et al* (1974), Mantai (1974), Wood (1975), Mantai & Haase (1977) and Graham *et al* (1982).

In response to conditions of nuisance growth of *Cladophora* in the Great Lakes, a three year program of ecological studies was recently completed under the auspices of the United States Environmental Protection Agency, Large Lakes Research Station at Grosse Ile, Michigan, and the University of Michigan. A series of papers was prepared covering field studies (Auer *et al* 1982), phosphorus kinetics (Auer & Canale 1982a, b), photosynthesis and respiration (Graham *et al* 1982), and mathematical model development and verification (Canale & Auer 1982a, b; Canale *et al* 1982). The following general conclusions are derived from laboratory experimentation, field observations and model simulations: 1) seasonal variation in *Cladophora* growth rate is regulated by the effects of light and temperature on photosynthesis and respiration, 2) onset of growth, summer die-off, and fall resurgence of *Cladophora* are consistent with energy balance concepts developed in the mathematical model, 3) spatial variation in *Cladophora* growth is regulated by the local supply of phosphorus where lakewide nutrient levels are not sufficient to support growth 4) optimal depth for growth is determined by the balance between enhanced respiration in shallow water and light limitation in deeper water and varies seasonally as environmental conditions change, and 5) where high levels of biomass are produced, growth can be reduced by carrying capacity (self-shading) limitations.

To what extent do these conclusions for *Cladophora* also apply to *Ulothrix zonata*? We would hypothesize that the temperature optimum for photosynthesis in *Ulothrix* would be consistent with its geographic and seasonal distribution; thus optimum photosynthesis would occur at colder temperatures for *U. zonata* than for *C. glomerata*. The dominance of *Ulothrix* in the littoral zone of Lake Superior, even where phosphorus is abundant, strongly suggests that temperature is a major factor regulating its distribution. Because *Ulothrix* normally occurs in shallow water and in the splash zone, we could also hypothesize that optimum photosynthesis occurs at higher light levels than for *Cladophora* [which had an optimum depth of one meter at Harbor Beach on Lake Huron (Canale *et al*, 1982)]. It is clear that light and temperature may play a major role in determining seasonal and vertical abundance of *Ulothrix* and *Cladophora* at a specific site. Nutrient abundance in the well-mixed nearshore region probably has little role in regulating vertical distribution patterns. In lakes exhibiting marked seasonal patterns of phosphorus and nitrogen, nutrients may partially regulate species succession.

Nutrient availability is extremely important in determining the geographical distribution of *Ulothrix* and *Cladophora* within the Great Lakes. In lakes Superior, Huron and Michigan, where phosphorus levels are insufficient to support extensive algal growth, the geographic distribution of *Cladophora* is restricted to local sites with an abundant supply of phosphorus. In Lake Superior, *Cladophora* growth is additionally restricted by low water temperatures (midsummer maximum about 10° C). If *Ulothrix* does possess more efficient phosphorus uptake kinetics, as implied by field observations, we could hypothesize that *Ulothrix* would dominate at phosphorus-poor sites in lakes Huron and Michigan. In the phosphorus-rich waters of lakes Erie and Ontario, light and temperature play the major role in determining seasonal distribution patterns. Our studies of the effects of light intensity, temperature, and phosphorus on the seasonal and spatial distribution of *Cladophora* now permit us to begin comparable research on the factors regulating the growth of *Ulothrix* in the Laurentian Great Lakes.

Materials and methods

Cladophora glomerata (L.) Kütz and *Ulothrix zonata* (Weber & Mohr) Kütz were isolated from Lake Huron at Harbor Beach, Michigan in spring, 1979. Filaments were freed of algal epiphytes by washing and dragging through agar and were cultured in a defined medium called D11 (Graham *et al* 1982). Stock cultures were maintained in a constant temperature room at 20° C in aerated 500-ml flasks containing 250 ml of D11 medium. Sylvania cool white fluorescent lamps (F96T12CW) provided a light intensity of 25 $\mu E/m^2 \cdot sec$ (PAR). The light-dark cycle was set at 16:8 for *Cladophora* and 12:12 for *Ulothrix*.

Experiments were conducted in a controlled environment facility (BIOTRON) at the University of Wisconsin-Madison. The crossed-gradients room in the BIOTRON is especially suited to carry out research programs requiring a wide range of light and temperature conditions. Measurements of net photosynthesis and respiration were carried out in the crossed-gradients room, permitting simultaneous operation of four temperature corridors at right angles to a gradient of light intensities. Temperatures examined were 1°, 5°, 10°, 15°, 20°, 25°, 30° and 35° C. The nominal values of light intensities used were 1200, 600, 300, 150, 75, 35, 25 and 10 $\mu E/m^2 \cdot sec$ measured with a spherical quantum sensor (Lambda Instruments). Experiments were conducted in two phases as the crossed-gradients room can maintain only four temperatures at one time. Fiberglass screening and cheesecloth were used to adjust incident light to desired experimental levels. A 16:8 light-dark cycle was used for all experiments, however, culture tubes containing *Ulothrix* were covered with boxes after 12 hours of light to inhibit zoospore production. Temperature was monitored hourly by six shielded thermocouples located in each temperature corridor and connected to a Honeywell strip chart recorder. Temperature control was within 1.5° C at each experimental level. Three replicate algal cultures in aerated test tubes containing 50 ml of medium were placed at every light-temperature combination for each algal species. The algae were preconditioned under experimental conditions in the crossed-gradients room for at least one week before measurements began.

Net photosynthesis and respiration were determined by measurement of oxygen evolution and consumption using an Orion Instruments oxygen electrode. Measurements were carried out in cylindrical Plexiglas chambers, filled with fresh culture medium and equilibrated to the corridor temperatures. The chambers are divided internally by a perforated plate and range in volume from 277 to 241 ml. The algae from one culture tube were transferred to one side of the chamber, the top sealed, and the oxygen electrode inserted through a small opening into the opposite side of the chamber. The chamber and electrode were placed on a magnetic stirrer, and a screen-covered frame was lowered over the unit to maintain the desired light intensity. The oxygen electrode was connected to an expanded scale pH meter calibrated to read directly in mgO_2/liter. Readings were made at 5-minute intervals for at least 30 min to obtain a regression line with slope in $mgO_2/L \cdot min$. Respiration rates were measured with the same apparatus after net photosynthesis rates were recorded by covering the chamber and electrode with a light-tight container. The algae were then removed from the chamber, wrapped in labelled foil packets, and plunged into liquid nitrogen. After freeze-drying, final dry weights were determined for all algal samples. The regression slopes for net photosynthesis and respiration were then corrected for chamber volume and final dry weight to yield rates in mgO_2/g dry weight· h. Gross photosynthesis was calculated as the sum of net photosynthesis and respiration.

Studies of phosphorus kinetics were made with field collections of *Ulothrix* and *Cladophora*. The primary field collection site was Harbor Beach, Michigan, on Lake Huron. Additional data for phosphorus kinetics in *Ulothrix* were obtained from samples collected at Eagle Harbor, Michigan, on Lake Superior. The internal phosphorus content of Lake Huron collections of *Cladophora* and *Ulothrix* varied predictably with distance from Spring Creek, the primary source of phosphorus at the Harbor Beach site (Auer & Canale 1982a). Algal material was collected at several sites to represent a range of internal phosphorus concentrations. Samples were removed from the substrate and transported to a field-site laboratory in beakers containing lake water from the sample location; travel time was less than 30 minutes. Algal material was examined microscopically for the presence of epiphytes and zoospores which might influence the results.

Experiments were conducted under constant conditions of light (309 μE/m$^2 \cdot$ sec) and temperature (19±2°C). Two-liter glass beakers, divided internally by a perforated Plexiglas plate and equipped with magnetic stir bars, were filled with a measured volume of filtered water from Lake Huron. The water was obtained from a site approximately 1 km offshore where soluble reactive phosphorus in the filtered lake water was less than 1 μgP/L. A standard addition of phosphorus as K_2HPO_4 was made at levels of 10, 20, 40, 60, 80, 100, 300, 500, 800, and 1200 μgP/L. Each beaker was thoroughly mixed and a preliminary sample taken. Algal material was set aside for analysis of initial internal phosphorus and then a portion was added to each beaker. The amount of algae varied from 30 to 150 mg dry weight depending on the phosphorus level in the beaker. Internal phosphorus was measured on dried (62–64µC) algal material after grinding and digestion according to the procedures of Gerloff & Fitzgerald (1976).

Aliquots of water were withdrawn from the beakers at 20-min intervals and analyzed for soluble reactive phosphorus using the ascorbic acid method (APHA, 1975). Soluble reactive phosphorus concentration was plotted against time to determine the uptake velocity (gP/L·h). The *initial* slope of the resulting phosphorus depletion curve for each substrate level was used to calculate uptake velocity because changes in substrate and internal phosphorus during the course of the experiment would alter rates. The calculated velocity was multiplied by the container volume and divided by the dry weight of algal material to yield the phosphorus uptake rate (gP/mgDW $\cdot$ h $\pi 10^{-8}$). The phosphorus uptake rate can be converted to % P/day by multiplying by $0.024 \pi \mathrm{x} 10^8$. Values of the maximum phosphorus uptake rate $\hat{\rho}$ and the half-saturation constant K_m at each level of internal phosphorus were calculated using a computer program designed for fitting hyperbolic functions (Bliss & James 1966).

Results

The BIOTRON experiments produced data on gross and net photosynthesis and respiration for *Cladophora* and *Ulothrix* at 64 different combinations of light intensity and temperature. A detailed presentation of the results for *Cladophora* has been reported previously (Graham *et al* 1982). The *Cladophora* data can be summarized graphically as a set of response surfaces in which the vertical axis represents gross photosynthesis, net photosynthesis, or respiration and the horizontal axes are tem-

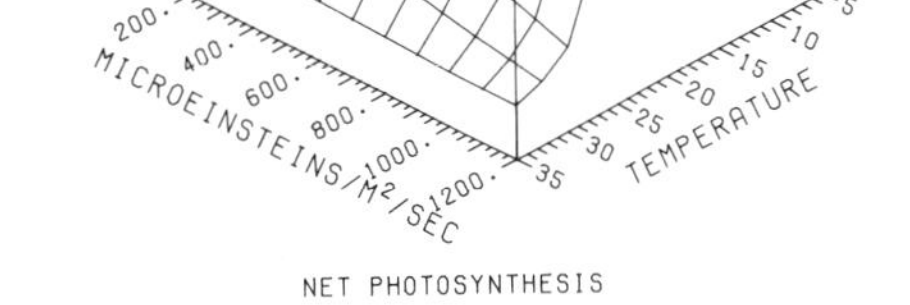

Fig. 1. Response surface for net photosynthesis (mgO_2/gDW $\cdot$ h) as a function of light intensity (μE/m$^2 \cdot$ s) and temperature (°C) in *Cladophora glomerata*.

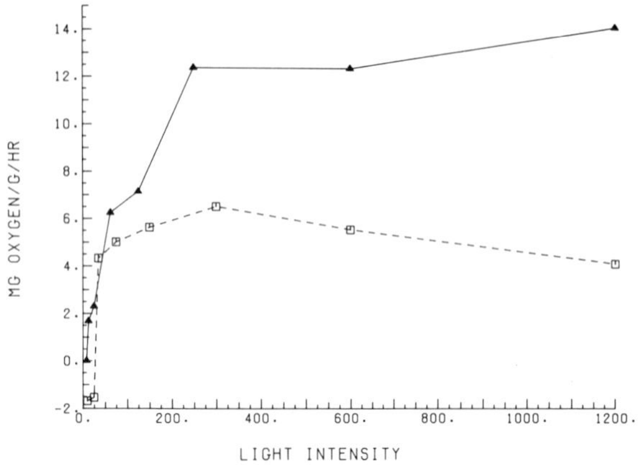

Fig. 2. Net photosynthesis (mg O_2/gDW · h) in *Cladophora* (□) and *Ulothrix* (▲) as a function of light intensity (μE/m^2 · s). Each data point is the average of measurements at 10°, 15° and 20° C.

perature and light intensity. The response surface for net photosynthesis in *Cladophora* is shown in Fig. 1. The response surface is seen from the direction of highest light intensity and temperature and is also tilted 30° toward the viewer. A line has been added at zero net photosynthesis as a guide. The lines running parallel to the light intensity axis represent net photosynthesis at each 2.5° C interval from 0° to 35° C. The lines paralleling the temperature axis represent net photosynthesis at each 100 μE/m^2 · sec interval from 1200 to 300 μE/m^2 · sec. Additional lines were added between 300 and 0 μE/m^2 · sec.

Optimal conditions for net photosynthesis in *Cladophora* occurred between 13° and 17° C and between 150 and 600 μE/m^2 · sec (Figs. 1, 2 & 3). Net photosynthesis drops off rapidly with decreasing light levels and with either an increase or decrease in temperature. Net photosynthesis is negative at all light levels at 35° C (Fig. 1). The net photosynthesis data in Fig. 2 were averaged over 10°, 15° and 20° C at each light intensity. Net photosynthesis in *Cladophora* is negative at 10 and 25 μE/m^2 · sec but reaches 4 mgO_2/gDW · h at 35 μE/m^2 · sec. The compensation point therefore lies between 25 and 35 μE/m^2 · sec. Maximum net photosynthesis occurs at 300 μE/m^2 · sec. In Fig. 3 the values of net photosynthesis at each temperature represent the average of measurements at 150, 300, and 600 μE/m^2 · sec. Net photosynthesis is negative at 1° C and reaches a maximum at 15° C.

A preliminary analysis of net photosynthesis data for *Ulothrix zonata* is presented with comparable data for *Cladophora* in Figs. 2 and 3. Values shown for *Ulothrix* are averages for 10°, 15° and 20° C (Fig. 2) and 125, 250, and 600 μE/m^2 · sec (Fig. 3). *Ulothrix* clearly had much higher average rates of net photosynthesis than *Cladophora* (Fig. 2). The maximum value of net photosynthesis for *Ulothrix* occurred at the highest light intensity used (1200 μE/m^2 · sec) in contrast to the *Cladophora* maximum at 300 μE/m^2 · sec. At the lowest light intensity (8 μE/m^2 · sec), net photosynthesis in *Ulothrix* was just barely positive while that in *Cladophora* was strongly negative. *Ulothrix* exhibited a higher net productivity than *Cladophora* at all light intensities. Net photosynthesis in *Ulothrix* as a function of temperature is shown in Fig. 3. At temperatures from 1° to 20° C, net photosynthesis in *Ulothrix* was distinctly higher than in *Cladophora*, especially at 1° and 5° C. Net photosynthesis was a maximum at 5° C but remained high up to 20° C. Between 20° and 25° C net photosynthesis in *Ulothrix* dropped below rates for *Cladophora* and became negative at 30° C. Figure 3 suggests that *Ulothrix* is a true cold-water alga.

A detailed discussion of phosphorus uptake kinetics for *Cladophora* has been reported by Auer &

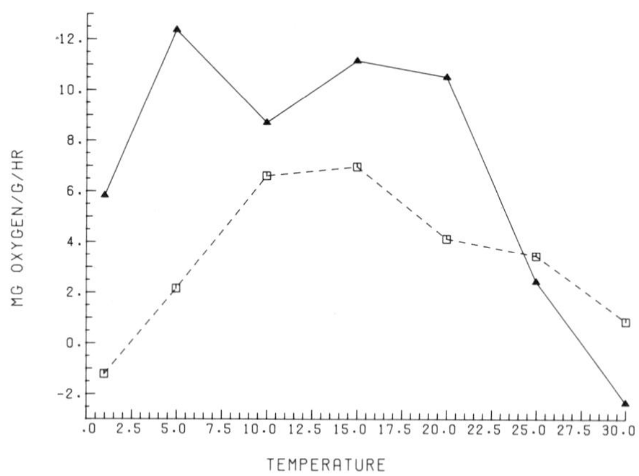

Fig. 3. Net photosynthesis (mg O_2/gDW · h) in *Cladophora* (□) and *Ulothrix* (▲) as a function of temperature. Each data point is the average of measurements made at 150, 300 and 600 μE/m^2 · s.

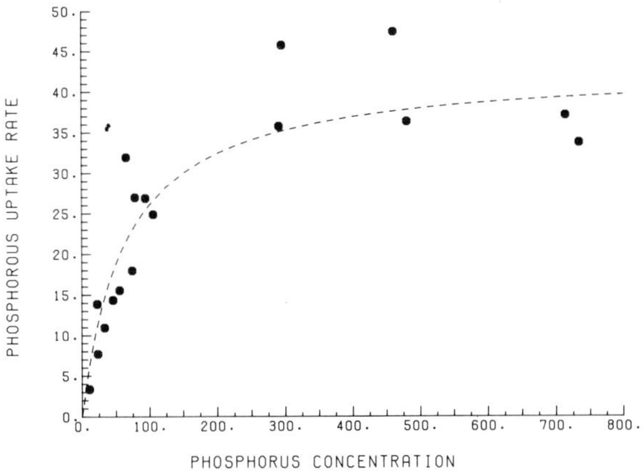

Fig. 4. Phosphorus uptake rate (10^{-8} gP/mgDW · h) as a function of external phosphorus concentration (μgP/L) in *Cladophora* at an internal phosphorus level of 0.23%P. The dashed line is the fitted hyperbolic uptake function with $\hat{\varrho} = 43 \times 10^{-8}$ gP/mg DW · h and $K_m = 65$ μg P/L.

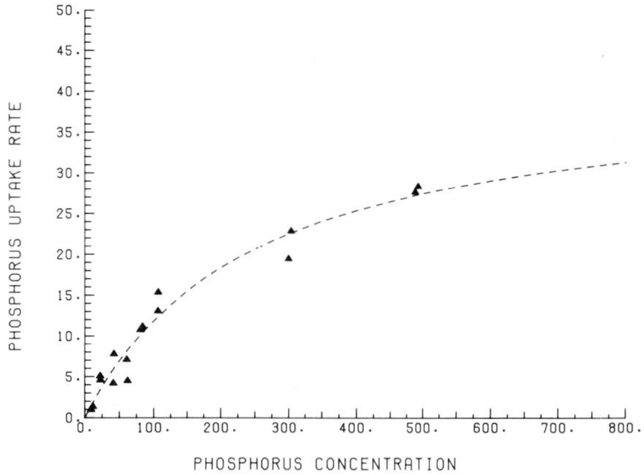

Fig. 5. Phosphorus uptake rate (10^{-8} gP/mgDW · h) as a function of external phosphorus concentration (μg P/L) in *Ulothrix* at an internal phosphorus level of 0.21% P. The dashed line is the fitted hyperbolic uptake function with $\hat{\varrho} = 41 \times 10^{-8}$ gP/mg DW · h and $K_m = 245$ μg P/L.

Canale (1982a, b). Preliminary phosphorus uptake data for *Ulothrix* is presented here for the first time. Phosphorus uptake experiments were conducted over a range of internal phosphorus levels. Both the maximum phosphorus uptake rate (ϱ) and the half-saturation constant (K_m) were found to vary with the level of internal phosphorus. High phosphorus uptake rates and high half-saturation constants were associated with low levels of internal phosphorus. Values of both uptake constants decreased as internal pool size increased. Phosphorus uptake rates ranging from 21 to 188 × 10^{-8} gP/mg DW · h were measured for *Cladophora* over an array of internal phosphorus levels. Rates for *Ulothrix*, measured over a more limited range of internal phosphorus levels, varied from 14 to 133 10^{-8} gP/mg DW · h. Values for the half-saturation constant ranged between 30 and 250 μgP/L for *Cladophora* and 35 and 349 μgP/L for *Ulothrix*.

A direct comparison of phosphorus kinetics in *Cladophora* and *Ulothrix* can be made by examining uptake coefficients at similar internal phosphorus levels. In Fig. 4, the phosphorus uptake rate is shown as a function of external phosphorus concentration for *Cladophora* cells with an internal phosphorus concentration of 0.23% P (0.23 gP/100 gDW). The value of the maximum phosphorus uptake rate ($\hat{\varrho}$) was 43 × 10^{-8} gP/mg DW · hr, and the half-saturation constant K_m was 65 μgP/L. Similar data for *Ulothrix* are illustrated in Fig. 5. The maximum phosphorus uptake rate was 41 × 10^{-8} gP/mgDW · h, and the half-saturation constant was 245 μgP/L. Although the values for $\hat{\varrho}$ were quite similar, *Ulothrix* exhibited a much higher half-saturation constant at this level of internal phosphorus than did *Cladophora*.

Discussion

In bodies of water where lakewide phosphorus levels are not capable of supporting extensive algal growth, the geographic distribution of *Cladophora* and *Ulothrix* appears to be governed by the supply of phosphorus. The data presented here indicate that the dominance of *Ulothrix* in the littoral zone of Lake Superior cannot be explained solely on the basis of phosphorus kinetics. Our original hypothesis based on field observations was that *Ulothrix* would possess more efficient phosphorus uptake mechanisms than would *Cladophora*. In contrast, the data indicate that phosphorus uptake rates and half-saturation constants for *Cladophora* and *Ulothrix* are quite similar over a range of internal phosphorus levels. Furthermore, the values of the maximum phosphorus uptake rate for *Ulothrix* and

Cladophora at a specific internal phosphorus level are essentially identical. The half-saturation constants of the two algae also indicate that *Cladophora* would be more effective at taking up phosphorus than *Ulothrix*. One aspect of phosphorus kinetics which has not been considered, however, is the potential for variation in uptake rate with temperature. Rhee and Gotham (1981) have shown that *Scenedesmus*, under conditions of nitrate limitation, has a clear uptake optimum at 15° C. We suspect that uptake temperature optima differ for *Ulothrix* and *Cladophora*, in a fashion similar to that for photosynthesis temperature optima.

Although phosphorus plays a major role in determining geographic distribution of these algae in the Great Lakes, it does not play a significant role in seasonal abundance. Phosphorus concentrations in the nearshore waters of the Great Lakes do not generally vary in a seasonal manner capable of influencing attached algal growth. In the eastern basin of Lake Erie, however, a seasonal decline in internal phosphorus in *Cladophora* reflects a reduction in soluble phosphorus levels during the summer (Millner *et al* 1982). The area occupied by *Ulothrix* and *Cladophora* is generally restricted to depths of less than two meters (Blum 1982; Garwood 1982). It is unlikely that nutrient gradients influence vertical zonation in turbulent nearshore waters as this zone is well mixed and essentially homogenous with respect to chemical constituents.

Patterns of vertical zonation and seasonal abundance appear to be primarily influenced by the impact of light and temperature on rates of photosynthesis and respiration. At Harbor Beach on Lake Huron, *Cladophora* had an optimum depth for growth of one meter (Canale *et al* 1982). In shallow water (<1 m) respiration was enhanced by high light levels creating an unfavorable energy balance. In deeper waters, light limitation became increasingly important. In contrast, *Ulothrix* has its highest rates of net photosynthesis at the highest light levels examined (Fig. 2). The measured light optima for the two algae are consistent with observed patterns of vertical zonation. *Ulothrix* is well adapted to high light levels and occurs in the splash zone, while *Cladophora* grows best at moderate light levels (300 $\mu E/m^2 \cdot sec$) and has an optimal depth which varies with season and water turbidity. To what extent desiccation tolerance may be a factor is unknown for any Great Lakes alga; the ability to survive periods of exposure and even

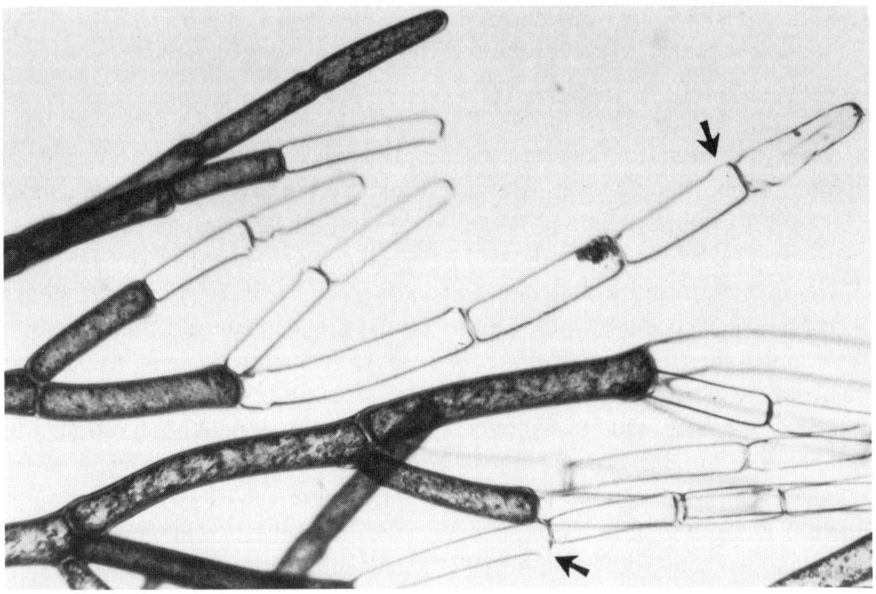

Fig. 6. Cladophora glomerata. Terminal cells or rows of cells in terminal branches are empty zoosporangia with small discharge pores (arrows). Note that proximal cells have not sporulated. × 230.

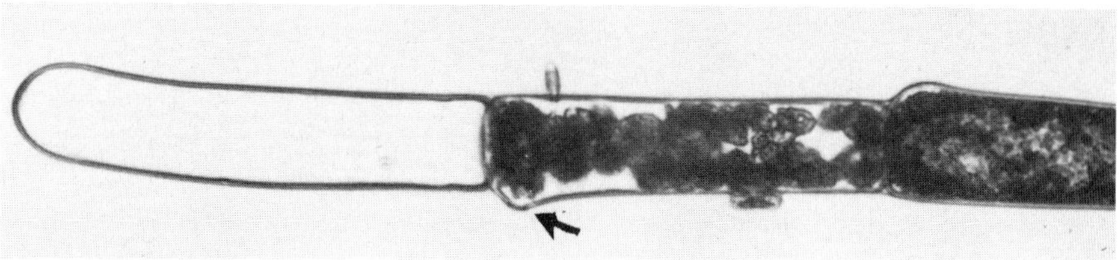

Fig. 7. Cladophora glomerata. A terminal branch showing, from right to left, a vegetative cell, a sporangium containing numerous developing zoospores (note developing discharge pore (arrow)) and a terminal, empty sporangium. × 510.

to carry on enhanced photosynthesis in air has been shown to a significant factor in algal ecology in the marine intertidal zone (Johnson *et al* 1974). Seasonal patterns of abundance of the two algae are consistent with the measured effects of temperature on net photosynthesis and respiration. *Ulothrix* has a temperature optimum for photosynthesis at only 5° C and a high rate even at 1° C (Fig. 3). In contrast the optimum temperature for growth of *Cladophora* is at 15° C. Thus *Ulothrix* occurs in the early spring and late fall in the lower Great Lakes and is dominant in Lake Superior where cold water temperatures exclude *Cladophora*.

Closer examination of the data in Figs. 2 and 3 raises some interesting questions about the seasonal occurrence of *Ulothrix*. *Ulothrix* exhibits higher rates of net photosynthesis than *Cladophora* at all light levels and temperatures through 20° C. Yet, field observations show *Cladophora* replacing *Ulothrix* at 10° C. Why does *Ulothrix* not remain dominant up to 20° C? Why also do we not see large masses of *Ulothrix* piled upon the beaches of the Great Lakes as we often see for *Cladophora*?

The difference in magnitude of biomass accumulation is related to contrasts in the reproductive strategies of these algae. Relatively little difficulty was encountered in growing large amounts of *Cladophora* biomass for BIOTRON studies. *Ulothrix*, however, formed large numbers of zoospores upon transfer to fresh culture medium. Lokhorst & Vroman (1974) stated that European isolates of *U. zonata* underwent zoosporogenesis on a short-day (8:16) regime and gametogenesis under a long day (16:8) photoperiod. Our Lake Huron isolates of *U. zonata* formed zoospores under long-day (16:8) conditions at 20° C and 35 $\mu E/m^2 \cdot sec$. Culture of *Ulothrix* under neutral day conditions (12:12) was found to reduce zoospore formation sufficiently to permit the growth of large amounts of biomass. It was concluded that control of reproduction in *Ulothrix* was the key to understanding its seasonal abundance [as suggested by Garwood (1982)] and therefore we undertook a comprehensive study of the effects of light, temperature, and photoperiod on reproduction. A similar comprehensive study of *Cladophora* reproduction had already been made (Hoffmann & Graham, in press). Zoosporogenesis in *Cladophora* (Figs. 6 and 7) occurred under conditions of short days (8:16), low light (62 $\mu E/m^2 \cdot sec$), and moderate temperatures (13°–20° C). Preliminary analysis of reproduction data for *Ulothrix* indicates that, compared with *Cladophora*, zoosporogenesis occurs at high rates at all combinations of light, temperature, and photoperiod; definite photoperiod and temperature effects do occur, however. In *Cladophora*, zoosporogenesis occurs under a rather restricted set of environmental conditions and is confined to terminal cells or rows of cells on terminal branches (Fig. 6); small discharge pores permit zoospore release with little mechanical disruption (Fig. 7). The main thallus of *Cladophora*, therefore, remains intact following the release of zoospores. In contrast, cells of *Ulothrix* form zoospores simultaneously all along the filament (Fig. 8). When the zoospores are released, the filament is left empty, and large discharge pores disrupt the cell walls causing the filaments to disintegrate. Large masses of *Ulothrix* do not accumulate because biomass is expended as zoospores. As water temperature and daylength increase in spring, the balance between vegetative growth and zoosporogenesis shifts further toward zoo-

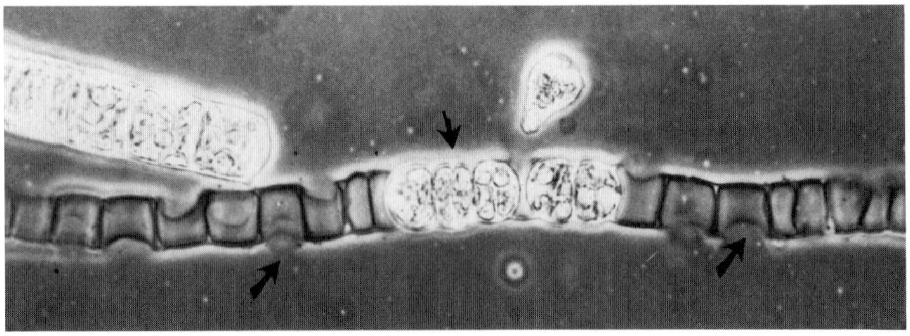

Fig. 8. Ulothrix zonata. The filament at upper left has not sporulated, but the cells in the lower filament have all been transformed into zoosporangia. Zoosporangia produce a variable number of quadriflagellate spores (straight arrow), depending on the size of the original vegetative cell. Zoospores have been discharged from empty cells. Note that discharge pores occupy relatively large areas of the cell walls (curved arrows). Phase-contrast optics. × 1300.

sporogenesis, and vegetative biomass declines.

The field and laboratory experiments described in this manuscript have shown how light, temperature, photoperiod and phosphorus act through photosynthesis and reproduction to regulate the observed patterns of seasonal, spatial, and geographical distribution of *Ulothrix* and *Cladophora* in the Laurentian Great Lakes. The complex nature of these interactions points to the need for comprehensive integrated studies of periphyton ecology in freshwater environments.

Acknowledgements

BIOTRON work was carried out at the University of Wisconsin and was supported under grants from the U.S. Environmental Protection Agency (R806600), the National Science Foundation (PCM 7912093) and the Office of Research, Michigan Technological University. Phosphorus kinetic studies were supported by a grant from the U.S. Environmental Protection Agency. The support of Mr. Nelson Thomas of the USEPA Large Lakes Research Station is gratefully acknowledged. The authors would like to thank Craig Meston, Claude Taylor, and Lee Wilcox for their assistance in the BIOTRON studies, and the staff of the BIOTRON for monitoring the crossed gradients room.

References

Adams, M.S. & W. Stone, 1973. Field studies on photosynthesis of Cladophora glomerata (Chlorophyta) in Green Bay, Lake Michigan. Ecology 54:853–862.

American Public Health Association, American Water Works Association and Water Pollution Control Federation. 1975. Standard Methods for the Examination of Water and Wastewater. 14th Edition, APHA, Washington, D.C., 1193 pp.

Auer, M.T. & R.P. Canale, 1980. Phosphorus uptake dynamics as related to mathematical modeling of Cladophora at a site on Lake Huron. J. Great Lakes Res. 6:1–7.

Auer, M.T. & R.P. Canale. 1982a. Ecological studies and mathematical modeling of Cladophora in Lake Huron: 2. Phosphorus uptake kinetics. J. Great Lakes Res. 8:84–92.

Auer, M.T. & R.P. Canale. 1982b. Ecological studies and mathematical modeling of Cladophora in Lake Huron: 3. The dependence of growth rates on internal phosphorus pool size. J. Great Lakes Res. 8:93–99.

Auer, M.T., R.P. Canale, H.C. Grundler, & Y. Matsuoka. 1982. Ecological studies and mathematical modeling of Cladophora in Lake Huron: 1. Program description and field monitoring of growth dynamics. J. Great Lakes Res. 8:73–83.

Bellis, C.I. & A.T. James, 1966. Fitting the rectangular hyperbola. Biometrics 22:573–602.

Bellis, V.J. & D.A. McLarty, 1967. Ecology of Cladophora glomerata (L.) Kütz in southern Ontario. J. Phycol. 3: 57–63.

Bliss, V.J., 1968. Unialgal cultures of Cladophora glomerata (L.) Kütz. I. Response to temperature. J. Phycol. 4:19–23.

Blum, J.L. 1982. Colonization and growth of attached algae at the Lake Michigan water line. J. Great Lakes Res. 8:10–15.

Canale, R.P., & M.T. Auer. 1982a. Ecological studies and mathematical modeling of Cladophora in Lake Huron: 5. Model development and calibration. J. Great Lakes Res. 8:112–125.

Canale, R.P., & M.T. Auer. 1982b. Ecological studies and mathematical modeling of Cladophora in Lake Huron: 7. Model verification and system response. J. Great Lakes Res. 8:134–143.

Canale, R.P., M.T. Auer, & J.M. Graham. 1982. Ecological studies and mathematical modeling of Cladophora in Lake Huron: 6. Seasonal and spatial variation in growth kinetics. J. Great Lakes Res. 8:126–133.

Evans, D. & J.G. Stockner. 1972. Attached algae on artificial and natural substrates in Lake Winnipeg, Manitoba. J. Fish. Res. Bd. Canada 29:31–44.

Garwood, P.E. 1982. Ecological interactions among Bangia, Cladophora, and Ulothrix along the Lake Erie shoreline. J. Great Lakes Res. 8:54–60.

Gerloff, G.C. & G.P. Fitzgerald. 1976. The Nutrition of Great Lakes Cladophora. U.S.E.P.A., Grosse Ile, Michigan. 111 pp.

Graham, J.M., M.T. Auer, R.P. Canale & J.P. Hoffmann. 1982. Ecological studies and mathematical modeling of Cladophora in Lake Huron: 4. Photosynthesis and respiration as functions of light and temperature. J. Great Lakes Res. 8:100–111.

Herbst, R.P. 1969. Ecological factors and the distribution of Cladophora glomerata in the Great Lakes. Amer. Midl. Nat. 82:90–98.

Hoffmann, J.P. & L.E. Graham. The effect of temperature, light intensity and photoperiod on growth and reproduction in the nuisance alga Cladophora glomerata (L.) Kütz. (Chlorophycophyta, Cladophorales). J. Phycol. (in press).

Jackson, D.F. 1966. Photosynthetic rates of Cladophora fracta from two sites in Lake Ontario under natural and laboratory conditions, pp. 44–50. In: Proceedings of the 9th Conference on Great Lakes Research, University of Michigan, Ann Arbor.

Johnson, W.S., A. Gigon, S.L. Gulman & H.A. Mooney. 1974. Comparative photosynthetic capacities of intertidal algae under exposed and submerged conditions. Ecology 55: 450–453.

Kirby, M.K. & W.E. Dunford, 1981. Attached algae of the Lake Erie shoreline near Nanticoke Generating Station. J. Great Lakes Res. 7:249–257.

Lester, W.W., M.S. Adams & E.H. Dettmann, 1974. Light and Temperature Effects on Photosynthesis of Cladophora glomerata (Chlorophyta) from Green Bay, Lake Michigan: Model Analysis of Seasonal Productivity. Masters Thesis, University of Wisconsin, 43 pp.

Lin, C.K., 1977. Accumulation of water soluble phosphorus and hydrolysis of polyphosphates by Cladophora glomerata (Chlorophyceae). J. Phycol. 13:46–51.

Lokhorst, G.M. & M. Vroman. 1974. Taxonomic studies on the genus Ulothrix (Ulotrichales, Chlorophyceae). III. Acta Bot. Neerl. 23:561–602.

Mantai, K.E. 1974. Some aspects of photosynthesis in Cladophora glomerata. J. Phycol. 10:288–291.

Mantai, K.E. 1978. The response of Cladophora glomerata to changes in soluble orthophosphate concentration in Lake Erie. Verh. Internat. Verein. Limnol. 20:347–351.

Mantai, K.E. & C. Haase. 1977. Some factors affecting photosynthesis and respiration measurements in Cladophora glomerata. J. Great Lakes Res. 3:191–195.

Michaelis, L. & M.L. Menten. 1913. Die Kinetik der Invertinwirkung. Biochem. Z. 49:333–369.

Millner, G.C., R.A. Sweeney & V.R. Frederick. 1982. Biomass and distribution of Cladophora glomerata in relation to some physical-chemical variables at two sites in Lake Erie. J. Great Lakes Res. 8:35–41.

Moore, L.F. 1978. Attached algae at thermal generating stations – the effects of temperature on Cladophora. Verh. Internat. Verein. Limnol. 20:1727–1733.

Parker, R.D.R. 1979. Distribution of filamentous green algae in nearshore periphyton communities along the north shore of Lake Superior. Amer. Mid. Nat. 101:326–332.

Parker, R.D.R. & D.B. Drown. 1982. Effects of phosphorus enrichment and wave simulation on populations of Ulothrix zonata from northern Lake Superior. J. Great Lakes Res. 8:16–26.

Rhee, G.Y., & I.J. Gotham. 1981. The effect of environmental factors on phytoplankton growth: temperature and the interactions of temperature with nutrient limitation. Limnol. Oceanogr. 26:635–648.

Smith, R.F. & H.F. Ludwig. 1968. Eutrophication mechanisms at Lake Tahoe – II. Source of an endogenous Ulothrix bloom on the south shore, July 1967. Water Res 2:625–630.

Storr, J.F. & R.A. Sweeney. 1971. Development of a theoretical seasonal growth response curve of Cladophora glomerata to temperature and photoperiod. In: Proceedings of the 14th Conference on Great Lakes Research: 119–127.

Taft, C.E. & W.J. Kishler. 1973. Cladophora as Related to Pollution and Eutrophication in Western Lake Erie. Water Resources Center, Ohio State University, Columbus. 103 pp.

Tilman, D. & S.S. Kilham. 1976. Phosphate and silicate growth and uptake kinetics of the diatoms Asterionella formosa and Cyclotella meneghiniana in batch and semicontinuous culture. J. Phycol. 12:375–383.

Whitton, B.A. 1970. Biology of Cladophora in freshwaters. Water Research 4:457–576.

Wood, K.G. 1975. Photosynthesis of Cladophora in relation to light and CO_2 limitation, and $CaCO_3$ precipitation. Ecology 56:479–484.

Authors' address:
M. T. Auer
Department of Civil Engineering
Michigan Technological University
Houghton, MI 49931, U.S.A.
J.M. Graham/L.E. Graham/J.A. Kranzfelder
Department of Botany
University of Wisconsin-Madison
Madison, WI 53706, U.S.A.

CHAPTER 18

Uptake and accumulation of some nutrient elements in relation to the biomass of an epilithic community

C. MÜLLER

Abstract. Seasonal biomass variations of an epilithic *Cladophora*-community in the littoral of an eutrophic lake were evaluated on six occasions, from May through September. The calcium-, magnesium-, phosphorus-, and chlorophyll *a*-content of the algal material were analyzed. A distinct seasonal variation existed in calcium content, and was coupled with variations in algal biomass with a maximum in July. Chlorophyll *a* was a good measure of biomass when the algal community was actively growing. Phosphorus content was very low when biomass was high, and thus was considered a limiting factor. The *Cladophora* community showed a bimodal growth pattern on the littoral substrata, with the first growth period from May to July and the second period from August to September/October. Low nutrient concentrations of lake water, the effect of competition from adhering epiphytic matrix, and thermal inhibition are believed to restrict growth of *Cladophora* in Krageholmssjön.

Introduction

The ecology of attached algae in lakes and rivers has been the subject of intensified research in connection with the monitoring of eutrophication processes during recent decades (Beeton 1969; Thomas 1969; Wetzel 1975; Sand-Jensen 1980). Many studies were made of *Cladophora glomerata* (L.) Kütz. in lentic and lotic environments (Bellis & McLarty 1967; Jackson & Lin 1968; Whitton 1970). Two distinct growth maxima of *Cladophora* have been explained in terms of factors limiting growth (temperature, nutrient availability, etc.). Various measures of biomass have been used (e.g. chlorophyll *a*, dry weight, carbon) for estimating photosynthesis, but the correlations have generally been poor (Pieczyńska & Szczepańska 1966; Goldman 1969). The importance of the thermal regime was stressed by Moore (1978). Critical levels of nutrients for growth of *Cladophora* were examined by Jackson & Lin (1968), Neil & Owen (1964), Bellis (1968), Adams & Stone (1973), and others.

Study area

Krageholmssjön is situated in the southern-most province of Sweden, Scania, *ca.* 10 km NW of Ystad. The lake was thoroughly investigated by Hamrin (1970) and the surroundings by Ahlberg *et al.* (1974).

Glacial clay gyttja is the dominating geological substratum. Beech (*Fagus silvatica* L.), oak (*Quercus robur* L.), and alder (*Alnus glutinosa* Gaertn.) dominate in the surrounding woodland. The phytoplankton communities consist mainly of centric diatoms, *Cyclotella* spp. in the spring, and a mixed green algae/diatom community occurs in summer (Hamrin 1970). Blue-green algae, e.g. *Aphanizomenon*, appear in late summer and persist until

Wetzel, R.G. (ed.), Periphyton of Freshwater Ecosystems.
 ISBN 90 6193 768 X.

Table 1. Morphometric and water chemistry data from Lake Krageholmssjön. Data from Hamrin (1970) and unpubl. data (summer aspect).

Lake surface (km^2)		2	
Catchment area (km^2)		15	
Mean depth (m)		4.4	
Maximum depth (m)		9.0	
Pelagial water, 0.5 m depth:			
Alkalinity	(meq/l)		2.4–2.5
pH			8.3–8.8
Ca	(mg/l)		56
Mg	(mg/l)		5.0
NO_3-N	(μg/l)		22–62
PO_4-P	(μg/l)		<10
total-N	(mg/l)		1.2–2.0
total-P	(mg/l)		0.04–0.10

November. The maximum gross production per unit lake surface in Krageholmssjön was calculated in 1969 at *ca*. 1950 mg C/m^2 (Hamrin 1970). Some morphometric and chemical data of Krageholmssjön are given in Table 1.

Materials and methods

Sampling was done at the tip of Point Hasslenos, an exposed peninsula in the central part of Krageholmssjön. Six sampling occasions were chosen during the growing season of *Cladophora* (May through September/October). Five replicate samples were taken on granite and sandstone substrata with a dense cover of filamentous algae. A metal cylinder (radius 4.8 cm) was held firmly against the substratum and the algal filaments were carefully removed from the stone surface with forceps. The water depth was 0.3–0.4 m. Two samples were used for chlorophyll extraction and three samples were used for biomass and chemical analyses. The chlorophyll extraction was done in 100-ml bottles with 90% acetone plus magnesium carbonate. After a 48-h extraction in a dark, cool (+4°C) place, the algal material was ground with a glass rod and chlorophyll *a* content was measured in a Carl Zeiss M4 Q III spectrophotometer at 663 and 750 nm. Previous studies proved this technique useful for extraction of filamentous algae (Müller, unpub.). Chlorophyll *a* content calculations were made according to Golterman & Clymo (1969). The biomass samples were dried to constant weight at 105°C and ashed at 550°C for 3 h. Two of the dried samples were then digested in a perchloric + nitric acid mixture (ratio 1:4) and analyzed for calcium, magnesium, and phosphorus content in a Perkin-Elmer 560 AAS or a Technicon Auto-Analyzer.

Results

The biomass of *Cladophora* as dry weight is given in Table 2. Net algal biomass was fairly constant during June, but in July a maximum of *ca*. $1.74 \cdot 10^{-3}$ kg/100 cm^2 was recorded. In August and September the standing crop decreased to $0.136 \cdot 10^{-3}$ kg/100 cm^2 at the end of September. Chlorophyll *a* content was relatively high in May, *ca* 6.24 mg/100 cm^2, but decreased in June, while the standing crop was still almost constant. The maximum biomass value in July was also reflected in the chlorophyll *a*-content maximum of *ca*. 6.97 mg/100 cm^2. In August, the chlorophyll value declined to approximately a third, and from late August to September the content was only *ca*. 1.65–1.70 mg/100 cm^2.

The calcium content in and on *Cladophora* increased in May and June, with a maximum of ca. 170 mg/g dry weight in July (Fig. 1). Parallel to the decrease in biomass, the amount of calcium was reduced to approximately the May level, *ca*. 50–55 mg/g DW. In September the calcium content increased slightly, about a 31% increase per gram dry weight alga.

Phosphorus content varied somewhat more per

Table 2. Mean dry weight of the epilithon May through September. Confidence limits (C.L.) are given at 95% probability level.

Date 1981	Mean weight ± C.L. (g/100 cm^2)		Ash content (%)	n
28/5	0.582	0.171	51–65	4
22/6	0.566	0.177	59–73	4
18/7	2.72	1.38	33–39	4
12/8	0.515	0.308	61–69	4
28/8	0.399	0.107	57–66	4
24/9	0.302	0.0892	55	5

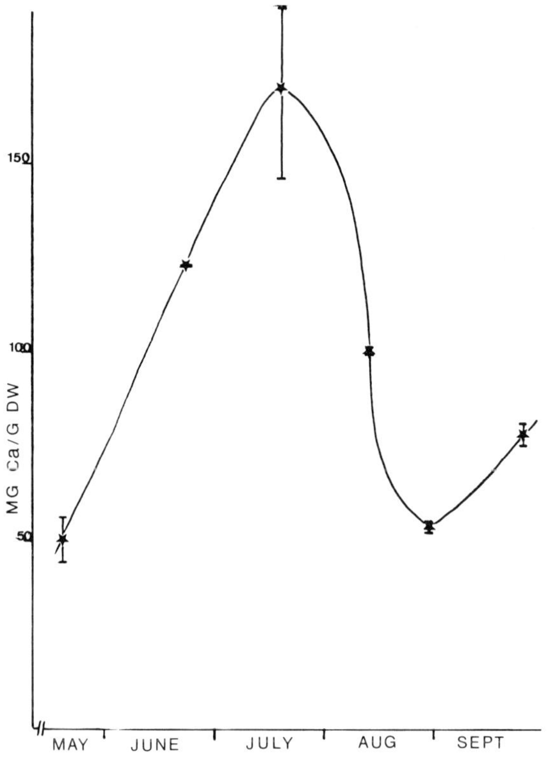

Fig. 1. Changes in the calcium content of the epilithic algal community from May to September in Krageholmssjön. 95% confidence intervals are given (n = 2).

sampling unit, but in general there was an inverse relationship with biomass (Fig. 2); the minimum content was registered in July, *ca.* 0.6 mg P/g DW, and the maximum P-content was found in late September, *ca.* 1.45 mg/g DW.

The magnesium content (Fig. 3) did not show any significant variations from May to September. The amount of magnesium varied between 0.12 and 0.15% of DW.

Discussion

The importance of the nutrient status with aquatic macrophytes and algae as the growth controlling factor was discussed by Gerloff & Krombholz (1966), Adams & Stone (1973), and Wong & Clark (1976). Eyster (in Jackson 1968) listed phosphorus and magnesium as macro-nutrients for green plants, and calcium as a micro-nutrient for algae in general. Phosphorus is known to be stored in excess by algal cells when available (Lewin 1962; Fogg 1966; Wong & Clark 1976). The role of magnesium and calcium in the physiology of algae is somewhat more obscure, apart from chlorophyll synthesis where magnesium is an essential component. Fogg (1966) and others suggested the existence of an optimum ratio between monovalent ions and divalent ions, especially favorable for growth of algae in synthetic media. Jackson (1968), studying rates of photosynthesis and respiration of *Cladophora* in laboratory media, found that 100 mg/l $CaCl_2$ was inhibiting to photosynthesis, probably due to the high salt concentration in the solution. Bellis (1968) found that *Cladophora* favored hard to very hard waters, and required an absolute minimum of 1.2 mg/l Ca and 0.7 mg/l Mg for survival. The deposition of calcium in and on cell walls of *Cladophora* was observed by Bellis & McLarty (1967) and Müller (1978). A high photosynthetic rate in the cells may cause shifts in the equilibrium of bicarbonate-carbonate system and pH-fluctuations, thus causing precipitations of calcium carbonate onto

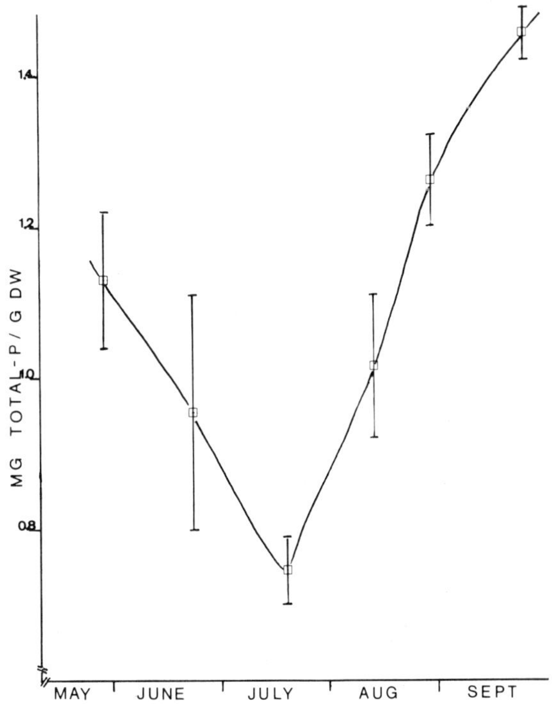

Fig. 2. Changes in the phosphorus content of the epilithic algal community, May to September. 95% confidence intervals are given (n = 2).

the cell walls of the algae.

The calcium content of *Cladophora* in Krageholmssjön rapidly increased during June and July as the standing crop attained a maximum. Although biomass was low in August, fresh *Cladophora* appeared at the water-line of the substratum. Thus, the algal community started a second growth phase, with precipitation of carbonate by the alga. It is assumed that this part of the algal community was actively growing in the lake during the autumn months, since the water temperature was still favorable for growth and nutrient

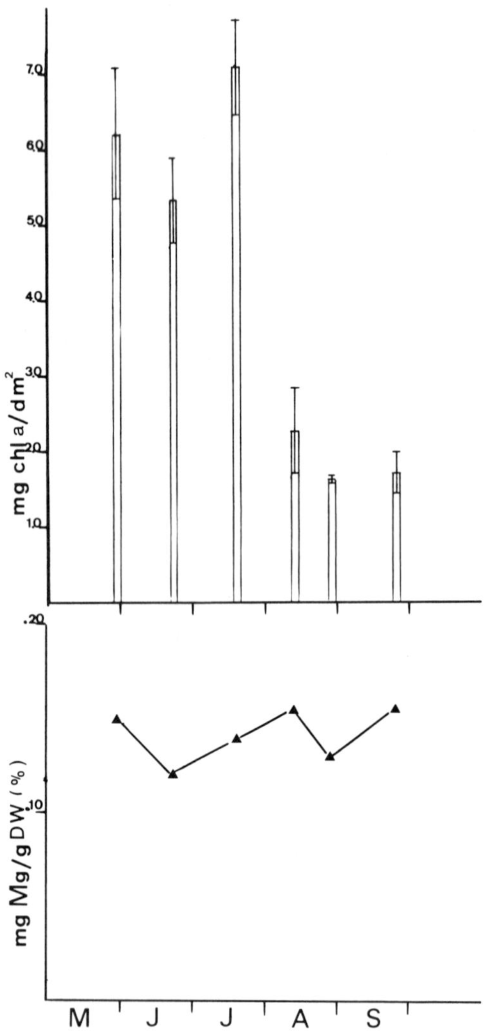

Fig. 3. Top: The chlorophyll *a* of *Cladophora* per 100 cm^2 substrate surface May to September. Mean values (± range) are given. *Bottom:* Magnesium content in percent of dry weight (DW) of *Cladophora*.

concentrations (phosphate, nitrate) increase (Hamrin 1970). The active growth period of *Cladophora* is generally considered to start when the lake temperature attains *ca.* 10° C (Moore 1978), which gives an active growth period for *Cladophora* in Krageholmssjön from May to the end of September/beginning of October.

The magnesium content of *Cladophora* was fairly constant during the investigation (Fig. 3). No significant correlation with chlorophyll *a* was found.

Much information has been gathered on the phosphate limitation for algal growth. Whitton (1970) reviewed the papers on phosphate as a limiting factor for growth of *Cladophora* in the Great Lakes, and Jackson (1968) concluded phosphorus to be a limiting factor in relation to the photosynthetic rate when phosphate concentrations were below 10 μg/l in media. Rosemarin (1982) showed that field samples of *Cladophora* from Lake Ontario were extremely phosphorus deficient at the end of June (2–4 μg P/mg C biomass). Wong & Clark (1976) concluded that the critical level for P in specific growth rate measurements of *Cladophora* was approximately 0.06 mg P/l. They found a significant correlation between phosphorus in plant tissue and dissolved phosphorus in river systems ($r^2 = 0.87$). The critical P level for growth in algal tissue would be approximately 1.6 mg P/g DW. Gerloff & Krombholz (1966) proposed the critical level to be 1.3 mg P/g DW for aquatic plants. The P content of *Cladophora* in Krageholmssjön is generally below 1.3 mg P/g DW from May to August, but attained a level above this critical limit in September (*ca.* 1.45 mg P/g DW). The phosphate-phosphorus in Krageholmssjön increased during autumn and was about 0.06 mg PO_4-P/l at the beginning of November (Hamrin 1970).

Another important factor to consider on the microlevel is the periphyton community on *Cladophora*. Microscopic examination revealed a matrix of bacteria, detritus, diatoms, mineral grains, protozoans, chironomids, etc. covering the algal filaments of *Cladophora*. *Cocconeis placentula* Ehr. and *C. pediculus* Ehr. were the most frequent epiphytes, and covered in some cases up to 80% of the *Cladophora* filaments. Whitton (1970) made the assumption that a dense covering of diatoms interfered with the nutrient uptake of *Cladophora glomerata.* Moore (1978) drew the same conclusion based on his observations of a rapid proliferation of epiphytes on *Cladophora* parallel to the cessation of growth. The functional significance of the described matrix on *Cladophora* in the littoral has been discussed by Wetzel & Allen (1970) and Allen (1971).

It appears that the uptake and accumulation of nutrients by *Cladophora* in temperate lakes are affected by different environmental conditions. A biomass increase affects the calcium carbonate deposition in the growth phase of the algae. Phosphorus availability is probably one of the limiting factors restricting the length of the first growth period (May to July). Temperature is certainly another limiting factor during the second growth period (August to late September). Further research is needed to quantify the effects of the *Cladophora* epiphytic matrix and the biological, chemical, and physical influences on the dynamics of *Cladophora glomerata.*

References

Adams, M.S. & Stone, W., 1973. Field studies on photosynthesis of Cladophora glomerata (Chlorophyta) in Green Bay, Lake Michigan. Ecology 54:853–862.

Ahlberg, E., Andersson, J., Bogren, A., Hernestrand, E., Hjelm, C., Korley, J., Larsén, B., Persson, A., Robertsson, I., Schlyter, F., Stibe, L. & von Wachenfeldt, J., 1974. Environmental Documentation of the Krageholm Area (in Swedish). Mimeogr. 72 pp.

Allen, H.L., 1971. Primary productivity, chemo-organotrophy and nutritional interactions of epiphytic algae and bacteria on macrophytes in the littoral of a lake. Ecol. Monogr. 41: 97–127.

Beeton, A.M., 1969. Changes in the environment and biota of the Great Lakes. In: Eutrophication: Causes, Consequences, Correctives. Nat. Acad. Sciences, Washington. pp. 150–187.

Bellis, V.J., 1968. Unialgal cultures of Cladophora glomerata (L.) Kütz. II. Response to calcium-magnesium ratio and pH of the medium. Proc. 11th Conf. Great Lakes Res:11–15.

Bellis, V.J. & McLarty, D.A., 1967. Ecology of Cladophora glomerata (L.) Kütz. in southern Ontario. J. Phycol. 3:57–63.

Fogg, G.E., 1966. Algal Cultures and Phytoplankton Ecology. Univ. Wisconsin Press, Madison. 126 pp.

Gerloff, G.C. & Krombholz, P.H., 1966. Tissue analysis as a measure of nutrient availability for the growth of angiosperm aquatic plants. Limnol. Oceanogr. 11:529–537.

Goldman, C.R., (ed.) 1969. Primary Productivity in Aquatic Environments. Univ. California Press, Berkeley. 457 pp.

Golterman, H.L. & Clymo, R.S., 1969. Methods for Chemical Analysis of Freshwaters. IBP Handbook 8. Blackwell Scientific Publ., Oxford.

Hamrin, S., 1970. Temporal Variations and long-term Changes in Lake Krageholmssjön and Lake Ellestasjön (south Scania). In Swedish. Inst. Limnology, Univ. Lund. Mimeogr. 53 pp.

Jackson, D.F. & Lin, S.D., 1968. Ecology of Cladophora fracta and Cladophora glomerata. Final report for Project WP 00782, Fed. Water Pollution Control Administration. 127 pp.

Lewin, R.A., (ed.) 1962. Physiology and Biochemistry of Algae. Academic Press, New York. 929 pp.

Moore, L.F., 1978. Attached algae at thermal generating stations – the effect of temperature on Cladophor. Verh. Internat. Verein. Limnol. 20:1727–1733.

Müller, C., 1978. On the productivity and chemical composition of some benthic algae in hard-water streams. Verh. Internat. Verein. Limnol. 20:1457–1462.

Neil, J.H. and Owen, G.E., 1964. The distribution, environmental requirements and significance of Cladophora in the Great Lakes. Publ. Great Lakes Research Div., Univ. of Mich. 11:113–121.

Pieczyńska, E. & Szczepańska, W., 1966. Primary production in the littoral of several Masurian lakes. Verh. Internat. Verein. Limnol. 16:372–379.

Rosemarin, A.S., 1982. Phosphorus nutrition of two potentially competing filamentous algae, Cladophora glomerata (L.) Kütz. and Stigeoclonium tenes (Agardh) Kütz. from Lake Ontario. J. Great Lakes Res. 8.66–72.

Sand-Jensen, K., 1980. The equilibrium among autotrophic components in temperate lakes with different nutritional loadings. (In Danish) Vatten 36/2:104–115.

Thomas, E.A., 1969. The process of eutrophication in central European lakes. In: Eutrophication: Causes, Consequences, Correctives. Nat. Acad. Sciences, Washington. pp. 29–49.

Wetzel, R.G., 1975. Limnology. W. B. Saunders, Philadelphia. 743 pp.

Wetzel, R.G. & Allen, H.L., 1970. Functions and interactions of dissolved organic matter and the littoral zone in lake metabolism and eutrophication. In: Z. Kajak and A. Hillbricht-Ilkowska (eds.), Productivity Problems of Freshwaters. Warsaw, PWN Polish Scientific Publishers. p. 333–347.

Whitton, B.A., 1970. Biology of Cladophora in freshwaters. Water Res. 4:457–476.

Wong, S.L. & Clark, B., 1976. Field determination of the critical nutrient concentrations for Cladophora in streams. J. Fish. Res. Board Can. 33:85–92.

Author's address:
C. Müller
Institute of Limnology
University of Lund
Box 3060
S-220 03 Lund, Sweden

CHAPTER 19

Dynamics of epiphytic communities on *Equisetum fluviatile* L.

Response to short-term variation in environmental conditions

TIMO KAIRESALO

Abstract. The response of epiphytic communities to short-term variations in environmental conditions was measured at the outer zone of an *Equisetum fluviatile* stand for an 18-day period. The epiphytic biomass was maximal after a period of bright and calm weather, but then during a week of cloudy weather two thirds of the epiphytic material disappeared. The epiphytic material was first quite evenly distributed along the macrophyte shoots to a depth of 50–60 cm, but a small change in the wind direction seemed to detach the diatom, *Fragilaria capucina* Desm., near the water surface. When light conditions worsened the algal dominance changed; *Fragilaria capucina* was replaced by filamentous green algae, *Oedogonium* spp., which were more tightly attached to their substrate. This change in the algal community composition and biomass was also reflected in their contribution to the total primary production of the macrophyte zone. The total number of herbivores (oligochaetes + chironomid larvae) in the epiphytic communities remained quite constant. The results of this study also reveal the importance of sampling intervals when measuring periphyton accumulation on plates.

Introduction

The outer zone of a littoral helophyte stand is a site of interaction between the dense part of the macrophyte stand and the open lake area. In this zone light conditions are generally good for epiphytic algal growth after the emergence of the macrophytes, whereas inside the stand underwater photosynthesis may become limited primarily by light (e.g. Straskraba & Pieczynska 1970; Hillbricht-Ilkowska *et al* 1972; Kairesalo 1980a). On the other hand, the environmental factors related either directly (as surface waves) or indirectly (as temperature and nutrient concentrations) with the force and direction of wind may greatly affect the species succession, biomass accumulation and productivity of the epiphytic communities. Therefore, when interpreting the succession of epiphytic communities over a longer period, knowledge of their response to short-term variation in the environmental conditions is very important.

This study aimed to measure intensively both quantitative and qualitative succession of epiphytic communities at the outer zone of an *Equisetum fluviatile* stand during a 2–3 week period. Since that period has commonly been used in the sampling of periphyton on artificial substrates, the results of this study are highly relevant to the validity of such experiments.

Materials and methods

The study area was located at the outer zone of an *Equisetum fluviatile* stand in the western end of Lake Pääjärvi, an oligotrophic, brown-water lake

Wetzel, R.G. (ed.), Periphyton of Freshwater Ecosystems.
 ISBN 90 6193 768 X.

in southern Finland (61° 04′ N, 25° 08′ E) (general description in Ruuhijärvi 1974). A permanent grid-like pier system with removable boards had been built in the macrophyte stand in 1978, at the beginning of 'the littoral project' (see Sarvala *et al* 1982), and it made possible repeated sampling without disturbance.

Sampling took place within a zone of about 1-m breadth and 5-m length near the outer edge of the *Equisetum fluviatile* stand. In this zone the macrophyte density was 200–300 stems in m^{-2} (Kairesalo 1982a), and changes in wind direction were quickly reflected there in the water temperature and in nutrient concentrations or ratios (Kairesalo 1982b).

Samples for determining epiphytic biomasses were taken at two- or three-day intervals using transparent polyethylene tubes (20 cm in diameter) which had many small holes (0.5–1 mm in diameter) in a row on one side. One macrophyte shoot was gently inserted into the tube and the shoot was then cut at the sediment surface with a sharp metal plate. The shoot and the epiphytes on it were raised to the water surface and while gently lifting the tube in the horizontal position the water drained out from the tube through the small holes and both ends, while the epiphytes adhered on the walls of the soft plastic tube and remained well in their position. At least four parallel samples (i.e. shoots) were taken in each biomass sampling. In addition, at least two samples for identification and enumeration of algae and animals were taken separately.

In the laboratory the shoots (in their tubes) were cut into 10-cm pieces starting from the level of the water surface. Epiphytic material was then removed directly into aluminium foil cups using a thin and flexible rubber plate. These biomass samples were dried at 60° C, and their total carbon contents were then determined with a CHN-analyser (Hewlett Packard). The algal species were identified with a Nikon research microscope, and the algal dominances in the homogenized samples were estimated as covering percentages in the microscopic fields. Nomenclature for the diatoms was based on Hustedt (1930). Oligochaetes and chironomid larvae were counted using a stereomicroscope (Wild M 5) with a grooved disc (Hakala 1971) and neutral red staining. Nomenclature for the oligochaetes was based on Brinkhurst (1971) and for the chironomid larvae on Oliver *et al* (1978).

On the first and the last date of the study period total community production and respiration were measured using an acrylic plastic chamber (volume 160 l, height 1 m) in which there was a continuous air and water circulation during the measurement. Primary production and respiration were determined as the total change in CO_2 content inside the chamber (see Kairesalo 1982a). The contribution of the epiphytic and planktonic algal communities and the macrophyte to the total primary production was measured as ^{14}C-fixation simultaneously with the measurement of total primary production (see Kairesalo 1982a), incubation time being 30 minutes. The ^{14}C-epiphyte samples were treated as the biomass samples. For the measurement of radioactivity the evolved $^{14}CO_2$ in the outflow gas of the CHN-analyser was trapped into alkylamine solution (Carbo-Sorb II, Lumac) in a gas-liquid exchange column connected with a scintillation vial. Radioactivity was measured in xylene based scintillation cocktail (Carbo-Luma, Lumac) using a Wallac Rackbeta liquid scintillation counter. The plankton samples (after the incubation) were taken from the chamber using an acrylic plastic tube (4.4 cm in diameter). Two parallel samples (500 ml) were filtered through Whatman GF/C-filters. The filters were totally combusted in the CHN-analyser, the evolved $^{14}CO_2$ was trapped into alkylamine solution and the radioactivity measured as for the corresponding epiphyte samples. The submerged and aerial parts of the macrophyte shoots were treated separately. On the final occasion the offshoots were analysed separately (at the beginning of the study period there were only few and tiny offshoots). The total carbon and radiocarbon contents of the macrophytes were determined from dried and homogenized plant material. The total carbon was determined with the CHN-analyser, and the radiocarbon with a Junitek-Oxidizer (see Kairesalo 1982a) in which larger subsamples could be combusted. Tests confirmed that both analysers gave identical ^{14}C-yields.

Air and water temperature and irradiance were measured continuously in the study area. Air temperature was measured with a sensor placed in the shade about 1 m above the water surface and water temperature with a sensor placed to the middle of the water column. Irradiance was measured using a Kipp & Zonen-solarimeter. Underwater radiation was measured with an underwater quantum sensor (LI-COR).

Results

Figure 1 shows light, temperature and wind conditions together with the development of the corresponding epiphytic biomass in the study area during an 18-day period in July, 1981. The epiphytic algae had their peak occurrence after a period of

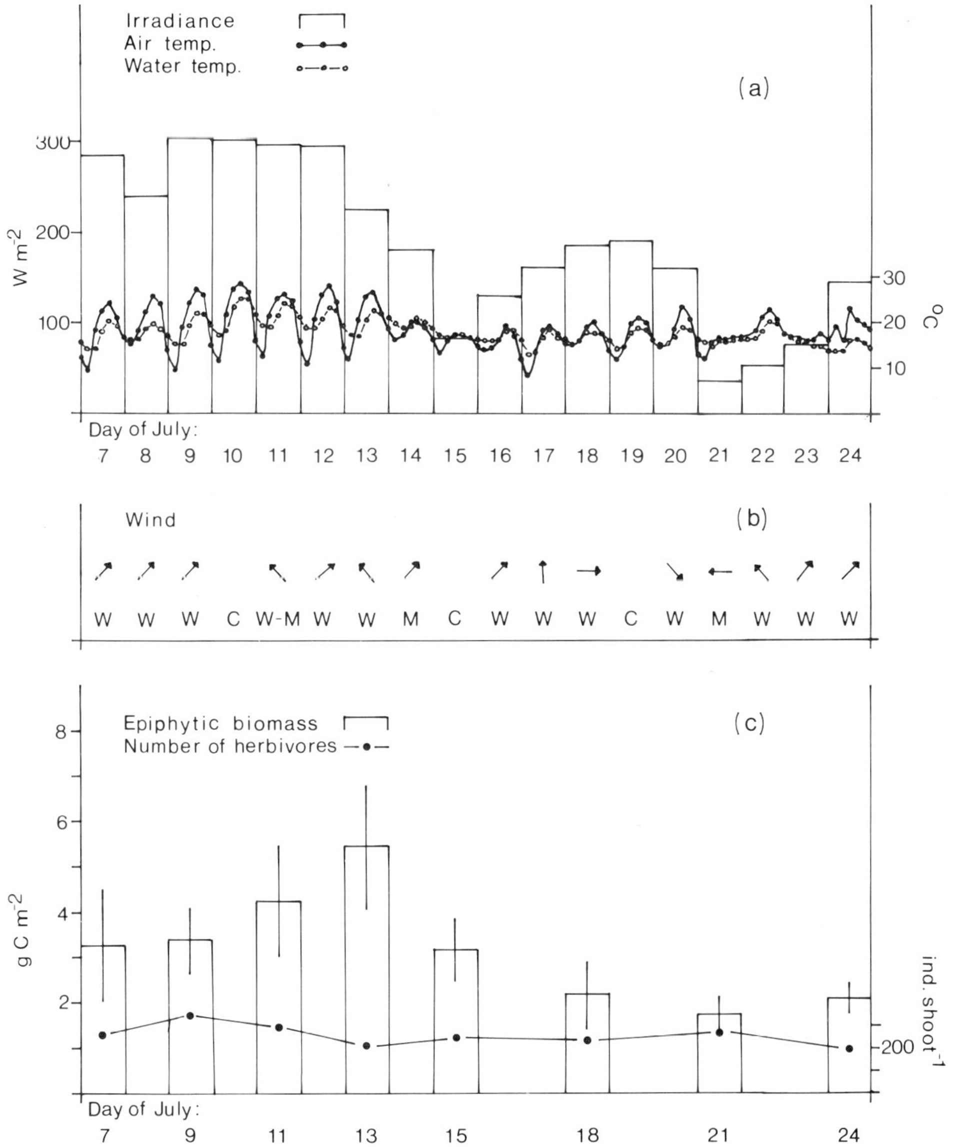

Fig. 1. (a) Mean daily irradiance, and air and water temperatures during the sampling period, 7–24 July 1981; (b) prevailing wind direction (north upward) and its relative power; C = calm, W = weak, M = moderate; (c) Total epiphytic biomass ($\bar{x} \pm$ S.D.) and the mean number of oligochaetes and chironomid larvae in the epiphytic material on one *Equisetum fluviatile* shoot between water and sediment surfaces.

Table 1. Number of oligochaetes and chironomid larvae in the epiphytic material on 10 cm pieces of *Equisetum fluviatile* shoots and per one milligram of total carbon biomass (in parentheses).

Length (cm)	Date (July)							
	7	9	11	13	15	18	21	24
Oligochaetes								
0–10	49	74	86	41	42	34	27	16
	(27)	(49)	(36)	(27)	(25)	(23)	(27)	(23)
10–20	64	70	44	56	41	38	36	26
	(36)	(37)	(15)	(22)	(27)	(29)	(36)	(26)
20–30	50	73	30	24	29	35	36	32
	(20)	(43)	(11)	(6)	(18)	(23)	(40)	(25)
30–40	54	55	22	35	40	36	38	38
	(30)	(23)	(12)	(6)	(21)	(30)	(29)	(18)
40–50	35	50	90	35	45	38	40	36
	(22)	(24)	(39)	(11)	(13)	(21)	(22)	(18)
Total number (0–50 cm)	252	322	272	191	197	181	177	148
Chironomid larvae								
0–10	2	8	7	3	8	8	16	3
	(1)	(5)	(3)	(2)	(5)	(5)	(16)	(4)
10–20	4	4	2	3	9	10	16	7
	(2)	(2)	(.8)	(1)	(6)	(8)	(16)	(7)
20–30	2	6	2	1	9	12	18	10
	(1)	(4)	(.8)	(.5)	(6)	(8)	(20)	(8)
30–40	3	7	3	3	12	11	23	12
	(2)	(3)	(2)	(.5)	(6)	(9)	(18)	(6)
40–50	3	4	10	3	15	17	27	15
	(2)	(2)	(4)	(1)	(4)	(9)	(15)	(8)
Total number (0–50 cm)	14	29	24	13	53	58	100	47

bright and calm weather (wind, if blowing, was weak and off from the shore) (Fig. 1a). The algal biomass then consisted almost completely of a diatom, *Fragilaria capucina,* which formed long, loosely attached cell chains on and around the macrophyte shoots. Among these chains there occurred only few other diatom species (*Synedra ulna* (Nitzsch) Ehr., *Tabellaria fenestrata* (Lyngb.) Kütz., *T. flocculosa* (Roth) Kütz. and *Gomphonema constrictum* Ehr.). During the maximal biomass (July 13), however, filamentous green algae, *Oedogonium* spp., appeared on the macrophyte shoots, especially on their uppermost parts. While the epiphytic biomass then decreased with cloudy weather the contribution of the *Oedogonium*-filaments increased in the algal communities, and at the end of the study period the algal biomass consisted mainly of the *Oedogonium*-filaments, which were tightly attached to their substrate, and of minute diatoms, *Achnanthes minutissima* Kütz. and *Gomphonema parvulum* (Kütz.) Grunow, which covered the filaments of the larger green algae. During the study period, the wind was blowing against the macrophyte stand on four days; weakly on the 11th, 13th and 22th, and more strongly on the 21th of July (Fig. 1b). The total number of herbivorous animals in the epiphytic communities seemed to be quite constant (Fig. 1c). Oligochaetes, however, had their peak occurrence simultaneously with the mass occurrence of

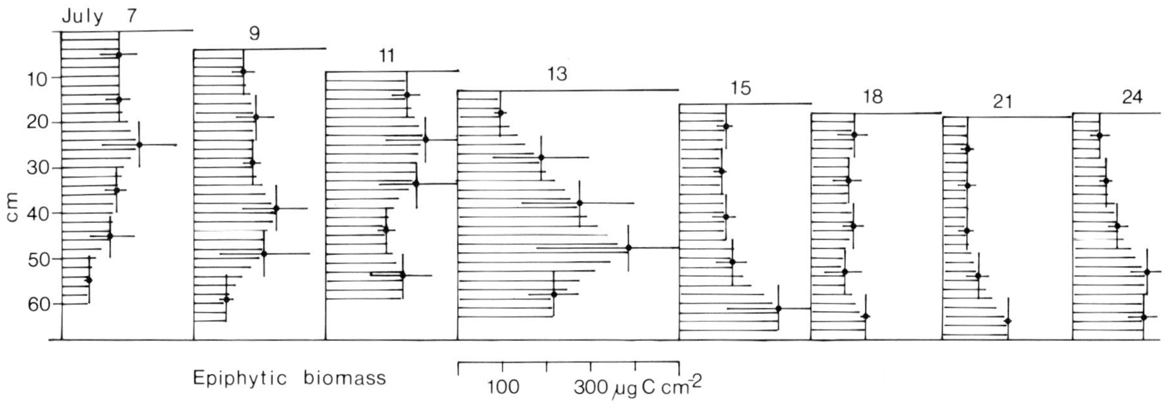

Fig. 2. Vertical distribution of epiphytic material ($\mu g\ C\ cm^{-2}$) on the *Equisetum fluviatile* shoots ($\bar{x} \pm$ S.D.). Alteration of water depth at the sampling site is also presented in the figure.

the diatom, *Fragilaria capucina,* whereas the chironomid larvae were relatively more abundant among the filamentous green algae (*Oedogonium* spp.) and the minute diatoms (*Achnanthes minutissima, Gomphonema parvulum*) (Table 1). The most abundant species of oligochaetes were *Stylaria lacustris* Linn. and *Tubifex tubifex* Müller, and of the larger chironomid larvae *Psectrocladius* sp. and *Microtendipes pedellus* Deg., and of the smaller chironomid larvae *Cricotopus (Isocladius) sylvestris* Fabr., *Cladotanytarsus* sp. and *Diamesa* sp. The number of small chironomid larvae, particularly, increased at the end of the sampling period (Table 1).

The epiphytic material was quite evenly distributed along the macrophyte shoots during the first three sampling dates (Fig. 2), but thereafter the epiphytic material seemed to accumulate near the sediment surface. This decrease and change in the distribution of the epiphytic biomass happened simultaneously with the change in algal dominance; the loosely attached diatoms were replaced by the more tightly attached green algae starting from the water surface.

There was a distinct difference in the photosynthetic activity of the epiphytic algal communities between the first and the last day of the sampling period (Fig. 3). This difference could mostly be explained by the threefold higher irradiance during the first incubation period (cf. Fig. 4). When the diatom, *Fragilaria capucina,* dominated in the algal communities the photosynthetic activity seemed to be closely related to the available light down to a depth of 40 cm, but nearer the sediment surface it increased. This increase in the photosynthetic activity, however, was not caused by the change in the algal species composition since the same species, *Fragilaria capucina,* dominated equally at all depths. No surface inhibition in the photosynthetic activity was evident on the bright day (July 7) whereas such inhibition was measured on the less bright day (July 24).

The contribution of the epiphytic algae to the total primary production in the macrophyte zone seemed to be very important although their biomass was minute in comparison with the corresponding macrophyte biomass (Fig. 4). The contri-

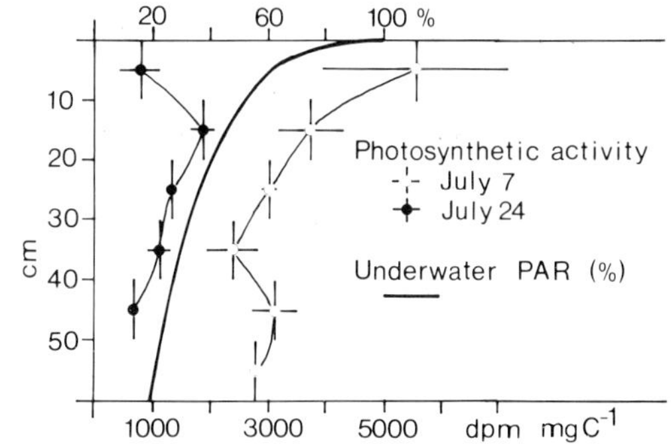

Fig. 3. Photosynthetic activity of the epiphytic communities at different depths on two dates (7 and 24 July 1981) together with underwater penetration (%) of photosynthetically active radiation (PAR), measured on the 24 July 1982.

bution of plankton to the total primary production was negligible since planktonic algae were almost totally lacking, the plankton biomass consisting mostly of the zooplankton species, *Bosmina* spp. and *Polyphemus pediculus* L.

Discussion

The results of this study show the possibility for a rapid change in the quantity and quality of the epiphytic communities when light and/or wind conditions alter. Particularly, in the outer zones of dense helophyte stands the epiphytic communities may undergo great fluctuations. In the dense part of the macrophyte stand heterotrophic processes generally prevail in the water after the emergence of the macrophytes (Kairesalo 1982a), producing very high nutrient concentrations; for instance, phosphorus concentration (both total and phosphate phosphorus) might be over ten times higher inside the *Equisetum fluviatile* stand than in the surrounding open lake water (Kairesalo 1982b). These 'extra' nutrients then at least partly are fixed into algal biomass in the outer macrophyte zone where light conditions are more favourable for the algal photosynthesis. The epiphytic algal growth and accumulation in such circumstances may be very striking, especially in oligotrophic lakes where light-saturated algal production in open lake areas is often mostly limited by phosphorus (cf. Schindler *et al* 1978). In addition, the contribution of epiphytic communities to the total metabolism may become very significant, the primary production in the water of that zone being then mainly attributed to the epiphytic algae. However, this 'scum flora', which mostly consists of loosely attached, metaphytic species, is very susceptible to stronger water currents, and may rapidly disappear from the substrates under such conditions. The detached epiphytic algae then either settle down to the sediment surface or disperse into the surrounding water, thus

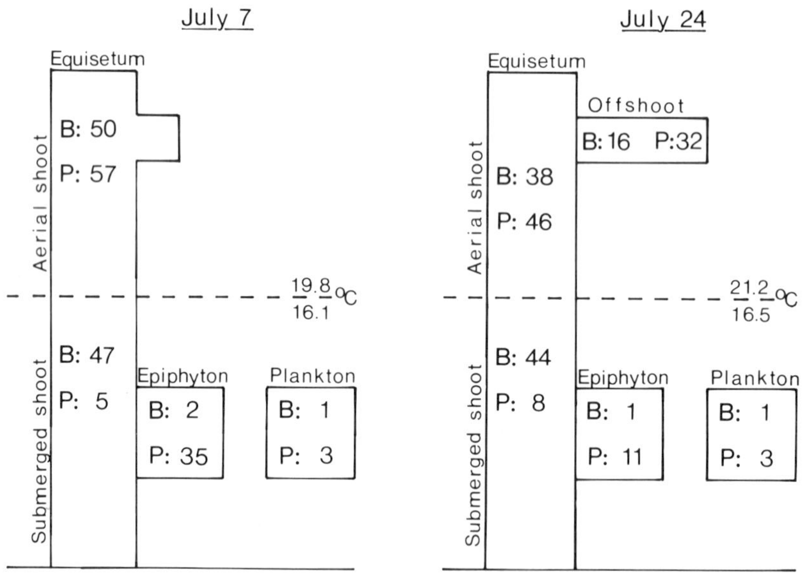

	July 7	July 24
I_o ($W m^{-2}$)	644	190
Tot. B ($g C m^{-2}$)	119	139
Tot. P ($mg C m^{-2} h^{-1}$)	455	113
Tot. R ($mg C m^{-2} h^{-1}$)	149	113

Fig. 4. Percentage contributions of different autotrophs to total aboveground biomass (B) and ^{14}C fixation (P) on two dates. Air and water temperature, surface irradiance (I_o), total aboveground biomass (Total B), net production (Total P) and respiration (Total R) are also given.

increasing the algal biomass, and perhaps the productivity, of those communities (cf. Wetzel 1975; Ilmavirta *et al* 1977; Sheath & Hellebust 1978). Of the common epiphytic diatom species, *Fragilaria capucina* and *Achnanthes minutissima* are very abundant also in the sediment and in the littoral plankton (Ilmavirta *et al* 1978; Kairesalo 1980a, b), *Synedra ulna* being abundant also in the pelagial plankton of Lake Pääjärvi (e.g. Ilmavirta 1974).

True epiphytic algae, on the other hand, may have 'enough' time to adapt to different light intensities, this adaptation perhaps being the main reason for the increase in the photosynthetic activity near the sediment surface (cf. Fig. 3). In cultures of diatoms, physiological adaptation from one light intensity to another may have caused by changes in the content of photosynthetically active pigments or by changes in the photosynthetically active enzymes, or both (Jørgensen 1977). Microscopic examination revealed that the benthic cells of the diatom, *Fragilaria capucina,* were darker coloured and thus presumably contained more pigments than the surface cells. In addition, the photosynthesis of the diatom communities in the surface layers of the water seemed not to have been inhibited by the high light intensity, a phenomenon almost always found in measurements of phytoplankton production.

The epiphytic animals seemed to have a close connection with the benthic populations since all the epiphytic species were also present in the sediment (Hiisivuori, pers. comm.). Their contribution to the rapid decrease of the epiphytic algal biomass soon after the light conditions had worsened for algal production remains uncertain. Although the total number of herbivores seemed to be constant throughout the sampling period (suggesting that grazing of algae also remained constant), at least qualitative changes occurred in the grazing. Diatoms, generally, have been considered more desirable food than green or blue-green algae (e.g. Kajak & Warda 1968; Nicotri 1977; Moore 1981). The minute diatoms, *Achnanthes minutissima* and *Gomphonema parvulum*, may have been a preferable food for the small chironomid larvae than the larger *Fragilaria capucina*-cells, which were more exploited by the larger oligochaetes.

In conclusion, long sampling intervals (2–3 weeks) may lead to false conclusions concerning the development of epiphytic and all periphytic communities if light and/or wind conditions have changed considerably during that period. For the same reason comparisons between different sampling sites may be invalid. To get more accurate figure for development of periphytic communities on exposed substrates intensive sampling periods rather than separate samplings on fixed dates with long intervals are, therefore, suggested.

Acknowledgements

I wish to thank Dr. Roger Jones for his comments on the manuscript. He also checked the English of the manuscript. Christa Hiisivuori, M.Sc., identified the epiphytic animals.

References

Brinkhurst, R.O., 1971. A Guide for the Identification of British Aquatic Oligochaeta. Freshwater Biological Association, Scientific Publication No. 22, 2nd edition. 55 pp.

Hakala, I., 1971. A new model of the Kajak bottom sampler, and other improvements in the zoobenthos sampling technique. Ann. Zool. Fennici 8:422–426.

Hillbricht-Ilkowska, A., Kowalczewski, A. & Spodniewska, I., 1972. Field experiments of the factors controlling primary production of the lake plankton and periphyton. Ekol. Pol. 20:315–326.

Hustedt, F., 1930. Bacillariophyta (Diatomeae). Die Süsswasser-Flora Mitteleuropas Heft 10. JENA, Verlag von Gustav Fischer. 466 pp.

Ilmavirta, V., 1974. Diel periodicity in the phytoplankton community of the oligotrophic lake Pääjärvi, southern Finland. I. Phytoplanktonic primary production and the related factors. Ann. Bot. Fennici 11:136–177.

Ilmavirta, V., Jones, R.I. & Kairesalo, T., 1977: The structure and photosynthetic activity of pelagial and littoral plankton communities in Lake Pääjärvi, southern Finland. Ann. Bot. Fennici 14:7–16.

Jørgensen, E.G., 1977. Photosynthesis. pp. 150–168. In: D. Werner (ed.), The biology of diatoms. Bot. Monogr. 13. Blackwell Scientific Publications, Oxford. 498 pp.

Kairesalo, T., 1980a. Comparison of in situ photosynthetic activity of epiphytic, epipelic and planktonic algal communities in an oligotrophic lake, southern Finland. J. Phycol. 16: 57–62.

Kairesalo, T., 1980b. Diurnal fluctuations within a littoral plankton community in oligotrophic Lake Pääjärvi, southern Finland. Freshwat. Biol. 10:533–537.

Kairesalo, T., 1982a. Photosynthesis and respiration within an Equisetum fluviatile L. stand in Lake Pääjärvi, southern Finland. Arch. Hydrobiol. 96:317–328.

Kairesalo, T., 1982b. The seasonal succession of epiphytic communities within an Equisetum fluviatile L. stand in Lake Pääjärvi, southern Finland. Int. Revue ges. Hydrobiol. (in press).

Kajak, Z. & Warda, J., 1968. Feeding of benthic non-predatory Chironomidae in lakes. Ann. Zool. Fennici 5:57–64.

Moore, J.W., 1981. Inter-species variability in the consumption of algae by oligochaetes. Hydrobiologia 83:241–244.

Nicotri, M.E., 1977. Grazing effects of four marine intertidal herbivores of the microflora. Ecology 58:1020–1032.

Oliver, D.R., McClymont, D. & Roussel, M.E., 1978. A Key to some Larvae of Chironomidae (Diptera) from Mackenzie and Porcupine River Watersheds. Fisheries & Marine Service, Technical Report No. 791. 73 pp.

Ruuhijärvi, R., 1974. A general description of the oligotrophic Lake Pääjärvi, southern Finland, and the ecological studies on it. Ann. Bot. Fennici 11:95–104.

Sarvala, J., Kairesalo, T., Koskimies, I., Lehtovaara, A., Ruuhijärvi, J. & Vähä-Piikkiö, I., 1982. Carbon, phosphorus and nitrogen budgets of the littoral Equisetum belt in an oligotrophic lake. Hydrobiologia 86:41–53.

Schindler, D.W., Fee, E.J. & Ruszczynski, T., 1978. Phosphorus input and its consequences for phytoplankton standing crop and production in the experimental lakes area and in similar lakes. J. Fish. Res. Bd. Canada 35:190–196.

Sheath, R.G. Hellebust, J.A., 1978. Comparison of algae in the euplankton, tychoplankton, and periphyton of a tundra pond. Can. J. Bot. 56:1472–1483.

Straskraba, M. & Pieczynska, E., 1970. Field experiments on shading effect by emergents on littoral phytoplankton and periphyton production. Rozpravy Ceskosl. Akad. Ved, Rada Matem. Prir. Ved 80:7–30.

Wetzel, R.G., 1975. Limnology. W.B. Saunders Co., Philadelphia. 743 pp.

Author's address:
T. Kairesalo
Department of Limnology
University of Helsinki
Viikki
00710 Helsinki 71, Finland

CHAPTER 20

Littoral zone production of oligotrophic lakes

The contributions of phytoplankton and periphyton

STANFORD L. LOEB, JOHN E. REUTER and C.R. GOLDMAN

Abstract. Littoral zone primary production can be expressed as the summation of the integral production of its phytoplanton and benthic components:

$$\text{littoral prod. } (\text{mgC}\cdot\text{m}^{-1}\cdot\text{time}^{-1}) = \int_0^z\int_0^x \text{phyto prod, dzdx} + \int_0 \text{benthic prod. dx.}$$

This expression represents a 'wedge' of the littoral zone, 1 m wide, extending from the shore a distance, x, to where the depth of the water, z, is equal to the maximum depth of the littoral zone (i.e. 1% of surface light intensity). The periphyton community contributed 63% of the total littoral production (0–60 m) in Lake Tahoe at a site where the benthic substratum was rock.

Evaluation of littoral productivity on a water column basis (i.e. m^{-2} of lake surface) illustrated that periphyton contributed varying amounts to the total productivity depending upon the depth of the water column, the time of year, the degree bottom of slope, and the benthic substratum type. Periphyton (epilithic) contributed its maximum to 2, 8 and 16 m littoral water column productivity at Lake Tahoe during the spring (>84%) and its minimum during the summer and winter (>60%). A survey of five oligotrophic lakes indicated that periphyton often contributed the majority of the littoral zone production when the substrata were rocks or organic sediments.

Introduction

Littoral zone production has often been found to dominate total lake production especially in shallow lakes where a large part of the lake's surface area lies within the littoral zone (Wetzel 1964; Hargrave 1969; Davis 1970; Wetzel *et al* 1972). Deep lakes usually have only a small part of their surface area within the littoral zone; therefore, the relative contribution of littoral zone production to total lake production is much reduced. Lake Tahoe ($\bar{z} = 313$ m), for example, has less than 19% of its surface area between the shore and the surface projection of the 60 m contour (i.e., the depth of 1% surface light intensity). The littoral zone of Lake Tahoe has been estimated to contribute about 10% of the lake's total annual primary production (Goldman & deAmezaga 1975; Goldman 1981).

The littoral zone of lakes serves as the interface or buffer zone between the watershed and the main body of the lake. As such, it responds more quickly and more site-specifically to pollutant inputs which are toxic or biostimulatory. The littoral zone also serves as the primary habitat for many secondary producers (e.g. insects, crayfish, and fish) and energy transfers within this area of the lake are important to the functioning of the whole system.

Wetzel, R.G. (ed.), Periphyton of Freshwater Ecosystems.
 ISBN 90 6193 768 X.

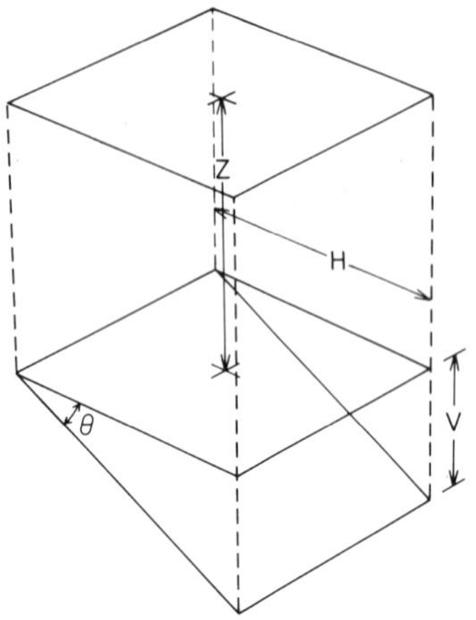

Bottom Slope (θ°)	Area of bottom substrate below 1.00 m² of lake surface	
	Smooth Slope (epipelic or epipsammic)	Step Function (epilithic)
15°	1.04	1.27
25°	1.10	1.47
35°	1.22	1.70
45°	1.44	2.00

Fig. 1. The area of benthos below 1 m² of lake surface for different bottom slopes (θ°). Z represents water depth; H = horizontal rock surface; V = vertical rock surface.

This paper discusses methods used to express littoral zone productivity of lakes. Partitioning of the major primary producers within the littoral zone (i.e. phytoplankton, periphyton, macrophytes) can facilitate an understanding of their individual production cycles and their interactions as they relate to the total littoral zone production. In the lakes investigated for this paper, macrophytes were either not present or insignificant; therefore, only phytoplankton and periphyton are discussed.

Methods

Periphyton productivity (^{14}C) was measured *in situ* following the method of Loeb (1981) and phytoplankton productivity according to Goldman *et al* (1963). Periphyton and phytoplankton incubations were conducted simultaneously (~1000–1300) with the exception of the deeper (20–60 m) phytoplankton productivity measurements at Lake Tahoe which were made within a week of the periphyton and shallow phytoplankton productivity measurements. Estimates of periphyton productivity at depths from 20–60 m were made from a regression equation with biomass (chl *a* and particulate nitrogen); these two parameters have been shown to be significantly correlated in Lake Tahoe (Loeb 1980; Loeb & Reuter 1981). Particulate nitrogen was determined using a Carlo Erba Elemental Analyzer (Model 1104). Chlorophyll *a* was calculated using the trichromatic equations of Strickland and Parsons (1968) following extraction of the samples in 90% acetone (24 h in the dark at 4° C after being homogenized in a tissue grinder).

A detailed description of the study lakes has been presented elsewhere (Loeb & Reuter 1981). These lakes are all subalpine, oligotrophic, and nutrient deficient. The comparative investigations of these five lakes were conducted during August and early September 1979. The other data presented for Lake Tahoe were collected between January 1978 and January 1979.

Methods to express littoral production

Water column phytoplankton production is commonly expressed as the integrated rate of primary productivity using values for discrete depths:

$$\text{water column phytoplankton production } (\text{mg C} \cdot \text{m}^{-2} \cdot \text{time}^{-1}) = \int_0^z \text{phyto. prod. } (\text{mg C} \cdot \text{m}^{-3} \cdot \text{time}^{-1})\, dz. \qquad (1)$$

The sum of water column phytoplankton production and benthic production from a depth equivalent to the depth of the water column is one way to evaluate littoral zone production:

$$\text{total littoral water column prod.} = \int_0^z \text{phyto. prod. } dz + \text{benthic prod.}_z \qquad (2)$$

An assumption in equation (2) is that the increase in benthic area below a square meter of lake surface resulting from the slope of the lake bottom is insignificant. This assumption may be valid for very shallow lakes or shallow areas of the lake littoral zone where the slope is slight; however, the error increases with increasing degree of slope (Fig. 1). For smooth sloping lake bottom (e.g. epipelic and epipsammic communities), the benthic production below one square meter of lake surface can be calculated using the equation:

$$\text{secant } \theta \text{ benthic prod. } (\text{mg C} \cdot \text{m}^{-2} \cdot \text{time}^{-1}). \quad (3)$$

If the benthic substrate is rock (epilithic community) a 'step function' is more appropriate; i.e. a combination of the horizontally (H) and vertically (V) oriented surfaces:

$$\text{H-benthic prod.} + \tan \theta \text{ V-benthic prod.} \quad (4)$$

Equation (4) requires measurements or estimates of periphyton productivity on both surface orientations (H and V).

Total littoral zone production can be expressed as the sum of the integrations of a series of total water column production measurements extending away from the shore a distance, x, to where the depth of the water is equal to the maximum depth

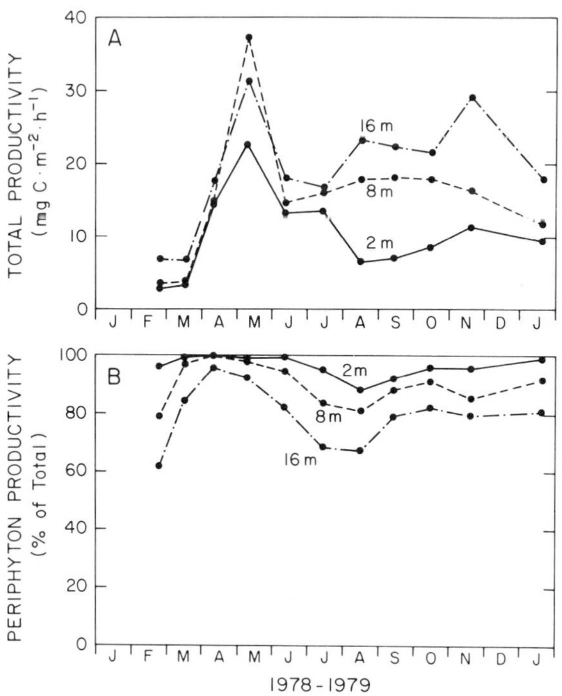

Fig. 2. Conceptual representation of the littoral zone expressed by Equation (5) (see text). X is the distance from shore to the surface projection of the depth contour equal to the maximum depth, Z, of littoral zone.

of the littoral zone (i.e. 1% of surface light intensity):

$$\text{total littoral zone production}$$
$$(\text{mg C} \cdot \text{m}^{-1} \cdot \text{time}^{-1}) = \int_0^x \int_0^z \text{phyto. prod. } dz dx + \int_0^x \text{benthic prod. } dx. \quad (5)$$

Equation (5) represents a 'wedge' of the littoral zone, 1m wide, extending from the shore out a distance to where the water depth is equal to the maximum depth of the littoral zone (Fig. 2). The benthic production term would be expressed as either equation (3) or (4) depending upon the type of substratum present. Macrophyte production, when significant, would be included in the benthic production term.

Results and discussion

In the shallow littoral zone of Lake Tahoe (2, 8 and 16 m), maximum annual productivities occurred in May while minimum productivities occurred during February or March (Fig. 3). The epilithic periphyton community dominated the littoral water column production throughout the year ($z \leq 16$ m; light intensity approximately 35% of surface light intensity at 16 m); maximum and minimum periphyton productivities occurred at the same time as did the total littoral water column productivities. Littoral phytoplankton, however, reached maximum production rates during August when periphyton production was decreasing and was near its annual minimum during the spring (March–May) when the periphyton community was nearing its maximum productivity. Dissimilar patterns of epilithic periphyton and littoral phytoplankton have also been reported by Kairesalo (1976). Seasonal differences in the production rates of periphyton and phytoplankton can, therefore, affect the total littoral zone production. Geomorphology and the composition of the substrata (e.g. rock, sand, or organic material) are also important factors affecting total littoral production.

The different seasonal patterns of productivity for these two algal communities observed during

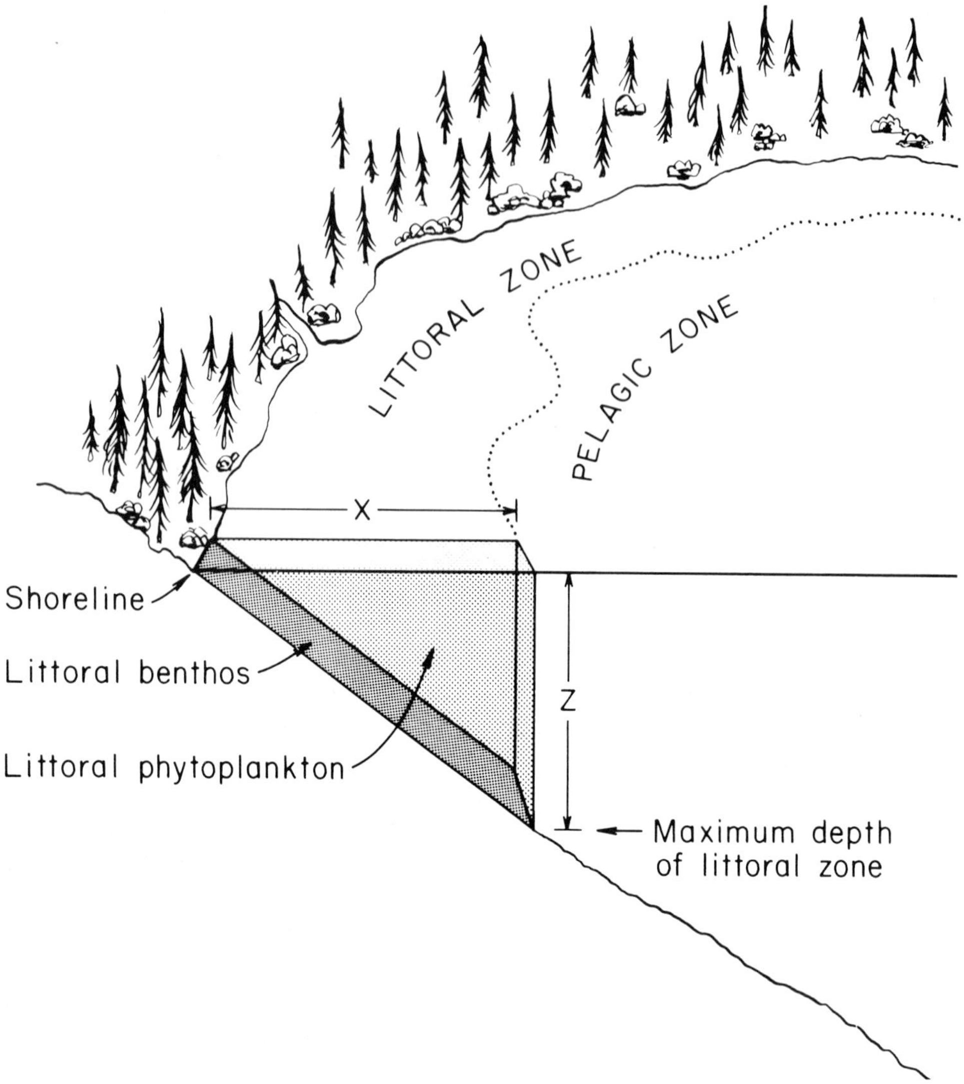

Fig. 3. A: Total littoral water column productivity at 2, 8, and 16 m from February 1978 to January 1979 at Lake Tahoe. B: The contribution of epilithic periphyton productivity to the total water column productivity (percent total).

1978 in Lake Tahoe suggest fundamental differences in factors regulating primary productivity (e.g. different sources of nutrients, differing physiological abilities to utilize available nutrients). Periphyton and phytoplankton, for example, respond differently to similar light regimes, particularly at high light levels. Several investigations have demonstrated that periphyton productivity is not suppressed at high light intensities which inhibit phytoplankton productivity (Goldman *et al* 1963; Hargrave 1969; Gruendling 1971; Hillbricht-Ilkowska *et al* 1972; Stanley & Daley 1976). Utilization of low ambient concentrations of dissolved nutrients (e.g. nitrate, ammonium) can also be different for these two algal communities. Differences in the kinetics of nitrate and ammonium assimilation by periphyton and phytoplankton have been shown (Loeb & Reuter 1981; Axler *et al* 1982a, b; Reuter 1983). Half-saturation constants (K_t) for dissolved inorganic nitrogen uptake are about two orders of magnitude lower for phytoplankton than for periphyton. These differences suggest that these two communities are adapted to different environments: the open lake water which has relatively low

Table 1. A comparison of littoral zone productivity of five oligotrophic lakes. In all lakes, the benthic substratum was rock and productivity on horizontal and vertical surfaces was assumed to be equal.

Lake	Secchi depth (m)	Slope ($\theta°$)	Depth (m)	Phyto. prod. dz ($mg\ C\ m^{-2}h^{-1}$)	Peri. prod. ($mg\ C\ m^{-2}h^{-1}$)	Total prod. ($mg\ C\ m^{-2}h^{-1}$)	Peri. as % total (%)
Crater Lake	28.0	45	2	0.23	7.02	7.25	97
			8	0.95	5.18	6.13	85
			16	2.39	10.48	12.87	81
Lake Tahoe	24.5	45	2	1.34	4.06	5.40	75
			8	5.90	8.02	13.92	58
			16	11.05	12.24	23.29	53
Fallen Leaf Lake	16.0	35	3	2.07	1.56	3.63	43
			7	5.02	3.66	8.68	42
			13	10.09	1.82	11.91	15
Donner Lake	13.5	30	2	3.39	6.47	9.89	66
			6.5	12.15	14.26	26.41	54
			17.5	25.63	0.43	26.06	2
Castle Lake	11.5	45	3	10.74	16.86	27.60	61
			7.8	27.37	13.02	40.39	32
			12	40.75	12.72	53.47	24

concentrations of nitrate and ammonium and benthic environments which has higher concentrations or alternative sources of nitrogen. The periphyton of the eulittoral (splash) zone represent a third type of environment and are distinctly different from the sublittoral periphyton both in species composition and nitrogen uptake kinetics (Ax ler *et al* 1982b; Reuter 1983).

Another striking difference between the phytoplankton and periphyton communities in certain oligotrophic lakes is the absence of nitrogen-fixing blue-green algae in the plankton and the dominance of these organisms in the epilithic periphyton (Jansson 1979; Loeb & Reuter 1981; Reuter *et al* 1982). N fixation accounted for over 50% of the total N uptake by the sublittoral epilithic periphyton community in the five oligotrophic lakes discussed in this paper (Loeb & Reuter 1981). This contrasting situation between the epilithic periphyton and phytoplankton communities in certain oligotrophic lakes may provide a framework for investigating factors which regulate the production of nitrogen fixing blue-green algae in open waters.

Regenerative nutrient processes are also likely to be different in the littoral benthic environment than in the overlying waters. Nutrient concentration gradients are sharpest at the interface of the benthos and the overlying waters (e.g. Mortimer 1941–1942). Benthic algae, in particular epipelic communities, may, therefore, have access to and be partially regulated by this pool of nutrients which is not as readily available to the phytoplankton. Subsurface groundwater inflow entering through the littoral sediments can also contain much higher concentrations of certain dissolved nutrients than surface runoff (Behnke 1975; Lee 1977; John *et al* 1978; Loeb & Goldman 1979). Benthic algae are first exposed to this source of nutrients and ongoing sampling programs have demonstrated increased periphyton growth adjacent to urban shoreline developments which are believed to contribute relatively higher groundwater nutrient loads than undeveloped areas (Loeb 1980; Leonard & Goldman 1981; Goldman *et al* 1982).

Comparison of the littoral water column production of five oligotrophic lakes further demonstrates the relatively large contribution of the periphyton community (Table 1). The trend of decreasing contribution of periphyton to the total production with increasing depth of the water column is also apparent from these data.

Expression of littoral production in terms of lake surface area is of limited use. Equation (5) was derived to allow the total production of the littoral zone to be expressed as a single term. Using this expression, the total productivity of the littoral zone (0–60 m) of Lake Tahoe was determined at a site (Rubicon Pt.) where the benthic substratum was entirely rock and the slope was 45°. This production rate was 1624 mg C m^{-1} h^{-1}. Of this total, 63% or 1022 mg C m^{-1} h^{-1} was contributed by epilithic periphyton and 37% or 602 mg C m^{-1} h^{-1} was contributed by littoral phytoplankton.

The utility of this conceptualization of the littoral zone of Lake Tahoe has aided our designing research programs to evaluate the impact of increased nutrient loading of the lake from urban development. Accelerated eutrophication of Lake Tahoe has resulted in the doubling of pelagial phytoplankton production over the past two decades (Goldman 1981). Ongoing studies indicate periphyton production has also greatly increased and that the spatial distribution of periphyton production is positively correlated with the distribution of urban development around the lake's shore line (Loeb 1980). These changes and relationships are all believed to be associated with increased nutrient loading. The periphyton community and the littoral zone of lakes are both valuable experimental areas for research related to nutrient dynamics, primary production, and the impact of human activities on these sensitive nearshore waters.

Acknowledgements

We thank R.P. Axler for his critical review of this manuscript and his assistance with field work. J.P. Zehr also reviewed this manuscript. We also thank M. Otto, A.P. Giannini, R.C. Richards and D. Fellows for their help with SCUBA diving and laboratory work. G. Malyj, E. deAmezaga, M. Smith and D. DuMont assisted with the preparation of this manuscript. Research support was provided by N.S.F. grants DEB 78-23824 and DEB 80-19918 to C.R. Goldman.

References

Axler,R.P., Gersberg, R.M. & Goldman, C.R., 1982a. Inorganic nitrogen assimilation in a subalpine lake. Limnol. Oceanogr. 27:53–65.

Axler, R.P., Goldman, C.R., Reuter, J.E., Loeb, S.L., Priscu, J.C. & Carlton, R.G., 1982b. Comparative studies of the nitrogen metabolism of phytoplankton and periphyton in oligotrophic lakes. FAO/IAEA, Rome, Italy.

Behnke, J., 1975. A summary of the biochemistry of nitrogen compounds in groundwater. J. Hydrol. 27:155–167.

Davis, G.S., 1970. Productivity of macrophytes in Marion Lake, British Columbia. J. Fish. Res. Board Can. 27:71–81.

Goldman, C.R., 1981. Lake Tahoe: Two decades of change in a nitrogen deficient oligotrophic lake. Verh. Internat. Verein. Limnol. 21:45–70.

Goldman, C.R. & deAmezaga, E., 1975. Primary productivity in the littoral zone of Lake Tahoe, California-Nevada. Symp. Biol. Hung. 15:49–62.

Goldman, C.R., Leonard, R.L., Axler, R.P., Reuter, J.E. & Loeb, S.L., 1982. Interagency Tahoe Monitoring Program – Second Annual Report: Water Year 1981. Tahoe Research Group, Univ. of Calif., Davis, June 1982. 193 p.

Goldman, C.R., Mason, D.T. & Wood, B.J.B., 1963. Light injury and inhibition in Antarctic freshwater phytoplankton. Limnol. Oceanogr. 8:313–322.

Gruendling, G.K., 1971. Ecology of the epipelic algal communities in Marion Lake, Britisch Columbia. J. Phycol. 7: 239–249.

Hargrave, B.T., 1969. Epibenthic algal production and community respiration in the sediments of Marion Lake. J. Fish. Res. Board Can. 26:2003–2026.

Hickman, M., 1971. Standing crops and primary productivity of epipelon of two small ponds in North Somerset, U.K. Oecologia (Berl.) 6:238–253.

Hilbricht-Ilkowska, A., Kowalczewski, A. & Spodniewska, I., 1972. Field experiment on the factors controlling primary production of the lake plankton and periphyton. Ekologia Polska 20(25):315–326.

Jansson, M., 1979. Nutrient budgets and the regulation of nutrient concentrations in a small sub-Arctic lake in northern Sweden. Freshwater Biology 9:213–231.

John, P.H., Gibbs, M.M. & Downes, M.T., 1978. Groundwater quality along the eastern shores of Lake Taupo, 1975–1976. N.Z. J. Mar. Freshwater Res. 12:59–66.

Kairesalo, T., 1976. Measurements of production of epilithiphyton and littoral plankton in Lake Pääjärri, Finland. Ann. Bot. Fennici 13:114–118.

Lee, D.R., 1977. A device for measuring seepage flux in lakes and estuaries. Limnol. Oceanogr. 22:140–147.

Leonard, R.L. & Goldman, C.R., 1981. Interagency Tahoe Monitoring Program – First Annual Report: Water Year 1980. Tahoe Research Group, Univ. of Calif., Davis, May 1981. 82 p.

Loeb, S.L., 1980. The Production of the Epilithic Periphyton Community of Lake Tahoe, California-Nevada. Ph.D. thesis. Univ. of California, Davis. 165 p.

Loeb, S.L., 1981. An in situ method for measuring the primary productivity and standing crop of the epilithic periphyton community in lentic systems. Limnol. Oceanogr. 26:394–399.

Loeb, S.L. & Goldman, C.R., 1979. Water and nutrient transport via groundwater from Ward Valley into take Tahoe. Limnol. Oceanogr. 24:1146–1154.

Loeb, S.L. & Reuter, J.E., 1981. The epilithic periphyton community: a five-lake comparative study of community productivity, nitrogen metabolism and depth-distribution of standing crop. Verh. Internat. Verein. Limnol. 21:346–352.

Mortimer, C.H., 1941–1942. The exchange of dissolved substances between mud and water in lakes. J. Ecol. 29:280–329; 30:147–201.

Reuter, J.E., 1983. Studies of Nitrogen Assimilation of the Epilithic Periphyton Community in Lake Tahoe, California – Nevada. Ph.D. thesis, Univ. of California, Davis.

Reuter, J.E., Loeb, S.L. & Goldman, C.R., 1983. Nitrogen fixation in oligotrophic Lake Tahoe. In: Periphyton in Freshwater Ecosystems, R.G. Wetzel (ed.). Developments in Hydrobiology 17. pp. 101–109.

Sanders, F.S., 1976. An Investigation of Carbon Flux in the Sediments of Castle Lake, California. Ph.D. thesis, Univ. of California, Davis, 131 p.

Stanley, D.W. & Daley, R.J., 1976. Environmental control of primary productivity in Alaskan tundra ponds. Ecology 57:1025–1033.

Strickland, J.D.H. & Parsons, T.R., 1968. A practical handbook of seawater analysis. Bull. Fish. Res. Board Can. 167. 311 p.

Wetzel, R.G., 1964. A comparative study of the primary productivity of higher aquatic plants, periphyton and phytoplankton in a large, shallow lake. Int. Rev. ges. Hydrobiol. 48:1–61.

Wetzel, R.G., Rich, P.H., Miller, M.C. & Allen, H.L., 1972. Metabolism of dissolved and particulate detrital carbon in a temperate hard-water lake. Mem. Ist. Ital. Idrobiol. (Suppl.) 29:185–243.

Authors' address:
S.L. Loeb/J.E. Reuter/C.R. Goldman
Division of Environmental Studies
University of California
Davis, CA 95616, U.S.A.

CHAPTER 21

Biomass and production of periphyton attached to dead reed stems in Lake Maarsseveen

JOHN T. MEULEMANS and FLOOR HEINIS

Abstract. Seasonal changes of periphytic biomass attached to dead reed stems were measured during two successive years. Biomass was compared at the bankside, the lakeside, and the center of a reed-belt. The biomass was high in winter; subsequently it decreased and remained low during summer (maximum and minimum values being, respectively, 49.52 ± 5.61 and 5.22 ± 2.88 g ash-free dry weight m^{-2} substratum). Preliminary results of periphytic production measurements *in situ* (O_2-method) are discussed.

The response of carbon fixation rate to light was studied in the laboratory (^{14}C-method). The ^{14}C-fixation response curves showed an inhibition of C-fixation rate under high light intensities. In summer, when the periphyton was 'shade-adapted', this inhibition was less in comparison with responses found in spring and late summer. These results may indicate that factors other than light are responsible for the observed inhibition.

Introduction

Limnological research was begun in Lake Maarsseveen I in the seventies. The littoral region occupies about 25% of the total surface area of the lake. More than 90% of the emergent littoral zone is covered with *Phragmites australis* (Cav.) Trin. ex Steudel. The submerged parts of these plants represent a large substratum for periphyton, which is dominated in winter by diatoms of genera *Diatoma, Fragilaria*, and *Tabellaria* and in summer by the genera *Achnanthes, Gomphonema*, and *Synedra*. The role of this periphyton was investigated in the littoral zone and in the entire lake. Several aspects of this periphyton, notably its dynamics and architecture, were investigated by Roos *et al* (1981). In the present study the seasonal changes in periphytic biomass attached to dead *Phragmites* stems were analyzed during two subsequent years. Some preliminary *in situ* measurements of the periphytic primary production (O_2-method) are given and the effects of light intensity on carbon fixation rates (^{14}C-method) are demonstrated.

Materials and methods

The study was made in Lake Maarsseveen I, in the center of the Netherlands. This artificial lake is oligo-mesotrophic; it is situated at 52°10′N, 05°06′E and has a surface area of 70 ha. The physical and chemical characteristics of this lake have been summarized by Ringelberg (1981).

Sampling sites. Two reed-belts situated at the SW- and NE-sides of the lake were chosen for the experiments. The samples within a reed-belt were collected at the bankside, the lakeside, and in the center of the stand. The mean shoot densities per m^2 were estimated, which averaged 53 for the living

Wetzel, R.G. (ed.), Periphyton of Freshwater Ecosystems.
 ISBN 90 6193 768 X.

ones, and about 150 for the submerged dead stems.

Determination of biomass. To estimate the periphytic biomass, three dead stems were collected in triplicate at every sampling station. Stems were cut off just above the bottom, transported to the laboratory, and the periphyton was removed using a scalpel. Biomass was expressed as ash-free dry weight/substratum area (g m^{-2}). The periphyton was dried at 70° C to constant weight; part of the dry matter was weighed and combusted in alumina combustion boats for 2 h at 550° C and reweighed.

Net production measurements. Net production of the periphyton community was estimated using the oxygen method. Each time 2 dead reed stems of about 20 cm length and its adhering periphyton were cut and incubated in a perspex chamber (volume 0.5 l), which was filled with filtered lake water (0.45 μm pore size Sartorius membrane filter). Oxygen evolution was measured using an YSI (Model 5739) oxygen electrode, built into the incubator. The water was stirred continuously. The measurements were performed *in situ* during midday and lasted 4 h. The chlorophyll a concentrations were determined after Nusch (1980) in a slightly modified way. The solar irradiation above the reed-belt was measured with a irradiance meter (LI-COR Quantum Radiometer type Li 185).

The effects of light intensity on ^{14}C-fixation. The periphyton was carefully scraped from the stems, and gently treated for 5 min by means of a Vibromixer to get a homogeneous periphyton mixture, diluted with filtered lake water ($\pm$ 8 μg chl a l^{-1}). The periphyton was stored for one night in dark at the same temperature as the lake water until the experiment was started. Bottles were filled with 100 ml of the diluted periphyton and incubated with $NaH^{14}CO_3$ (0.02 μCi ^{14}C ml^{-1}) for 3 h at different light intensities ranging from 0 to 1500 μE m^{-2} s^{-1}. The incubation was done according to the method of Fee (1973) as modified by Flik & Keyser (1981).

Results

Seasonal changes in the periphytic biomass at the center of the reed-belts are given in Fig. 1. From

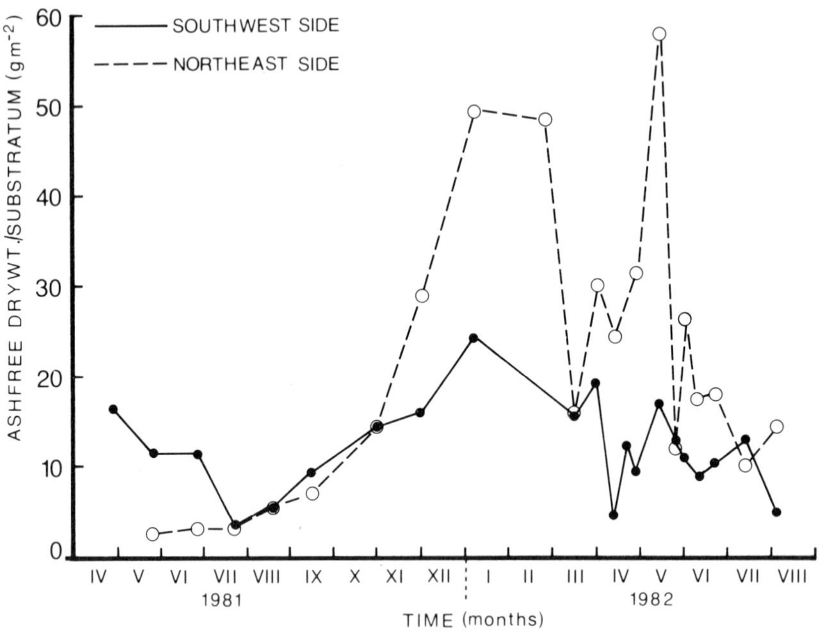

Fig. 1. Changes in periphytic biomass attached to dead reed stems. The sampling stations were in the center of a reed-belt, situated at the S.W.- and NE-side of the lake. Biomass is expressed as ash-free dry weight/substratum area (g m^{-2}).

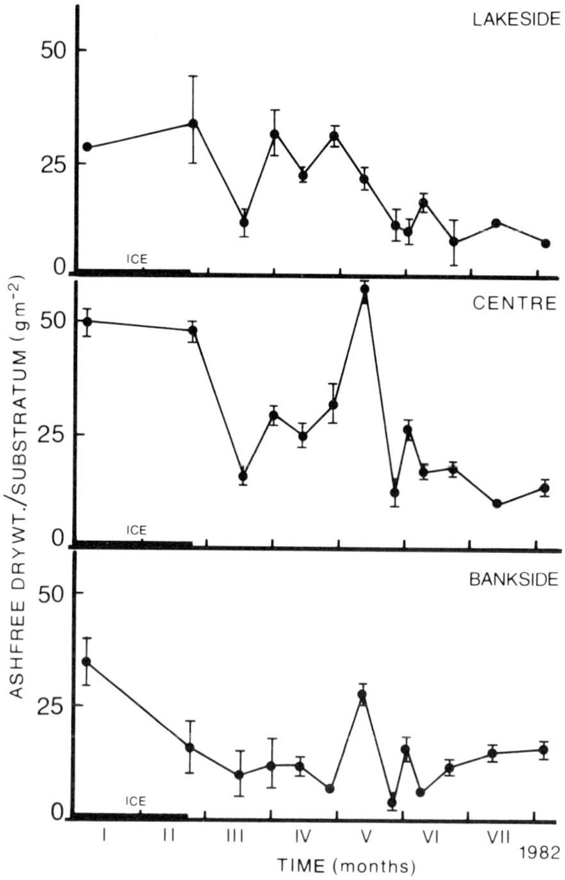

Fig. 2. Changes in periphytic biomass attached to dead reed stems. The sampling stations were at the bankside, the lakeside, and the center of a reed-belt situated at the NE-side of the lake. Biomass is expressed as ash-free dry weight/substratum area ($g m^{-2}$).

mid-May 1981 onwards the biomass at the NE-side of the lake was relatively low, and it remained low until the end of the summer (2.25–7.05 g ash-free dry weight m^{-2} substratum). Subsequently an increase in biomass occurred in autumn with maximum values during winter (49.52 ± 5.61 g ash-free dry weight m^{-2} substratum). The biomass then decreased except for a short sharp increase at mid-May. The ash-free dry weight consisted in autumn and winter mainly of epiphytic algae, while in spring and summer the algal component was actually smaller because of the contribution of small animals (nematodes, chironomids, ciliates, and a number of epiphytic protozoans, especially vorticellids). The detritus component within the periphyton was negligible. The same tendency in the dynamics of periphytic biomass was observed at the SW-side of the lake, but was less pronounced (Fig. 1). In Fig. 2 the differences in seasonal biomass within one reed-belt (NE-side) at several sites are shown. These differences were substantial and seasonally dependent. In winter and spring the biomass at the bankside was lower than at the other sampling stations.

Several net production measurements of the periphytic community showed that in May O_2-production was higher on cloudy days than on sunny days (Table 1). In July, when the reed leaves were fully expanded, the net production of the periphytic community was almost zero. This net production rate could be enhanced under higher light intensity (outside the reed-belt) *in situ*.

The ^{14}C-experiments revealed (Fig. 3) that in spring the C-fixation rate at a light intensity of 1500 $\mu E\ m^{-2}\ s^{-1}$ was only 44% of the maximum rate which occurred at 300 $\mu E\ m^{-2}\ s^{-1}$. In early summer, however, less inhibition by high light intensities was found. The C-fixation rate in July reached its maximum even at an intensity of 1500 $\mu E\ m^{-2}\ s^{-1}$, whereas subsequently in late summer light inhibition at high light intensities (800–1500 $\mu E\ m^{-2}\ s^{-1}$) occurred again.

Discussion

In general the periphytic biomass was high in winter and low in summer. This difference pertained in

Table 1. Net oxygen production of the periphytic community attached to dead reed stems. The *in situ* incubations (O_2-method) were done during midday and lasted 4 h.

Date	Oxygen production ($mg\ O_2\ mg^{-1}$ chl-a h^{-1})	Mean PHAR ($\mu E\ m^{-2}\ s^{-1}$)
27 April	0.22	275
	0.25	
6 May	0.23	605
	0.43	
9 May	0.08	1410
13 May	0.14	1716
	0.11	

particular to the biomass of the epiphytic algae, since in winter the ash-free dry weight consisted mainly of the organic material of epiphytic algae. In summer, however, the periphyton consisted partly of animal material. A possible explanation for the decline of the epiphytic algae in summer was the low available light intensity which was reduced by a well-developed leaf canopy of the reed stand by that time, which caused a low periphytic production. Less is known about other factors like the status of nutrients, water temperature, or the grazing-pressure on the periphyton in Lake Maarsseveen. That grazing can be important as a regulating factor for periphytic production in littoral zones has been shown for instance by Higashi *et al* (1981) in Lake Biwa. Although in general low periphytic biomass levels are found in summer, Sommer (1977) reported in contrast to this a high periphytic biomass in summer and low biomass in winter in the Neusiedler See. However, he expressed biomass on dry weight basis. In Lake Maarsseveen ± 50% of the dry weight consists of $CaCO_3$, which indicates that dry weight is not suitable as a measure of periphytic biomass.

Net production of the periphytic community was low in spring. Since the values represented the net result of heterotrophic and autotrophic activity within the periphyton community, the low values in that period may have been caused by high heterotrophic activity. The possible inhibition of pri-

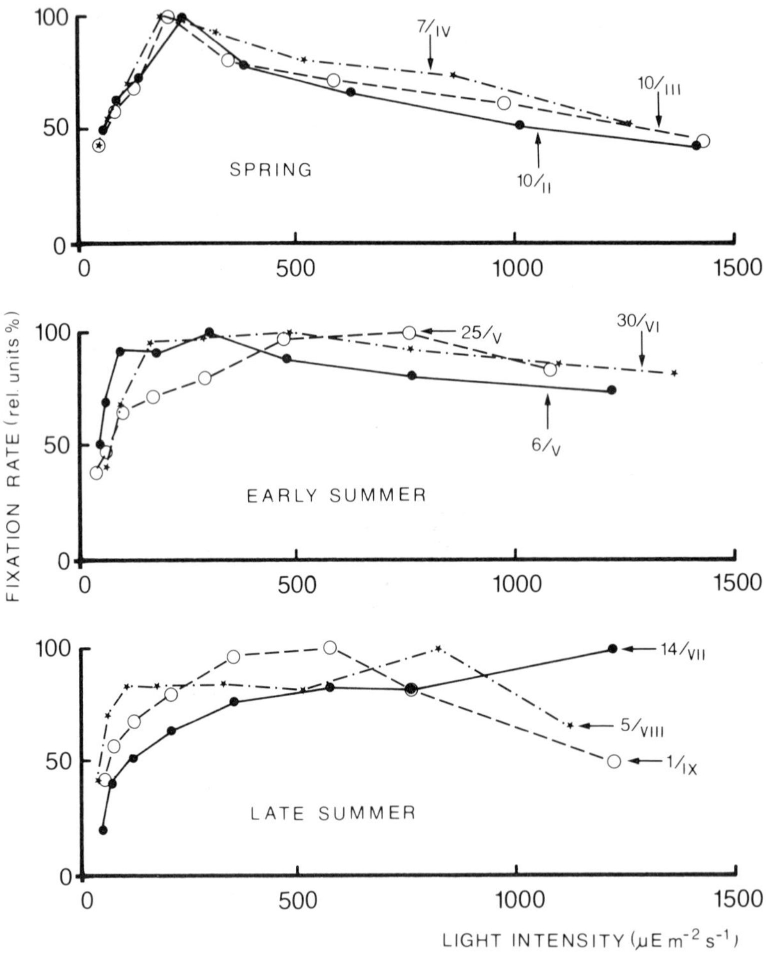

Fig. 3. The effect of light intensity of the ^{14}C-fixation rate periphyton, sampled from spring through late summer. The C-fixation rate is expressed as a percentage of the highest response for each incubation. The incubation temperatures during spring, early-, and late summer were, respectively, 4°, 15°, and 22° C.

mary production by high light intensity in spring, as indicated by the ^{14}C-response curves to light, will be studied further. The latter curves showed an inhibition of C-fixation in spring under high light intensity. In summer, this inhibition was less in comparison with curves found in spring and late summer. In that period, periphyton was adapted to low light intensity (strong attenuation of light under a well-developed leaf canopy). Most studies to the effect of light on the photosynthetic rate pertain to phytoplankton. Penhale (1977), however, examined the epiphyton attached to *Zostera marina* in this respect at three temperatures, notably 15°, 22°, and 29° C (^{14}C-method). No inhibition was observed at 15° and 29° C, while at 22° C the photosynthetic rate reached a maximum and declined slowly as the light intensity exceeded the 600 μE m^{-2} s^{-1}. Tanimizu *et al* (1981) failed to observe photoinhibition in winter during incubation (O_2-method) of periphyton attached to reed stems under several light intensities (maximum light intensity used was 20 K Lux). Our experiments, in which the 'shade adapted' periphyton showed almost no photoinhibition in summer under high light intensities, may indicate that factors other than light, e.g. nutrient limitation, may have been responsible for this phenomenon (see Flik, in press).

Acknowledgements

The authors thank Dr. E.P.H. Best, Dr. P.J. Roos, and Prof. Dr. J. Ringelberg for critically reading the manuscript. Thanks are due to Eveline de Hollander and Marco Heykoop for their skillful technical assistance. This investigation was supported by the foundation for Fundamental Biological Research of the Netherlands (B.I.O.N.), subsidized by the Netherlands Organization for the Advancement of Pure Research (Z.W.O).

References

Fee, E.J., 1973. A numerical model for determining integral primary production and its application to Lake Michigan. J. Fish. Res. Board Can. 30:1447–1468.

Flik, B.J.G. & Keyser, Á., 1981. Estimation of the primary production in Lake Maarsseveen I with an incubator technique. Hydrobiol. Bull. 15:41–50.

Higashi, M., Miura, T., Tanimizu, K. & Iwasi, Y., 1981. Effects of the feeding activity of snails on the biomass and productivity of an algal community attached to a reed stem. Verh. Internat. Verein. Limnol. 21:590–595.

Nusch, E.A., 1980. Comparison of different methods for chlorophyll and phaeopigment determination. Arch. Hydrobiol. Beih. Ergebn. Limnol. 14:14–36.

Penhale, P.A., 1977. Macrophyte-epiphyte biomass and productivity in an eelgrass (Zostera marina L.) community. J. exp. mar. Biol. Ecol. 26:211–224.

Ringelberg, J., 1981. Introduction to the research area. Hydrobiol. Bull. 15:5–9.

Roos, P.J., Post, A.F. & Revier, J.M., 1981. Dynamics and architecture of reed periphyton. Verh. Internat. Verein. Limnol. 21:948–953.

Sommer, U., 1977. Productionsanalysen am Periphyton im Schilfgürtel des Neusiedler Sees. Sitzber. Österr. Akad. Wiss. Math.-naturw. Kl. Abt. I. 186:219–246.

Tanimizu, K., Miura, T. & Higashi, M., 1981. Effect of water movement on the photosynthetic rate of an algal community attached to reed stems. Verh. Internat. Verein. Limnol. 21:584–589.

Authors' address:
J.T. Meulemans/F. Heinis
Department of Aquatic Ecology
University of Amsterdam
Kruislaan 320
1098 SM Amsterdam, The Netherlands

Seasonal changes in organic matter and dark oxygen uptake of epiphyton and epipelon in relation to seston deposition in Lake Vechten (The Netherlands)

HERMAN J. GONS and RONNY VAN KEULEN

Abstract. Epiphyton and epipelon, i.e. the particulate matter associated with submerged macrophytes, and on the sandy lake bottom, respectively, were quantitatively collected from the substrates in order to study their role in the carbon cycle of Lake Vechten (The Netherlands). Both the epiphyton and epipelon were mainly detrital, and the exchange with the seston principally determined their seasonal and spatial distribution. The epipelon was maximal in mid-summer, by which time it may contain more carbon that the seston on a lake-wide basis. The epiphyton continued to increase until the autumnal turnover. Dark oxygen uptake by the epipelon was equivalent to one third of the limnetic production of particulate organic matter. In years with extensive macrophyte development the contribution of the lower littoral region to the decomposition of organic matter produced in the epilimnion may be substantially higher than during this study. The role of hydrodynamics in the distribution of epiphyton and epipelon in relation to the deposition of seston is discussed.

Introduction

Regarding the significance of exchange of particulate organic matter for the metabolic interactions in macrophytic-epiphytic and epipelic associations, Wetzel & Allen (1972) emphasized the simultaneous roles of autochthonous and allochthonous materials entering the littoral of lakes. It is conceivable, however, that autochthonously produced planktonic organic matter is very important. In the phycological literature there are numerous accounts of massive exchange between periphyton and phytoplankton (e.g. Brown & Austin 1973; Cattaneo & Kalff 1978; Moss & Abdel Karim 1969).

Little is known about the input of total sestonic matter into the periphyton, since it is difficult to quantify the total epiphytic and epipelic matter. Indirect measurements of seston deposition, i.e. by using sediment traps, have seldom been made in the littoral region of lakes, and, in this case, estimating 'true' deposition meets with many uncertainties (Blomqvist & Håkanson 1981). Davis & Brubaker (1973) demonstrated that finer size pollen was deposited more in the littoral than in the deeper lake parts during stratification. The question was put forward whether this might apply also to the seston.

Gons (1982) studied the epiphyton associated with submerged macrophytes, mainly *Ceratophyllum demersum*, and the epipelon on the sandy bottom in Lake Vechten. Both epiphyton and epipelon could be quantitatively collected from the substrates, and were defined as all particulate matter between 0.2 and 125 μm in size. Throughout the year the epiphyton and epipelon consisted mainly of detrital aggregates; among the algae motile pennate diatoms generally predominated, but also

Wetzel, R.G. (ed.), Periphyton of Freshwater Ecosystems.
 ISBN 90 6193 768 X.

phytoplanktonic species were often abundant. Oxygen concentrations *in situ* indicated that heterotrophy might exceed autotrophic production. This was confirmed by the oxygen exchange during laboratory incubations. These findings were explained by import of sestonic organic matter.

The periphyton research is a part of ecosystem studies on Lake Vechten with the main aim to quantify the pool sizes and fluxes of carbon (see Gons 1981). In the present paper the quantitative importance of the epipelon- and epiphyton-seston interrelationships in Lake Vechten is described. Of interest for lakes in general are the parts played by hydrodynamic conditions and trophic state, which are discussed.

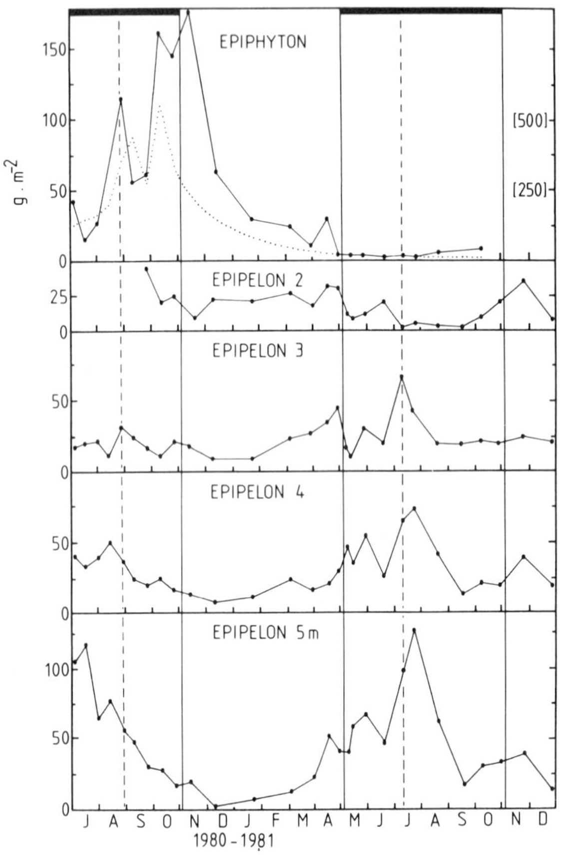

Fig. 1. Seasonal distribution of organic matter contained in epiphyton from sampling station 2 m, and in the epipelon from the stations 2–5 m. The epiphyton was associated with *Ceratophyllum demersum*, for which the total dry weight is represented by the dotted line (scale in brackets). The periods of summer stratification are indicated by the black bars, and the temperature maximum by the broken lines.

Study site

A recent description of Lake Vechten (4.7 ha; $z_m = 12$ m) is given by Steenbergen & Verdouw (1982). For species composition and distribution of the aquatic macrophytes, and physico-chemical characteristics of the littoral, see Best (1981) and Gons (1982). Based on the occurrence of the macrophytes, the littoral region extends down to the 5-m depth contour, which is 38% of the total lake area. During recent years only a small part of the shore was lined with emergent vegetation and on an areal basis the zone of the floating-leaved macrophytes was greatly exceeded by that of the submerged species. During 1979–1981 the latter zone was small compared with the area covered only by epipelon.

Most of the results refer to the samples collected in 1980–1981 along a fixed transect through the SE part of the lake (Gons 1982). All along the transect the bottom was sandy, as applies to the major part of Lake Vechten's littoral region. During 1980 a dense *Ceratophyllum* bed developed at 2 m depth (sampling station 2 m), but the macrophytic biomass was an order of magnitude lower in 1981. At the sampling station 3, 4, and 5 m the sand was covered only by epipelon.

Materials and methods

The methods used were described previously (Gons 1982); they are only outlined briefly below.

Sampling. Macrophytes and epiphyton were collected by SCUBA-diving. The loss of the loosely associated part of the epiphyton could not be completely prevented. Epipelon was collected using a suction method, which gave good results at the greater depths, but was inadequate for the quantitative collection of epipelon in the dense vegetation at station 2 m in 1980. This was mainly due to the macrophytic parts bent over the bottom: these prevented the isolation of a known area.

In the laboratory epiphyton was quantitatively separated from its substrate using vigorous stirring by a Vibromixer (Chemap, Zürich) and lowering the pH to 5 for removal of the tightly adhered fraction.

Gravimetric and chemical analyses. Aliquots of the suspensions of epiphyton and epipelon were centrifuged and the dry weights determined after freeze-drying. Ignition loss was determined after ashing at 450° C. For assessing the contribution of $CaCO_3$ and organic carbon content, elemental analysis was carried out on both total dry matter and the residue after ignition. Organic weight was assumed to be identical with the ignition loss. Percentage organic matter and $CaCO_3$ refer to the total dry weight. Chlorophyll *a* was extracted in methanol and computed according to Iwamura *et al* (1970).

Dark oxygen uptake. Aliquots of the suspensions were incubated in BOD-bottles in the laboratory at *in situ* temperature. Start and end concentrations were measured using a YSI meter and self-stirring BOD-bottle probe (Yellow Springs, Ohio).

Results

Seasonal distribution of epiphyton

The submerged macrophytes in Lake Vechten generally are heavily coated by epiphyton, which, including carbonates and ash, may exceed the macrophytes several times on the basis of dry weight. As a rule, the percentage organic matter is close to 20 (Gons 1982). Seasonal changes in epiphytic organic matter at sampling station 2 m and in the macrophyte total dry weight are given in Fig. 1.

A dense stand of *Ceratophyllum demersum* developed in 1980. The macrophyte standing crop maximum occurred late in the season; this was likely due to the exceptionally low irradiation during the second half of June and July. In October the macrophytes collapsed and partly began to disintegrate. The plant stand being small in area, sampling was limited to 3 shoots per date, and the integrated values for the macrophyte and the epiphyton were obtained after counting the shoot number per unit area. Starting from the end of October onward, this method could no longer be used. This later course was computed assuming that a quarter of the maximum standing crop, i.e. averaged September and October values, remained by the year end (cf. Best 1982). It was further assumed that the macrophyte dry weight continued to decline gradually to the low values of early Summer 1981.

During the summer 1980 integrated epiphytic organic matter was highest in August. After comparatively low values in September, the epiphyton increased rapidly, reaching maximum density in November, and the epiphytic organic matter surpassed that of the substrate. These events are by no means unique for 1980 and 1981, when the lake's submerged vegetation was poorly developed. In 1978 similar high epiphyton:macrophyte ratios occurred in autumn (Gons 1982).

The low macrophyte standing crop in 1981 implied, of course, that the epiphytic organic matter was much lower than in the previous year. The great difference in the macrophyte population density at station 2 m in the two years did not result in significant change in the epiphyton:substrate weight ratios. For total dry weight these fluctuated around 1 during July and August and reached 3 in autumn. During the study period the highest ration, i.e. 5.5, occurred in April 1981. Also, during the two summers the chemical composition of the epiphyton was much the same (Gons 1982).

Seasonal distribution of epipelon

At the sampling stations 3, 4, and 5 m macrophytes were absent. At these depths (Fig. 1) marked seasonal changes occurred in the particulate matter covering the sand; all year around the conditions remained aerobic. In winter as well, changes in the particulate matter at greater depth, down to 6–7 m, could be determined using the same method, but from spring onwards the sand was covered by at least several cm of anaerobic mud.

Epipelon values from 3–5 m depth were maximal in July or August. Thereafter the values declined rapidly until October during both years. The minima were observed towards the year end, but these were not as pronounced in 1981 compared with 1980.

The epipelon increased from January until the onset of the summer stratification. High gains in organic weight occurred in April. Accelerated accumulation followed the thermocline formation prior to the mid-summer temperature maximum.

The changes in epipelon at 2 m depth represented a special case, which was a result of the growth of *Ceratophyllum*. During the summer 1980, the population density of the macrophyte was so high that the sampler was considered to yield unreliable results (see Methods). Starting from January 1981 only could sampling be carried out with confidence; thus the 1980 data must be regarded as no more than indicative. In 1981 the seasonal changes at this depth often deviated from the above description for 3–5 m, notably in summer (see Discussion).

Depth distribution of epipelon

Throughout the stratification period epipelic organic matter increased with depth. Exceptions, however, were the 2-m values in 1980, which indicate that the epipelon in the vegetation was greater even than at the 5-m depth. This result agreed with visual impressions when diving. Accumulation of particulate matter, which probably arose for a great part from dissociated epiphyton, was also apparent from the steep decline in dissolved oxygen towards the bottom (Gons 1982). In early September 1980, the oxygen even reached zero in a night sample.

From December to April the concentrations did not differ greatly. Depth dependency relationships tended to be reversed from those during the summer: until early April, the 3 and 4-m values were higher than those for 5 m, but lower than those for 2 m. Only as late as mid-April did the epipelon from 5 m again reach a higher concentration than that from the lesser depths; this followed a rapid warming of the lake water.

At times, for instance around the beginning of the summer stratification, the changes in organic matter at the various depths were out of phase. This phenomenon as well as the differences with the course of the epiphytic organic matter is discussed below.

Epipelic vs. *sestonic organic carbon on lake basins*

The region between the 2 and 5-m depth contour occupies 20.3% of the area of Lake Vechten. Should the transect of the present study be representative for the lower littoral, the changes in particulate organic carbon contained in the epipelon per unit total lake area are represented in Fig. 2. The average sestonic POC-values measured during 1973–1978 (Gulati & Siewertsen, pers. comm.) are also given; see Gulati *et al* (1982) for more details on the seston dynamics.

The general conclusion is that increments to the epiphyton and epipelon cannot be accounted for by their own autotrophic production, but can be explained by imports of sestonic matter (Gons 1982). In this way, one obvious prerequisite is fulfilled: the seston concentrations are compatible.

From January to June, sestonic POC was higher than that of the epipelon and the changes for seston and epipelon were approximately parallel, including the clear water phase, which in 1981 occurred in the first half of May. For the seston a second peak generally occurred in September. The epipelon reached its maximum about two months earlier, also in 1980, and by then contained more organic

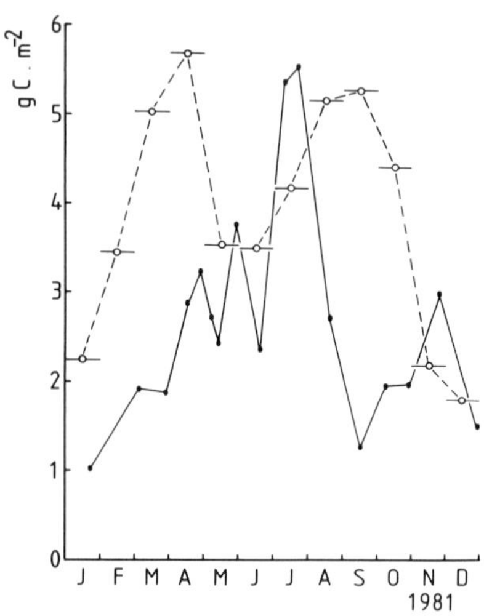

Fig. 2. Time courses of lake-wide concentrations of organic carbon contained in the epipelon from 2–5 m depth and in the seston (broken lines) – excluding zooplankton – above the oxygen zero line. The sestonic values are the monthly averages for the years 1973–1978 (Gulati & Siewertsen, pers. comm.).

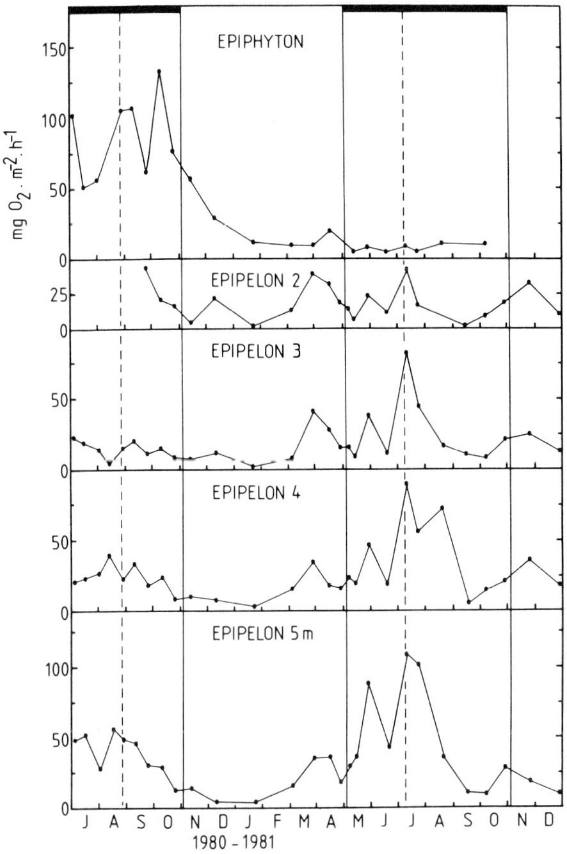

Fig. 3. Seasonal changes in dark oxygen uptake by epiphyton and epipelon. See Fig. 1.

carbon than the seston generally included in that time of the year. It might be anticipated that the epipelon lost from July onwards was mixed over the water column. In early September 1981, however, the lake water was even more transparent than in May. In the first part of September 1980 also, the vertical extinction of light reached a low value. It is very likely that the epipelon was transported down to the metalimnion, and together with vertical transport and local production contributed to the very high concentrations in that lake region during summer (Steenbergen & Korthals 1982). These concentrations were at least partly included in the areal values established by Gulati & Siewertsen (Fig. 2).

Epiphytic and epipelic oxygen consumption

The organic portion of the deposited sestonic matter is utilized by a great variety of heterotrophic organisms, of which the respiration is for the greater part of the year, including the summer months, the main component of the epiphytic and epipelic oxygen consumption.

The oxygen consumption rates (Fig. 3) were correlated primarily with the organic matter concentrations (Fig. 1). Thus the epipelic oxygen uptake increased with depth in summer, while in winter a slightly opposite trend was found.

From September 1980 to April 1981 the epiphyton and epipelon expressed similar weight-specific consumption rates. In the first part of the summer, the epiphytic specific oxygen uptake was markedly enhanced, but that of the epipelon remained comparatively low. This difference may reflect the contribution of newly deposited matter to both associations (Gons 1982).

For two reasons, (1) enhanced specific oxygen consumption during May-September, and (2) high organic matter concentrations up to autumn, the epiphyton may be as important for the mineralization of organic matter as the epipelon, despite a proportionately small area covered by submerged

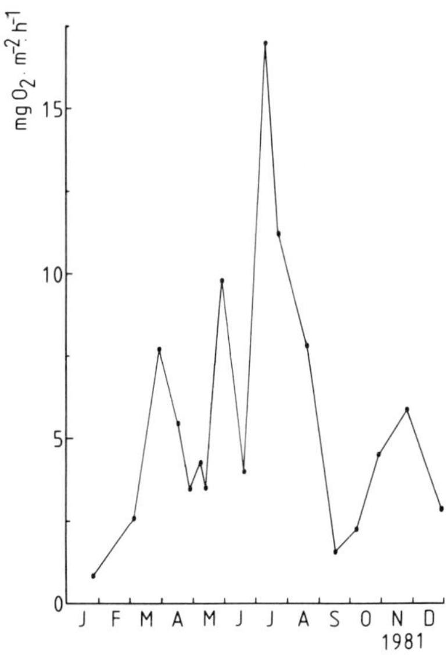

Fig. 4. Seasonal changes in lake-wide dark oxygen uptake by epipelon from 2–5 m depth.

vegetation. In 1981 though, the distribution of the macrophytes was so reduced that most of the oxygen consumption in the lower littoral was epipelic. The oxygen consumption rates for the epipelon on lake basis are shown in Fig. 4.

Epiphytic and epipelic dark oxygen uptake vs. *limnetic carbon fixation*

Carbon fixation by the phytoplankton was not measured during the greater part of 1981, but on the basis of that in previous years (de Kloet 1982) and taking into account of the night-time respiratory carbon loss, the lake-wide production of 50 g m^{-2} y^{-1} of particulate organic carbon should be a reasonable estimate.

Also on lake-wide basis, the epipelon from 2–5 m depth consumed 46 g O_2 m^{-2} y^{-1}, which is equivalent to one third of the phytoplanktonic production. For the following reasons a precise estimate of the part of the limnetically-produced organic matter being mineralized in the epipelon cannot be made.

Firstly, oxygen consumption due to the respiration of the epipelic algae, and that in the course of the mineralization of macrophytic and allochthonous matter, must be subtracted. Using artificial substrata it was demonstrated that active algal growth occurred up to at least the 4-m depth. But, relating dark oxygen uptake to the maximum gross photosynthesis of epipelon, it was established that as a rule 40% of the respiration of algal origin was the upper limit (Gons 1982). A complication is the respiration of deposited phytoplanktonic species; this may contribute significantly to the algal part of the oxygen uptake, before these cells lose their viability and are decomposed. Regarding the timing of the seasonal maximum in epipelic oxygen consumption (Fig. 4), a substantial contribution of mineralization of macrophytic and allochthonous matter to the annual epipelic oxygen uptake may be excluded. Allochthonous particulate input occurs mainly in the form of leaf-fall. Considering the tree species and the lake morphology, this particulate input represents no more than 5–10% of the annual limnetic production on the basis of the relationship derived by Gasith & Hasler (1976).

Secondly, the estimate was based on the one transect through the eastern depression of the lake. In 1982 a study was started on the distribution of epipelon along 4 transects; preliminary results indicate that epipelic oxygen uptake in the western part is lower than in the eastern lake part (Kromkamp, pers. comm.).

All these factors do not detract from the conclusion that the lower littoral is of utmost importance in the recycling of carbon fixed by the planktonic algae. In 1981 the major share of the mineralization in the littoral was epipelic. In years with extensive growth of the submerged macrophytes, the decomposition of sestonic matter undoubtedly will occur for a still greater part in the littoral region, since the vegetation both enhances and prolongs net deposition and, consequently, oxygen uptake. Of the phytoplankton particularly in summer and autumn, a greater percentage will be decomposed under aerobic conditions.

Discussion

For deposition of sestonic matter in general, and for import into the epiphyton and epipelon in the present study, several factors may be important. Since sufficiently detailed data are lacking on both hydrodynamical consequences of wind velocity and direction and on settling velocities of the particles in the epilimnion of Lake Vechten, the discussion is confined to phenomena relating to the thermal cycle. For Lake Vechten, being a temperate, holomictic lake, at least five seasonal phases in the thermal cycle may be distinguished (see Mortimer 1974).

In phase (1), turbulent mixing occurred over the entire water column from November 1980 until March 1981; ice cover was insignificant. Both seston and epipelon were low in mid-winter (Fig. 2). Their increase during the subsequent months correlated with the recurrent diatom spring bloom (Blaauboer 1982) and high carbon fixation (de Kloet 1982). Verdouw and Dekkers (1982), using sediment traps at the 8-m depth, observed low deposition rates in winter and increasing rates until April.

In both the epiphyton and epipelon planktonic

diatoms, mainly *Stephanodiscus* spp., predominated during January-March. It was already established in 1977 that the algal species composition in seston and epiphyton was similar during this phase (Gons 1982). That the epipelic matter was comparatively homogeneously distributed (Fig. 1) also appeared from the chemical composition (Table 1A); differences with the epiphyton were slight as well.

From December to March, the epipelic concentration was somewhat higher at the lesser depth, thus closer to the shore (Fig. 1). These results indicate that horizontal differences in exposition to the prevailing wind force were more important during this time than decreases in the wind-induced action going from 2 to 5 m depth.

Phases (2) and (3) in the thermal cycle are the initial and main warming periods, during which wind stress is counteracted by buoyancy forces. In Lake Vechten the transition from (2) to (3) involves the change from the 'spring' to the 'summer' phytoplankton (Blaauboer 1982) and low sestonic values in May and June (Gulati *et al* 1982; Fig. 2). During the two phases calm sunny periods result in weak density gradients which are destroyed by spells of turbulent weather. In 1981 significant temperature gradients (4–5° C over 5 m depth) occurred in mid-April, early May, the end of May, and during the first half of July. In the intervening periods these gradients did not occur, e.g. at the onset of May, or were not more than 2–3° C, as in the remaining part of May and in June (Gons 1982).

Table 1. Chemical characteristics of epiphyton and epipelon.

	% organic matter	% $CaCO_3$	mg chlorophyll *a*/g organic matter
A. November 1980–March 1981 (n = 5)			
epiphyton	18.4 ± 3.1	18.5 ± 3.8	1.01 ± 0.57
epipelon 2 m	20.5 ± 2.1	12.2 ± 1.0	1.31 ± 0.60
epipelon 3 m	15.6 ± 4.3	11.9 ± 0.8	1.00 ± 0.33
epipelon 4 m	13.3 ± 4.7	12.0 ± 0.7	1.17 ± 0.55
epipelon 5 m	12.1 ± 7.1	14.0 ± 2.4	1.15 ± 0.39
B. June–September 1981 (n = 5)			
epiphyton	16.6 ± 3.5	36.0 ± 4.9	2.38 ± 1.65
epipelon 2 m	14.6 ± 3.5	12.9 ± 2.0	0.98 ± 0.36
epipelon 3 m	10.0 ± 4.5	13.9 ± 3.5	0.72 ± 0.28
epipelon 4 m	6.6 ± 1.2	12.9 ± 1.1	0.71 ± 0.30
epipelon 5 m	7.8 ± 1.6	14.1 ± 1.5	0.54 ± 0.12

Moss (1969) attributed changes in the phytoplankton composition to similar temperature gradients. For the case of exchange between seston and epiphyton and epipelon, the above fluctuations in the thermal gradient distributions also explain the time course and depth distribution of organic matter (Fig. 1).

Firstly, the temperature gradients mark the times favorable for seston deposition: rapid increases in epiphyton and epipelon usually coincided. A notable exception was in early May 1981. Then, however, the seston concentration in the upper strata was very low compared with that given for May in Fig. 2.

Secondly, the resulting density gradients reduced the influence of wind at greater depth; consequently net deposition increased downwards, as is evident from the epipelic depth distribution (Fig. 1). Besides, this may explain the differences in the time course at the various depths, since increases in deposition rate will start at the deeper places, but losses due to wind mixing will occur first at the lesser depths. The sampling frequency was too low to demonstrate this relationship clearly.

It is debatable what happens when epiphyton and epipelon diminish. This reduction generally occurs at rates which cannot be accounted for by mineralization. Moreover, this reduction involves not only organic matter, but also the carbonate and ash fractions, since changes in the chemical composition were slight (Gons 1982). Probably, during the summer months, resuspension over the complete water column was less important than downward transport along the bottom. This transport became apparent in early May when deposition was low due to the sestonic concentrations. At the lesser depths the epipelon decreased markedly, but over the 2–5-m interval as a whole no significant changes occurred (Figs. 1 & 2; see also Gons 1982). Transport along the bottom was consistent with the results of Verdouw & Dekkers (1982): more material was collected in sediment traps near the shore, and its chemical composition was markedly different from that in traps in the central lake part. In the more turbulent periods the epiphytic loose fraction may become easily dissociated and settle down. In places with dense macrophyte stands this fraction may accumulate (epipelon from 2 m in 1980, Fig. 1), while the epipelon at the same depth but unsheltered by vegetation can be transported downwards.

Since during the summer months the changes due to photosynthetic production and redistribution of organic matter vary greatly with depth, the chemical composition of the epiphytic and epipelic matter, and of epipelon from the various stations, showed considerable differences (Table 1B).

Phases (4) and (5), respectively the mid-summer stratification and autumnal cooling, brought about only slight temperature differences in the upper strata. During phase (4), starting mid-July in 1981, epipelon and epiphyton declined dramatically; in Fig. 1 this change was more clear for the epiphyton in August 1980. By this time the seston was concentrated in the metalimnion (Steenbergen & Korthals 1982), and the phytoplankton of the upper strata was dominated by the motile dinoflagellate *Ceratium hirundinella* (Blaauboer 1982). The decrease in epiphytic and epipelic matter probably resulted from both increased wind mixing and low deposition rates due to quantitative and qualitative changes in the seston. The active form of *Ceratium*, for instance, was virtually unknown from epiphytic and epipelic samples, but its resting stage was encountered frequently.

During phase (5) the seston concentrations increased again, but high net deposition rates no longer occurred in the major part of the lower littoral. In these lake regions, however, with still comparatively low turbulence, accumulation rates were high. During this time of the year, Verdouw & Dekkers (1982) measured the highest deposition rates in the lower strata, and at the same time the epiphyton reached its seasonal maximum (Fig. 1).

Summarizing, on annual basis the lower littoral of Lake Vechten is a zone of transportation. Although significant amounts of deposited matter may accumulate periodically, the net deposition over the years is zero. Two principal reasons for this are: resuspension due to wind stress, and smooth steep slopes of the sandy bottom – 11° along the transect of this study. During the phase of turbulent mixing the deposition of sestonic matter was rather evenly distributed over the lower littoral region, and presumably over the whole lake area. Prior to and during the stratification period there were great horizontal differences. The results indicated that the bottom from 2–5 m depth in the SE part of the lake collected more sestonic matter than may be expected on the basis of the lake morphometry alone. In case of both epiphyton and epipelon, the imports were so massive compared to the local production that the specific interactions between periphytic organisms are their substrata were concealed (Gons 1982).

Lake Vechten may qualify to occupy the lower end of the scale for eutrophic lakes. It may be proposed that in both more eutrophic and oligotrophic temperate lakes of similar morphometry, the proportions in the spatial deposition of sestonic matter remain the same. In more eutrophic situations the seston deposition, in addition to reducing water transparancy, may be a main cause for the decreased distribution of littoral producers. In the oligotrophic case, the role of the sestonic organic matter deposited into the littoral may be obscured by the local production and import of allochthonous materials.

Acknowledgements

We are deeply indebted to R.D. Gulati and K. Siewertsen for the use of their data. P.H. Best, R.D. Gulati and S. Parma are thanked for critically reading the manuscript.

References

Best, P.H., 1982. The aquatic macrophytes of Lake Vechten. Species composition, spatial distribution and production. Hydrobiologia 95:65–77.

Blaauboer, M.C.I., 1982. The phytoplankton species composition and its seasonal periodicity in Lake Vechten from 1956–1979. Hydrobiologia 95:25–36.

Blomqvist, S. & Håkanson, L., 1981. A review on sediment traps in aquatic environments. Arch. Hydrobiol. 91:101–132.

Brown, S.-D. & Austin, A.P., 1973. Spatial and temporal variation in periphyton and physico-chemical conditions in the littoral of a lake. Arch. Hydrobiol. 71:183–232.

Cattaneo, A. & Kalff, J., 1978. Seasonal changes in the epiphyte community of natural and artificial macrophytes in Lake Memphremagog (Que. & Vt.). Hydrobiologia 60:135–144.

Davis, M.B. & Brubaker, L.B., 1973. Differential sedimentation of pollen grains in lakes. Limnol. Oceanogr. 18:635–646.

Gasith, A. & Hasler, A.D., 1976. Airborne litterfall as a source of organic matter in lakes. Limnol. Oceanogr. 21:253–258.

Gons, H.J., 1981. Project carbon cycle in Lake Vechten. In: Parma, S. & Gulati, R.D. (Eds.) Prog. Rep. Limnol. Inst. 1980. Verh. K. Ned. Akad. Wet. Afd. Natuurkd. Tweede Reeks 77. 66 pp.

Gons, H.J., 1982. Structural and functional characteristics of epiphyton and epipelon in relation to their distribution in Lake Vechten (The Netherlands). Hydrobiologia 95:79–114.

Gulati, R.D., Siewertsen, K. & Postema, G., 1982. The zooplankton: its community structure, food and feeding, and role in the ecosystem of Lake Vechten (The Netherlands). Hydrobiologia 95:127–163.

Iwamura, T., Nagai, H. & Ichimura, S., 1970. Improved methods for determining contents of chlorophyll, protein, ribonucleic acids and deoxyribonucleic acid in planktonic populations. Int. Rev. ges. Hydrobiol. 55:131–147.

Kloet, W.A. de, 1982. The primary production of phytoplankton in Lake Vechten (The Netherlands). Hydrobiologia 95:37–57.

Mortimer, C.H., 1974. Lake hydrodynamics. Mitt. Internat. Verein. Limnol. 20:124–197.

Moss, B., 1969. Vertical heterogeneity in the water column of Abbot's Pond. II. The influence of physical and chemical conditions on the spatial and temporal distribution of the phytoplankton and a community of epipelic algae. J. Ecol. 57:397–414.

Moss, B. & Abdel Karim, A.G., 1969. Phytoplankton associations in two pools and their relationships with associated benthic flora. Hydrobiologia 33:587–600.

Steenbergen, C.L.M. & Korthals, H.J., 1982. Distribution of phototrophic microorganisms in the anaerobic and microaerophilic strata of Lake Vechten (The Netherlands). Pigment analysis and role in primary production. Limnol. Oceanogr. 27:883–895.

Steenbergen, C.L.M. & Verdouw, H., 1982. Lake Vechten (The Netherlands): aspects of its morphometry, climate, hydrology and physico-chemical characteristics. Hydrobiologia 95:11–23.

Verdouw, H. & Dekkers, E.M.J., 1982. Nitrogen cycle of Lake Vechten (The Netherlands): role of sedimentation. Arch. Hydrobiol. 94:251–263.

Wetzel, R.G. & Allen, H.L., 1972. Functions and interactions of dissolved organic matter and the littoral zone in lake metabolism and eutrophication. In: Kajak, Z. & Hillbricht-Ilkowska, A. (Eds.) Productivity Problems of Freshwaters. PWN, Warszawa. pp. 333–347.

Authors' address:
H.J. Gons/R. van Keulen
Limnological Institute
Vijverhof Laboratory
Rijksstraatweg 6
3631 AC Nieuwersluis, The Netherlands

CHAPTER 23

Studies on periphyton community dynamics of acidic streams using track autoradiography

HAMISH C. DUTHIE and PAUL B. HAMILTON

Abstract. Track autoradiography was used in 1980 to measure species-specific production rates in periphyton communities in acidic streams near Sept-Iles, Quebec. Diatoms were the most important spring colonizers of natural rock surfaces; e.g., *Tabellaria flocculosa* reached maximum populations in early June, had a maximum specific production rate of 0.014 pg C $cell^{-1}$ 1000 $\mu E\ m^{-2}\ s\text{–}h^{-1}$, and formed 40% of total community production Although *Tabellaria* growth rate remained high in the summer, its contribution to community production dropped to 4%. *Mougeotia* sp. was the dominant community producer in the summer. In contrast, the cyanophyte *Calothrix parva* had a very low specific growth rate and formed less than 1% of the community production. The rhodophyte *Batrachospermum moniliforme* flourished late in the season, had a maximum specific growth rate of 1.06 pg C $cell^{-1}$ 1000 $\mu E\ m^{-2}\ s\text{–}h^{-1}$, and contributed 12% of community production. Community trends were largely controlled by the growth and loss rates of the major species.

Introduction

In many respects our knowledge of periphyton ecology has lagged behind that of phytoplankton and other aspects of botanical limnology. This situation is mainly because of problems related to the lack of standardized sampling techniques and the paucity of studies using sound experimental approaches. When in 1979 a long-term project began on the ecosystem dynamics of boreal salmonid streams in NE Quebec, the opportunity was afforded to initiate a comprehensive study on the stream periphyton. The initial phase of the research included the development of a standard radiocarbon technique for the measurement of *in situ* primary production, the standardization of quantitative sampling and enumeration techniques for estimating cell numbers, the application of scanning electron microscopy to study spatial relations and species succession on natural rock surfaces, and the use of track autoradiography to estimate species-specific production rates. This contribution addresses only the last of these and has two main objectives; to describe the track autoradiographic technique, and to show how the technique may be used in determining the growth strategies of individual periphyton species.

The streams studied are situated in the Matamek Research Watershed about 25 km east of Sept-Iles, Quebec. This Precambrian Shield location is characterized by low gradient streams (<2% slope), humic acid coloured waters (5–15 mg C L^{-1}), approximately 110 cm precipitation evenly distributed throughout the year, and a mean annual temperature of 1° C. The area has a short growing season of about 1025 growing degree days. Stream pH is around 5.5. The physical and chemical characteristics of the streams are more fully described in Naiman (1983) and Naiman & Sedell (1981).

Most of the research was carried out during the

Wetzel, R.G. (ed.), Periphyton of Freshwater Ecosystems.
 ISBN 90 6193 768 X.

summer of 1980 in the Matamek River, a sixth order stream. Other sites investigated were First Choice Creek, a first order stream; the Muskrat River, a fifth order stream; and the Moisie River, a ninth order stream.

Methods

Community primary productivity was estimated by the ^{14}C method. Two naturally colonized rocks representative of the local community were placed in an 18-L recirculating plexiglas metabolic chamber (Bott *et al* 1978) and incubated for 4 h with 50 μCi ^{14}C sodium bicarbonate under natural light. Illumination during the experiment was recorded using a Licor LI-192S PAR photocell and integrator. At the end of the incubation period the rocks were submerged for three short periods in 0.1 N HC1 and quantitative scrape samples taken with a Douglas sampler (Douglas 1958) for scintillation counting.

Samples for track autoradiography were taken from the natural rocks after incubation with ^{14}C and were preserved in Lugol's iodine. After checking for a suitable dilution, cells were sedimented onto slides using the sliding chamber sedimentation apparatus described by Knoechel & Kalff (1976). The track autoradiography development procedure followed the methods described by Knoechel & Kalff with the following modifications: 1) A minimum of four turbulent rinses were used to remove inorganic ^{14}C remaining in the sample. 2) After rinsing the sedimented samples the remaining water was removed using minimum heat from overhead lamps. 3) Instead of a Coplin jar, a modified glass container was used to hold the nuclear emulsion thereby reducing unnecessary wastage (R. Knoechel, pers. comm.). 4) Dry film exposure periods ranged from 2 h to 4 days. 5) Fixed ^{12}C was expressed in picograms C cell^{-1} 1000 μE m^{-2} s-h^{-1}. 6) At least two counts were done at different incubation times. These were checked for Poisson distribution and compared with each other before a value of the fixed carbon rate was determined. If the counts were not in agreement, repeated counts were made until reproducible results were obtained. Both old and new slides were used during the recounts. All counts were done on slides with hidden labels. Each result reported here is the mean of several estimates.

Cell counts were made by the inverted microscope method on samples obtained from natural rock surfaces using the Douglas sampler. Details of community primary productivity and species populations will be published elsewhere. Only those results relevant to the track autoradiographic measurements are given here.

Results and discussion

Periphyton productivity studies generally fall into one of two categories: those concerned with a holistic approach to stream metabolism (e.g. Naiman 1983), and those examining primary production under controlled experimental conditions (e.g. Cattaneo & Kalff 1979). In both categories the periphyton community is regarded as an entity, the contributions of the individual species populations are not resolved. Population studies give insight into the community structure at the time of sampling, but, except possibly with very close interval sampling, give no indication of the growth and loss rates of the individual species. The use of autoradiographic methods to estimate individual species rates of carbon fixation gives a finer resolution of community dynamics (Knoechel & Kalff 1978). The results presented here are insufficient to allow a detailed discussion of the seasonal dynamics of the Matamek periphyton but do allow an overview of the major trends.

By the end of winter most rock surfaces were scoured bare. Observations by us (unpublished) suggest that colonization was effected from stream drift and from residual populations in natural depressions in the rocks. Diatoms were the first important colonizers and appeared to photosynthesize well at low temperatures. In this regard their response is similar to planktonic diatoms (Fogg 1975). The first colonizers of bare rock surfaces following ice break-up in mid-May 1980 were *Tabellaria flocculosa, Eunotia pectinalis,* and *Gomphonema angustatum* (Table 1). The first of these to reach a large population was *G. angustatum* with a density of 85×10[8] cells m-2 on 27 May in First Choice Creek. The major colonizer in the Matamek River and other higher order streams was *Tabellaria flocculosa;* between 20 May and 2 June its population increased from 2.7 to 91.5 × 10[8] cells m-2 and its share of community primary productivity rose to nearly 40%. However, on 2 June the rate of carbon fixation per cell was the lowest measured. Presumably the period of growth was coming to an end and there was increased competition for space, light, or nutrients. Over the remainder of the season the cellular growth rate was higher than in early spring but the share of community production never exceeded 4%. The high cell growth rate coupled with low population density implies a high loss rate, presumably either sloughing off or grazing by invertebrates. We saw little evidence of parasitism. The large diatom *E. pectinalis* also had a large population on 2 June coinciding with a low rate of individual cell growth. *Achnanthes minutissima* was a rare diatom in all the rivers studied, but it had the highest measured cellular growth rate. While further measurements are needed, we can only infer that this diatom must suffer very high loss rates.

Several species of Chlorophyceae were important in the periphyton community of all the sites studied. The major colonizer in the Matamek and Moisie rivers over much of the growing season was an unidentified species of *Mougeotia*. Unfortunately, it had a cellular carbon fixation rate very much greater than the measuring capacity of the experimental design and is therefore not included in the results. We estimate, by subtracting known values, that *Mougeotia* fixed up to 80% of the total community production in June and early July. Two other green algae, *Cylindrocapsa conferta* and *Tetraspora cylindrica,* were important species in spring colonization, the former in the Matamek River and the latter in First Choice Creek, Both formed around 20% of community production in

Table 1. Cellular rates of carbon fixation, standardized for light, and percentage of total community primary productivity realized by 10 species of algae in streams in NE Quebec, Canada. All estimates are for the Matamek River except where indicated: FC = First Choice Creek, MU = Muskrat River, MO = Moisie River.

	Date (1980)	Cell population (10^8 m^{-2})	pg C fixed $cell^{-1}$ (1000 μE m^{-2} s–h^{-1}	% of total community primary productivity
Bacillariophyceae				
Tabellaria flocculosa	20 May	2.7	0.026	0.9
Tabellaria flocculosa	2 June	91.5	0.014	39.7
Tabellaria flocculosa	5 August	12.4	0.075	4.0
Tabellaria flocculosa	10 August	6.1	0.047	0.9
Tabellaria flocculosa	30 August	3.6	0.058	1.6
Eunotia pectinalis	20 May	0.1	0.016	0.03
Eunotia pectinalis	2 June	17.2	0.013	6.9
Gomphonema angustatum	27 May (FC)	85.0	0.485	33.2
Achnanthes minutissima	23 August (MO)	0.63	0.150	0.02
Chlorophyceae				
Cylindrocapsa conferta	2 June	51.5	0.013	20.5
Cylindrocapsa conferta	10 August	10.1	0.147	0.06
Tetraspora cylindrica	27 May (FC)	705.6	0.039	21.9
Tetraspora cylindrica	27 June (FC)	5.6	0.012	1.8
Scenedesmus species a	23 August (MO)	0.03	1.18	0.008
Cyanophyceae				
Calothrix parietana	2 June	1.14	0.002	0.07
Calothrix parietana	10 August	0.60	0.014	0.03
Merismopedia tenuissima	23 August (MO)	0.01	0.56	0.001
Rhodophyceae				
Batrachospermum moniliforme	1 August (MU)	7.61	1.06	12.0

late May-early June. *C. conferta* persisted in low to moderate numbers in the Matamek River over the ramainder of the season but *T. cylindrica* was strictly a spring form, all but disappearing from First Choice Creek in the summer. As was the case with *T. floculosa*, the population maximum of *C. conferta* coincided with minimal cellular growth values. Growth per cell was ten times higher in August but the contribution to total community production was minimal and the population low implying high loss rates. The only other measurement made of a green alga was on *Scenedesmus* sp. *a* (related to *S. abundans*). This species had the highest cellular growth rate of any alga we measured, but the species was always very rare in the community.

The Cyanophyceae never formed significant populations in these acidic waters and contributed very little to the total community productivity. *Calothrix parietana* had a very low cellular fixation rate but a single measurement on *Merismopedia tenuissima* in the Moisie River in late August was surprisingly high.

Late in the growing season the rhodophyte *Batrachospermum moniliforme* reached maximum population density in the community. Our observations found that this species overwintered as small basal cell clusters in minute depressions on the rocks, grew slowly in the spring, then rapidly with warmer summer temperatures. Our single measurement on 1 August showed a high cellular growth rate contributing 12% to the total community productivity. This species appears to be ungrazed.

There are sufficient results presented here to conclude that the species studied have various strategies for growth and survival. The diatoms are the most successful colonizers in early spring although their highest growth rates are in summer. Possibly other algae are better competitors in summer or have lower loss rates. *Tetraspora cylindrica* was dominant in First Choice Creek in early spring but is apparently shade intolerant. *Calothrix* and *Batrachospermum* are probably perennial forms, the latter growing best in warm waters. Considerably more observations are required for a more comprehensive understanding of the ecosystem dynamics of lotic periphyton. Preliminary observations by us showed little correlation of species populations with dissolved nutrients; physical factors and growth and loss rates appear to be more important.

In conclusion, track autoradiography presents an attractive new tool for experimental work involving periphyton. We are currently carrying out autecological studies of *Tabellaria flocculosa, Batrachospermum moniliforme*, and *Mougeotia* sp. These three species are of major importance in the Matamek system and have widely different strategies for growth. We are investigating the colonization of these species on natural substrates and are using a variety of experimental approaches including track autoradiography and scanning electron microscopy. Track autoradiography should also facilitate work in the areas of alga-substrate and alga-herbivore interactions. It may be possible to resolve, in greater detail than at present, the response of individual periphyton algae to fertilization, or to stresses such as pollution, toxicity, or acidification.

Acknowledgements

This research was founded by a grant from Woods Hole Oceanographic Institution and is publication No. 64 of the Matamek Research Station. We gratefully acknowledge the assistance and advice of R. Knoechel in using track autoradiography.

References

Bott, T.L., Brock J.T., Cushing, C.E., Gregory, S.V., King, D. & Peterson, R.C., 1978. A comparison of methods for methods measuring primary productivity and community respiration in streams. Hydrobiologia 60:3–12.

Cattaneo, A. & Kalff, J., 1979. Primary production of algae on natural and artificial aquatic plants. Limnol. Oceanogr. 24:1031–1037.

Douglas, B., 1958. The ecology of attached diatoms and other algae in a small stony stream. J. Ecol. 46:295–322.

Fogg, G.E., 1975. Algal Cultures and Phytoplankton Ecology. Second edition. University of Wisconsin Press, Madison. 175 pp.

Knoechel, R. & Kalff, J., 1976. Track autoradiography: a

method for the determination of phytoplankton species productivity. Limnol. Oceanogr. 21:590–596.
Knoechel, R. & Kalff, J., 1978. An in situ study of the productivity and population dynamics of five freshwater planktonic diatom species. Limnol. Oceanogr. 23:195–218.
Naiman, R.J., 1983. The annual pattern and spatial distribution of aquatic oxygen metabolism in boreal forest watersheds. Ecological Monographs 53:73–94.
Naiman, R.J. & Sedell, J.R., 1981. Stream ecosystem research in a watershed perspective. Verh. Internat. Verein. Limnol. 21:804–811.
Naiman, R.J. & Sedell, J.R., 1981. Stream ecosystem research in a watershed perspective. Verh. Internat. Verein. Limnol. 21:804–811.

Authors' address:
H.C. Duthie/P.B. Hamilton
Department of Biology
University of Waterloo
Waterloo, Ontario, Canada N2L 3G1

CHAPTER 24

A geomorphic approach for examining the role of periphyton in large watersheds

ROBERT J. NAIMAN

Abstract. Geomorphic analysis of watersheds, when combined with biotic studies of specific ecosystem components, place these components in a proper ecosystem perspective. In undisturbed boreal forests of Quebec, Canada, annual periphyton production can be reliably predicted from stream order and extrapolated to the watershed level when combined with some simple geomorphic analyses. At the watershed scale periphyton contribute only 11 to 27% to total ecosystem metabolism, depending on the size of the watershed. The most metabolically active components are fine particulate organic matter (0.5 μm–1 mm) and mosses.

Introduction

Predictable changes occur in the structure of periphyton assemblages and their metabolism as streams gradually merge to form rivers (Vannote *et al* 1980; Naiman & Sedell 1980; Naiman 1983). This fact makes the role of periphyton difficult to evaluate in general terms for running waters since, as watershed area increases, the role and character of periphyton assemblages constantly change relative to other biotic components.

To understand the basic ecosystem level properties of watersheds, it is essential to quantify biotic components relative to their proportional occurrence. Concomitantly, since watersheds act as a unit, they therefore must be treated as one if the basic ecological properties are to be perceived properly. This philosophy avoids difficulties commonly associated with emphasing biotic components without regard to their proportional occurrence.

To achieve this perspective I have combined standard geomorphological analyses of drainage networks with biotic measurements of specific ecosystem components. Partial results of a project using this approach have been presented in detail elsewhere (Naiman & Sedell 1981; Naiman 1982, 1983). In this article I describe the methods used and discuss how the results may increase our understanding of the metabolic role of periphyton in various sized watersheds.

Study sites

All the watersheds are located about 25 km east of Sept-Iles, Quebec, Canada (50′ 15″ N; 66′ 00″ W). This area is Precambrian Shield with a completely undisturbed boreal forest. Running water here are brown colored, generally have a gentle slope (<2%), and high concentrations of dissolved organic carbon (5–15 mg C/L). Four of the five watersheds (First Choice Creek, Beaver Creek, Muskrat River, and the Matamek River) are within a reserve protected by the Province of Quebec for scientific research only (Table 1). The fifth site (Moisie River) is adjacent to the Matamek River watershed. Detailed descriptions of each site can be found in Naiman (1982, 1983).

Wetzel, R.G. (ed.), Periphyton of Freshwater Ecosystems.
 ISBN 90 6193 768 X.

Methods and materials

Geomorphic analyses

The drainage network of each watershed was characterized for stream orders, number and length of streams of each order, and gradient from Canadian Department of Energy, Mines, and Resources topographic maps (scale 1:50 000) following procedures of Horton (1945), Leopold *et al* (1964), Morisawa (1968) and Dunne & Leopold (1978). Results were field checked for accuracy with the conclusion that no first order streams were indicated on the maps but >95% of the permanent second order streams were represented (Naiman 1983). The total number and length of first order streams were determined by regressing available data on number and length of streams greater than or equal to second order, as a function of stream order, and extrapolating back to first order. The mean width of each stream order was measured for a large number of transects ($n > 50$) either in the field or from aerial photographs (scale 1:15 000) provided by the Quebec Ministry of Lands and Forests. Basin area and total lake surface were measured planimetrically from topographic maps.

Validity of the Horton (1945) system of fluvial morphology depends upon the concept that conditions of surface relief, slope, and channel configuration reach a time-independent steady state in which channel morphology is adjusted to the quantity of debris and excess water characteristically produced under the controlling climatic and physical regime. Thus when several of watersheds are studied it is essential to examine them from geomorphological uniformity. To do this some simple parameters should be examined, including:

1. The ratio of segments of a given order (N_u) to the number of segments of the next highest order (N_{u+1}) is termed the bifurcation ratio (R_b). The bifurcation ratio will not be precisely the same from one order to the next due to chance variations in watershed geometry, but will tend to be a constant throughout the series. Bifurcation ratios characteristically range between 3.0 and 5.0 for watersheds where geological structures do not distort the drainage pattern.
2. The constant of channel maintenance (C) is the area required to sustain one linear kilometer of channel. It is calculated by regressing the logarithm of basin area (km^2) against the logarithm of cumulative stream length (km). When the logarithm of the intercept is read at $\log_{10}$ [stream length] = 0 and the antilog of this intercept is taken, the constant of channel maintenance (C) for the watershed is obtained. If the resulting C's for each watershed are similar, then this is one indication that the watersheds are essentially uniform.

Representative sections, either the whole stream if it was short, or reaches between 300 m and 10 000 m if the stream was long, were mapped using stan-

Table 1. A general characterization of the Quebec watersheds (modified from Naiman 1982, 1983).

Parameter	First Choice Creek	Beaver Creek	Muskrat River	Matamek River	Moise River
Stream order	1	2	5	6	9
Watershed area (km^2)	0.25	1.83	207	673	19 871
Mean width (m)	0.3	2.0	21.9	51.7	208.7
Mean depth (m)	0.05	0.3	2.0	1.8	2.5
Mean annual discharge (m^3/s)	0.013	0.033	8.4	13.7	466.1
Gradient (%; lower reaches)	1.5	1.0	0.14	1.33	0.16
Substrate	Detritus	Sediment/ cobble	Sediment/ cobble	Gravel/ cobble	Sand/ cobble
Forest canopy development	Closed	Mostly closed	Mostly open	Open	Open
Temperature range (°C)	0.4–11.0	0.0–19.0	0.0–19.5	0.1–19.8	0.1–21.8
Annual degree days (°C/y)	1702	2069	2157	2225	2219

Table 2. Fluvial geomorphological characterization of the Moisie River (19 871 km^2) drainage network (modified from Naiman & Sedell 1981).

Stream order	No. of streams	% Total	Total length of streams (km)	% Total	Mean width (m)	Estimated surface area ($m^2 \times 10^6$)	% Total	Streams km^2	km/km^2	Bifurcation ratio	Mean length (km)
1	38 770	78.2	16 142	49.9	0.3	4.8	2.3	1.957	0.815		0.42
2	8 400	16.9	8 249	25.5	2.1	17.3	8.4	0.424	0.416	1:4.76	0.98
3	1 855	3.7	3 842	11.9	6.7	25.7	12.5	0.094	0.194	1:4.53	2.07
4	415	0.84	1 879	5.8	15.6	29.3	14.3	0.021	0.095	1:4.46	4.83
5	95	0.19	1 072	3.3	30.1	32.3	15.8	0.005	0.054	1:4.37	11.29
6	19	0.04	471	1.5	51.3	24.2	11.8	0.001	0.024	1:5.00	24.80
7	7	0.014	340	1.1	80.6	27.4	13.4	0.000	0.017	1:2.71	48.57
8	2	0.004	292	0.9	119.1	34.8	17.0	0.000	0.015	1:3.50	146.00
9	1	0.002	54	0.2	168.2	9.1	4.4	0.000	0.003	1:2.00	54.00
Total	49 564		32 341			204.9		2.502	1.633		$\bar{x} = 0.653$

dard survey methods. Pools, riffles, backwater areas, and large debris were delineated for each site. This information was then used to calculate habitat percentages for representative stream orders and was incorporated into the metabolic measurements for final calculation of site-specific metabolism.

Metabolic measurements

These measurements have been described by Naiman & Sedell (1980, 1981) and Naiman (1983). In general, natural periphyton on rocks was collected from each stream during the ice-free season, placed in recirculating 17-L Plexiglas chambers in streams from which the rocks were taken, and oxygen changes monitored over 24 h. Metabolic parameters were calculated using the equations of Bott *et al* (1978) and Naiman (1983).

Results

Geomorphology

Usefulness of the stream order system depends upon the premise that given a sufficiently large sample, stream order will be directly proportional to watershed area, to channel dimensions, and to discharge. After streams in a drainage network have been assigned order numbers, the number of segments of each order are counted and the lengths measured. Obviously the number of stream segments of any given order will be more numerous than for the next highest order (Horton 1945). When the logarithm of number of lengths of streams is plotted against order, watersheds in which geological structures do not distort the drainage pattern characteristically show a linear relationship with only a small deviation from a straight

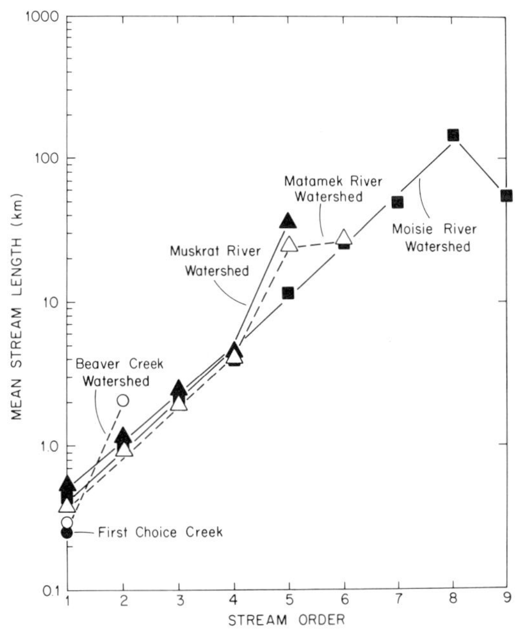

Fig. 1. Mean stream length is plotted as a function of stream order for all streams of specified order in each watershed.

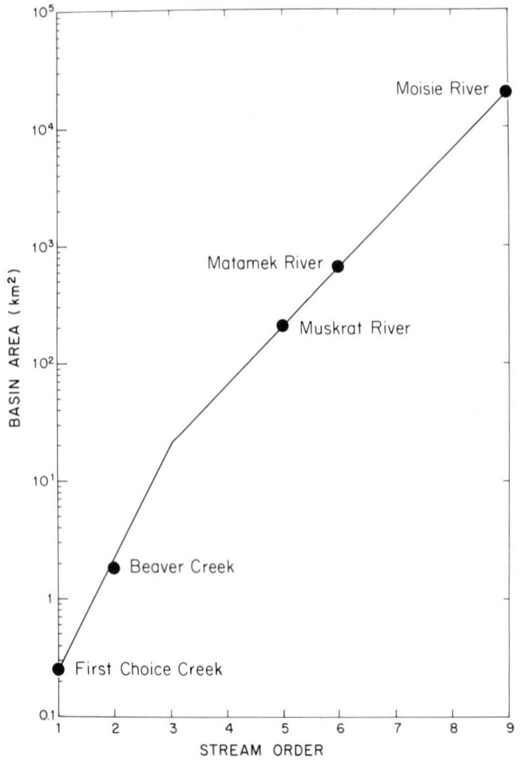

Fig. 2. Basin area is regressed as a function of terminal stream order for the five watersheds used for this study.

line. In watersheds chosen for this study, the relationship for lengths is slightly distorted at higher orders (Naiman 1983). The large rivers are either slightly truncated (e.g., Moisie and Matamek rivers) or elongated (e.g., Muskrat River), due to insufficient time since the recent glaciation for an equilibrated drainage pattern to completely develop (Power *et al* 1973). However, the number and length of 1st order streams can be reliably estimated ($r^2 > 0.95$) from the resulting relationships between number, lengths and stream order for these watersheds (Naiman 1983).

The Moisie River watershed contains an estimated 49 564 streams and rivers with a combined length of 32 341 km (Table 2), resulting in a density of 2.5 streams km^{-2}. Of the total number of streams, 98.9% are small (orders 1–3) making up 87.3% of the total length in the drainage network. In contrast, stream orders 7–9 make up <0.1% of the total number and 2.2% of the total length. The Matamek and Muskrat River watersheds show a similar pattern, while the Beaver Creek and First Choice Creek watersheds are too small to provide a meaningful analysis.

Stream width is reliably predicted ($r^2 = 0.99$) from order, and can be estimated from the equation: Log_{10} (width) = 2.93 Log_{10}(order) −0.57, where width is in meters. When stream-width data are combined with the number and lengths of each stream order, water surface area in the drainage network can be estimated (Table 2). Stream surface area in the Muskrat River watershed accounts for only 1.0% of the basin area; it is similar for the Matamek (1.2%) and Moisie basins (1.0%). In the Moisie watershed the numerous 1st–3rd order streams contribute only 23.2% to the total stream water surface area of 204.9 km^2, the middle order (4–6) streams contribute 41.9%, while 7th–9th order streams comprise 34.8% of the surface area.

Each of the larger watersheds contain numerous lakes. In the Muskrat River basin lakes encompass 7 km^2, in the Matamek River basin 49.6 km^2, and in the Moisie River basin they cover 1227.3 km^2. These lake areas represent 3.4%, 7.4%, and 6.2% of the basin areas, respectively. Lake surface area can be reliably estimated from basin area ($r^2 = 0.98$) by the equation: Log_{10} (Lake surface area) = −1.56 + 1.09 Log_{10} (Basin area), where both have units of km^2. In the Moisie River water-

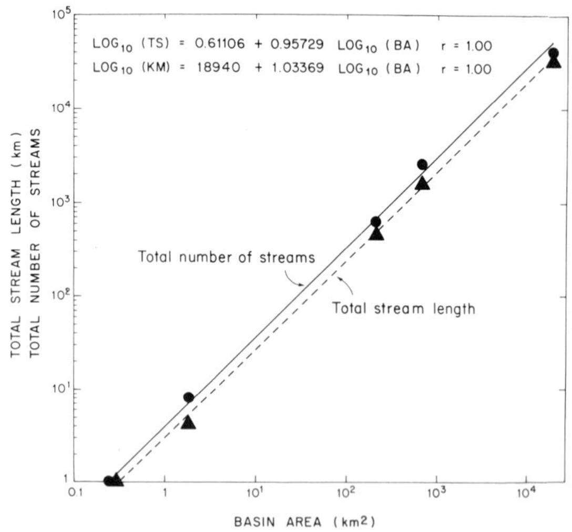

Fig. 3. The total number of streams and the total length of all streams in a watershed are plotted as a function of basin area.

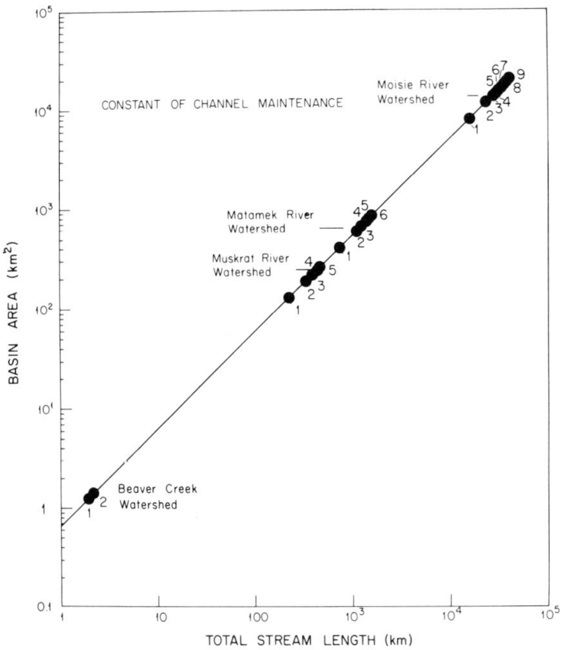

Fig. 4. The constant of channel maintenance is calculated by regressing basin area against the total length of all streams in that watershed and determining the intercept at a stream length of 1 km. Numbers refer to specific stream orders within a particular watershed.

shed the aquatic surface area represented by lakes is 6.2 times that of stream surface area. As age of the watershed increases the ratio of lake surface area to stream surface area would be expected to decrease.

The bifurcation ratios calculated for the Moisie River watershed indicate that geological structures do not distort the drainage pattern to any great extent (Table 2). Most values are between 3.0 and 5.0; those ratios that are not between these two values are based on the rarer large rivers. Either the large rivers have not yet had sufficient time to reach a geologically stable configuration or the sample size (e.g. number of rivers) is not large enough to provide a meaningful analysis as the terminal stream order for the basin is approached.

Horton (1945) postulated that the mean lengths of stream segments of each successive order of an equilibrated watershed approximate a direct geometric sequence. A plot of logarithm of mean stream length as a function of order normally yields a set of points lying essentially along a straight line (Fig. 1). However, at higher orders the relationship for our sites is distorted with the larger rivers being either slightly truncated (Moisie River) or elongated (Muskrat River).

Assuming the validity of the laws of stream lengths and basin areas, in which both properties are related in an exponential function with order (Fig. 2), length should be related to area by a power function (Strahler 1957). For our study sites both the total stream length and the number of streams in each watershed are perfectly correlated ($r = 1.00$) with basin area (Fig. 3).

The constant of channel maintenance for each watershed forms a straight line of 45° slope, indicating that all watersheds have a uniform character (Fig. 4). For the Moisie River $C = 0.66$, for the Matamek River it is 0.66, and for the Muskrat River it is 0.60. This means that, on the average, 0.60–0.66 km^2 is needed to support each linear km of stream channel.

These various simple geomorphic analyses indicate that the watersheds are remarkably uniform, except for the large river segments. Therefore, some caution must be utilized when extrapolating the biotic data from the large river sites to the

Table 3. Annual periphyton production can be estimated for each stream order from these equations. SO = steam order; NS = not significant (modified from Naiman 1983).

Metabolic parameter	Equation	Correlation coefficient (r)	Significance (p)
Gross production	GP = 62.25 + 7.60 (SO)	0.89	<0.05
Net community production	NCP = 19.93 + 6.62 (SO)	0.97	<0.01
Diel (24-h) respiration	$\text{Log}(R_{24}) = 1.84 + 0.05$ (SO)	0.30	NS
Net daily metabolism	NDM = 4.85(SO) − 0.83	0.95	<0.01

watershed level. However, overall, the sites chosen for this study represent a typical continuum for this region.

Metabolic measurements

Certain metabolic parameters associated with periphyton can be reliably predicted from stream order in undisturbed watersheds (Naiman 1983). Annual gross production, net community production, and net daily metabolism are significantly correlated with stream order, while diel respiration tends to remain relatively constant throughout the stream orders studied (Table 3). Similar analyses for other major metabolic components [e.g., moss, macrophytes, fine particulate organic matter (FPOM; 0.5 μm–1 mm), and coarse particulate organic matter (CPOM; >1 mm)] generally resulted in significant positive correlations between some metabolic components and stream order, but for others metabolism tended to remain constant as stream size increased (Naiman 1983).

Comparison of annual gross production by periphyton to other metabolic components gives an indication of the relative role of periphyton in running waters. On an areal basis (g $O_2/m^2/y$), periphyton contributes only 11 to 27% to the total metabolism for a particular reach of stream (Table 4). Surprisingly, the largest contribution of periphyton to total metabolism is in the small, heavily shaded headwater streams (20–27%). In small streams most metabolism is associated with FPOM, while in large rivers most metabolism is associated with FPOM and the moss community.

Equations predicting metabolism as a function of stream order, in combination with geomorphic analyses of each watershed, provide an estimate of total watershed metabolism (g O_2/y; Table 5). Again the results are similar concerning the relative metabolic role of periphyton. Only 15 to 27% of total watershed metabolism can be attributed to periphyton with the highest percentages (25–27%) in small watersheds. FPOM is the dominant metabolic component, regardless of basin size.

Discussion

An appropriate starting point for river zonation studies or watershed studies is Lotspeich's (1980) outline for watershed classification which utilizes geology, macro- and micro-climate, soil type, and terrestrial vegetation as heirarchical classification criteria. As suggested by Culp & Davies (1982), by coupling longitudinal zonation studies with Lotspeich's hierarchical system and the river continuum framework, the biological comparison of stream components in different geographical regions could be enhanced, and the understanding of biological processes significantly aided. Additionally, adaptations to specific regions must be made (e.g., alpine regions, deserts), as well as to areas where the drainage network has been variously altered by damming, pollution, or grazing. In these instances further classification of streams is essential.

For this study I used stream order as a classification system. For the region investigated stream order is significantly correlated with other physiochemical parameters such as discharge, channel

Table 4. Comparison of gross production by the major metabolic components for each site. Data are in units of g $O_2/m^2/y$. FPOM = fine particular organic matter; CPOM = coarse particulate organic matter; NP = not present (data extracted from Naiman 1983).

Site	Periphyton	Moss	Macrophytes	FPOM	CPOM	Total	Percent periphyton
First Choice Creek	54.7	NP	NP	177.9	41.1	273.7	20.0
Beaver Creek	91.6	NP	NP	198.4	45.0	335.0	27.3
Muskrat River	96.7	160.6	NP	327.3	30.8	615.4	15.7
Matamek River	120.1	288.7	7.9	206.4	49.6	672.7	17.9
Moisie River	123.1	340.4	57.8	531.0	29.0	1081.3	11.4

Table 5. Comparison of gross production by the major metabolic components for each watershed. Data are in g $O_2/y \times 10^8$. FPOM = fine particulate organic matter; CPOM = coarse particulate organic matter; NP = not present (data extracted from Naiman 1983).

Watershed	Periphyton	Moss	Macrophytes	FPOM	CPOM	Total	Percent periphyton
First Choice Creek	3.4	NP	NP	7.1	2.1	12.6	27.0
Beaver Creek	16.8	NP	NP	39.5	9.5	65.8	25.5
Muskrat River	98.3	111.3	NP	273.3	44.3	527.2	18.6
Matamek River	124.4	169.8	1.9	356.2	53.4	704.7	17.7
Moisie River	210.7	386.1	28.2	648.6	78.5	1352.1	15.6

width, temperature regime, and water chemistry. In effect, stream order provided an index integrating the physiochemical environment. However, for studies of specific ecosystem components, other researchers may find that certain parameters, such as the chemical regime, are better predictors of processes than is stream order.

By examining periphyton metabolism in a watershed perspective I have shown that periphyton contributes only a small percentage to the annual metabolism of the drainage network. Surprisingly, periphyton seems most important in canopied headwater streams where it contributes nearly a quarter of the total annual metabolism. In larger rivers, where water remains relatively clear and shallow, periphyton is conspicuous but accounts for only 11 to 19% of the total annual metabolism. It can only be concluded that in streams of undisturbed boreal forest watersheds the role of periphyton is minor compared to FPOM and moss. It is clear that future studies of stream metabolism in the boreal forest of Quebec should emphasize the role of FPOM and mosses rather than periphyton.

To understand and evaluate ecosystem dynamics in a proper perspective, an approach similar to the one described here must be used. This approach provides a framework for evaluating the contribution of any biological component to the ecosystem, thus avoiding problems inherent with site-specific and component-specific studies. While this study was conducted in a region of uniform geomorphology, climate and flora it demonstrated salient features about the role of periphyton. Future studies will require an examination of the influence of geology, climate, soil, and flora (Lotspeich 1980) and longitudinal zonation (Vannote *et al* 1980) on biological features if new unifying concepts in watershed research are to emerge, especially for large basins.

Acknowledgements

I thank J.M. Teal and R. Morin for helpful suggestions towards improving the manuscript. This is Contribution No. 5253 from the Woods Hole Oceanographic Institution and No. 65 from the Institution's Matamek Research Station. Funds were provided by the Matamek Research Program.

References

Bott, T.L., Brock, J.T., Cushing C.E., Gregory S.V., King, D. & Petersen, R.C., 1978. A comparison of methods for measuring primary productivity and community respiration in streams. Hydrobiologia 60:3–12.

Culp, J.M. & Davies, R.W., 1982. Analysis of longitudinal zonation and the river continuum concept in the Oldman-South Saskatchewan River system. Can. J. Fish. Aquat. Sci. 39:1258–1266.

Dunne, T. & Leopold, L.B., 1978. Water in Environmental Planning. W.H. Freeman. 818 pp.

Horton, R.E., 1945. Erosional development of streams and their drainage basins; hydrophysical approach to quantitative morphology. Bull. Geol. Soc. Am. 56:275–370.

Leopold, L.B., Wolman, M.G., Miller, J.P., 1964. Fluvial Processes in Geomorphology. W.H. Freeman. 522 pp.

Lotspeich, F.B., 1980. Watersheds as the basic ecosystem: this conceptual framework provides a basis for a natural classification. Water. Res. Bull. 16:581–586.

Morisawa, M., 1968. Streams: Their Dynamics and Morphology. McGraw-Hill. 175 pp.

Naiman, R.J., 1982. Characteristics of sediment and organic carbon export from pristine boreal forest watersheds. Can. J. Fish. Aquat. Sci. 39:1699–1718 (in press).

Naiman, R.J., 1983. The annual pattern and spatial distribution of aquatic oxygen metabolism in boreal forest watersheds. Ecol. Monogr. 53:73–94.

Naiman, R.J. & Sedell, J.R., 1980. Relationships between metabolic parameters and stream order in Oregon. Can. J. Fish. Aquat. Sci. 37:834–847.

Naiman, R.J. & Sedell, J.R., 1981. Stream ecosystem research in a watershed perspective. Verh. Internat. Verein. Limnol. 21:804–811.

Power, G., Pope, R.F. & Coad, B.W., 1973. Postglacial colonization of the Matamek River by fishes. J. Fish. Res. Bd. Canada 30:1586–1589.

Strahler, A.N., 1957. Quantitative analysis of watershed geomorphology. Trans. Am. Geophys. Union 38:913–920.

Vannote, R.L., Minshall, G.W., Cummins, K.W., Sedell J.R. & Cushing, C.E., 1980. The river continuum concept. Can. J. Fish Aquat. 37:130–137.

Author's address:
R.J. Naiman
Woods Hole Oceanographic Institution
Woods Hole, MA 02543, U.S.A.

CHAPTER 25

Adaptations to benthic freshwater herbivory

C. OTTO

Abstract. Contrary to periphyton, fresh macrophytes are not ingested by many freshwater invertebrates. Different adaptations that serve to maximize net energy gain in grazers utilizing a food resource which is heterogeneously distributed in space and time are discussed. Data presented support the hypothesis, that aquatic macrophytes often protect themselves against herbivores by producing repellent substances that make them unpalatable. However, some elaborate plant-snail associations suggest these plants in one way or another profit from attracting snails in spite of the grazing damage. Long periods of exposure, lack of alternative food for herbivores, and a vegetative mode of reproduction are factors suggested to enhance the adaptive value of the production of secondary substances by aquatic macrophytes.

Introduction

In freshwater ecosystems a variety of different taxa such as gastropods, crustaceans and insect larvae graze upon periphyton. Periphyton is sometimes heavily fed upon and the grazers may be exposed to food shortage (e.g. Flint & Goldman 1975; Hart 1981). Although the periphyton quantity may be reduced due to grazing, quality may improve (Hunter 1980). Increased grazing may also increase periphyton production through a) increased availability of nutrients brought about by the presence of grazers, or b) altered competitive interactions between various species of algae (cf. Doremus & Harman 1977). In contrast to periphyton, fresh macrophytes are seldom used as food by many animals (e.g. Soszka 1975; Cummins & Klug 1979). In a lake littoral Soszka (*op. cit.*) only found aquatic lepidopteran larvae and some caddis larvae to ingest fresh macrophyte tissue to a considerable extent. Vascular plant tissue periodically also seems to be important to some snails (Pip & Stewart 1976), and crayfish has been shown to be responsible for the control of submerged pond vegetation (Abrahamsson 1966).

The food ingested by benthic herbivores is structurally very heterogenous. Thus the height of plants ranges from below one millimeter to more than one meter. The more obvious adaptations of herbivores consequently concern their mouth parts. The scoop-shaped mandibles in insects and the rasping radula of gastropods represent appropiate tools for those herbivores scraping periphyton. Herbivores shredding the plant tissue of fresh macrophytes have sharp and stout mandibles, while the piercers, imbibing plant fluid, e.g. some Hydroptilidae (Trichoptera), Corixidae (Hemiptera), and Chrysomelidae (Coleoptera) have mouth parts specialized for sucking. Aside from these morphological adaptations directly related to feeding behaviour, adaptations that arose to maximize the net energy gain also exist in herbivores.

Plants are the targets of herbivores, and some results indicate that macrophytes often protect

Wetzel, R.G. (ed.), Periphyton of Freshwater Ecosystems.
 ISBN 90 6193 768 X.

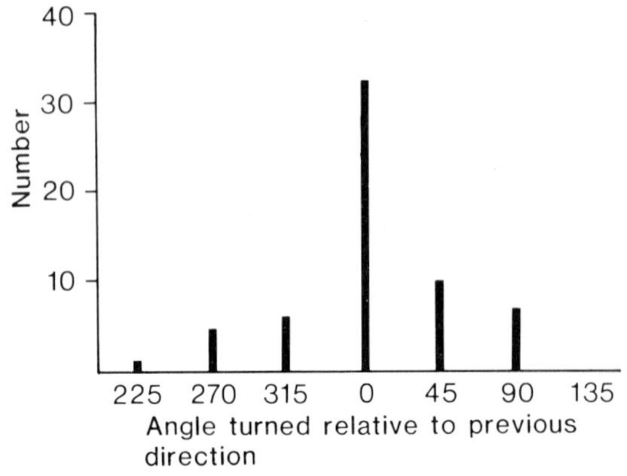

Fig. 1. Angle turned relative to previous bearing by *Dicosmoecus gilvipes* (Trichoptera) larvae when moving between food patches. Redrawn from Hart 1981.

themselves by producing secondary substances rendering them unpalatable to herbivores. Thus lauric, myristic and palmitoleic acids in *Cladophora* sp. have been ascribed an inhibitory effect to a variety of invertebrates (Lalonde *et al* 1979). Also in *Nuphar* sp. and *Nelumbo* sp. the presence of strange alkaloids has been attributed a defensive function (Hutchinson 1981). Algal populations making up the periphyton community are selectively grazed upon (e.g. Bovbjerg 1965; Patrick 1970; Calow & Calow 1975), and snails tend to prefer algae which are easiest to digest (Calow & Calow 1975). However, evidence as to the cause for the rejection of some algal species is still lacking. One hypothesis is that those algae which are easiest to assimilate are also most palatable. On the other hand, compensation due to low assimilation of green algae in terms of high consumption has been documented in a grazing naidid (*Stylaria lacustris* L.) (Streit 1978). Thus, at least to this species, assimilation efficiency seems unrelated to the palatability of its food.

Adaptations in herbivores grazing periphyton

Apart from the overall positive correlation between annual mean biomass of periphyton and macroinvertebrate abundance (e.g. Ward & Dufford 1979), individual grazers show a number of behavioural adaptations to periphyton heterogenously distributed in space and time. Active transportation demands energy, and since tranportation movements usually are not consistent with other activities such as feeding and mating, it should serve an adaptive function when it occurs (see Hart & Resh 1980). In a grazing limnephilid caddis larva (*Dicosmoecus gilvipes* Hagen) Hart (1981) analyzed the relationship between movement, feeding, and resource patchiness. A time-activity budget showed, that feeding was the predominant activity (97.9%), while movement and inactivity only took place during 1.9 and 0.2 per cent of the time, respectively. Further, it was demonstrated, that in a patchy environment the larvae spent a disproportionately high amount of time in those patches where food was plentiful (approximately 20 times longer than they did in recently grazed patches). Individuals moved more slowly when they entered a patch, whereas the rate of mandibular scraping increased. Also, the larvae often turned back when they reached the patch boundary. The ability to distinguish among quantitative levels of food has also been demonstrated for the grazing snail *Physa gyrina* Say (Warner 1976). When a caddisfly larva moved between food patches, it preferably continued straight ahead (Fig. 1), thereby reducing the probability of encountering a patch recently grazed by itself. Calow (1974) investigated the effect of starvation on the limpet *Ancylus fluviatilis* Müll. in terms of pattern

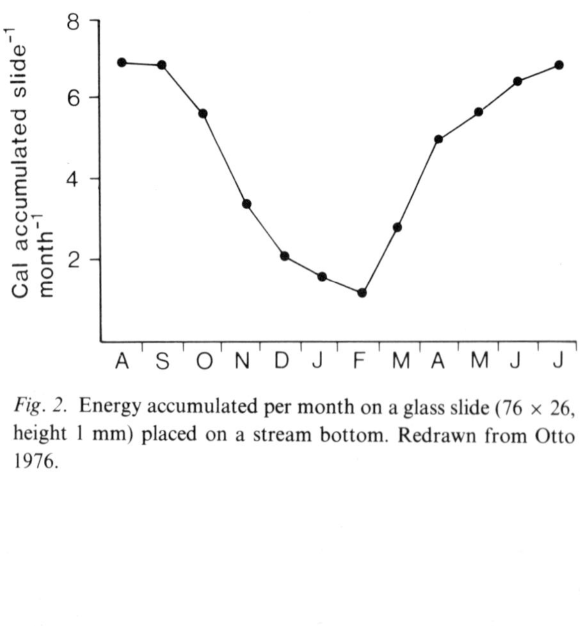

Fig. 2. Energy accumulated per month on a glass slide (76 × 26, height 1 mm) placed on a stream bottom. Redrawn from Otto 1976.

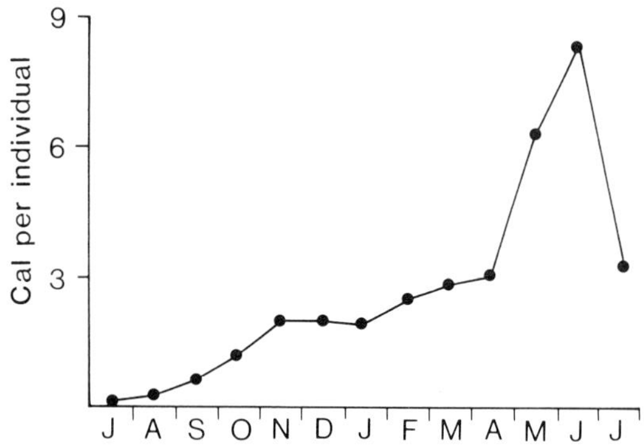

Fig. 3. Growth curve of an average *Ancylus fluviatilis* individual. Redrawn from Otto 1976.

and rate of movement. He found that distance moved by satiated snails did not differ whether food was present or absent. In contrast, some variation occurred in the patterns of movement in that those produced in the absence of food were haphazard. Furthermore, as starvation time increased the rate of movement also increased up to some optimum level and then declined. The observed changes in pattern and rate of movement at starvation were expected to increase the probability of finding new food. Calow (1974) also concluded that *A. fluviatilis* makes random movements between, and concentrated efforts within, food patches. However, the result by Streit (1981) indicates that *A. fluviatilis* preferably moves against the current when searching for food. These studies clearly demonstrate that food discontinuities in space influence the behaviour of grazers.

Aside from effects of short-term variations in periphyton availability just mentioned, seasonal variations occur and are often correlated to growth patterns and dynamics of invertebrates. Relative food availability to the snail *A. fluviatilis,* measured as energy accumulated per month on submerged glass slides, varied fourfold over the year (Fig. 2). Low amounts of energy accumulated during winter, while the opposite was found during summer. Hatching occurred in July, i.e. at a time when food was plentiful. Growth rate was comparatively high during autumn. During winter growth rate was low or non-existant, but in April it resumed, and the limpets reached their final size in June (Fig. 3), after which the generation rapidly died off. Also the energy content of the limpets showed a seasonal pattern correlated to food availability (Fig. 4). Peak densities of *A. fluviatilis* occurred in August. The decline in abundance was comparatively fast from August to September and from May onwards, but during the rest of the year the rate of decline seemed fairly constant with the exception of the time December to January i.e. a period of poor food availability (Fig. 5). Of course the decline in abundance during winter did not necessarily mean that it was caused by poor food supply, but it coincided with a poor condition of the limpets (Fig. 4) and low food availability (Fig. 2). Besides, experimentally, food availability has been shown to greatly influence growth, fecundity and survivorship in a number of species (e.g. Eisenberg 1966, 1970; Brown 1982).

In insects seasonal variation in food availability may affect the duration of emergence and flight periods. Different guilds of lotic caddis larvae have differential access to food during the year (Otto 1981). Thus filter feeders seem to be relatively evenly supplied with food during the year leading to long flight periods because of long emergence periods. To grazers feeding on periphyton, seasonal variations in food supply seem to have a synchronizing effect on growth, pupation and emergence. On the other hand, grazers are often small and often make cases of mineral particles, which means that transportation costs are high. Therefore it seems probable that the feeding area of each larva is small, which would imply that local differences in food availability (cf. Summer & Fisher 1979) tend to decrease this synchronizing effect.

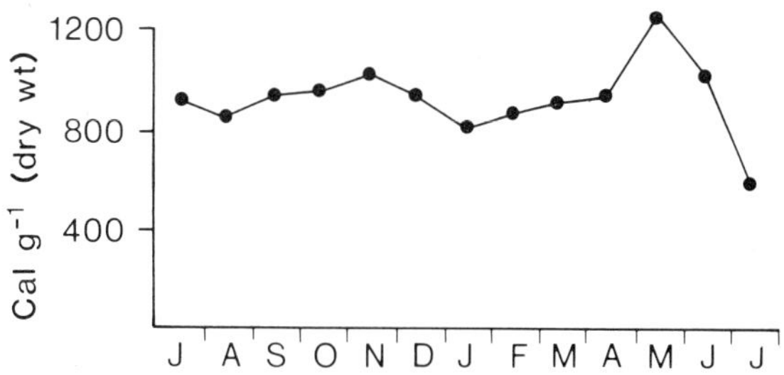

Fig. 4. Seasonal changes in energy content during life cycle in *Ancylus fluviatilis*. Redrawn from Otto 1976.

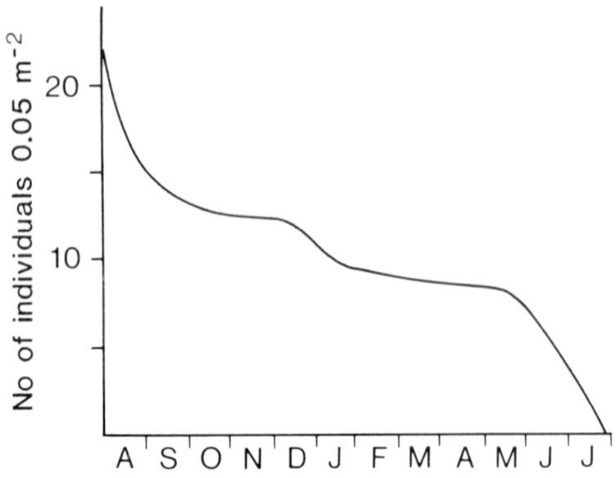

Fig. 5. Seasonal changes in density of *Ancylus fluviatilis*. Redrawn from Otto 1976.

Since it may also happen that their continously renewed food resource (periphyton) is in short supply (see above), temporal variation in pupation and emergence would be the logical effect. This variation leads to relatively long flight periods caused by potentially long periods of emergence. In summary, it appears that conflicting adaptive forces are acting on this larval group. Contrary to filter feeders and grazers, life cycles of shredders, dependent on the allochthonous input during autumn, seem to be highly synchronized to this non-renewable food source. Short emergence and flight periods characterize this group (Fig. 6).

Adaptations in plants

White (1978) argued that a plant species which is adequately nutritious and freely available to a herbivore throughout its life cycle would soon be exterminated. Plants have evolved different ways of protection against herbivores. Thick cell walls may render diatoms invulnerable to grazing stonefly nymphs (B.L. Madsen, pers. comm.) or mayfly nymphs such as *Cloeon dipterum* L. (Brown 1960). Another way of protection may be by the production of secondary substances rendering the plant unpalatable and/or toxic (see above). Thus Pennak (1973) found *Daphnia rosea* Sars to be strongly repelled by substances secreted by aquatic macrophytes, and extracts of freshwater macrophytes have been shown to be toxic against mosquitoe (*Aedes aegypti* L.) larvae (Angerilli 1980). On the other hand, certain snails and plants are often associated, and the snails are known to inflict 'extensive grazing damage' on the plants (Pip & Stewart 1976). Pip & Stewart (*op. cit.*) speculated that the significant amounts of soluble exudates excreted by macrophytes (cf. Wetzel 1969) may act as specific attractants. If so, it may be suspected that the plants in one way or another profit from attracting the snails in spite of the grazing damage (cf. Owen 1977).

Several factors indicate aquatic macrophytes to be much more exposed to herbivores than are terrestrial ones. Aquatic plants or at least parts of them are exposed during a long time of the year or even the year round. In lotic waters a very high pressure must be exerted by the herbivores, since most of them have their most intense period of growth during spring, when alternative food (dead leaves) is scarce. In addition many lotic herbivores are confined to areas of low current, i.e. the only places where many macrophytes can become rooted, and thus plants and herbivores are forced together through the action of the heterogeneity of the abiotic environment. Because of the restricted area available for colonization and a vegetative mode of reproduction many aquatic macrophytes often begin their life not very far from their parents, hence increasing the risk of being discovered by potential herbivores. On the other hand truly terrestrial plants often start growing at long distances from their parents and thus may escape herbivores by not being found at the new location.

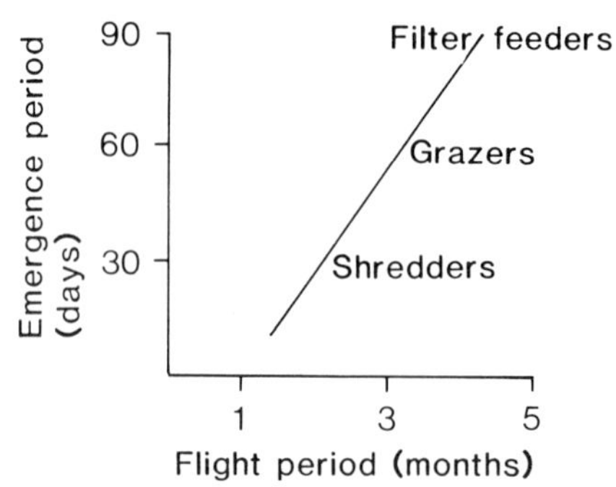

Fig. 6. Relationship between emergence period (days) and flight period (months) in lotic caddisflies whose larvae have different feeding behaviour. Redrawn from Otto 1981.

Besides, aquatic plants are often confined to borders between aquatic and terrestrial environments, making attacks from both directions possible, and also they should be easily encountered at the border of an uninhabitable area. All these factors contribute to a high exposure to herbivores i.e. a situation where secondary plant substances are likely to be selected for.

To investigate if plants growing in or associated (periodically in contact) with freshwater generally are unpalatable compared to terrestrial ones, Otto & Svensson (1981) compared the consumption rate by a polyphagous herbi- and detritivorous caddis larva (*Potamophylax cingulatus* Steph.) on terrestrial and 'aquatic' plants. Nine plant pairs, each of which was made up of one terrestrial and and one 'aquatic' species belonging to the same genus (one exception), were tested for consumption by larvae. In 7 out of 9 plant pairs consumption was highest on the terrestrial plant. Nitrogen content was highest in the 'aquatic' plant in all pairs but two. The consumption rate is normally inversely related to nitrogen content of the food (e.g. House 1965; Taylor & Bardner 1968; Baker 1975), and thus the low consumption on 'aquatic' plants was according to expectation. Texture has been shown to influence consumption by herbivores, i.e. high toughness inhibits consumption (e.g. Feeny 1970). In the field *P. cingulatus* larvae mainly fed on shed leaves, including those of beech (*Fagus silvatica* L.) which, though wet, are much tougher than fresh macrophytes. The larvae are sometimes observed feeding on alder (*Alnus glutinosa* L.) roots, indicating a good biting capacity. A high water content, as found in the rejected aquatic plants, would presumably facilitate biting, and thus texture seems difficult to accept as a factor of significance in explaining the different consumption rates in this case.

Feeding on plants of low nitrogen (N) content implies high consumption rates in order to achieve enough N for normal development (cf. Fox & Macauley 1977). These authors also showed that on plants (*Eucalyptus* leaves) of very low N contents (0.5–1.9%), the accumulation of N in larvae of a chrysomelid beetle increased linearly with leaf N content. However, on more 'normal' N content plants (>1.9% N), and here we find almost all plants tested in the cited study, N accumulation and N content of food do not seem to be related (Slansky & Feeny 1977). Thus, under normal feeding conditions, where consumption rate is inversely related to N content of food, from consumption rate and N content of plant alone nothing could be postulated concerning N accumulation. Although knowledge about N accumulation in relation to N content of food in *P. cingulatus* larvae is lacking, we feel that the N consumption as calculated from consumption rates and N content of macrophytes should show similar variations in terrestrial and aquatic plants. To put it more precisely: N accumulation with respect to N content of food should be the same for both groups of plants, and if the mechanism regulating N use efficiency is the same for both groups, then N consumption regression lines should be similar. However, as seen in Fig. 7, the regression lines differ markedly and, irrespective of N use efficiency, the low N consumption rates of the aquatic plants would imply these to be of low nutritional value as compared to the terrestrial plants.

Consumption of the most primitive plant pairs (genera *Scirpus* and *Rumex*) was unexpectedly high for the aquatic plants as compared to the terrestrial ones. The N content of these primitive aquatic plants was about as high as in more advanced plants where consumption rates were very low. This observation led us to assume that high N

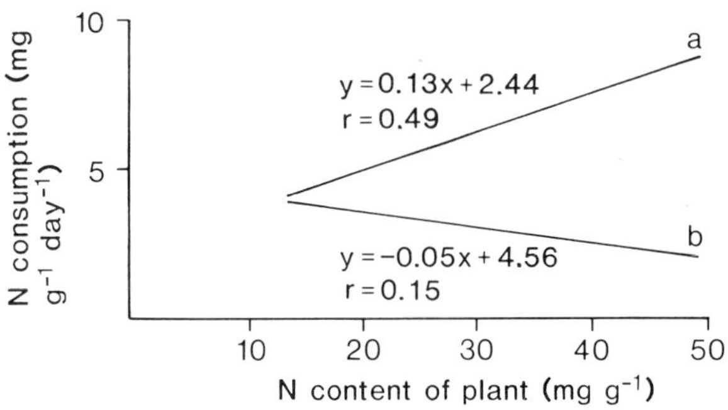

Fig. 7. N consumption of plant (mg $g^{-1}day^{-1}$, 60° C dry wt) in relation to N content (mg g^{-1}) of plant species by *Potamophylax cingulatus* (Trichopera) larvae. Line *a* refers to 'terrestrial' plants while *b* refers to 'aquatic' macrophytes. Redrawn from Otto & Svensson 1981.

content of the aquatic food plants was not the only factor responsible for the low consumption rates on the more advanced aquatic plants. From previous studies (Otto 1974) we also knew that the consumption of the advanced aquatic plants was definitively too low for a normal development of *P. cingulatus* larvae. These facts support the hypothesis that many aquatic plants have evolved secondary substances to reduce consumption by aquatic herbivores. Some secondary plant substances are known to form indegistible complexes with proteins, and the assimilation of dietary N is then inhibited (Feeny 1970). In the cited study, where most aquatic plants were rejected as food, it was concluded that at least these plants primarily rely upon repellent substances reducing palatability.

Acknowledgments

I thank B.S. Svensson, University of Lund, for comments on the manuscript. Support was received from the Swedish Natural Science Research Council.

References

Abrahamsson, S.A.A., 1966. Dynamics of an isolated population of the crayfish Astacus astacus Linné. Oikos 17:96–107.

Angerilli, N.P.D., 1980. Influences of extracts of freshwater vegetation on the survival and oviposition by Aedes aegypti (Diptera: Culicidae). Can. Ent. 112:1249–1252.

Baker, J.E., 1975. Protein utilization by larvae of the black carpet beetle, Attagenus megatoma. J. Insect Physiol. 21: 613–621.

Bovbjerg, R.V., 1965. Feeding and dispersal in the snail Stagnicola reflexa (Basommatophora: Lymnaeidae). Malacologia 2:199–207.

Brown, D.S., 1960. The ingestion and digestion of Cloeon dipterum L. Hydrobiologia 16:81–96.

Brown, K.M., 1982. Resource overlap and competition in pond snails: an experimental analysis. Ecology 63:412–422.

Calow, P., 1974. Some observations on locomotory strategies and their metabolic effects in two species of freshwater gastropods, Ancylus fluviatilis Müll. and Planorbis contortus Linn. Oecologia (Berl.) 16:149–161.

Calow, P. & Calow, L.J., 1975. Cellulase activity and niche separation in freshwater gastropods. Nature 255:478–480.

Cummins, K.W. & Klug, M.J., 1979. Feeding and ecology of stream invertebrates. Ann. Rev. Ecol. Syst. 10:147–172.

Doremus, C.M. & Harman, W.N., 1977. The effects of grazing by physid and planorbid freshwater snails on periphyton. Nautilus 91:92–96.

Eisenberg, R.M., 1966. The regulation of density in a natural population of the pond snail, Lymnaea elodes. Ecology 47:889–906.

Eisenberg, R.M., 1970. The role of food in the regulation of the pond snail, Lymnaea elodes. Ecology 51:680–684.

Feeny, P., 1970. Seasonal changes in oak leaf tannins and nutrients as a cause of spring feeding by winter moth caterpillars. Ecology 51:565–581.

Flint, R.W. & Goldman, C.R., 1975. The effects of a benthic grazer on the primary productivity of the littoral zone of lake Tahoe. Limnol. Oceanogr. 20:935–944.

Fox, L.R. & Macauley, B.J., 1977. Insect grazing on Eucalyptus in response to variation in leaf tannins and nitrogen. Oecologia (Berl.) 29:145–162.

Hart, D.D., 1981. Foraging and resource patchiness; field experiments with a grazing stream insect. Oikos 37:46–52.

Hart, D.D. & Resh, V.H., 1980. Movement patterns and foraging ecology of a stream caddisfly larva. Can. J. Zool. 58: 1174–1185.

House, H.L., 1965. Effects of low levels of the nutrient content of a food and of nutrient imbalances on the feeding and the nutrition of a phytophagous larva, Celerio euphorbiae (Linnaeus) (Lepidoptera: Sphingidae). Can. Ent. 97:62–68.

Hunter, R.D., 1980. Effects of grazing on the quantity and quality of freshwater Aufwuchs. Hydrobiologia 69:251–259.

Hutchinson, G.E., 1981. Thoughts on aquatic insects. BioScience 31:495–500.

Lalonde, R.T., Morris, C.D., Wong, C.F., Gardiner, L.C., Eckert, D.J., King, D.B. & Zimmerman, R.H., 1979. Response of Aedes seriatus larvae to fatty acids of Cladophora. J. Chem. Ecol. 5:371–381.

Otto, C., 1974. Growth and energetics in a larval population of Potamophylax cingulatus (Steph.) (Trichoptera) in a South Swedish stream. J. Anim. Ecol. 43:339–361.

Otto, C., 1976. Production of Ancylus fluviatilis Müller (Gastropoda) in a South Swedish stream. Pol. Arch. Hydrobiol. 23:421–429.

Otto, C., 1981. Why does duration of flight periods differ in caddisflies? Oikos 37:383–386.

Otto, C. & Svensson, B.S., 1981. How do macrophytes growing in or close to water reduce their consumption by aquatic herbivores? Hydrobiologia 78:107–112.

Owen, D.F., 1977. Are aphids really plant pests? New Scientist 76:76–77.

Patrick, R., 1970. Benthic stream communities. Am. Sci. 58: 546–549.

Pennak, R.W., 1973. Some evidence for aquatic macrophytes as repellents for a limnetic species of Daphnia. Int. Revue ges. Hydrobiol. 58:569–576.

Pip, E. & Stewart, J.M., 1976. The dynamics of two aquatic plant-snail associations. Can. J. Zool. 54:1192–1205.

Slansky, F. Jr. & Feeny, P., 1977. Maximization of the rate of oxygen accumulation by larvae of the cabbage butterfly on wild and cultivated foodplants. Ecol. Monogr. 47:209–228.

Soszka, G.J., 1975. Ecological relations between invertebrates and submerged macrophytes in the lake littoral. Ekol. pol. 23:393–415.

Streit, B., 1978. A note on the nutrition of Stylaria lacustris (Oligochaeta: Naididae). Hydrobiologia 61:273–276.

Streit, B., 1981. Food searching and exploitation by a primary consumer (Ancylus fluviatilis) in a stochastic environment: nonrandom movement patterns. Revue suisse Zool. 88: 887–895.

Summer, W.T. & Fisher, S.G., 1979. Periphyton production in Fort River, Massachusetts. Freshw. Biol. 9:205–212.

Taylor, W.E. & Bardner, R., 1968. Leaf injury and food consumption by larvae of Phaedon cochleariae (Coleoptera: Chrysomelidae) and Plutella maculipennis (Lepidoptera: Plutellidae) feeding on turnip and radish. Ent. exp. appl. 11:177–184.

Ward, J.V. & Dufford, R.G., 1979. Longitudinal and seasonal distribution of macroinvertebrates and epilithic algae in a Colorado springbrook-pond system. Arch. Hydrobiol. 86:284–321.

Warner, J.S., 1976. Choice among food levels by an aquatic grazer (Physa gyrina Say). Behav. Biol. 16:379–383.

Wetzel, R.G., 1969. Excretion of dissolved organic compounds by aquatic macrophytes. BioScience 19:539–540.

White, T.C.R., 1978. The importance of a relative shortage of food in animal ecology. Oecologia (Berl.) 33:71–86.

Author's address:
C. Otto
Department of Animal Ecology
University of Lund
Ecology Building
Helgonav. 5, S-223 62 Lund, Sweden

CHAPTER 26

Attached algal-substrata interactions: fact or myth, and when and how?

ROBERT G. WETZEL

Abstract. Much controversy has evolved in recent times over the relationships between attached algal and other microfloral populations and communities and the substrata upon which they grow. A critical review of the multitude of types of substrata and what is known about the interdependency of the microflora on the substrata permits several generalizations. (1) Nearly all substrata are highly dynamic in their physical characteristics and chemical contributions to attached microflora. (2) Functional interactions between substrata and attached microflora are greatly influenced by water turbulence both in lakes as well as streams. (3) Adaptations of epipelic algae and advantages of steep nutrient gradients towards the algae and overlying water compensate for the disadvantages of living on unstable organic-rich sediments in close proximity to toxic endproducts of anaerobic metabolism. (4) The many physical disadvantages of epiphytic algae associated with a growing, metabolically-demanding macrophytic substratum are greatly outweighed by the advantages gained in access to required physical (particularly light) and chemical factors. Each of these advantages and disadvantages is discussed in relation to existing knowledge and dogma. It is concluded that some contemporary generalizations are premature and misleading in relation to variable and complex reciprocal metabolic interactions existing between attached algae and substrata. Much of the incongruous information emanates from the lack of application of essential metabolic experimentation at the micro-level of investigation of attached communities and from the poor understanding of metabolism in associated substrata.

Introduction

Renewed controversy has evolved recently concerning the relationships between periphytic algae growing on substrata within fresh waters. Based upon limited evidence for concrete statements of either viewpoint, claims divide into (a) those statements that no significant interaction exists between algae and the substrata upon which they rest, and (b) those who believe that very complex metabolic relationships, including symbiosis, exist between attached microflora and the substrata, whether they are relatively inorganic or whether they are living either as a tissue (macrophyte) or metabolically active as in sediments. It is useful to address these two positions, because the first veiwpoint can in effect direct research away from investigation of periphyton-substrata interrelationships. In the following summary, I argue that the non-functional viewpoint is based upon insufficient information and is a premature conclusion.

Periphytic communities

Attached periphytic communities are adnate to

Wetzel, R.G. (ed.), Periphyton of Freshwater Ecosystems.
 ISBN 90 6193 768 X.

substrata, and these substrata are living or dead, or often a combination of the two components. Algal communities fall into two general types that are useful for discussion. (a) Those algae associated with largely inorganic substrata, which may be epipelic on rigid or loosely aggregated substrata, or on inorganic sediments with varying amounts of particulate organic matter. (b) That periphyton associated with living substrata, usually submersed portions of angiosperms, which consists of algae, bacteria, particulate detritus, inorganic products, (e.g. $CaCO_3$), and organic complexes of polysaccharides and proteins which originate from both the substratum as well as the periphytic community.

Soon after the turn of the century, statements were made that all biotic substrata in freshwater ecosystems could be replaced with inert substances and the lake would not be changed (Shelford 1918). Of course those statements are simply untrue, not only among epiphytic algae but also among epipelic communities. Most lakes of the world are shallow and the primary productivity of the attached algae often exceeds that of the phytoplankton. This ratio of attached versus planktonic productivity, of course, varies with basin morphometry, substrata, and nutrient loading (cf. discussion below).

Secondly, it is important to recognize that most of the attached microfloral communities live, grow, and die. During the more or less continuous senescence of portions of the community, much of the synthesized organic matter is released as dissolved organic compounds. Most of the particulate organic matter also enters the detrital organic matter pool for subsequent degradation and mineralization. On the average less than 10 per cent of the total organic matter produced is consumed by animals (cf. Wetzel 1983).

Several recent articles have appeared renewing the controversy of any potential relationship between the periphytic algae and the substrata (e.g. Cattaneo & Kalff 1979; Gough & Gough 1981). The essence of the argument of the former workers is that (a) growth and productivity of algae is the same whether upon natural or 'inert'* substrata, and (b) that no metabolic relationship exists between the algae and the 'host' substratum. My thesis is that this opinion of a lack of functional relationships is based upon inadequate information obtained at macrocommunity levels of analysis that are unable to analyze the complex exchanges (some positive, others negative) between the components of the attached communities and the substrata.

It is instructive to briefly review the general environmental conditions to which the periphytic communities are exposed. For purposes of discussion, the major communities are separable into those communities epipelic upon sediments (including largely inorganic substrata such as sand and rock) and those growing upon living substrata, primarily submersed macrophytes but also organic sediments that contain active bacteria and other microflora.

Physical environmental characteristics (Table 1)

Water turbulence is an important feature of the littoral zone. It is clear that excessive turbulence of shallow zones of lakes and streams is generally detrimental to attached algae. The molar action of moving sediments abrades and injures many attached algae. Epipsammic algae, although generally small and adapted to attachment within crevices of sand grams (Meadows & Anderson 1966, 1968; Steele & Baird 1968), are subject to displacement out of adequate light by churning of littoral sand sediments. Disturbance of organic sediments in shallow littoral areas is great and frequent, commonly with a net export by transport to deeper areas and subsequent overlying deposition of sediments (cf. Wetzel 1983).

Turbulence decreases rapidly with increasing depth in lakes. As a result, even loosely aggregated organic-rich sediments can serve as substrata for epipelic algae below the zone of active turbulence. Because of the unstable nature of loosely-aggregated sediments and often high rates of deposition of

* For example, glass or plastic substrata. Characteristics of texture or chemical composition of these artificial substrata are usually not stated; any nutrient or toxicity characteristics of the substrata are also usually not evaluated.

Table 1. Summary of responses of epipelic and epiphytic algae to primary physical, chemical and biotic environmental parameters.

Environment	Epipelic	Epiphytic
Physical		
1. Water movement	1. High → low depth gradient	1. High exchange
2. Substrata	2. Unstable, constant	2. Stable, increasing, changing
3. Temperature	3. High → low depth gradient	3. Commonly high
4. Light	4. Poor	4. Moderate to good
a. Quantity	a. High → low depth gradient	a. Good → poor → good gradient with changing depths-densities
b. Quality	b. Decreasing with increasing depth	b. Selective attenuation
		i. By water
		ii. By epiphytes
Chemical-biotic		
1. Conditions	1. Reducing	1. Aerobic: Diurnal variations
2. Inorganic nutrients	2. High	2. Low
a. Phosphorus	a. High	a. Low; variable
b. CO_2 (HCO_3^-)	b. High	b. Low; high competition
c. Others	c. High	c. Low
3. Organic substrates/nutrients	3. High	3. Moderate
a. Organic micronutrients	a. High	a. Low; high competition
b. Heterotrophy	b. High potential	b. Low; variable; high bacterial heterotrophy
4. Primary productivity	4. Low	4. High
5. Symbiotic interactions	5. Low	5. High

new overlying sediments from settling seston, epipelic algae risk burial. Many species are capable of migration in rhythmic fashion to compensate for light attenuation by sediment (reviewed by Round 1981 and Wetzel 1983).

Epiphytic algae adnate to growing tissues of submersed portions of aquatic macrophytes are generally exposed to greater water movements than are many algae sessile on sediments. The movements offer the potential for high exchange of water and disruption of metabolically-induced chemical gradients at the surfaces of the periphyton. As the macrophytes grow, new surfaces are created and rapidly colonized by epiphytes (see, for example, Rogers & Breen 1983). The alterations in substrata occur at several levels, both spatially and temporally. Epiphytic algae are often exposed to entirely different chemical and physical conditions of the environment on older tissues than on newly formed tissue that is higher in the water column. As discussed below, the metabolic characteristics of the supporting tissue substrata are also greatly different in actively growing macrophytes; variations from exponential growth versus often lower, senescing tissue on the same macrophyte provide entirely different substrata conditions from beneath as well as above the periphytic community.

Water temperatures to which epipelic periphyton are exposed are variable but in thermally stratified lakes are often low within the metalimnion and hypolimnion. In contrast, epiphytic periphyton are often exposed to warmer epilimnetic temperatures, since the macrophytes are commonly concentrated in shallow, quiescent waters and grow upwards.

Underwater irradiance has been widely studied in pelagial regions of lakes. Relatively little is known of the light quantity and quality in the microhabitat conditions of periphyton communities. The exponential and selective attenuation of light by primarily the molecular characteristics of water itself and dissolved organic compounds of water is well known and is applicable to the overall light available to periphyton along depth gradients, especially to those algae attached to sediments. As substrata change position spatially, as, for example, growing macrophytes, epiphytic communities are exposed to constantly changing light quantity and quality. Since the macrophytic growth form (increased dissection and surface area) has evolved to maximize light reception, changing growth positions can work to advantage of epiphytes by growth upward and to disadvantage of earlier colonizers on lower tissue that is subsequently shaded by the plant overstory. Similarly, epiphytes attached to the basal portions of emergent macrophytes, and epipelic algae growing among emergent vegetation, are subject to greatly reduced light availability as the canopy develops seasonally. Very little is known of the adaptations of algae to this selective light attenuation by means of chromatic adaptation and other physiological compensatory mechanisms.

Within the periphyton itself, the growth of the microfloral community markedly affects light quantity and quality reaching algal populations as well as the supporting substratum. Algal pigments within the periphyton, particularly chlorophyll *a*, are foremost in selective light attenuation (Losee & Wetzel 1983). Bacteria, particulate detritus, and calcium carbonate within the periphyton complex are relatively non-selective to light attenuation.

Chemical-biotic relationships (Table 1)

As we evaluate the nutrient/biotic relationships, the dichotomy between epipelic and epiphytic periphyton increases. A major distinction between the epipelic and epiphytic habitats is the condition of the substratum. Rather than relating to the macrohabitat of the overlying water, it is important that we understand the conditions at the microhabitat level to which the algae and bacteria are exposed.

The epipelic habitat is nearly always reducing, even if the overlying water is aerobic, immediately below (*ca.* 1 mm) and at times within the epipelic community. Very steep microgradients exist vertically within the surficial sediment. Fermentative conditions occur within a few millimeters or micrometers of epipelic algae. Fermentative end products of sediment bacteria (CO_2, H_2S, CH_4, volatile fatty acids) are produced, and dissolved oxygen of overlying water or produced in photosynthesis can be rapidly consumed by high oxidative and respiratory demands. Diffusion gradients function in both directions across the periphytic community, with most fluxes in the direction of the overlying water through the periphyton. Migratory abilities of epipelic algae are important not only to poise the community within light gradients but likely also for movement to minimize the effects of toxic metabolic end products.

Epipelic periphyton can be exposed to high concentrations of essential nutrients, in particular CO_2 and HCO_3^-, inorganic phosphorus, ammonium ions, and inorganic micronutrients. Concentrations of these nutrients tend to be high in the interstitial water of sediments undergoing active mineralization, and the concentration gradients encourage upward diffusion through the periphytic community. Bacterial productivity is very high in the sediments at the sediment-water interface. Synthesis of essential organic micronutrients, such as the water-soluble vitamins (e.g. B_{12}, thiamine, biotin; e.g. Kurata *et al* 1979a, 1979b), required to varying degrees by algae, is high by benthic bacteria in close juxtaposition to be epipelic algae. Some bacterial competition for organic micronutrients is likely but production interstitially is high and excessive amounts are probably readily available to the attached algae.

Production of high energy-containing organic substrates, e.g. acetate, can be very large at the sediment-water interface. Although in general algal assimilation of organic substrates by chemoorganotrophy or photoorganotrophy is low when in competition with the more efficient bacterial active permease uptake mechanisms, if significant algal heterotrophy occurs in freshwater ecosystems, this mechanism would be expected by epipelic algae. Very little is known of the significance of such mechanisms by epipelic algae under *in situ* conditions.

In contrast, the epiphytic habitat is generally aerobic. However, the microstructure within the epiphytic community is highly structured. The epiphytic complex consists of living algae, bacteria, detrital particulate organic matter, dead cells, as well as, in bicarbonate waters, particulate $CaCO_3$, all of which is held together with an organic matrix of polysaccharides and proteinaceous compounds (e.g. Wetzel & Allen 1970; Allanson 1973). This epiphytic complex is *stratified*, in which new cell aggregates are constantly forming over the older cells. Another important characteristic of the epiphytic community is the *close juxtaposition* of all cells and components, so that diffusion distances are small. Although the close juxtaposition is an advantage for the microflora, the proximity of the photosynthesizing tissue of the supporting submersed macrophyte can have both advantages and disadvantages. The macrophyte tissue can effectively compete for nutrients from the water, e.g. CO_2, HCO_3-, other inorganic nutrients, even though many nutrients are assimilated by the macrophytes from the sediments (reviewed by Wetzel 1983). This epiphyte-macrophyte competition, however, varies diurnally with the daily photosynthetic, respiratory, and photorespiratory cycles of the larger plant itself. The result can be a dynamic, variable access of the periphyton to nutrients, both from the water and from the macrophyte substratum. As photosynthesis proceeds diurnally, a microenvironment occurs at the surfaces with rapidly increasing pH and reduced availability of many nutrients, in particular inorganic CO_2. Following light availability, the most frequent limitation to macrophytic photosynthesis is the

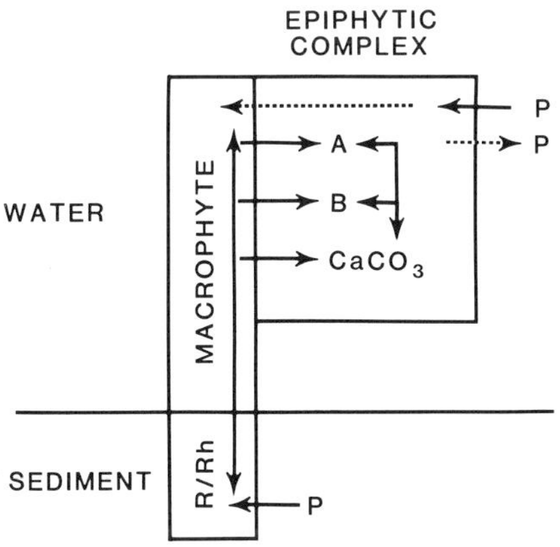

Fig. 1. Diagrammatic representation of the potential fluxes of phosphorus within a submersed macrophyte and the epiphytic complex consisting of algae (A), bacteria (B), and monocarbonate deposition (see text).

availability of inorganic carbon (cf. Wetzel & Grace 1983). Despite many adaptations in submersed angiosperms to conserve respired CO_2, release of CO_2 by the macrophytes from respiration and photorespiration is much less or nil in light than in darkness.

The availability and sources of phosphorus to epiphytic algae and bacteria have been a subject of much recent controversy. Most submersed angiosperms assimilate phosphorus from the sediments where interstitial concentrations are manifold times greater than in the water. Phosphorus is translocated to the foliage in angiosperms and even in certain macroalgae. Phosphorus from the water must pass through the epiphytic community in order to reach the macrophyte tissue (Fig. 1). Competition for phosphorus results from utilization by algae, bacteria, and by adsorption to $CaCO_3$ encrustations produced by photosynthetic metabolism of CO_2 by both the macrophyte and the periphytic algae. Very little phosphorus usually reaches the macrophyte until concentrations within the water are very large. Some exchange within the periphytic complex towards the macrophyte may occur, although competition is intense and it is probable that phosphorus within the epiphytic community will be conserved.

Much of the controversy focused on the release of phosphorus from the macrophyte to the water and thus through the epiphytic community (Fig. 1). Some studies have indicated small quantities (*ca.* 5%) of phosphorus release from macrophytes (e.g. Carignan & Kalff 1982), particularly from submersed macrophytes of oligotrophic freshwaters. Other workers show extracellular release of phosphorus translocated from the sediments of up to 15 per cent of the total (Moeller & Wetzel, in preparation). In all cases, it is apparent that little, if any, phosphorus of macrophytic origin passes through the epiphytic complex to the surrounding water, as long as the plants and periphyton are actively growing. When the macrophytes are senescing, a somewhat greater release of phosphorus to the water may occur. However, among perennial and 'evergreen' angiosperms, most residual phosphorus is translocated to the root/rhizome system. In addition, at the end of a growth phase, much of the macrophyte detritus and residual epiphytic complex settles to the sediments, with relatively little release of phosphorus to the water. Phosphorus of the macrophyte-epiphyte complex is thus largely conserved and short-circuited back to the sediments and is continually being recycled between the sediments and the plant-epiphytic complex.

The availability of organic micronutrients to epiphytic algae is essentially unknown. The proximity of epiphytic algae to bacterial sources of organic micronutrients (e.g. vitamins) is probably an advantage for the algae (Table 1). A high competition for these substances exists, however, from the macrophytes and by chemical adsorptive and inactivation processes.

Chemoorganotrophic and photoheterotrophic utilization of simple organic substrates can occur by epiphytic algae, especially at low light intensities. The results of Pip & Robinson (1982) and in my own laboratories indicate, however, that heterotrophic utilization of organic substrates by epiphytic algae is very low. Only about one percent or less of carbon fixation was by heterotrophy; the remainder was by light-mediated photosynthetic fixation. Marked differences were found in the algal species utilizing the organic substrates. As in the phytoplankton (e.g. McKinley & Wetzel 1978; Ellis & Stanford 1981), competition for organic substrates released by bacteria and from other sources is high. Simple compounds (<700 Daltons) released from macrophytes are rapidly utilized and mineralized by bacteria (e.g. Søndergaard 1982). Even though much is known about the quantities and types of organic substrates released by submersed macrophytes, relatively little is known about the utilization of these compounds by the epiphytic community or indirect effects of altered bacterial metabolism (e.g. increased production of CO_2 or organic micronutrients).

It is at this microlevel of community interactive utilization and release of organic substrates and inorganic nutrients that a marked potential for symbiosis exists between epiphytic community components and between these epiphytes and the substratum. Because of the close proximity of the community components, the potential for reciprocation and exchange of essential nutrients and organic growth factors is high. Although antibiotic effects of organic compounds released by the macrophytic substratum are known (e.g. Anthoni & Christophersen 1982; Wium-Andersen *et al* 1982), this antibiotic reduction of epiphytic microflora is probably an exception rather than the rule. In many cases the epiphytic microflora is likely enhanced by metabolic products (O_2, organic compounds) released by the macrophyte. In turn, the macrophyte may obtain required nutrients, in particular CO_2, from the periphyton. My hypothesis is that the microflora complex on the plant surface can function as a highly integrated system metabolically. Whether such an epiphytic community was evolutionarily advantageous for the macrophytes as a means of diverting grazing away from the macrophyte tissue (Hutchinson 1975) is conjectural and very difficult to verify. The advantages of the epiphytic community spatially within a habitable environment are obvious; the advantages of the epiphytes on the macrophytes are less clear. Under oligotrophic conditions the reciprocity of the metabolic exchanges are likely more advantageous than the disadvantages associated with light reduction to the macrophytes by the epiphytes.

Comparison of the rates of primary productivity by epipelic and epiphytic algae indicates that the primary productivity of epiphytic algae is often much greater, often by a factor of 10, than that of epipelic algae per unit area of littoral zone (cf. Wetzel 1983). As discussed earlier, much of this difference is associated simply with the much larger surface area of macrophytes for epiphytic colonization within a well-lighted zone than is the case with much of the epipelic habitat.

The hypothesis has been expressed (Phillips *et al* 1978), that the nutrient concentrations of the water are of major importance, in combination with light regulation, in controlling the development of the epiphytic periphyton (Fig. 2). Under low nutrient loading conditions to the water, characteristic of oligotrophic and mesotrophic lakes, epiphytic algal development is low because, by implication, of a competition for nutrients of the water *with phytoplankton*. Light conditions for the macrophytes would be good because of modest light attenuation by both the epiphytic and macrophytic communities. The hypothesis further states that as nutrient loading to the water increases, increased growth of both the phytoplankton but particularly the epiphytic algae causes severe light reduction and the eventual demise of the supporting macrophytes (Fig. 2). Evidence for the release of excreted organic compounds that suppress the phytoplankton is very poor (reviewed by Wetzel 1983). More likely any suppression results from competition for available nutrients. This hypothesis assumes (a) that concerning the epiphytic algae, the macrophyte functions as an inert substratum on which the sessile algae grow, and (b) that 'suppressants' released by the macrophyte inhibit phytoplankton algae but not epiphytic algae.

I argue that, based on existing evidence, it is premature to state what and how much the 'actors' in this complex community are doing and effecting each other. We simply do not know. Most analyses addressing these problems are performed at the whole community level. Resolution of the regulating interactions among the metabolism of each component (algae, bacteria, supporting substrata) must be analyzed separately but simultaneously. We know from the few analyses performed at the necessary microlevel, at which the interactions are occurring, that specific relations exist among species of both the epiphytes and the substrata types. The *in situ* metabolism at the microlevel must be carefully analyzed; quantification of the metabolic pathways must be analyzed by direct experimental

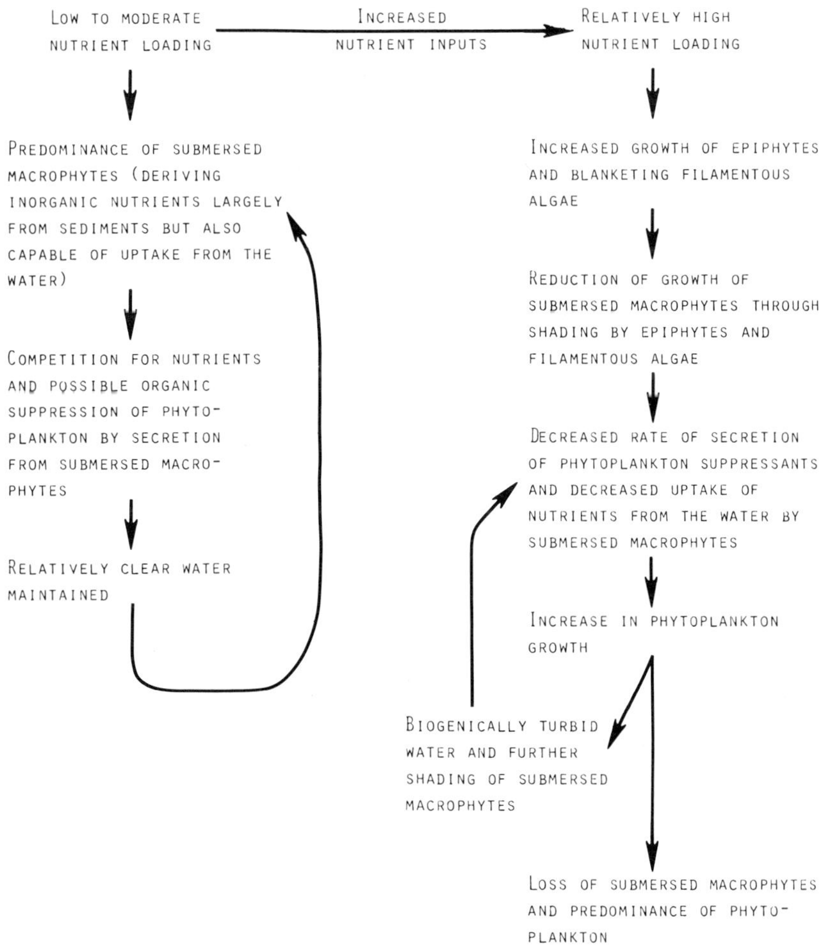

Fig. 2. Primary phases of one hypothesis of the interrelationships between macrophytes, epiphytic periphyton, and nutrient loading (see text). (Modified from Phillips *et al*, 1978.)

methods in this heterogeneous community. Much of the required methodology to do this is presently available, although it is tedious and requires much patience. Developments in this area are required with vigor; hence my belief, as I encouraged G. Robinson (1983) to emphasize, that methodology is the key to advances in periphyton research.

To delineate, however, the metabolic relationships in the attached periphyton community is only half of the story. The substratum itself is a very dynamic metabolic system that can be coupled to the metabolism of the periphyton. That statement is true whether one considers the tissue of the submersed portions of a macrophyte or sediment, which is also a very dynamic metabolic system. In order to evaluate the real interactions between epiphytes and substrata, we must recognize (a) that the epiphytic community consists of a community of algae, bacteria and detritus, and (b) the rapidity with which conditions change *below* the periphyton within the substratum.

Acknowledgements

Portions of these analyses were supported by the National Science Foundation (DEB-8001190) and the U.S. Department of Energy (EY-76-02-S-1599, COO-1599-225). Contribution No. 504, W. K. Kellogg Biological Station of Michigan State University.

References

Allanson, B.R., 1973. The fine structure of the periphyton of Chara sp. and Potamogeton natans from Wytham Pond, Oxford, and its significance to the macrophyte-periphyton metabolic model of R. G. Wetzel and H. L. Allen. Freshwat. Biol. 3:535–541.

Anthoni, U. & Christophersen, C., 1982. Synthesis of 4-methylthio-1,2-dithiolane and 5-methylthio-1,2,3-trithiane. Two naturally occurring bioactive compounds. Tetrahedron 38:2425–2427.

Carignan, R. & Kalff, J., 1982. Phosphorus release by submerged macrophytes: Significance to epiphyton and phytoplankton. Limnol. Oceanogr. 27:419–427.

Cattaneo, A. & Kalff, J., 1979. Primary production of algae growing on natural and artificial aquatic plants: A study of interactions between epiphytes and their substrate. Limnol. Oceanogr. 24:1031–1037.

Ellis, B.K. & Stanford, J.A., 1982. Comparative photoheterotrophy, chemoheterotrophy, and photolithotrophy in a eutrophic reservoir and an oligotrophic lake. Limnol. Oceanogr. 27:440–454.

Hutchinson, G.E., 1975. A Treatise on Limnology. Vol. III. Limnological Botany. New York, J. Wiley & Sons. 660 pp.

Kurata, A., Saraceni, C. & Kadota, H., 1979a. The status of B group vitamins in macrophyte and pelagic zones of Lake Biwa. Mem. Ist. Ital. Idrobiol. 37:63–85.

Kurata, A., Saraceni, C. & Kadota, H., 1979b. Diurnal changes of concentration in water of B group vitamins in macrophyte and pelagic zones of Lake Biwa. Mem. Ist. Ital. Idrobiol. 37:87–103.

Losee, R.F. & Wetzel, R.G., 1983. Selective light attenuation by the periphyton complex. In: Wetzel, R.G. (ed.) Periphyton of Freshwater Ecosystems. Developments in Hydrobiology 17:89–96.

McKinley, K.R. & Wetzel, R.G., 1979. Photolithotrophy, photoheterotrophy, and chemoheterotrophy: Patterns of resource utilization on an annual and a diurnal basis within a pelagic microbial community. Microbial Ecol. 5:1–15.

Meadows, P.S. & Anderson, J.G., 1966. Microorganisms attached to marine and freshwater sand grains. Nature 212: 1059–1060.

Meadows, P.S. & Anderson, J.G., 1968. Microorganisms attached to marine sand grains. J. Mar. Biol. Assoc. U.K. 48:161–175.

Phillips, G.L., Eminson, D. & Moss, B., 1978. A mechanism to account for macrophyte decline in progressively eutrophicated freshwaters. Aquat. Bot. 4:103–126.

Pip, E. & Robinson, G.G.C., 1982a. A study of the seasonal dynamics of three phycoperiphytic communities, using nuclear track autoradiography. I. Inorganic carbon uptake. Arch. Hydrobiol. 94:341–371.

Pip, E. & Robinson, G.G.C., 1982b. A study of the seasonal dynamics of three phycoperiphytic communities, using nuclear track autoradiography. II. Organic carbon uptake. Arch. Hydrobiol. 96:47–64.

Robinson, G.G.C., 1983. Methodology: The key to understanding periphyton. In: R.G. Wetzel, (ed.) Periphyton of Freshwater Ecosystems. Developments in Hydrobiology 17: 245–252.

Rogers, K.H. & Breen, C.M., 1983. An investigation of macrophyte/epiphyte interactions. In: Wetzel, R.G. (ed). Periphyton of Freshwater Ecosystems. Developments in Hydrobiology 17:217–226.

Round, F.E., 1981. The Ecology of Algae. Cambridge, Cambridge University Press. 653 pp.

Shelford, V.E., 1918. Conditions of existence. In: Ward, H.B. & Whipple, G.C. (ed.) Freshwater Biology. New York, J. Wiley & Sons, Inc. pp. 21–60.

Søndergaard, M., 1981. Kinetics of extracellular release of ^{14}C-labelled organic carbon by submerged macrophytes. Oikos 36:331–347.

Steele, J.H., & Baird, I.E., 1968. Production ecology of a sandy beach. Limnol. Oceanogr. 13:14–25.

Wetzel, R.G., 1983. Limnology. 2nd Ed. Philadelphia, Saunders College Publishing. 860 pp.

Wetzel, R.G. & Allen, H.L., 1970. Functions and interactions of dissolved organic matter and the littoral zone in lake metabolism and eutrophication. In: Kajak, Z. & Hillbricht-Ilkowska, A. (ed). Productivity Problems of Freshwaters. Warsaw, PWN Polish Scientific Publishers. pp. 333–347.

Wetzel, R.G. & Grace, J.B., 1983. Aquatic plant communities. In: Lemon, E. (ed.) CO_2 and Plants: The Response of Plants to Rising Levels of Atmospheric Carbon Dioxide. Boulder, Colorado, Westview Press, Inc. pp. 223–280

Wium-Andersen, S., Anthoni, U., Christophersen, C. & Houen, G., 1982. Allelopathic effects on phytoplankton by substances isolated from aquatic macrophytes (Charales). Oikos 39:187–190.

Author's address:
Robert G. Wetzel
W.K. Kellogg Biological Station
Michigan State University
Hickory Corners, MI 49060, U.S.A.

CHAPTER 27

An investigation of macrophyte, epiphyte and grazer interactions

K.H. ROGERS and C.M. BREEN

Abstract. Hutchinson (1975) and Eminson & Moss (1980) proposed that a delicate belance develops between macrophyte and epiphyton, such that the epiphyton provides maximum benefit to the host by diverting grazing without becoming so dense as to interfere with light and CO_2 reception. This study investigated the interaction between *Potamogeton crispus* L. from a tropical lake, its epiphyton and snail grazers. The host plant was resistant to consumption by snails until it was invaded by epiphytic, necrotrophic bacteria which 'condition' the host tissues. Grazing of epiphytes by the snails, however, reduced the rate of development of the bacteria and probably extended the life of the plants. Once mature, the plants are uprooted by feeding waterfowl causing a reduced resistance to necrotrophic attack and permitting increased invasion and conditioning. Conditioned plants were rapidly consumed by snails and converted to a fine detritus. This study indicates an additional factor, the necrotrophic bacteria/grazer interactions, in the interpretation of the response of host/epiphyte associations to nutrient enrichment, and demonstrates that epiphyte/grazer interactions can play an important role in determining the fate of submerged macrophyte production.

Introduction

The elucidation of the chemo-organotrophic relationship between epiphytes and their host plant (Allan 1971) has stimulated considerable interest in the nature of the epiphyte/host association. A general concept which has emerged is that dissolved organic matter (DOM) and nutrients released by the macrophyte host are utilized by bacterial and algal epiphytes (Wetzel 1969; Wetzel & Allen 1970; Allen 1971; Allanson 1973; Harlin 1973; McRoy & Goering 1974) and that the rates of transfer may affect epiphyte diversity (Hossell & Baker 1979; Eminson & Moss 1980) and productivity (Allen 1971; Rogers 1981) especially in systems not enriched with nutrients.

Hutchinson (1975) suggested that the epiphyton growth provides an easily available and nutritious food source for grazing invertebrates, thereby diverting them from consuming and damaging the host plant. He postulated that epiphytes play the same role in aquatic systems as do chemical adaptations which prevent excessive grazing in terrestrial systems. More recently Eminson & Moss (1980) developed the more broad concept that a delicate balance develops between macrophyte and epiphyton such that it provides maximum benefit to the host by diverting grazing without becoming so dense as to interfere with light and CO_2 reception. Studies of epiphytes on *Potamogeton crispus* L. in a South African lake (Rogers & Breen 1981) have shown that some epiphytic bacteria invaded the host epidermis. This led us to suggest that secretion of organic acids by the invading organisms promoted senescence of the host thereby facilitating consumption by invertebrates.

Wetzel, R.G. (ed.), Periphyton of Freshwater Ecosystems.
 ISBN 90 6193 768 X.

These bacteria were therefore considered necrotrophic.

This study examines the effects of grazing by the snail *Bulinus* (*Bulinus*) *natalensis* Küst. on the epiphyte/host relationship of *Potamogeton crispus*.

Study area

Potamogeton crispus, a winter producing annual, forms dense stands in the small shallow lakes of the Pongolo river floodplain, South Africa (27° 30′ S; 32° 15′ E). A full description of the floodplain is given by Furness & Breen (1980) and of the annual growth cycle of *P. crispus* by Rogers & Breen (1980).

Methods

Epiphyte colonization and snail grazing. Mature plants of *P. crispus* bear leaves of different ages with the youngest at the shoot apex and oldest at the base (Rogers & Breen 1981). Shortly after the *P. crispus* population attained maximum standing crop in 1978, approximately 100 mature plants were cut at the sediment and gently transferred to a trough of lake water. Two leaves were removed from each of four zones on the plant: 'Youngest' leaves, which were not fully expanded, near the apex; 'young' leaves, fully expanded near the apex; 'mature' leaves midway down the stem showing no evidence of senescence; 'oldest' leaves from the base of the stem were yellowing and senescent.

All snails and other invertebrates were removed from the leaves which were immediately transferred to the laboratory. Despite care in handling, loose 'detrital aggregate' (Bowen 1979) was dislodged and this study therefore considers only the more securely attached epiphyton.

The differences in epiphyton composition and the ultrastructure of the epiphyton/host interface with leaf age, at the time of sampling, were assessed by examination of five leaves of each age class using scanning and transmission electron microscopy, respectively (Rogers & Breen 1981).

A further 4 leaves of each age class were transferred to each of 5 glass dishes, containing unfiltered lake water and 8 snails of uniform size (shell height 2.5–3.5 mm). The effects of snail grazing on the epiphyton and leaf surface were assessed by a combination of electron microscopy and determination of the rate of consumption of leaves by snails. The leaves of 1 dish (4) were removed for microscopical analysis after 24 h; the remaining leaves (16) of each age were examined every second day to determine the time required for disappearance of each leaf of each age class.

Additional leaves of each age class were incubated in unfiltered, aerated lake water to permit invasion by necrotrophic bacteria to continue in the absence of grazers. The change in edibility, determined as the rate of consumption (Nicotri 1977) was estimated, as described above, at 2, 4, 6, 8, 12 and 18 day intervals.

Seasonal changes in P. crispus *and snail populations.* Sequential estimates of standing crop (oven dry weight m^{-2}) were used to follow the seasonal changes in snail populations and the growth pattern of *P. crispus.* Samples were taken with an electromechanical sampler (Rogers & Breen 1980) and the snails separated from the plant material during the washing process. Samples were oven dried (60° C) and weighed and snail numbers counted.

The effects of snail grazers and epiphytes on detritus formation. Potamogeton crispus produces vegetative propagules (turions) when the plants mature and intensive grazing of turions by waterfowl (Rogers 1980) results in most plants being uprooted and broken. Since these plants disappear rapidly (Rogers & Breen 1982) and their fate is unknown, *in vitro* experiments were conducted to assess the effects of necrotrophic bacteria and snails on them.

Uprooted plants were collected from the field, washed gently in lake water to remove sediment, and all invertebrates, roots and reproductive structures separated out. The excess water was then removed by centrifugation of the material in a commercial spin drier.

Sixty-four subsamples of 30 g each (about 3 g dry weight) were weighed rapidly and transferred to plastic bags before being transported to the laboratory. The subsamples were placed in glass jars containing 2 *l* of aerated unfiltered water and divided into four treatments of sixteen jars each:

1. No snails: *P. crispus* plants incubated without snails as described above.
2. Snails: Forty snails of 0.35 g fresh weight were added to each jar. This represented the density and mass of snails per gram of *P. crispus* in the field at the time.
3. No snails plus EDA: Since preparation of subsamples would have dislodged much of the epiphyton, 20 ml (0.025 g dry weight) of epiphytic detrital aggregate (EDA) was added to act as a source of bacteria during incubation. The EDA was washed from other plants.
4. Snails plus EDA: Both snails and EDA were added to the jars.

The effects of the treatments were assessed by following mass loss of the plant material and detritus accumulation over a period of 24 days. Four replicates of each treatment were harvested at 6-day intervals and the remaining plant material separated from the detritus by pouring the contents of the jars through a 1 mm^2 sieve. The water was filtered through pre-weighed glass-fibre filters to remove the detritus and both detritus and remaining plant material were oven dried (60° C) and weighed.

Results were fitted to an exponential function ($y = a^{-bk}$, where y = mass of plant material remaining after time interval b, a is the initial mass and k the rate constant) which was then linearised to $\ln y = a - bk$ (where a is then the natural log of the initial mass). Differences in rates of mass loss were tested for significance by following the 'Simultaneous Test Procedure' for differences between regression co-efficients (k) (Sokal & Rohlf 1969).

Results

Epiphyton colonization and snail grazing

The youngest leaves collected from the field were devoid of epiphyton (Fig. 1a) and the epidermal cells had a thin cuticle overlying a layered cell wall (Fig. 1b). After 24-h exposure to snails the leaf surface remained apparently unchanged. Thus if the snails had been grazing on the leaves they apparently caused little or no damage.

A sparse epiphyton, consisting mostly of *Cocconeis placentula* (Ehrenberg) Cleve was evident on young leaves (Fig. 1c and d) but did not appear to disrupt the structure of the epidermal wall (Fig. 1d). After 24 h, snails had removed most of these epiphytes, leaving debris and the ventral valves of diatoms behind (Fig. 2a). The surface of these young leaves was also unaffected by the grazing action of the snails (Fig. 2b).

Mature leaves supported a more dense and diverse epiphyton (Fig. 1e) and the epidermal wall had been invaded by bacteria which were evident below the cuticle (Fig. 1f). Although the cuticle was generally intact, electron translucent areas surrounding the bacteria indicated enzymatic degradation of the cell wall. Grazing snails had, within 24 h, removed the epiphyton and parts of the outer walls of some epidermal cells. This exposed numerous canals (Fig. 2d, arrowed) which are thought to have been caused by the necrotrophic action of the bacterial epiphytes within the cell walls. The incidence of canals increased in the vicinity of damaged cells, indicating a correlation between heavy localised infections of bacteria and cell damage caused by snail grazing.

Senescent leaves supported the most dense and diverse epiphyton (Fig. 1g). The epidermal walls were almost entirely electron translucent as a result of bacterial degradation (Fig. 1h) and the leaves were extensively damaged by snail grazing within a period of 24 h (Figs. 2e and f).

After 6 days in the presence of snails, all the youngest leaves (Fig. 3a) and most young leaves (Fig. 3b) were still intact and showed no macroscopic evidence of damage. Furthermore, while most mature leaves had been damaged (Fig. 3c), the few senescent leaves remaining were extensively damaged (Fig. 3d). Microscopical examination of the leaf surfaces showed them to be devoid of epiphyton and while the youngest leaves were undamaged, damage increased with leaf age.

Grazing snails therefore fed primarily on epiphytes but in older leaves, those parts which had been degraded by epiphytic, nectrotrophic bacteria

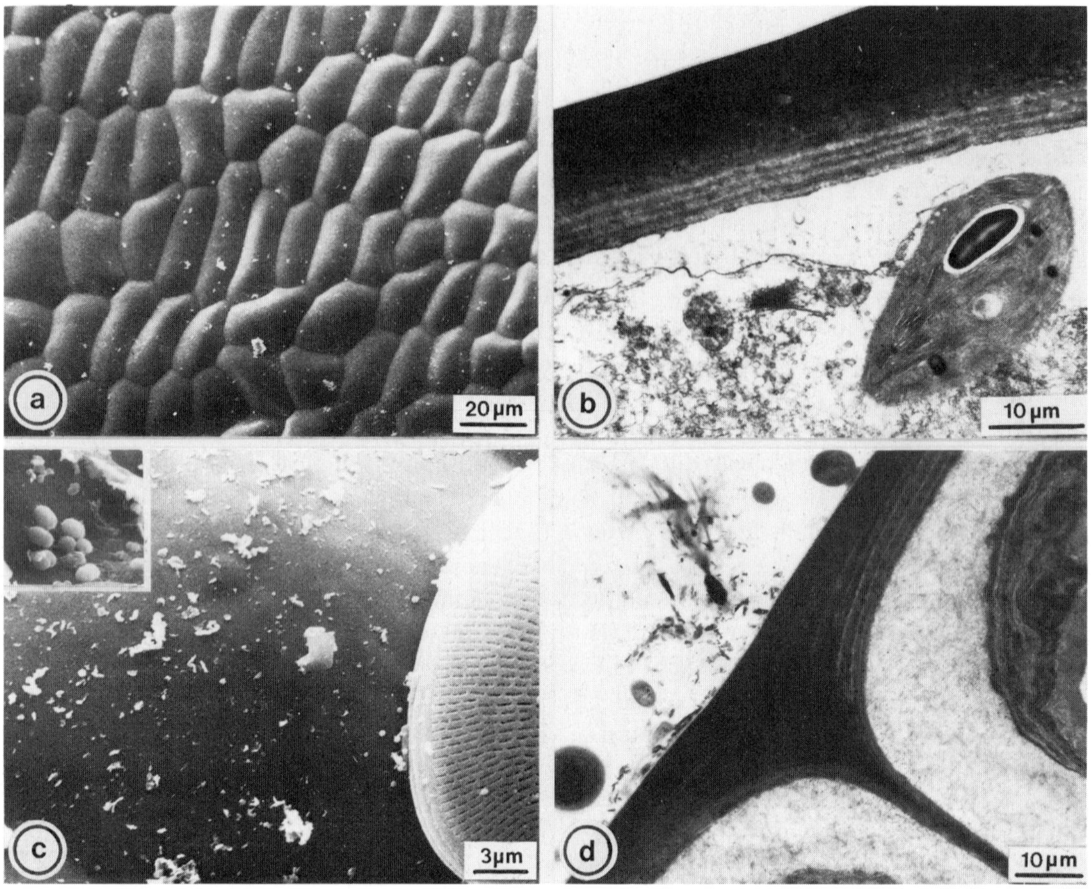

Fig. 1. Changes in the epiphyton community (left), and invasion by necrotrophic bacteria (right), of *P. crispus* leaves of different ages: (a + b) youngest; (c + d) young; (e + f) mature; (g + H) senescent.

were also consumed. Preparation of plant tissues for consumpton by invertebrates in this way has been noted for autumn shed leaves in streams and termed 'conditioning' by Cummins (1974). Grazing *in vitro* also, however, controlled the extent of recolonization by epiphytes. It is suggested that this would have the effect of reducing conditioning by necrotrophs which would retard consumption by snails. This hypothesis was tested in the laboratory by comparing the change in 'edibility' (Nicotri 1977) of leaves of different ages with increasing time allowed for conditioning before grazers were introduced.

'Conditioning' and 'edibility' of leaves

When leaves of the two youngest categories (youngest and young) were presented to snails immediately after collection, they were inedible and remained so until the experiment was terminated 20 days later (Fig. 4a and b). Even when these leaves were incubated with detrital aggregate in the absence of snails for 2, 4 and 8 days before being introduced to snails, they remained apparently unconditioned and inedible.

Mature and senescent leaves, however, had been conditioned *in situ* before collection and edibility, which was always greatest in the oldest senescent leaves, increased with time of conditioning in the laboratory (Fig. 4a and b).

After 18 days conditioning all leaves were similarly edible and most had been consumed by the 20th day (Fig. 4b). Thus, the number of leaves remaining after 20 days was greatest when con-

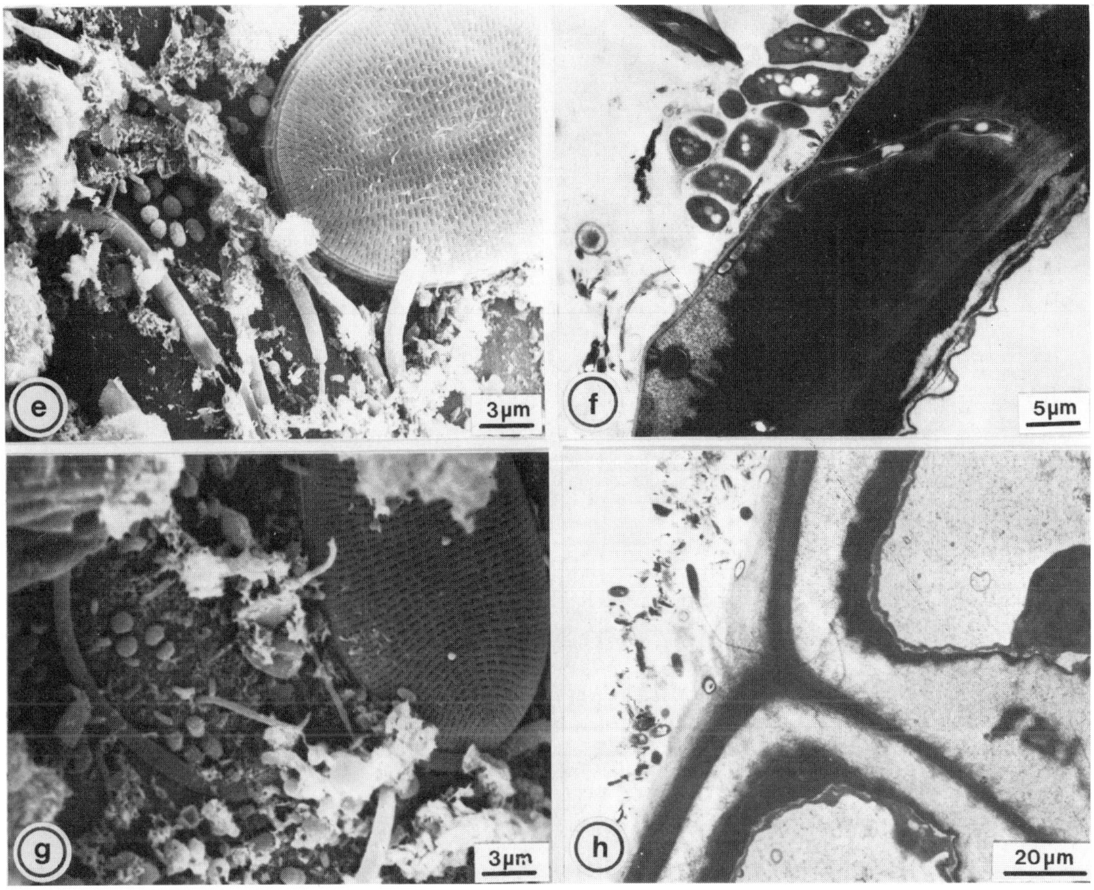

ditioning was least and when snail grazing continuously removed the developing epiphyton. The results show that although the degree of conditioning of leaves controls edibility and thus consumption by snails, grazing, by controlling the development of epiphyton, arrests conditioning and curtails consumption of the leaves.

In the field, however, the number and mass of snails per gram of plant tissue (Fig. 5) was less than that used in the above experiments and did not remain constant but increased to a maximum on senescing plants which support the most epiphyton (Rogers 1981). Thus it might be expected that the effects of epiphytes and snail grazers on the host plant in the field would be most pronounced at this time when feeding waterfowl uprooted the plants.

The effects of snail grazers and epiphytes on detritus formation

Uprooted plants, incubated without snails, lost *c*.38% of their mass in 12 days (Fig. 6a) at which stage they were brown, flaccid and apparently dead. Thereafter mass loss continued exponentially ($k = -0.033$) (Fig. 6a, Table 1) such that 50% loss ($t\,^1/_2$) occurred within 20 days. The pattern and rate of mass loss was not significantly ($p>0.05$) affected by the presence of EDA when grazers were absent (Fig. 6a, Table 1). In contrast, plants incubated with snails, although showing the same exponential pattern of mass loss, lost mass significantly ($p>0.05$) more rapidly ($k = 0.056$, $t\,^1/_2 = 13$) and this was greatly enhanced by addition of EDA.

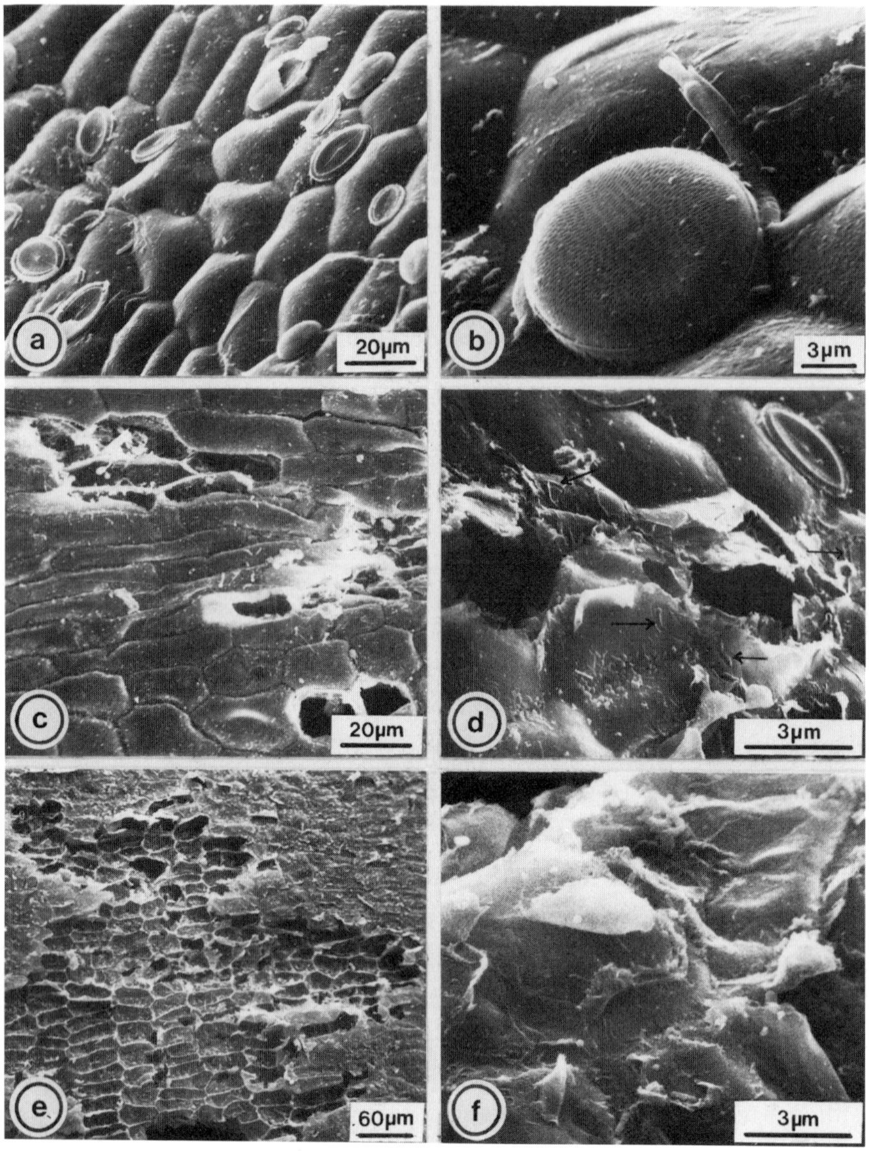

Fig. 2. The effects of snail grazing on (a) youngest, (b) young, (c) mature and (d) senescent *P. crispus* leaves, 6 days after introduction of the snails.

Despite the greater rates of mass loss in the presence of snails the plants did not become brown and flaccid.

Accumulation of a fine detritus ($<1\,mm^2$) consisting mainly of snail faeces, was also significantly more rapid (Fig. 6b) when EDA was added (1095 mg in 24 days) than in the presence of snails alone (600 mg). In the absence of snails, however, most of the dead plant material remained unfragmented and both with and without EDA less than 200 mg of fine detritus accumulated.

These results provide further evidence that snail grazing, by removing epiphytes, controls the spread of necrotrophic organisms which promote senescence in *P. crispus*. Grazing of conditioned plant parts, however, leads to a rapid breakdown and loss of mass.

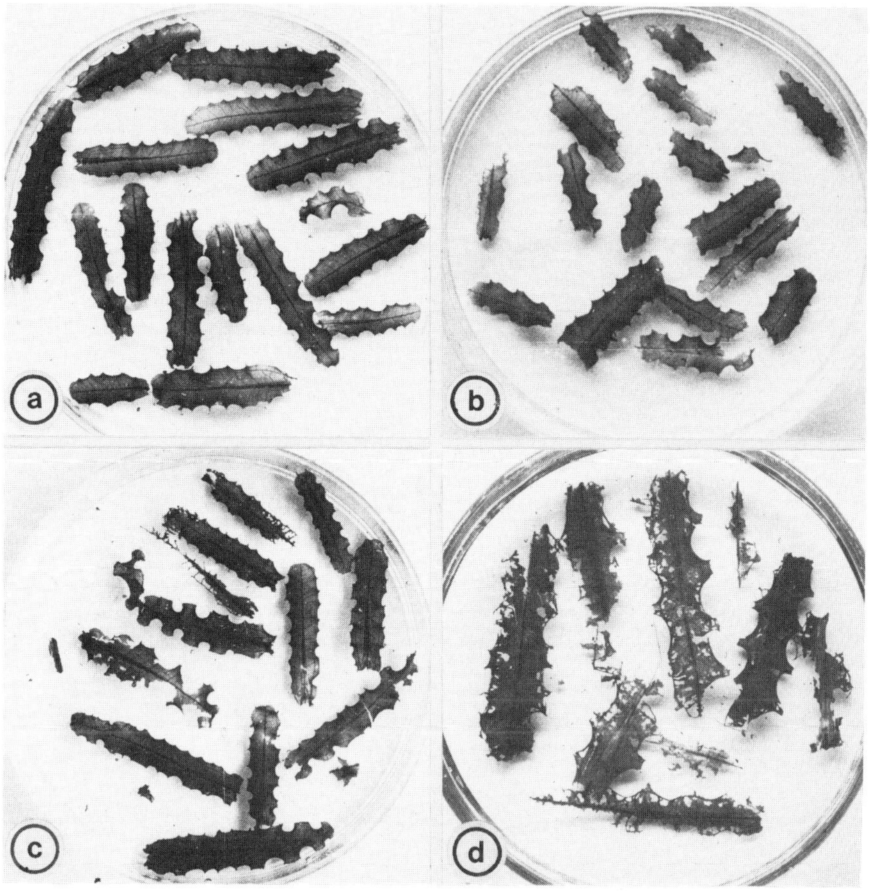

Fig. 3. The effects of snail grazing on the epiphyton and surface structure of *P. crispus* leaves of different ages: (a) youngest; (b) young; (c) mature; (d) senescent.

Discussion

The 'delicate balance', which Eminson & Moss (1980) suggest develops between epiphyton and the host plant, is thought to be controlled by fluxes of nutrients and DOM at the host/epiphyte interface. It is thus dependent on host metabolism and topography, and manifested in a host specific epiphyton. Although their hypothesis is based on algal epiphytes, the mechanisms they proposed might also apply to bacteria, especially as they are also universal contributors to the epiphyton.

The epiphyton developing on artificial substrates in the lakes of the Pongolo system shows distinct differences in species composition, biomass and nutrition from those on *P. crispus* (Rogers 1981), indicating that host metabolism does influence the epiphyton. However, in contrast to the situation proposed by Eminson & Moss (1980) the epiphyton does not prevent damage to the host by diverting grazing since the host cells are resistant to grazing until conditioned by necrotrophic bacterial epiphytes. It may be correct that snail grazing prevents excessive growth of epiphyton on *P. crispus* but, perhaps of greater significance is the reduction in the rate of attack by necrotrophs which appears to extend the life span of the plant. The presence of necrotrophic bacteria in *P. pectinatus* from a south temperate lake (Allanson, pers. comm.*) suggests

* Professor B.R. Allanson, Institute of Freshwater Studies, Rhodes University, Grahamstown, South Africa.

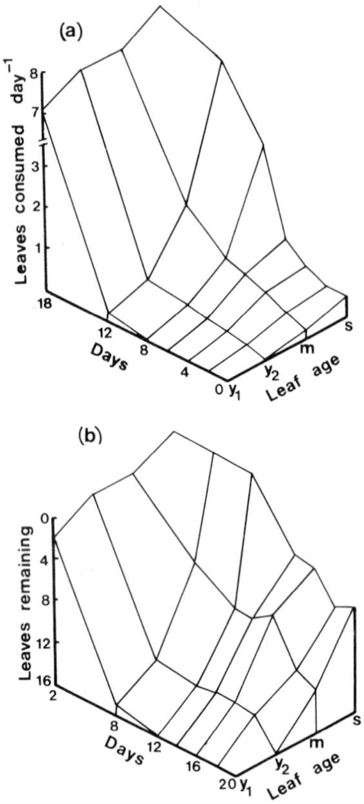

Fig. 4. The relationship between age, leaf conditioning and snail grazers in *P. crispus*. a: The effects of leaf age and time allowed for conditioning in the absence of grazers, on leaf edibility (leaves consumed day^{-1}). b: The effects of leaf age and time permitted for grazing, after conditioning, on the number of leaves remaining at the end of the experiment (20 days). Y_1 = Youngest leaves; Y_2 = Young leaves; M = Mature leaves; S = Senescent leaves).

that this interaction may be fairly widespread.

The epiphyton of *P. crispus* therefore has a dual role; it does provide grazing for snails but, when once the bacterial component has gained entry into the cell wall, it also serves to enhance grazing of the macrophyte rather than divert it. However, the rapid growth and very short life cycle of *P. crispus* (4–5 months; Rogers & Breen 1980) ensures that reproduction can take place before the plants are consumed. The question, however, arises as to how easily the natural balance between epiphyte and host can be upset.

Eminson & Moss (1980) consider that the balance between epiphyton and host may be upset when the fertility of the water increases to the point where it has a greater infuence on the epiphyton than the relatively small fluxes at the host/epiphyte interface. The result being that growths of epiphyton become so dense as to outcompete for nutrients, inorganic carbon and light and a decline in macrophyte populations in highly eutrophic waters follows (Phillips *et al* 1978). Our studies indicate an additional factor in this balance, the necrotrophic bacteria/grazer interactions. As lake fertility increases and the influence of the host on epiphytes decreases the rate of invasion by pathogenic or necrotrophic micro-organisms might exceed the rate of grazing earlier in the host's life cycle. Both growth and reproduction would thus be impaired resulting in a decline of the macrophyte population.

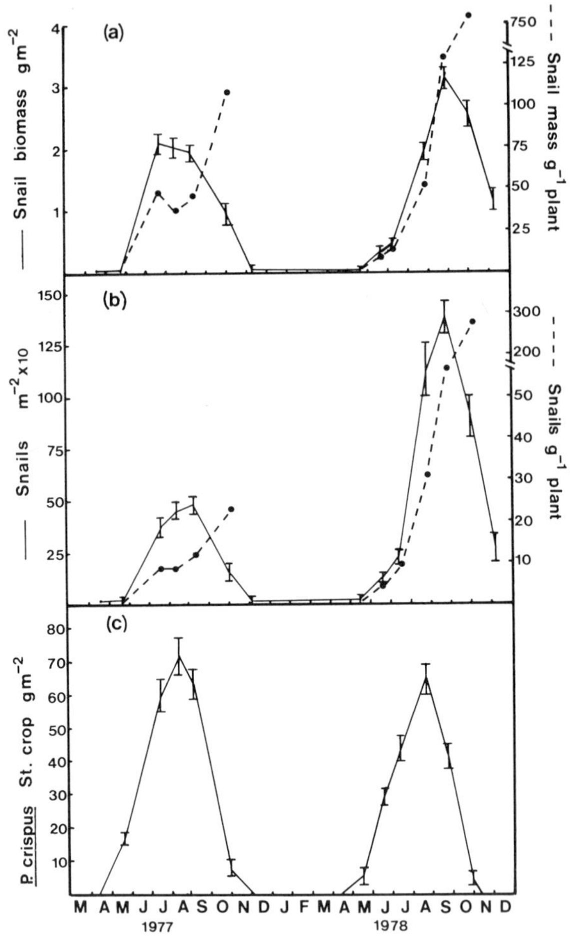

Fig. 5. Changes in (a) snail biomass (g dry wt. m^{-2}) and (b) snail numbers in relation to (c) changes in *P. crispus* standing crop in a Pongolo Floodplain lake during 1977 and 1978.

Table 1. The effects of snails and epiphytic detrital aggregate (EDA) on mass loss of uprooted *P. crispus* plants. Described by a simple exponential function ($y = ae^{-kt}$) where r^2 = coefficient of determination, k = rate constant, $t_{\frac{1}{2}}$ = time taken for 50% mass loss. Significance between regressions (Sokal & Rohlf 1969): NS = not significant; * = $p<0.05$; ** = $p<0.01$.

Treatment	r^2	k	$t_{\frac{1}{2}}$	Significance			
No snails	0.981	−0.033	22	NS	*	**	
No snails & EDA	0.877	−0.034	20	NS	*	**	
Snails	0.987	−0.056	13		*	**	**
Snails & EDA	0.999	−0.091	6			**	**

Under natural conditions on the Pongolo floodplain, however, the effects of epiphytes and snails on the host plant are most important in determining the fate of plant production at the time of senescence, after the plants have been uprooted by waterfowl.

Before the plants are uprooted the invading bacteria are restricted to the outer epidermal wall even in the oldest leaves (Rogers & Breen 1981). Once uprooted, however, the plant's defenses against, or control of, epiphytes appears to break down and the invading bacteria spread rapidly. This situation supports the suggestion that host metabolism influences the epiphyton.

Conditioning of the cell walls of *P. crispus* clearly permits the snails to consume the plants before they die and before the readily utilizable organic matter is leached out. This conditioning permits the rapid formation of a fine, nutritious (unpublished data) detritus and so influences the rates and patterns of nutrient and energy transfer within the ecosystem.

Most studies of the fate of aquatic plant production during decomposition indicate that the factors most important in determining the rates and patterns of decay exert their influence after death of the plant and are either environmental (temperature, O_2 and nutrients) or related to the structure of the plant itself (refractility) (Godshalk & Wetzel 1978; Carpenter & Adams 1979; Howard-Williams & Davies 1979). This study, however, shows that the biological interactions occurring between epiphytes, grazers and the host can also be important and should be taken into account when considering the importance of plant and epiphyte production to ecosystem functioning.

We would also suggest, on the basis of our studies, (Rogers & Breen 1980, 1981, 1982) that even in systems where important grazers are absent, the influence of epiphytes, particularly bacteria, on host senescence and decay will have important implications in energy and nutrient transfer from submerged macrophytes.

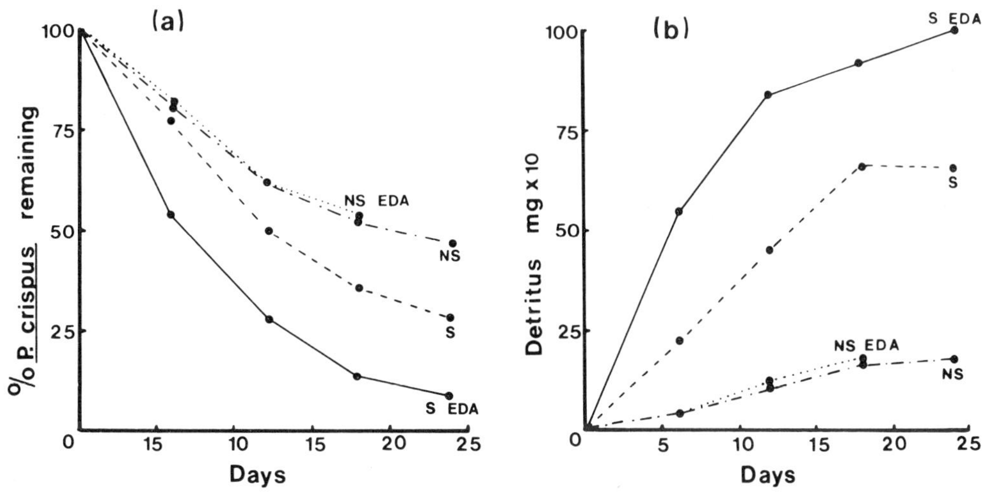

Fig. 6. The effects of snails and epiphytic detrital aggregate (EDA) on (a) loss of mass and (b) detritus accumulation from senescent *P. crispus* plants which had been uprooted by waterfowl. (NS = no snails; S = snails).

Acknowledgements

This work was supported by a grant from the Inland Waters Ecosystems Section of the National Programme for Environmental Sciences, Council for Scientific and Industrial Research. Fiona Rogers provided invaluable assistance and encouragement throughout the study.

References

Allanson, B.R., 1973. The fine structure of the periphyton of Chara sp. and Potamogeton natans from Wytham Pond, Oxford, and its significance to the macrophyte-periphyton metabolic model of R.G. Wetzel and H.L. Allen. Freshwat. Biol. 3:535–541.

Allen, H.L., 1971. Primary productivity, chemo-organotrophy and nutritional interactions of epiphytic algae and bacteria on macrophytes in the littoral of a lake. Ecol. Monogr. 41: 97–127.

Bowen, S.H., 1979. Determinants of the chemical composition of periphytic detrital aggregate in a tropical lake (Lake Valencia, Venezuela). Arch. Hydrobiol. 87:166–177.

Carpenter, S.R. & Adams, M.S., 1979. Effects of nutrients and temperature on decomposition of Myriophyllum spicatum L. in a hard water eutrophic lake. Limnol. Oceanogr. 24: 520–528.

Cummins, K.W., 1974. Structure and function of stream ecosystems. BioScience 24:631–641.

Eminson, D. & Moss, B., 1980. The composition and ecology of periphyton communities in freshwaters. 1. The influence of host type and external environment on community composition. Br. phycol. J. 15:429–446.

Furness, H.D. & Breen, C.M., 1980. The vegetation of seasonally flooded areas of the Pongolo River Floodplain. Bothalia 13:217–230.

Godshalk, G.L. & Wetzel, R.G., 1978. Decomposition of aquatic angiosperms. III. Zostera marina L. and a conceptual model of decomposition. Aquat. Bot. 5:329–354.

Harlin, M.M., 1973. Epiphyte – host relations in seagrass communities. Aquat. Bot. 1:125–131.

Hossell, J.C. & Baker, J.H., 1979. A note on the enumeration of epiphytic bacteria by microscopic methods with particular reference to two freshwater plants. J. Appl. Bacteriol. 46: 87–93.

Howard-Williams, C. and Davies, B.R., 1979. The rates of dry matter and nutrient loss from decomposing Potamogeton pectinatus in a brackish south-temperate coastal lake. Freshwater Biol. 9:13–21.

Hutchinson, G.E., 1975. A Treatise on Limnology. Vol. III. Limnological Botany. John Wiley and Sons, Inc., New York. 660 pp.

McRoy, C.P. & Goering, J.J., 1974. Nutrient transfer between the seagrass Zostera marina and its epiphytes. Nature 248: 173–174.

Nicotri, M.E., 1977. Grazing effect of four marine intertidal herbivores on the microflora. Ecology 58:1020–1032.

Phillips, G.L., Eminson, D.F. & Moss, B., 1978. A mechanism to account for macrophyte decline in progressively eutrophicated freshwaters. Aquat. Bot. 4:103–125.

Rogers, F.E.J., 1981. Studies on the Epiphyton of Potamogeton crispus L., MSc Thesis, University of Natal, Pietermaritzburg. 172 pp.

Rogers, K.H., 1980. The vegetation of the Pongolo floodplain: Distribution and utilization. In: Bruton, M.N. & Cooper, K.H. (eds.) Studies on the Ecology of Maputaland. Rhodes University Press, Grahamstown. 560 pp.

Rogers, K.H. & Breen, C.M., 1980. Growth and reproduction of Potamogeton crispus L. in a South African lake. J. Ecol. 68:561–571.

Rogers, K.H. & Breen, C.M., 1981. Effects of epiphyton on Potamogeton crispus L. leaves. Microb. Ecol. 7:351–363.

Rogers, K.H. & Breen, C.M., 1982. Decomposition of Potamogeton crispus L.: The effects of drying on the pattern of mass and nutrient loss. Aquat. Bot. 12:1–12.

Sokal, R.R. & Rohlf, F.J., 1969. Biometry: The Principles and Practice of Statistics in Biological Research. W.H. Freeman and Co., San Francisco. 776 pp.

Wetzel, R.G., 1969. Factors influencing photosynthesis and excretion of dissolved organic matter by aquatic macrophytes in hard-water lakes. Verh. Int. Ver. Limnol. 17:72–85.

Wetzel, R.G. & Allen, H.L., 1970. Functions and interactions of dissolved organic matter and the littoral zone in lake metabolism and eutrophication. In: Kajak, Z. & Hillbricht-Ilkowska, A. (eds.) Productivity Problems of Freshwaters. PWN Polish Scientific Publishers, Warsaw. pp. 333–347.

Authors' address:
K.H. Rogers*/C.M. Breen
Pongolo River Research Group
Department of Botany
University of Natal
Pietermaritzburg
Natal, South Africa

* Present address:
Department of Botany
Witwatersrand University
I Jan Smuts Avenue
Johannesburg, South Africa

CHAPTER 28

Potential effect of *Myriophyllum spicatum* on the primary production of phytoplankton

H. GODMAIRE and D. PLANAS

Abstract. The effect of the presence of *Myriophyllum spicatum* on natural phytoplankton populations was evaluated in a semi-natural environment (limnocorral) and in laboratory assays. Phytoplankton primary production and physicochemical parameters were followed through a growing season in limnocorral containing *M. spicatum* and their epiphytic algae, other macrophytes, or plastic control macrophytes. In the laboratory, where nutrients, light, and grazing were controlled, natural assemblages of phytoplankton were grown in enriched water taken from aquaria with and without *M. spicatum*, and primary production was measured.

Phytoplankton primary production was stimulated during the early stages of growth of *M. spicatum*. The stimulation did not seem to be related to nutrient conditions or the light climate. Results suggest that an interaction other than nutrient or light competition exists between *M. spicatum* and the surrounding algae.

Introduction

A strong inhibition of phytoplankton production was demonstrated in the littoral zone when the macrophyte stand was quite dense (Stangenberg 1968; Goulder 1969; Dokulil 1973). This reduction was principally attributed to the competition for light and nutrients (Brandl *et al* 1980). Nevertheless, the possibility that stimulating or inhibiting substances are liberated by the macrophytes cannot be dismissed (Fitzgerald 1969; Søndergaard 1981; Hellebust 1974; Anthoni *et al* 1980). Such substances could also influence the availability of micronutrients (Allen 1971) or act as supplements of labile products and carbon sources for heterotrophic algal populations (Brylinsky 1977).

Although the liberation of metabolic products by algae, and their effects on phytoplankton are relatively well known (Hellebust 1970, 1974; Anthoni *et al* 1980; Carlucci & Bowes 1979), information concerning angiosperms is scarce. Planas *et al* (1981) have shown that phenolic compounds extracted from *Myriophyllum spicatum* could inhibit the growth of algae in unialgal cultures and in natural assemblages in the laboratory. In natural environments, Davis *et al* (1973) and Paerl (1979) have observed an absence of algal blooms in the presence of *M. spicatum*. Conversely, Cattaneo & Kalff (1979) measured a higher specific production of epiphytes with *M. spicatum* than with macrophytes such as *Vallisneria americana* and *Potamogeton richardsonii.* Wile & Hitchin (1977) observed a reduction of the annual phytoplankton biomass following extensive harvesting of *M. spicatum* in a lake where this plant was dominant. These observations suggest that *M. spicatum* may inhibit or stimulate phytoplankton. The purpose of this study was to verify the possible interaction between *M. spicatum* and the phytoplankton, and to explore the mechanisms.

Wetzel, R.G. (ed.), Periphyton of Freshwater Ecosystems.
 ISBN 90 6193 768 X.

Two approaches were used to achieve these objectives. The first involved limnocorrals in a typical lake of the Canadian Shield. Seasonal observations of physicochemical parameters and of phytoplankton primary production in limnocorrals with *M. spicatum*, and in controls with and without other macrophytes, permitted us to analyze the potential effect of *M. spicatum* on the phytoplankton. A second approach used assays on natural populations of phytoplankton in aquaria with and without *M. spicatum*, which permitted us to investigate further this type of interaction.

Materials and methods

1. Limnocorrals. The limnocorrals experiment was carried out in Lake St. Charles, a dystrophic lake in the Canadian Shield, where *M. spicatum* has never grown. The limnocorrals had a capacity of 135 l, a diameter of 0.7 m, and were supported by 4 styrofoam floats to prevent any contact with surrounding water and to conform to varying water levels. The limnocorrals were filled with 25 l of homogenized sediments and with water from the littoral zone. Some physicochemical and morphometric characteristics of the lake are presented in Table 1. After one month of settling, in order to verify the replication and water-tightness of the limnocorrals, 5 cm shoots of *M. spicatum* were planted in two limnocorrals at a density of one plant per 50 cm^2. Shoots of dominant floating-leaved macrophytes of the lake, *Sparganium* sp. and *Nuphar microphyllum*, were planted in two other limnocorrals at the same density. The two remaining limnocorrals contained only plastic plants.

From the end of July to mid-October, an integrated water sample (peristaltic pump) from 0 to 1 m was collected at two-week intervals for chemical and biological measurements.

The following parameters were measured *in situ*: conductivity, measured with an Horizon Ecology meter (Model 1484), pH with a Metrom Herisau (E488) meter, and light with a Licor underwater quantum sensor.

Nitrite ($N\text{-}NO_2$) and nitrate ($N\text{-}NO_3$) concentrations were determined according to Strickland & Parsons (1972). Total phosphorus (P-TP), total dissolved phosphorus (P-TDP), and reactive soluble phosphorus (P-RSP) were determined using a modified Johnson's method (1971), in which persulfate was used to oxidize the organic matter. Ammonia ($N\text{-}NH_4$) and silica (SiO_2) were determined as in Stainton *et al* (1974). Chlorophyll a was extracted with acetone, and calculated as in Vollenweider (1969). The carbon-14 method of Steemann Nielsen (1952) was used to measure primary production.

All samples were incubated in the lake at a depth of 0.5 m in 350-ml bottles (2 clear, 1 black) for 2 to 4 h. After incubation, 50-ml aliquots were filtered on Sartorius 0.45-μm porosity filters. The radioactivity of added carbon-14 and of the filters was determined in a Beckman LS 8000 Scintillation Counter (Schindler 1966; Weimer *et al* 1975).

2. Aquarium assays of natural phytoplankton populations. M. spicatum was grown in 20 l aquaria. Two types of controls were used: one with *Elodea canadensis*, the other with plastic plants. The aquaria were filled with 3 l of a rich organic sediment from a lake where *M. spicatum* is absent and with water from the same lake. Five centimeter shoots were planted at a density of one per 15 cm^2. The aquaria were aerated.

Since the previous limnocorral studies showed that the influence of *M. spicatum* changed the phy-

Table 1. Some morphometric and physicochemical characteristics of Lake St-Charles.

Characteristics	Range
Ice cover period	Dec.–Apr.
Maximum depth (m)	19.0
Mean depth (m)	7.8
Light extinction coefficient (littoral zone)	0.45–1.37
pH	5.6–6.2
Alkalinity (meq l^{-1})	0.04–0.10
Conductivity (μmhos cm^{-1})	20–25
Total dissolved N (μg-at N l^{-1})	0.05–8.00
Total P (μg-at P l^{-1})	0.03–0.60
SiO_2 (μg-at SiO_2 l^{-1})	0.10–50.0

toplankton production according to the life cycle of the plant, the assays were done at three different stages of macrophytic growth, namely when the rate of stem elongation was 0.8 cm/day, 0.2 cm/day, and senescing at, respectively, 2, 5, and 8 weeks after planting.

The assays were carried out as follows: 2 l of water sampled from each aquarium were filtered (Whatman GF/C filters) and enriched with potassium nitrate and potassium phosphate at concentrations of 15 μg-at N l^{-1} and of 1 μg-at P l^{-1}. Each sample was inoculated with an equal volume of phytoplankton concentrate, obtained from the same lake as the water used for the aquaria. The concentrate was obtained by centrifugation of 15 l of water from the epilimnion, previously filtered through a 200 μm mesh to eliminate the zooplankton. The flasks were continuously agitated and exposed to natural light. After 60 h, primary production was determined with the carbon-14 method. Preliminary assays showed that between 48 and 60 h were needed to avoid primary production depression after nutrient enrichment (Healey 1979) and to obtain an optimal response. Samples were exposed to 300 μE m^{-2} sec^{-1} in an incubator (Shearer 1976). Temperature was maintained at 25° C ± 2. The entire experiment was repeated three times beginning on 24 July, 14 August and 11 September 1981.

3. Statistical analysis. A discriminant analysis (discriminant, direct method, SPSS) was applied to the physicochemical results from limnocorrals for the seven first samplings. The following parameters were considered: temperature, light extinction coefficient, light intensity at 0.5 m, conductivity, pH, alkalinity, dissolved oxygen, P-TP, P-TDP, P-SRP, N-NO_3, N-NO_2, N-NH_4, and SiO_2. Groups formed a priori for the analysis were: (1) the two limnocorrals with *M. spicatum*, (2) the two limnocorrals with other macrophytes, and (3) the two limnocorrals without macrophytes. Missing values were approximated following mean values. The data did not respect completely the multinormality condition; nevertheless the discriminant analysis is robust to deviations of that condition (Legendre & Legendre 1979).

An analysis of variance was carried out on the primary production data according to the date and on the assays.

Results

1. Limnocorrals. The variance analysis of different dates shows that *M. spicatum* significantly modified the primary production ($p \leq 0.01$) of the phytoplankton when this macrophyte is 4 to 6 weeks old (Fig. 1). The maximum difference occurred on the

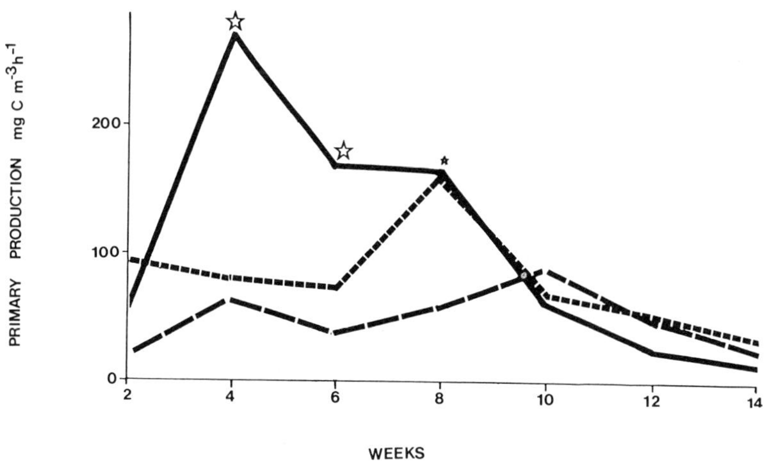

Fig. 1. Primary production of limnocorrals from July to October. Each point is the mean of four replicates. The large ($p \leq 0.05$) and small ($p \leq 0.01$) stars indicate the probability levels of the significant differences among the three treatments, *M. spicatum* (solid line), no macrophytes (dashed line), other macrophytes (dotted line).

4th week when primary production in the presence of *M. spicatum* was 4.3 times greater than in controls without macrophytes, and 3.5 times greater than in the presence of other macrophytes. In the 8th week the difference between treatments was significant at a probability level of 0.05, but subsequently no differences were observed.

A discriminant analysis was applied to the physicochemical results to look for the parameters which distinguish the pairs of limnocorrals having the same treatment, and which could influence primary production. Results of the parameters used in the discriminant analysis which might influence directly the primary production are presented in Table 2.

The percentages of classification of the samplings (per limnocorral, per date) in relation to the groups in which they were placed, a priori, were as follows: those of group 1 were well classified at 100%, those of group 2 at 85.7% and those of group 3 at 78.6%. The analysis shows that pH ($p \leq 0.005$) and conductivity ($p \leq 0.001$) were significant for the discrimination of each group. Light and nutrients were not significant. Only the first cannonic discriminant function was significant ($p \leq 0.001$); it explained 81.8% of the variability. These results reflected that the groups created a priori were valid for the whole experiment. We could then assume that each treatment i.e. the presence of *M. spicatum*, the presence of other macrophytes, and the absence of macrophytes entailed a particular relation to the measured parameters.

2. Aquarium assays of natural phytoplankton populations. The three assays at 2, 5 and 8 weeks included phytoplankton populations from different stages of the annual succession. For the three experiments, a significantly higher primary production was observed in the presence of water taken from the aquarium with *M. spicatum* when the macrophyte was in the early stages of seasonal growth (Fig. 2). At other stages of *M. spicatum*'s growth, primary production was not stimulated by the medium taken from this aquarium. In general, the primary production associated with the medium with *M. spicatum* was lower than that of the control without macrophytes, but the differences between the medium with *M. spicatum,* the control, and the medium with *E. canadensis* were not significant. At no stage of growth of *E. canadensis* did the medium produce any significant stimulation of primary production (Fig. 2).

Offsetting of experiments (Fig. 2) permitted us to use the same phytoplankton populations simultaneously for different stages of the macrophyte growth. For example, the assay of week 5 for experiment 2 was carried out at the same time as the assay of week 2 of experiment 3 and the assay of week 8 of experiment 1. Assays for the same population of phytoplankton at different stages of *M. spicatum* growth gave similar results as above.

Discussion

The two approaches used, i.e. the limnocorrals and

Table 2. Results of the parameters used in the discriminant analysis for the three samplings where the differences of primary production values between the three treatments were significant.

Weeks	With *M. spicatum*	With macrophytes	Without macrophytes
Extinction coefficient			
4	0.67 ± 0.12	1.60 ± 0.50	0.81 ± 0.33
6	2.66 ± 0.80	2.45 ± 0.93	0.78 ± 0.24
8	4.42 ± 0.17	1.83 ± 0.51	2.82 ± 1.26
pH			
4	5.4 ± 0.1	5.9 ± 0.2	5.9 ± 0.1
6	5.3 ± 0.1	6.0 ± 0.0	6.0 ± 0.1
8	5.3 ± 0.1	5.9 ± 0.2	5.8 ± 0.1
Conductivity (μmhos cm^{-1})			
4	37.0 ± 1.4	30.0 ± 2.0	33.0 ± 2.5
6	38.0 ± 0.7	32.0 ± 1.5	31.0 ± 1.3
8	36.0 ± 1.2	26.0 ± 3.0	27.0 ± 3.0
P-SRP (μg-at P l^{-1})			
4	0.055 ± 0.008	0.050 ± 0.005	0.110 ± 0.020
6	≤0.03	≤0.03	≤0.03
8	≤0.03	≤0.03	≤0.03
P-TP (μg-at P l^{-1})			
4	0.150 ± 0.070	0.250 ± 0.000	0.165 ± 0.070
6	0.358 ± 0.000	0.344 ± 0.020	0.546 ± 0.090
8	0.511 ± 0.110	0.361 ± 0.100	0.544 ± 0.190
N-NO_2 + N-NO_3 (μg-at N l^{-1})			
4	≤0.05	≤0.05	≤0.05
6	0.063 ± 0.010	0.119 ± 0.010	0.116 ± 0.003
8	0.070 ± 0.020	0.141 ± 0.000	≤0.05

the assays, show that the medium in which *M. spicatum* grew stimulated primary production of phytoplankton at the beginning of the growth of this macrophyte. These results could explain the observations of Wile & Hitchin (1977) where the phytoplankton biomass decreased after extensive harvesting of *M. spicatum* in a lake where this plant was dominant. Our results are also in agreement with Nichols' (1973) observations that in relatively deep water heavily colonized by *M. spicatum* the production rate of the macrophyte and the algae peaked at about the same time. Nichols also remarked that the decrease in macrophyte biomass by harvesting led to a decrease in algal biomass.

The causes of this stimulation observed in our experiments did not seem to be related to the light climate conditions or to the nutrient concentration of the medium where this plant grows. In the limnocorrals, there were differences in light extinction coefficients and/or phosphorus and nitrogen concentrations (Table 2), although these differences were not significant in the discriminant analysis of the limnocorrals, and were nonexistant in the assays where phosphorus and nitrogen were added in excess. In the discriminant analysis of the limnocorrals, pH and conductivity were the most important factors in the grouping of the limnocorrals by treatment.

It is difficult to explain the low pH found in the limnocorrals with *M. spicatum* concurrent with higher primary production. Two hypotheses can be suggested: the low pH could be related to the preference of *M. spicatum* for ammonia as a nitrogen source (Nichols & Keeney 1976), or to the possible excretion of phenolic compounds such as gallic, ellagic and tannic acids that are abundant in tissues of *M. spicatum* (Planas *et al* 1981). Laboratory results show that a percentage of organic matter excreted by *M. spicatum* is effectively composed of phenolic compounds (reported as 'total phenols' by Sarhan, *et al* 1982).

The higher conductivity values may be attributed to increased cation release from the sediments at the lower pH. This phenomenon would be amplified by the enclosure effect in the limnocorrals. The buffering effect of the sediments, following a lowering of the pH, has been observed in laboratory experiments (Serodes, pers. comm.).

Since the general physicochemical conditions assayed did not explain the higher values of phytoplankton primary production in the presence of *M. spicatum*, biological factors can be invoked. Of

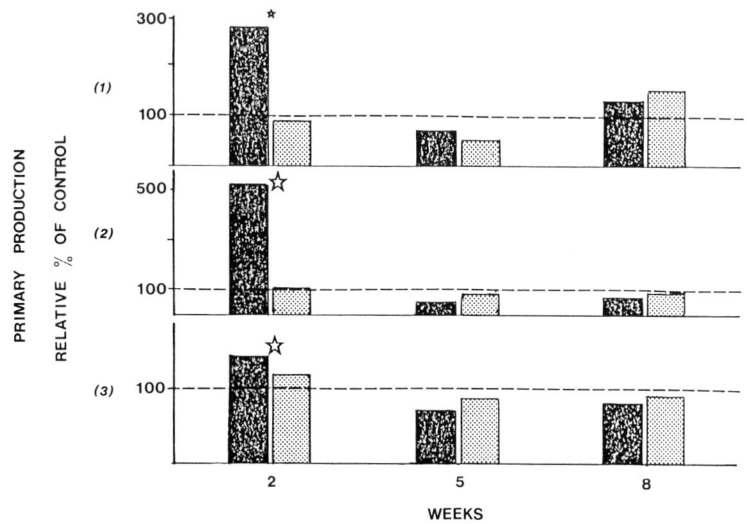

Fig. 2. Primary production of assays with *M. spicatum* medium, *E. canadensis* medium for the three experiments. The large ($p \leqslant 0.001$) and small ($p \leqslant 0.01$) stars indicate the probability levels of the significant differences among the three experiments. For each experiment assays were done at 2, 5 and 8 weeks. The results are expressed as percentage of the control, but the analysis of variance was done on the untransformed data.

these, the absence or decrease of grazing could be important. Pennak (1973) and Dorgelo & Koning (1980), among others, showed that *M. spicatum* repulses certain species of *Daphnia* and *Diaptomus*. However, in our assay experiments grazing was reduced and the stimulating effect was still observed.

The observed stimulation can be considered as the outcome of a factor induced by the interaction of *M. spicatum* with the periphyton, phytoplankton, zooplankton, and bacteria present in the environment. One possibility is that stimulation of primary production could be related to the release of extracellular organic matter (EOM) by *M. spicatum*, which could be used directly, or influence indirectly, the growth of the phytoplankton.

It is known that a significant portion of photosynthetically fixed carbon is released during the active growth of several angiosperms in the form of dissolved organic compounds (Wetzel 1969; Wetzel & Manny 1972; Hough & Wetzel 1975). For *M. spicatum*, Sarhan *et al* (1982) demonstrated that the excretion rate of extracellular organic carbon (EOC) did not exceed 10% for the first three weeks of growth, but subsequently increased with age, and could attain 35% of initial internal ^{14}C in plants. Furthermore, several studies on the heterotrophic use of these products by epiphytes show that 10% (Søndergaard 1981) to 40% (Brylinsky 1977) can be assimilated in short term experiments. Data on the effects of the absorption of these products by the phytoplankton are nonexistant.

Although the macrophyte-epiphyte proximity favours the absorption of these products by the epiphytes, the possibility exists that the products released by the macrophytes also affect the phytoplankton (Wetzel & Allen 1972).

The stimulation phenomenon could be related to the physiological state and/or the type of product excreted by the plant, since the effect of the stimulation was only observed at the beginning of the growth of *M. spicatum*. Several researchers (Khailov & Burlakova 1969; Brylinsky 1977; Søndergaard 1981) have shown higher excretion rates in active primary producers, i.e. the macrophytes at the beginning of growth should excrete more than later. Yet in *M. spicatum*, the excretion rate at the beginning of growth was at its minimum. The fact that the stimulation of the phytoplankton primary production appears at the beginning of the growth of *M. spicatum*, and that this stage of growth corresponds to a minimal excretion of EOM, could indicate that it is not the quantity, but the quality of the excreted products that is responsible for the stimulation.

Acknowledgements

This study was supported by grants from Doyen des Sciences de l'UQAM and from Environment Canada 1980 to D.P. and by postgraduate scholarships from NSERC and FCAC to H.G.

References

Allen, H.L., 1971. Primary productivity, chemo-organotrophy, and nutritional interactions of epiphytic algae and bacteria on macrophytes in the littoral of a lake. Ecol. Monogr. 41: 97–127.

Allen, H.L., 1977. Experimental studies of dissolved organic matter in a softwater lake. In: Cairns, J., Jr. (ed.), Aquatic Microbial Communities. Garland, New York. pp. 477–527.

Anthoni, U., Christophersen, C., Madsen, J.O., Wium-Andersen, S. & Jacobsen, N., 1980. Biological active sulfur compounds from the green alga Chara globularis. Phytochemistry 19:1228–1229.

Brandl, Z., Brandlová, J. & Poštolková, M., 1970. The influence of submerged vegetation on the photosynthesis of phytoplankton in ponds. Rozpravy Českosl. Akad. Věd, Řada Matem. Přír. Věd. 80:33–62.

Brylinsky, M., 1977. Release of dissolved organic matter by some marine macrophytes. Mar. Biol. 39:213–220.

Carlucci, A.F. & Bowes, P.M., 1970. Vitamin production and utilization by phytoplankton in mixed culture. J. Phycol. 6:393–400.

Cattaneo, A. & Kalff, J., 1979. Primary production of algae growing on natural and artificial plants: A study of interactions between epiphytes and their substrate. Limnol. Oceanogr. 24:1031–1037.

Davis, G.J., Jones, M.N. & Davis, D., 1973. Seed germination in Myriophyllum spicatum L. J. Elisha Mitchell Sci. Soc. 89:246–247.

Dokulil, M., 1973. Planktonic primary production within the Phragmites community of Lake Neusiedlersee (Austria). Pol. Arch. Hydrobiol. 20:175–180.

Dorgelo, J. & Koning, K., 1980. Avoidance of macrophytes and additional notes on avoidance of the shore by Acanthodiaptomus denticornis (Wierzejski, 1887) from Lake Pavin (Auvergne, France). Hydrobiol. Bull. 14:196–208.

Fitzgerald, G.P., 1969. Some factors in the competition or antagonism among bacteria, algae and aquatic weeds. J. Phycol. 5:351–359.

Goulder, R., 1969. Interactions between the rates of production of a fresh water macrophyte and phytoplankton in a pond. Oikos 20:300–309.

Healey, H., 1979. Short-term responses of nutrient-deficient algae to nutrient addition. J. Phycol. 15:289–299.

Hellebust, J.A., 1970. The uptake and utilization of organic substances by marine phytoplankton. In: Hood, D.W., (ed.) Organic Matter in Natural Water, Fairbanks: University of Alaska Institute of Marine Science Occasional Publications 1:225–256.

Hellebust, J.A., 1974. Extracellular products. In: Stewart, W.D.P. (ed.), Algal Physiology and Biochemistry. Blackwell, Oxford. pp. 838–863.

Hough, R. & Wetzel, R.G., 1975. The release of dissolved organic carbon from aquatic macrophytes. Diel, seasonal, and community relationships. Verh. Int. Ver. Limnol. 19: 939–949.

Johnson, D.L., 1971. Simultaneous determination of arsenate and phosphate in natural waters. Environ. Sci. Technol. 5:411–414.

Khailov, K.A. & Burlakova, Z.P., 1969. Release of dissolved organic matter by marine seaweeds and distribution of their total organic production to inshore communities. Limnol. Oceanogr. 14:521–532.

Legendre, L. & Legendre, P., 1979. Ecologie numérique 2. La Structure des Données Écologiques. Masson, Paris et les Presses de l'Université du Québec. 154 pp.

Nichols, D.S. & Keeney, D.R., 1976. Nitrogen nutrition of Myriophyllum spicatum uptake and translocation of ^{15}N, by shoots and roots. Freshwater Biol. 6:145–154.

Nichols, S.A., 1973. The effects of harvesting aquatic macrophytes on algae. Trans. Wis. Acad. Sci. Arts Lett. 61: 165–172.

Paerl, H.W., 1979. Study of the importance of nitrogen fixation to the growth of Myriophyllum spicatum. Preliminary report to the NWEI, Burlingon, Ontario.

Penhale, P.A. & Smith, Jr., W.O., 1977. Excretion of dissolved organic carbon by eelgrass (Zostera marina) and its epiphytes. Limnol. Oceanogr. 22:400–407.

Pennak, R.W., 1973. Some evidence for aquatic macrophytes as repellent for a limnetic species of Daphnia. Int. Rev. ges. Hydrobiol. 58:569–576.

Planas, D., Sarhan, F., Dubé, L., Godmaire, H. & Cadieux, C., 1981. Ecological significance of phenolics compounds of Myriophyllum spicatum. Verh. Internat. Verein. Limnol. 21:1492–1496.

Sarhan, F., Planas, D., Dubé, L. & Godmaire, H., 1982. Kinetics of extracellular release of ^{14}C labelled organic carbon by Myriophyllum spicatum and its effect on the phytoplankton. Preliminary report to the NWRI, Burlington, Ontario. 18 p.

Schindler, D.W., 1966. A liquid scintillation method for measuring carbon-14 uptake in photosynthesis. Nature 211: 844–845.

Shearer, J.A., 1976. Construction and operation of a portable incubator for phytoplankton primary production studies. Fish. and Marine Service Tech. Rapp. 638. 22 pp.

Søndergaard, M., 1981. Kinetics of extracellular release of ^{14}C-labelled organic carbon by submerged macrophytes. Oikos 36:331–347.

Stainton, M.P., Capel, M.J. & Armstrong, F.A.J., 1974. The Chemical Analysis of Freshwater. Misc. Spec. Publ. 25. 119 pp.

Strangenberg, M., 1968. Bacteriostatic effects of some algae- and Lemna minor extracts. Hydrobiol. 32:88–96.

Steemann Nielsen, E., 1952. Measurement of the production of organic matter in the sea by means of ^{14}C. Nature 167:684.

Strickland, J.D.H. & Parsons, T.R., 1972. A Practical Handbook of Seawater Analysis. 2nd ed. Bull. Fish. Res. Bd. Canada 167. 310 pp.

Vollenweider, R.A. (ed.), 1969. A Manual on Method for Measuring Primary Production in Aquatic Environments. International Biology Program Handbook 12. Oxford: Blackwell Scientific. 213 pp.

Weimer, C.W., Rodel, M.G. & Armstrong, D.E., 1975. Loss of ^{32}P-phosphate and ^{14}C-carbonate activity during liquid scintillation of aqueous sample. Envirn. Sci. Technol. 9:966–970.

Wetzel, R.G., 1969. Factors influencing photosynthesis and excretion of dissolved organic matter by aquatic macrophytes in hard-water lakes. Verh. Int. Ver. Limnol. 17:72–85.

Wetzel, R.G. & Manny, B.A., 1972. Secretion of dissolved organic carbon and nitrogen by aquatic macrophytes. Verh. Int. Verein. Limnol. 16:163–170.

Wetzel, R.G. & Allen, H.L., 1972. Functions and interactions of dissolved organic matter and the littoral zone in lake metabolism and eutrophication. In: Kajak, Z. & Hillbricht-Ilkowska, A. (ed.). Productivity Problems of Freshwaters. PWN Polish Scientific Publishers, Warsaw. pp. 333–347.

Wile, I. & Hitchin, G., 1977. An assessment of the practical and environmental implications of mechanical harvesting of aquatic vegetation in southern Chemung lake. Ontario Ministry of the Environment and Ontario Ministry of Natural Resources. 199 pp.

Authors' address:
H. Godmaire
Life Sciences Division
Scarborough College
University of Toronto
1265 Military Trail
West Hill, Ontario, Canada M1C 1A4

D. Planas
Dep. Sc. Biologiques
Université du Québec à Montréal
C.P. 8888 Succ. "A"
Montréal, Québec, Canada H3C 3P8

CHAPTER 29

Exchange of phosphorus between water, macrophytes and epiphytic periphyton in the littoral of Mikolajskie Lake, Poland

H.H. RIBER, J.P. SØRENSEN and A. KOWALCZEWSKI

Abstract. A major part of the phosphorus content of the submerged macrophyte-epiphyte complex is accumulated in the epiphytes. While little uptake of excess phosphorus could be measured by incubation of plankton, an uptake response of the macrophyte-epiphyte complex to addition of PO_4-P (200 μg L^{-1}) was detectable throughout the period of study, ranging from 2.5 to 16 μg P dm^{-2} h^{-1}. In unenriched water a net release of soluble reactive phosphorus was measured in all incubations. Detailed studies in enriched water showed that some release of phosphorus occurred when epiphytes were separated from their substratum. Filtration of water before incubation had little effect on the results. Net exchange rates of soluble reactive phosphorus of leaves and epiphytes of *Potamogeton perfoliatus* were of similar magnitude. Stems of *Phragmites australis* released enough phosphorus to account for a major part of epiphyte uptake. A turn-over time of a few days or less was indicated for phosphorus in epiphytic periphyton. The possible significance of this exchange for the phosphorus dynamics of lakes is discussed.

Introduction

The importance of the littoral zone for the phosphorus cycle of smaller lakes has been evident since the earliest field experiments with radio-phosphorus were published, although the role of the littoral as a source or sink for phosphorus could not be determined from these early studies (Rigler 1973). Recently, some studies have indicated a considerable internal load of phosphorus from the littoral (e.g. Prenkti *et al* 1979, Landers 1982) while Howard-Williams (1981) emphasizes the ability of the littoral to act as a nutrient filter.

A number of studies of the phosphorus nutrition of submerged macrophytes as summarized by Carignan (1982), indicate that the relative uptake of roots and leaves can be predicted from SRP (soluble reactive phosphorus) concentrations in sediment pore water and water surrounding the shoots. Release of phosphorus to the water from healthy macrophytes seems to be of little importance (Carignan & Kalff 1982), while a considerable release has been found from decaying macrophytes (Carpenter 1980).

Less attention has been paid to the significance of periphyton for the lacustrine phosphorus cycle. Since Wetzel & Allen (1972) emphasized the significance of the physical proximity between macrophyte and epiphytic periphyton, several authors (e.g. Harlin 1973, Penhale & Thayer 1980) have demonstrated transfer of phosphorus between macrophyte and epiphytes. Carignan & Kalff (1982) estimated that 3.4–9% of the phosphorus in the loosely attached fraction of epiphytes was derived from the macrophyte as an average for several species of macrophytes, while the rest originated from the surrounding water, and they concluded that in this respect macrophytes merely act

Wetzel, R.G. (ed.), Periphyton of Freshwater Ecosystems.
 ISBN 90 6193 768 X.

as a physical support for an active microbial community. From a review of existing literature Wetzel (1983) concluded that further research is needed in order to understand the significance of interactions between periphyton and its substratum as well as the role of this community in nutrient cycling of freshwaters.

In this investigation we studied some aspects of the exchange of soluble reactive phosphorus between the macrophyte-epiphyte complex and the surrounding water, in an attempt to evaluate the ability of this system to influence the phosphorus dynamics of a lake.

Materials and methods

The studies were carried out in Mikolajskie Lake, Northern Poland (53° 47′ N, 21° 36′ E), a euthopic, dimictic lake. Some morphometric and biological characteristics of the lake are given in Table 1. The numerous biological studies carried out in this lake have been summarized by Kajak (1978), while Pieczyńska (1976) gives a detailed account on its littoral zone. The main source of phosphorus for Mikolajskie Lake is municipal sewage from the town Mikolajki, at the north-western end of the lake.

Samples of macrophytes and epiphytic periphyton were taken from a bed of *Potamogeton perfoliatus* L. and an adjacent bed of *Phragmites australis* (Cav.) Trin., both located about 100 m to the south-east of the harbour of the PAN Ecological Field Station. Littoral water for analysis and incubation was taken from the harbour.

Table 1. Some morphometric, environmental and biological characteristics of Mikolajskie Lake (data from Kajak 1978).

Area (ha)	460
Max. depth (m)	27.8
Mean depth (m)	11.0
Area of littoral in % of total area	19.0
Water exchange rate (y^{-1})	3.0
Annual external load of phosphorus (g m^{-2})	78.0
Total net primary production (kcal m^{-2} y^{-1})	2483
Phytoplankton (%)	77
Macrophytes (%)	14
Periphyton and metaphyton (%)	9

Exchange of SRP was studied by incubating leaves of *Potamogeton* and of new and old (dead) *Phragmites* stem pieces in wide mouth glass bottles (230 ml) for 4 h followed by measurement of final concentrations of SRP. Incubations were done at natural light and temperature conditions in the harbour. Immediately after incubation the water was filtered (Whatman GF/C), epiphytes were removed from the macrophyte and the colonized area measured by planimeter (*Potamogeton*) or by slide gauge (*Phragmites*). Exchange was calculated by subtracting the concentration in control bottles, and expressed per unit area of substratum. For *Potamogeton* both sides of the leaves were thus considered in the calculations. In order to get representative samples of epiphytes leaves of *Potamogeton* were selected between the 8th and 20th visible leaf counted from the apex. Samples of *Phragmites* were taken about 30 cm below the water surface.

Experiments were carried out at both ambient SRP concentrations and in water enriched with 200 $\mu g\ PO_4 - P\ L^{-1}$. Detailed studies of the phosphorus exchange of *Potamogeton* were done in enriched water by simultaneous incubations in unfiltered and filtered (Whatman GF/C) water. Incubations of leaves with intact epiphyte cover were compared with separate incubations of leaves and the removed epiphytes. When epiphytes separated from *Phragmites* were incubated, submerged sheaths were included with the epiphytes, as separation of these from the already decomposing sheaths was impossible. All incubations were carried out in triplicate.

Incubations of *Phragmites* stems in bottles were disturbed by leakage of SRP from the cut vascular tissue. This leakage was estimated by comparison of bottle incubations with incubations done in stoppered perspex vessels, with the stems penetrating the stoppers, so no cuts were present, and the estimated correction was applied to the other experiments.

Measurements of SRP and chlorophyll a were started within two hours. SRP was analysed by the $SnCl_2$ version of the molybdenum blue method (Just & Hermanowicz 1964). Chlorophyll *a* of phytoplankton, periphyton and *Potamogeton* leaves was measured spectrophotometrically with correction for phaeopigments after extraction with 90% acetone (Golterman 1969).

Samples of littoral water were fixed with 4 M H_2SO_4 and samples of macrophytes and epiphytes were dried for later determination of total phosphorus by digestion with HNO_3:$HCLO_4$ (4:1) (Just & Hermanowicz 1964).

Results

Physical-chemical characteristics of littoral water, phytoplankton and epiphytic periphyton during the sampling period are summarized in Fig. 1. While a long lasting summer maximum of phytoplankton, culminating in August, is usually found in Mikolajskie Lake (Kajak 1978), only minor changes in phytoplankton chlorophyll *a* were observed during the sampling period. SRP was present in detectable amounts all the time and made up a rather high percentage of total P for a stratified lake in summer, mostly from 12 to 38% (range 2.7–70%).

Epiphytic biomass differed considerably between the host plants, being lowest on *Potamogeton* and highest on old *Phragmites* stems. A peak of epiphytic biomass was evident in September-October. The general magnitude of the biomass

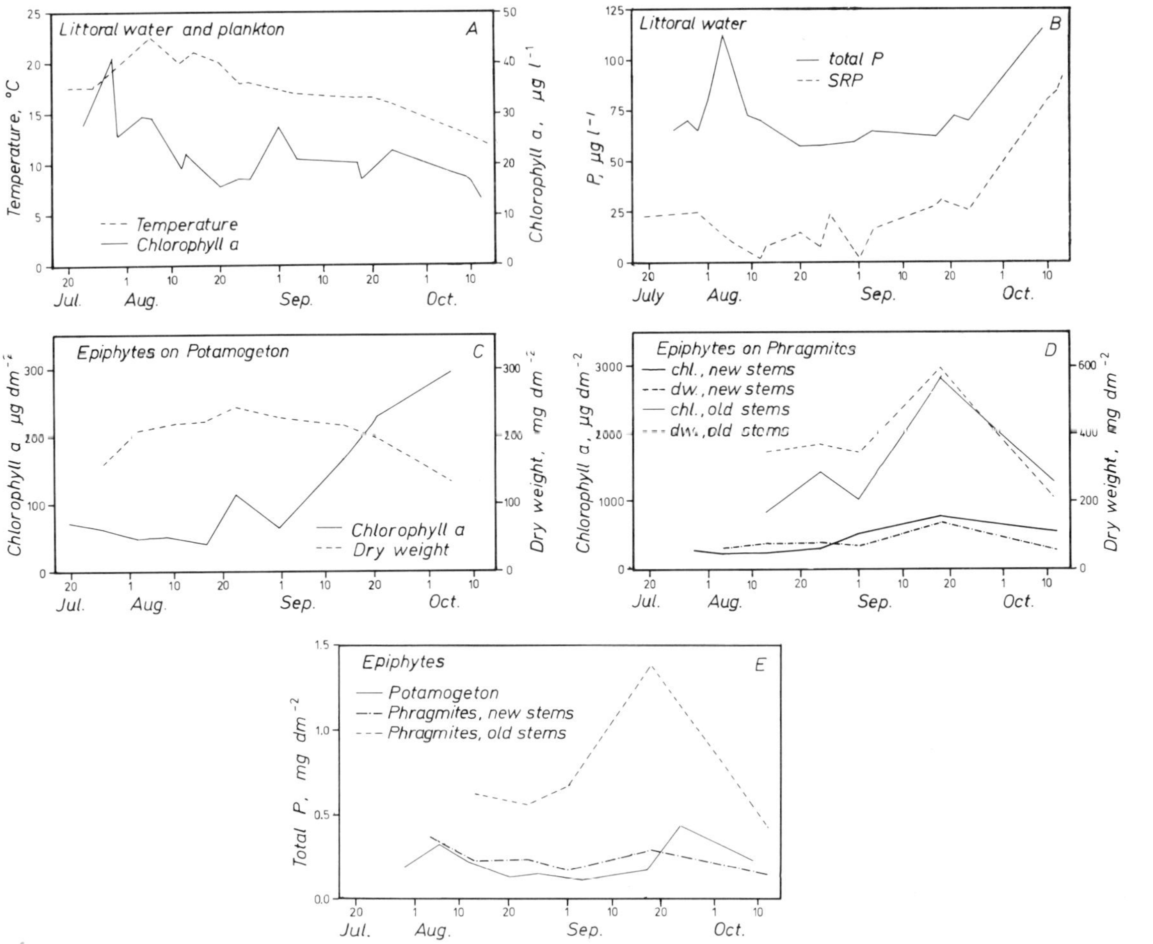

Fig. 1. Characteristics of littoral water, phytoplankton and periphyton during the sampling period. A. Water temperature and phytoplankton chlorophyll *a*. B. SRP and total P of water. C. Dry weight and chlorophyll *a* content of epiphytic periphyton on *Potamogeton perfoliatus*. D. Dry weight and chlorophyll *a* of epiphytic periphyton on *Phragmites australis*. E. Total P of periphyton.

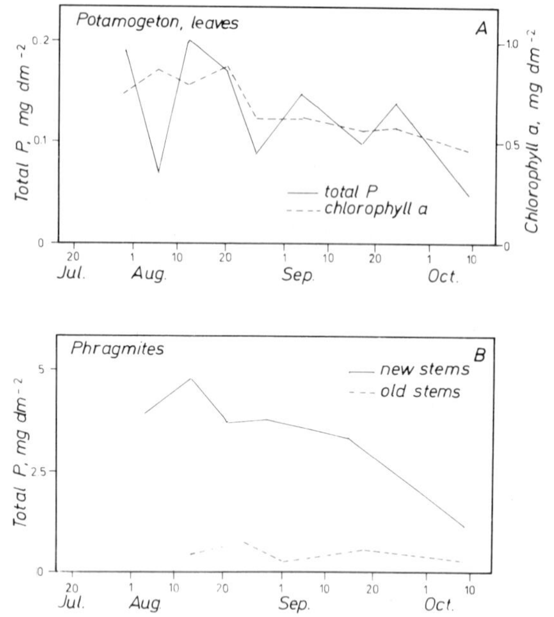

Fig. 2. A. Total P and chlorophyll *a* of *Potamogeton perfoliatus* leaves. B. Total P of new and old *Phragmites* stems.

and the seasonal pattern were in agreement with earlier data reported for Mikolajskie Lake (Szczepánska 1970, Kowalczewski 1975), although the biomass on new *Phragmites* stems was relatively low. A major part of the epiphytic dry weight found on *Potamogeton* consists of carbonate incrustations formed by macrophyte photosynthesis, while a smaller, but still considerable amount is found in the epiphytes on *Phragmites* (Wetzel 1960, Kowalczewski 1975, Szczepánska 1970). The amount of phosphorus present in the periphyton (Fig. 1e) generally followed the pattern observed for the biomass, when expressed per unit area of substratum. Average phosphorus content, expressed as a percentage of dry weight was: 0.11% on *Potamogeton,* 0.28% on new *Phragmites* stems and 0.19% on old *Phragmites* stems. For the zone of emergent vegetation the amount of phosphorus present in epiphytic periphyton can be roughly estimated for the whole lake. The zone covers 39 ha (Ozimek & Balcerzak 1976), and mean area of substratum for epiphytes is 0.74 m^2 per m^2 of littoral zone (Pieczyńska 1968). As the average amount of total epiphyte phosphorus was 0.24 mg dm^{-2} on new stems, the total amount of epiphytic phosphorus in the emergent zone is 6.7 kg, assuming that the amount of old stems remaining in late summer was negligible. In comparison the average total phosphorus in water and seston in the whole lake is 3900 kg (Kajak 1978).

Analyses of chlorophyll *a* and total P of *Potamogeton* leaves and *Phragmites* stems are shown in Fig. 2. The amount of phosphorus in *Potamogeton* leaves (mean value 0.129 mg P dm^{-2}) was usually smaller than the amount present in epiphytes on the leaves (mean value 0.216 mg P dm^{-2}). Senescence of the leaves was indicated by the decrease in chlorophyll *a* and total P in October. Total P of *Phragmites* stems has been expressed per unit area of stem surface in order to relate it to epiphyte data. While old stems contained a slightly smaller amount of phosphorus than their epiphytic cover, phosphorus content of new stems was higher than the content in epiphytes by an order of magnitude. Expressed as a percentage of dry weight, old stems contained on the average 0.03% P, new stems 0.16% P, and submerged sheaths of new stems 0.08% P (not included in Fig. 2). The decline in total P of new stems during September–October indicates the translocation of nutrients to the rhizome system.

Phytoplankton uptake of an excess of phosphorus (ΔP) as measured following incubations *in situ* of water samples enriched with 200 μg P L^{-1} is shown in Fig. 3. During most of the period ΔP did not exceed the confidence limits of zero, but a peak was observed at the beginning of August. Expressed on a biomass basis the highest value observed was 0.35 μg P μg Chl. a^{-1} h^{-1}.

Incubations of *Potamogeton* leaves with epi-

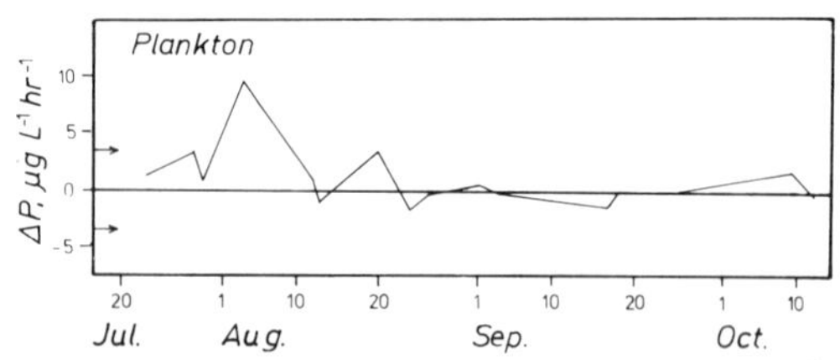

Fig. 3. SRP exchange (ΔP) of phytoplankton after enrichment with 200 μg L^{-1} PO_4-P. Arrows indicate 95% confidence limits for $\Delta P = 0$.

phytes in enriched water consistently showed a positive rate of P uptake, while all incubations in unenriched water showed a release of SRP (Fig. 4a). The possible release due to cutting the leaves is not known, but experiments with leaves cut into smaller pieces failed to show an increased release of SRP. Apart from this source of error the apparent release may be due to any combination of release and transformation of soluble or particulate unreactive phosphorus from the water into SRP. The difference between ΔP of enriched and unenriched water seems to be a relatively stable parameter, although it decreased steadily during most of the sampling period. All observed values were in the range of 5.3–14.8 μg P dm^{-2} h^{-1}. Similar incubations of *Phragmites* stem pieces with epiphytes showed a slightly different pattern, as ΔP in enriched water was negative in September (Fig. 4b). However, the seasonal tendency was similar to that of *Potamogeton*, and the difference between ΔP of enriched and unenriched water was of similar magnitude (2.5–16 μg P dm^{-2} h^{-1}). When *Potamogeton* leaves with epiphytes were incubated at different SRP values, ΔP of this material turned out to be concentration dependent with no indication of saturation up to 400 μg P L^{-1} under the given experimental conditions.

Results from incubations of *Potamogeton* leaves and epiphytes, intact or separated from each other, in unfiltered and filtered enriched water, are compared in Fig. 5. ΔP of separated leaves plus epiphytes usually differed little from ΔP of the intact complex, although the latter was almost consistently higher in September–October. This tendency was significant for incubations in unfiltered water ($P<0.001$, 2-way ANOVA). Control experiments with intact and removed periphyton from glass slides indicated a leakage of SRP due to scraping, of 0.4–7 μg P dm^{-2} h^{-1}, which would explain the observed tendency. Incubations in unfiltered and filtered water generally resulted in similar ΔP values. Only for the intact complex a significantly higher ΔP was observed in September/October ($P<0.01$, 2-way ANOVA). The relative contribu-

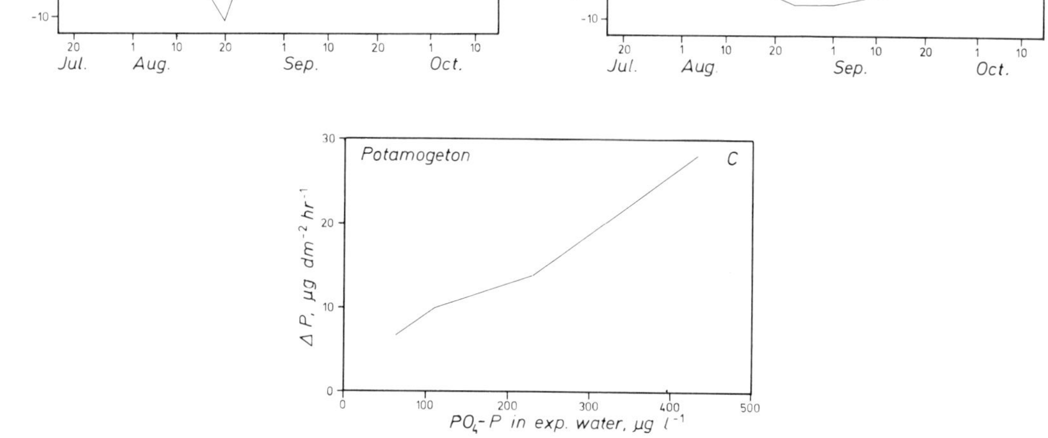

Fig. 4. ΔP of intact macrophyte–epiphyte complexes. A. Comparison of ΔP of *Potamogeton perfoliatus* in unenriched and enriched (200 μg L^{-1} PO_4-P) water. B. Same as A, but for *Phragmites australis*. C. ΔP of *Potamogeton*, measured at different concentrations of PO_4-P.

tions of leaves and epiphytes to the observed ΔP are shown in Fig. 5c and d. Although the influence of the removal process is not fully known, the result clearly indicated that the contributions of the two parts are of similar magnitude.

Epiphytes from *Phragmites* seemed to be little affected by the process of removal (Fig. 6a). ΔP of the separated epiphytes is shown in Fig. 6b together with ΔP of epiphytes from old stems. Though epiphyte biomass differed markedly between the two substrata, only minor differences in ΔP were observed. Also these values were similar to ΔP of epiphytes from *Potamogeton*. A release of SRP from new *Phragmites* stems was apparent, even after subtraction of leakage from cuts (Fig. 6c). Apart from the contribution of this release to the total ΔP of the complex, it probably presents nutritional advantages to the epiphytes of this substratum.

Discussion

Rates of phosphorus uptake by algae after enrichment cannot be interpreted as a measure of a long term uptake potential, but rather as an intermittent uptake indicating the nutritional state of the community. Thus intermittent uptake rates have been shown to be strongly related to the availability of phosphorus in algal cultures (Fuhs *et al* 1972), and in natural phytoplankton (Sproule & Kalff 1978). ΔP of plankton after enrichment as measured in this study should be largely comparable to the 'saturated rate of uptake' of Sproule & Kalff (1978). However, their values are typically more than an order of magnitude higher than the highest values of this investigation. Thus a phosphorus status with little if any deficiency is indicated by the phytoplankton data.

The seasonal pattern of ΔP in enriched water was similar in plankton and epiphytes, especially regarding the decrease during August–September. Leaves of *Potamogeton* showed a similar pattern, although it probably was influenced by the phenology of this plant. This fact and the similar magnitude of ΔP of epiphytes on *Potamogeton* and new and old *Phragmites* stems despite considerable differences in biomass, indicates that this parameter

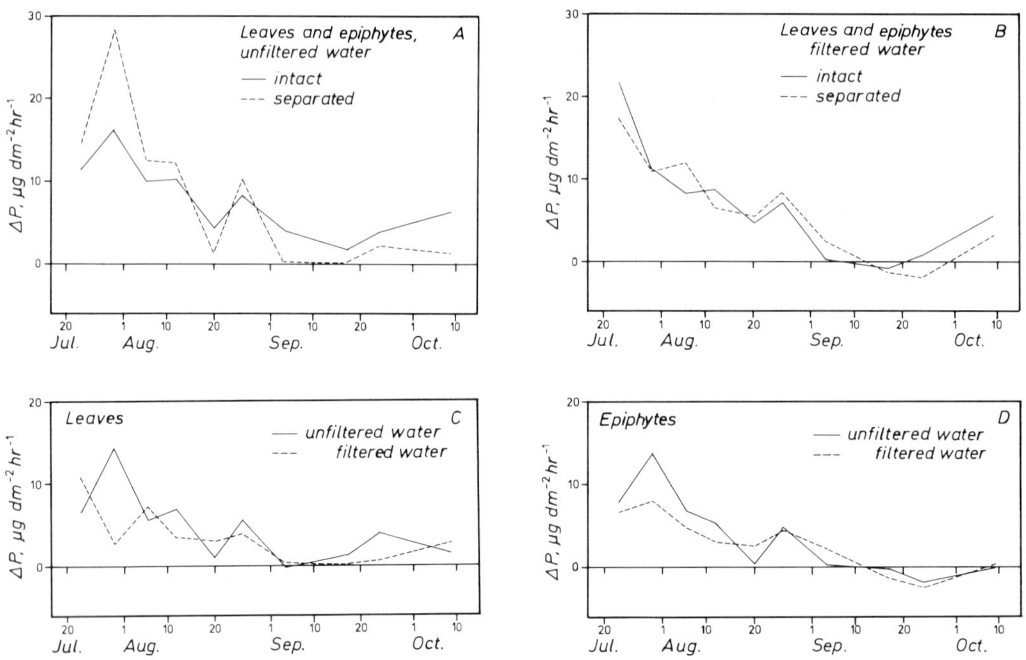

Fig. 5. Comparisons of different measurements of ΔP of *Potamogeton perfoliatus* and its epiphytes in enriched water. A. The intact complex *vs.* the sum of the separated components in unfiltered water. B. Same as A, but in filtered water. C. Leaves in unfiltered and filtered water. D. Epiphytes in unfiltered and filtered water.

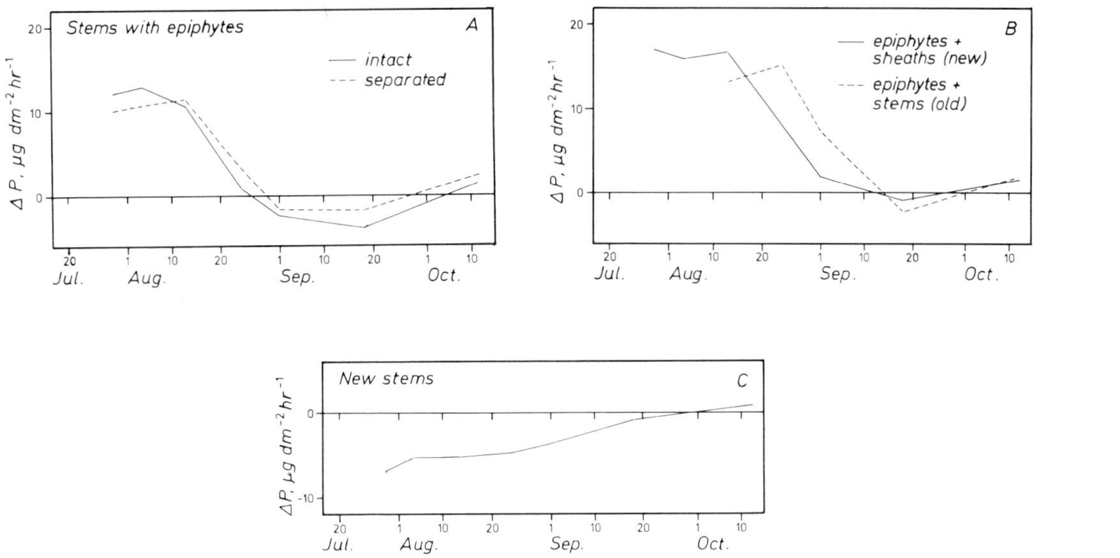

Fig. 6. Comparisons of different measurements of ΔP of *Phragmites australis* stems and their epiphytes. A. The intact complex *vs.* the sum of the separated components. B. Epiphytes from new *vs.* old stems. C. ΔP of new stems.

reflects community adaptation to the local nutritional status in the littoral, rather than individual properties of a given periphyton community.

All the ΔP values observed can be interpreted as the sum of (1) uptake of SRP, (2) release of SRP, and (3) transformation of unreactive into reactive phosphorus. It may be assumed that release of SRP and transformation of unreactive phosphorus are not immediately affected by an addition of PO_4-P, so the effect of enrichment is due to the uptake component. The observed uptake response of the macrophyte–epiphyte complex shows that some uptake of PO_4-P must have been in effect during the unenriched incubations, due to ambient PO_4-P. The value of the uptake component should be regarded as relative, since the process is strongly dependent on water movements. In an experiment with *Potamogeton perfoliatus* Riber (in press) found a threefold increase in ^{32}P uptake for both leaf and epiphytes, if the incubation bottle was rotated 12 times per min, which would correspond to realistic water movements for the littoral. The net release of SRP observed at natural concentration may thus partly be explained by an underestimation of uptake. Apart from that it must be balanced by uptake of unreactive phosphorus or supply from root uptake by the macrophyte. In the case of *Phragmites* it is clear that release of phosphorus from the stem, which in some cases accounted for more than the net release of the macrophyte–epiphyte complex without enrichment, was derived from root uptake. Epiphytes on *Phragmites* are situated favourably for utilization of this phosphorus output, although its contribution to their phosphorus nutrition cannot be estimated from the results.

Recent results by Carignan & Kalff (1982) show that only 9.9% of the total P release from *Myriophyllum spicatum* L. plus epiphytes was derived from the macrophyte. Recycling of phosphorus by the epiphytes is probably also the dominating process in the observed release from *Potamogeton* plus epiphytes in this investigation. Release to unenriched water may thus be considered a rough estimate of phosphorus exchange of the epiphytes, which indicates turnover times of 2 days or less, while turnover times for the whole biomass of periphyton typically lie between 14 and 60 days (Wetzel 1975).

Thus while the amount of phosphorus accumulated in periphyton appears to be small, as exemplified by the epiphytic phosphorus in the zone of emergent macrophytes, a considerable proportion of the phosphorus pool of a lake passes through the periphyton during the year. The absolute amount depends on a number of biotic factors, such as mass appearences of littoral algae, especially occurring in the eulittoral zone, which contribute decisively to algal primary production in this zone (Pieczyńska & Szczepańska 1966). This phenomenon may appreciably affect the phosphorus economy of the lake, by uptake and later release due to decomposition. Extremely high concentrations of SRP have been reported from such situations in isolated parts of the littoral (Pieczyńska 1972). However, the significance of this phenomenon is extremely difficult to assess because of its casual occurrence. Another important factor is the dependence of periphyton on a suitable substratum (Pieczyńska 1968). In Mikolajskie Lake the increasing eutrophication has restricted the depth limit of submerged macrophytes to *ca.* 2 m (Ozimek, unpubl.), while in 1970 they were abundant down to 6 m. This development clearly restricts the potential influence of epiphytic periphyton in this zone, although this community, where present, is enhanced by eutrophication (Cattaneo & Kalff 1980).

The role of periphyton in the phosphorus economy of lakes may be summarized as: (1) A temporary storage of phosphorus derived from epilimnetic water and from the sediment via macrophyte roots. More research is needed on the fate of this phosphorus, especially on the significance of sedimentation of sloughed periphyton. (2) Transformation of phosphorus, which is then returned to the water. Fee (1979) observed a positive correlation between phytoplankton primary production and the relative area of sediment exposed to the epilimnion, and he explained this by a faster recirculation of nutrients by microbial activity at the sediment surface. In this connection the macrophyte–epiphyte complex may be regarded as an expansion of colonized surface area.

Acknowledgements

Scholarships for Hans H. Riber and John P. Sørensen were provided under the Polish-Danish cultural agreement (DPWKN-D-54-11-80). Supplementary grants were received from the Danish Natural Science Research Council (J.nr. 11-0830) and from the Egmont H. Petersen Foundation.

We wish to thank Dr. M. Planter for doing the total P analyses, Prof. E. Pieczyńska, Prof. Z. Kajak and Dr. I. Spodniewská for inspiring discussions of the results, and Prof. R.G. Wetzel, Dr. A. Cattaneo and Dr. L.M. Jensen for critical comments on the manuscript.

References

Carignan, R., 1982. An empirical model to estimate the relative importance of roots in phosphorus uptake by aquatic macrophytes. Can. J. Fish. Aquat. Sci. 39:243–247.

Carignan, R. & Kalff, J., 1982. Phosphorus release by submerged macrophytes: significance to epiphyton and phytoplankton. Limnol. Oceanogr. 27:419–427.

Carpenter, S.R., 1980. Enrichment of Lake Wingra, Wisconsin, by submerged macrophyte decay. Ecology 61:1145–1155.

Cattaneo, A. & Kalff, J., 1980. The relative contribution of aquatic macrophytes and their epiphytes to the production of macrophyte beds. Limnol. Oceanogr. 25:280–289.

Fee, E.J., 1979. A relation between lake morphometry and primary productivity and its use in interpreting wholelake eutrophication experiments. Limnol. Oceanogr. 24:401–416.

Fuhs, G.W., Demmerle, S.D., Canelli, E. & Chen, M., 1972. Characterization of phosphorus-limited plankton algae. Amer. Soc. Limnol. Oceanogr. Spec. Symp. 1:113–133.

Golterman, H.L., 1969. Methods for chemical analysis of fresh waters. – IBP Handbook no. 8, Blackwell, Oxford.

Harlin, M.M., 1973. Transfer of products between epiphytic marine algae and host plants. – J. Phycol. 9:243–248.

Howard-Williams, C., 1981. Studies on the ability of a Potamogeton pectinatus community to remove dissolved nitrogen and phosphorus compounds from lake water. J. Appl. Ecol. 18:619–637.

Just, J. & Hermanowicz, W., 1964. Fizyczne i chemiczne badanie wody do picia i potrzeb gospodarczych (Physical and chemical examination of water for drinking and for economic purposes) – Warszawa, 320 pp.

Kajak, Z., 1978. The characteristics of a temperate eutrophic, dimictic lake (Lake Mikolajskie, Northern Poland. – Int. Rev. ges. Hydrobiol. 63:451–480.

Kowalczewski, A., 1975. Periphyton primary production in the zone of submerged vegetation of Mikolajskie Lake. – Ekol. Pol. 23:509–543.

Landers, D.H., 1982. Effects of naturally senescing aquatic macrophytes on nutrient chemistry and chlorophyll *a* of surrounding waters. Limnol. Oceanogr. 27:428–439.

Ozimek, T. & Balcerzak, D., 1976. Macrophytes. In: Pieczyńska, E., (ed), Selected problems of Lake Littoral Ecology, pp. 33–53. WUW, Warsaw.

Penhale, P.A. & Thayer, G.W., 1980. Uptake and transfer of

carbon and phosphorus by Eelgrass (Zostera marina L.) and its epiphytes. J. exp. mar. Biol. Ecol. 42:113–123.
Pieczyńska, E., 1968. Dependence of the primary production of periphyton upon the substrate area suitable for colonization. Bull. Acad. Pol. Sci. Cl.II. 16:165–169.
Pieczyńska, E., 1972. Ecology of the eulittoral zone of lakes. Ekol. Pol. 20:637–732.
Pieczyńska, E. (ed). 1976. Selected Problems of Lake Littoral Ecology. WUW, Warsaw, 238 pp.
Pieczyńska, E. & Szczepańska, W., 1966. Primary production in the littoral of several Mazurian lakes. – Verh. Int. Ver. Limnol. 16:372–379.
Prentki, R.T., Adams, M.S., Carpenter, S.R., Gasith, A., Smith, C.S. & Weiler, P.R., 1979. The role of submersed weedbeds in interception of allochthonous materials in Lake Wingra, Wisconsin, USA. – Arch. Hydrobiol./Suppl. 57: 221–250.
Rigler, F.H., 1973. A dynamic view of the phosphorus cycle in lakes. – In: Griffith, E.J. et al (eds), Environmental Phosphorus Handbook. Wiley, New York, pp 539–572.
Riber, H.H., 1983. Phosphorus uptake from water by the macrophyte-epiphyte complex in a Danish lake: Relationship to plankton. Verh. Int. Ver. Limnol. 22.
Sproule, J.L. & Kalff, J., 1978. Seasonal cycles in the phytoplankton phosphorus status of a north temperate lake (Lake Memphremagog, Qué-Vt), plus a comparison of techniques. – Verh. Int. Ver. Limnol. 20:2681–2688.
Szczepańska, W., 1970. Periphyton of several lakes of the Mazurian lakeland. – Pol. Arch. Hydrobiol. 17:397–418.
Wetzel, R.G., 1960. Marl incrustation on hydrophytes in several Michigan lakes. – Oikos 11:223–236.
Wetzel, R.G., 1975. Limnology. – Saunders, Philadelphia, 743 pp.
Wetzel, R.G., 1983. Attached algal-substrata interactions: Fact or myth, and when and how? In: Wetzel, R.G. (ed.) Periphyton in Freshwater Ecosystems. Developments in Hydrobiology 17: pp. 207–215.
Wetzel, R.G. & Allen, H.L., 1972. Functions and interactions of dissolved organic matter and the littoral zone in lake metabolism and eutrophication. In: Kajak, Z. & Hillbricht-Ilkowska, A. (eds.) Productivity Problems of Freshwaters, pp. 333–347. PWN, Warsaw–Krakow.

Authors' address:
H.H. Riber/J.P. Sørensen
Botanical Institute
University of Århus
Nordlandsvej 68
DK-8240 Risskov, Denmark

A. Kowalczewski
Department of Hydrobiology
Institute of Zoology,
University of Warsaw,
Nowy Swiat 67
00–046 Warsaw, Poland

CHAPTER 30

Methodology: the key to understanding periphyton

GORDON G.C. ROBINSON

Abstract. The present status of the development of methodology for periphyton research is reviewed with emphasis on needs for current and future research directions. Sampling procedures and methods of assessing the dynamics of biomass changes and population and community productivity are evaluated. The importance of future methodological improvements is stressed as a basis for quantitative determination of the interactions among periphytic components and regulating environmental parameters.

This is an attempt to review the present status of the development of methodology for periphyton research. It will attempt to place some emphasis on what may be perceived to be problematic areas, and respectfully suggest areas where there is a current need for new or improved procedures. This review should like to address, in order, sampling procedures, methods for the estimation of primary productivity and biomass, and steps that have and perhaps should be taken in the extremely complex area of the interaction of physical, chemical and biotic ecological factors with periphyton communities and their components. Although this review cannot be comprehensive, it is hoped that it may serve some function as methodological background.

Sampling procedures cannot for the most part be separated from the use and efficacy of use of artificial substrata, although epipsammic and epipelic communities of periphyton have not generally encouraged the use of artificial substrata. For the most part quantitative sampling of these communities has involved the removal of intact cores of sediment of various, but generally small, diameter, from which thin layers are removed for examination or experimentation (e.g. Hunding 1971; Swanson & Bachman 1976; Stanley 1976; van Raalte *et al* 1976; Darley *et al* 1976; Hargrave 1969). Hargrave (1969) has used a variant procedure in the design and use of a sampler with the ability to cut thin layers of sediment. An alternate procedure which is often used in shallow waters is that of Eaton & Moss (1966) in which surface material from a delineated area is aspirated into a collecting bottle. A hand held peristaltic pump as described by Hamala *et al* (1981) appears to be a useful alternative to an aspirator. Where such benthic samples are of a mixed nature, separation may be accomplished by serial washing in which firmly attached epipsammon is physically separated with its substrate from the epipelon and silt (Tai & Hodgkiss 1975). Ultrasonic treatment of sand grains has been shown to be an effective means of separating algal cells from sand grains to which they are firmly attached (Hickman 1969). For the harvesting of mobile epipelic algae trapping in lens paper and adherence to cover slips placed over sediment samples appear to be popular and quantitatively acceptable (Eaton & Moss 1966). Clearly

Wetzel, R.G. (ed.), Periphyton of Freshwater Ecosystems.
 ISBN 90 6193 768 X.

care must be taken with these procedures that vertical migrations of algae are accounted for.

Sampling of naturally occurring epilithic periphyton has very often used the well tried methods of Douglas (1958) or some modification of them. Delineated areas of colonized rock substrata which have been brought to the surface are often scraped and the scrapings used further. A sophistication of the half-bottle and brush technique is the bar-clamp sampler and ultrasonification as described by Gale (1975) and used by Gale *et al* (1979). Clearly, substrates may not always be brought to the surface nor, for some purposes, is it advisable (for example light shock may be deleterious for some experiments and indeed physical loss of loosely attached material may make substratum removal inappropriate). As a result the use of SCUBA or other diving procedures has led to some interesting *in situ* sampling – for example a syringe brush combination as used by Loeb (1981).

Artificial substrata are often used for collection of epilithic periphyton. The variety of types is large – smooth and roughened glass, a variety of plastics, slate, and granite have all been used (e.g. Rosemarin & Gelin 1978; Hudon & Bourget 1981; Castenholz 1960, 1961; Gale *et al* 1979). The advisability of using such substrata may at times be questionable. There are, however, clear advantages to their use: they may be readily manipulated into many positions. They may be adequately replicated and they may be readily sampled. Furthermore, exclusion of major species does not appear to be a problem in most uses. On the other hand, the times normally permitted for colonization of artificial substrata seem to be such that the periphyton which develops on them does not mimic that on natural substrata. This difference is particularly so for substrata positioned in oligotrophic situations. Is this a dilemma? I suspect in fact that it need not be. Suitable artificial substrata may be used successfully for many types of investigation – e.g. rates of colonization, community interactions, impact of environmental variables and comparison of environments. On the other hand, they should perhaps not be utilized in situations where they are intended to exactly mimic natural conditions, particularly where biomass and production estimates are important.

The sampling of epiphytic periphyton does of course often also involve the use of artificial substrata, and such use is complicated by the emerging fact that submerged and emergent macrophytes provide substrata which are not inert. This fact makes the use of artificial substrata a highly questionable procedure for some investigations. Many investigations have, however, been conducted on the epiphyton of natural substrates, where sampling has most often involved the removal of macrophytes from the water prior to the detachment of the periphyton. This removal may involve the loss of loosely attached material and is therefore perhaps best preceeded by the enclosure of the macrophyte under water, such as has been done by Hickman (1971) and some others. Subsequent detachment is then often accomplished by scraping or agitation. Woelkerling (1976) and Gough & Woelkerling (1976) report agitation in the presence of hydrolizing FAA solution to be effective, although Cattaneo & Kalff (1980) have reported that agitation may remove only a variable proportion of epiphytes (6–68%). This difference led these authors to consider loosely and tightly associated components of the epiphyton separately. Sampled epiphytes are often expressed in terms of the macrophyte surface area colonized, or, since this may be difficult to determine for some macrophytes, on the basis of dry weight of macrophyte colonized.

Glass, plastic, polystyrene, and wood have also been widely used to gather epiphytic periphyton. In these cases the length of time provided for colonization can be readily arranged to be quite comparable with that provided by natural substrata, particularly if these are annual macrophytes. The question must, however, be asked whether periphyton on natural substrata is different in biomass, composition, and productivity from that on artificial substrata, by virtue of the interaction between macrophytes and their epiphytes. There is evidence that such interaction does occur but evidence on whether or not this influences the periphyton in ways that are detectable with current methodology is conflicting. Some studies suggest real differences in periphyton on natural and artificial substrata while others do not. Part of this conflict may be due to methodological differences. For example, dissimilarity indices between communities of periphyton established on smooth and roughened cellulose acetate substrata and *Typha* indicate large differences in community composition (Shamess & Robinson 1982). On the other hand the similarity of productivity of periphyton on artificial and real substrata from a similar situation can also be demonstrated (Hooper & Robinson 1976). Recent work by Eminson (1978) and Eminson & Moss (1980) suggests a reconciliation of some of the conflict. They suggest that with progressive eutrophication, the role of the aqueous milieu of the periphyton progressively masks any controlling role of the macrophyte substratum. Moss (1981) goes further to suggest that, with extreme eutrophy, habitat preferences of phytoplankton and periphyton may themselves be eroded. If this is, indeed, a widely applicable phenomenon, then perhaps artificial substrata may be expected to be most effective in eutrophic waters and not as effective in oligotrophic situations. It should be added, however, that if macrophytes have the potential to influence periphyton at all, then the density of macrophyte stands may well be important in both oligotrophic and eutrophic situations, and there do exist highly eutrophic situations with very high densities of rooted aquatics. I think that before the influence of macrophytes on epiphytes in eutrophic situations is neglected there is a great deal more work to be done.

Not only may natural substrata influence their periphyton directly but also perhaps indirectly. For example, there is some evidence (Mason & Bryant 1975) that chironomid grazers may be excluded from solid artificial substrata by virtue of their inability to burrow into the substratum.

Clearly there are many advantages to the use of artificial substrata for the assessment of epiphytic periphyton, some of which I have already mentioned in relation to the epilithon, but which also include the ease with which their surface area can be determined, the elimination of contamination problems from host tissue, and the possibility of biomass, productivity, and community composition analysis techniques being applied without a prior removal of the periphyton from its substratum.

Before leaving these few comments on sampling, mention should be given to some peel procedures which may be extremely useful for analyses of colonization and early community dynamics. Margalef (1949) used a collodium peel of epilithic communities and Cattaneo (1978) successfully used a facial mask peel of epiphytes from *Potamogeton richardsonii*. I have found that adhesive tape followed by the selective removal of its backing can be utilized quite effectively for examination of diatom colonization. I should also mention that the epifluorescence technique described by Jones (1974, 1979) may be useful in some situations for direct observation of periphytic communities, and for the distinction between living and non-living components.

In all sampling, statistical reliability is, of course, important. This need may present severe problems, for periphyton development appears to be generally quite heterogeneous. This problem is so in early stages of community development where colonizing organisms spread from random clumps, and in late stages on community development, where variance appears to increase with community density. Eaton & Moss (1966) have provided a good analysis of sources of variation in epipelic communities and Hillebrand & van Dijk (1981) have described sources of variance at different scales. I think it is a safe generalization that most is to be gained in reducing variance by replication of primary sampling.

Considerable emphasis has been placed upon the determination of primary productivity of periphytic communities, so I will now review generally the methodology employed. For epipsammic and epipelic communities intact cores of sediment or separated algal populations have both been used, and these sometimes for *in situ* incubations and incubation under controlled environmental conditions. Epilithic production has been estimated from colonized rocks that have been enclosed within an incubation chamber of some sort (e.g. Schindler *et al* 1973) or by *in situ* placement of incubation chambers on colonized substratum (Wetzel 1964; Loeb 1981). This latter procedure has the clear advantage that samples are not brought to the surface prior to the commencement of a primary productivity determination. Productivity of epiphytic periphyton has also been estimated from incubations of detached and attached samples. Once again incubation chambers may be placed around communities without prior removal from the water (Cattaneo & Kalff 1979), which is clearly advantageous. Very often, however, incubations follow the removal of the periphyton and its substrate from the water.

Of course, periphyton colonizing artificial substrata have also been widely used for estimates of primary productivity. Most widely used methodologies are $^{14}CO_2$ uptake and O_2 evolution techniques. $^{14}CO_2$ uptake in light and dark incubation chambers has been measured either directly by Geiger Muller or liquid scintillation procedures or following combustion and collection of labelled carbon. Oxygen changes in light and dark incubations have been measured either by titration or by the use of respirometers. Disappearance of labelled carbon from surrounding water and changes in dissolved inorganic carbon are the bases of other procedures (Schindler *et al* 1973), which have had more limited use, and biomass accumulation rates have also on occasion been utilized as estimators of production (e.g. Cooper & Wilhm 1975; Adams & Stone 1973).

Comparison of different methods (mostly ^{14}C and O_2) have mostly shown very poor relationship between the methods (Schindler *et al* 1973), although overall temporal trends may be similar (Bott & Ritter 1981).

Generally speaking, the sources of error in these methodologies are such that none can be said to be entirely satisfactory. Although some error sources can be accounted for, some are still troublesome. Perhaps the most difficult problems to deal with in O_2 methodology are the effects of non-algal components of the community upon overall gaseous exchange, although antibiotic treatment as used by Hargrave (1967) may be useful in distinguishing bacterial effects, and the role of the photosynthetic quotient, which is of course likely to be variable and depend upon the degree of reduction of algal metabolites produced. Carbon methodology is pestered by the possible re-fixation of intracellular respiratory CO_2, and the resulting confusion of whether a method is providing an estimate of gross or net production or something in between. For the most part, determinations of primary productivity are actually measurements of community productivity. This is, of course, a problem with phytoplankton studies as well, but the possibility of separating non-algal material from periphytic communities (and particularly epiphytic and epilithic communities) is extremely remote. Despite this problem there are many very useful reports of *in situ* productivity, potential productivity, comparative productivity and specific productivity. Some special adaptations of periphytic organisms which influence their productivity have also been reported. Examples include the adaptations of epipsammic and epipelic organisms to darkness and diminishing O_2 levels (Moss 1977) and seasonal sun/shade adaptation to light in epiphytes (Jones & Adams 1982). We have also been able to demonstrate some degree of vertical light adaptation in communities which have grown at different depths of water and been incubated at different intensities (Robinson & Putt 1981).

Depending upon the purpose of an investigation, there are a number of indices of periphyton biomass that are available. Where spatial or temporal trends are desirable or where specific productivity is to be determined, chlorophyll a is a widely used index. This pigment is usually determined from spectrophotometric or fluorometric analysis of acetone or methanol extracts of periphytic communities. Periphyton communities, perhaps more than their planktonic counterparts, are likely to contain variable amounts of chlorophyll degradation products. Consequently, correction for phaeophytins is commonly incorporated into methodology.

Where energetic studies are conducted, or where turnover times are to be determined, particulate organic carbon content (as determined commonly by either wet or dry combustion), ash free dry weight, or simply dry weight may be used. Of course, with these procedures there may be considerable interference from the inclusion of non-algal material.

In many cases there is the need to know of the relative contribution of individual algal populations to community biomass. Cell numbers may make comparison of communities possible, but really do not yield a very useful measure of relative biomass. Consequently cell volumes (mostly derived from assumptions of classic geometric configurations) are often derived from measured cell dimensions. While useful, such measurements of biovolume may not always be accurate indicators of biomass, for as Sicko-Goad *et al* (1977) have shown from morphometric analyses of electron-micrographs of representative species, the variable extent of inert walls and cell vacuoles results in metabolizing biovolume being from 30–82% of the total biovolume of cells. This problem is considerably less if one is dealing with one group of organisms only (e.g. periphytic diatoms). When it is desirable to make comparisons between species in terms of assimilation rates, it may be advisable to use cell surface area or some standard unit of cell surface area. We have found, for example, that epiphytic periphyton cell surface area is better correlated with production values than is cell volume (Hooper-Reid & Robinson 1978). On the other hand there is good correlation between cell surface area and cell volume (Pip & Robinson 1982).

In all of the foregoing areas there has been much methodological progress, but also much repetition. There is still considerable latitude for advancement. Perhaps, one of the most challenging areas for the development of appropriate methodology is in the understanding of the interactions between whole periphyton communities and their physical, chemical, and biological environment, but more perhaps in the understanding of the ecology of individual components of communities and interactions between them.

Some progress has been made in examining the response of community productivity, biomass, and composition to environmental variables, but mostly by correlation or lack of correlation. There has also been some limited use made of physiological indicators of periphyton interaction with its environment. For example, phosphatase activity in periphyton as an indicator of phosphate limitation and organic phosphorus utilization, nitrogenase activity as an indicator of N-deficiency, and high silicon debt as an indicator of silicon-shortage (Cattaneo & Kalff 1979; Hooper-Reid & Robinson 1978).

Generally speaking, however, the complexity of periphyton communities and the nature of their environments appear to have been quite prohibitive to the analyses of the detailed interactions between components of communities and between communities and their environments. This problem may be in part due to the extreme difficulty in actually determining the true nature of environmental variables. How, for example, do we measure the light that is actually available to periphyton throughout its mass? How do we reconcile the extensive diurnal thermal gradients which may occur in shallow littoral waters that are populated with macrophytes (Dale & Gillespie 1977a, b, 1978)? How do we analyze the kinetics of supply and utilization of inorganic and organic nutrients available to periphyton – and particularly to epiphytic and epipelic periphyton in which the substratum is not inert?

Once again there has been some progress. For the type of detail that is necessary, we should have very specific information on processes of substratum colonization and community development. Scanning electron microscopy is providing some of this. Allanson (1973) provided a structural basis for the Wetzel & Allen model (Wetzel 1975), and more recently Hudon & Bourget (1981) and Hoagland *et al* (1982) have provided us with rather beautiful scanning micrographs demonstrating the progression of community development from organic film, to solitary and clumped bacteria, to filamentous bacteria and flat adherent diatoms, to clonal rosettes of *Gomphonema,* to the superstory of stalked and filamentous diatoms, and to filamentous greens. This type of information could be enhanced considerably with information on the metabolism of the individual populations involved. One approach to this may be in the use of quantitative autoradiography. So far this method can be used to monitor inorganic and organic carbon assimilation by individual species, and might well in the future have the potential to provide quantification of other nutritional processes. Autoradiography has permitted us to estimate inorganic and organic carbon assimilation by individual epiphytic taxa throughout a season on three species of *Potamogeton* (Pip & Robinson 1982) and to demonstrate differences between communities on the three macrophytes.

There is also progress on epiphyte/macrophyte studies of nutrient supply from host macrophytes. That such might occur was suggested by Prowse (1959) in his observation of the different communities which developed on previously uncolonized macrophytes. Transfer of nutrients has acutally been demonstrated by Allen (1971), Harlin (1973, 1975) and McRoy & Goering (1974) by the use of radioisotopes of carbon and phosphorus and heavy isotopes of nitrogen. The experiments of Twilley *et al* (1977) on the translocation of labelled P by compartmentalized *Nuphar* also suggest that this is possible.

As we see, progress is being made, but also that much more is required for a more complete understanding of periphyton. The methodological keys to such improved understanding lie, in my opinion, with the development of systems for the analyses of the fine detail of the highly compressed environmental niches of periphyton and for the analyses of the physiological ecology of the numerous components of these complex communities.

References

Adams, M.S. & Stone, W., 1973. Field studies on photosynthesis of Cladophora glomerata (Chlorophyta) in Green Bay, Lake Michigan. Ecology 54:853–862.

Allanson, B.R., 1973. The fine structure of the periphyton of Chara sp. and Potamogeton natans from Wytham Pond, Oxford, and its significance to the macrophyte-periphyton metabolic model of R.G. Wetzel and H.L. Allen. Freshwat. Biol. 3:535–542.

Allen, H.L., 1971. Primary productivity, chemo-organotrophy and nutritional interactions of epiphytic algae and bacteria on macrophytes in the littoral of a lake. Ecol. Monogr. 41: 97–127.

Bott, T.L. & Ritter, F.P., 1981. Benthic algal productivity in a Piedmount stream measured by ^{14}C and dissolved oxygen change procedures. J. Freshwat. Ecol. 1:267–277.

Castenholz, R.W., 1960. Seasonal changes in the attached algae of freshwater and saline lakes in the lower Grand Coulee, Washington. Limnol. Oceanogr. 5:1–28.

Castenholz, R.W., 1961. An evaluation of a submerged glass slide method of estimating production of attached algae. Verh. Int. Verein. Theor. Angew. Limnol. 14:155–159.

Cattaneo, A., 1978. The microdistribution of epiphytes on the leaves of natural and artificial macrophytes. Br. Phycol. J. 13:183–188.

Cattaneo, A. & Kalff, J., 1978. Seasonal changes on the epiphyte community of natural and artificial macrophytes in Lake Memphremagog (Que.-Vt.), Hydrobiol. 60:135–144.

Cattaneo, A. & Kalff, J., 1980. The relative contribution of aquatic macrophytes and their epiphytes to the production of macrophyte beds. Limnol. Oceanogr. 25:280–289.

Cooper, J.M. & Wilhm, J., 1975. Spatial and temporal variation in productivity, species diversity, and pigment diversity of periphyton in a stream receiving domestic and oil refinery effluents. The Southwestern Naturalist 19:413–428.

Dale, H.M. & Gillespie, T.J., 1977a. The influence of submersed aquatic plants on temperature gradients in shallow water bodies. Can. J. Bot. 55:2216–2225.

Dale, H.M. & Gillespie, T.J., 1977b. Diurnal fluctuations of temperature near the bottom of shallow water bodies as affected by solar radiation, bottom colour and water circulation. Hydrobiologia 55:87–92.

Dale, H.M. & Gillespie, T.J., 1979. Diurnal temperature gradients in shallow water produced by populations of artificial aquatic macrophytes. Can. J. Bot. 56:1099–1106.

Darley, W.M., Dunn, E.L., Holmes, K.S. & Larew, H.G., 1976. A ^{14}C method for measuring epibenthic microalgal productivity in air. J. Exp. Mar. Biol. Ecol. 25:207–217.

Douglas, B., 1958. The ecology of attached diatoms and other algae in a small stony stream. J. Ecol. 46:295–322.

Eaton, J.W. & Moss, B., 1966. The estimation of numbers and pigment content in epipelic algal populations. Limnol. Oceanogr. 11:584–595.

Eminson, D.F., 1979. A comparison of diatom epiphytes. Their diversity and density attached to Myriophyllum spicatum L. Br. Phycol. J. 13:57–64.

Eminson, D.F. & Moss, B., 1980. The composition and ecology of periphyton communities in fresh waters. I. The influence of host type and external environment on community composition. Br. Phycol. J. 15:429–226.

Gale, F.W., 1975. Ultrasonic removal of epilithic algae in a barclamp sampler. J. Phycol. 11:472–473.

Gale, F.W., Gurzynski, A.J. & Lowe, R.L., 1979. Colonization and standing crop of epilithic algae in the Susquehanna River. J. Phycol. 15:117–123.

Gough, S.B. & Woelkerling, W.J., 1976. Wisconsin desmids. I: Aufwuchs and plankton communities of selected soft water lakes, hard water lakes and calcareous spring ponds. Hydrobiologia 49:3–25.

Hamala, J.A., Duncan, S.W. & Blinn, D.W., 1981. A portable pump sampler for lotic periphyton. Hydrobiologia 80: 189–191.

Hargrave, B.T., 1969. Epibenthic algal production and community respiration in the sediments of Marion Lake. J. Fish. Res. Bd. Canada 26:2003–2026.

Harlin, M.M., 1972. 'Obligate' algal epiphyte: Smithora naiadum grows on a synthetic substrate. J. Phycol. 9:230–232.

Harlin, M.M., 1975. Epiphyte-host relationships in seagrass communities. Aquatic Bot. 1:235–131.

Hickman, M., 1969. Methods for determining the primary productivity of epipelic and epipsammic algal associations. Limnol. Oceanogr. 14:936–941.
Hickman, M., 1971. The standing crop and primary productivity of the epiphyton attached to Equisetum fluviatile L. in Priddy Pool, North Somerset. Br. Phycol. J. 6:51–59.
Hillebrand, H. & van Dijk, C.P.J., 1982. Patterns of freshwater algae on artificial substrates at different sampling scales (in press).
Hoagland, K.D., Roemer, S.C. & Rosowski, J.R., 1982. Colonization and community structure of two periphytic assemblages, with emphasis on the diatoms (Bacillariophyceae). Amer. J. Bot. 69:188–213.
Hooper, N.M. & Robinson, G.G.C., 1976. Primary production of epiphytic algae in a marsh pond. Can. J. Bot. 54: 2810–2815.
Hooper-Reid, N.M. & Robinson, G.G.C., 1978a. Seasonal dynamics of epiphytic algae growth in a marsh pond: productivity, standing crop, and community composition. Can. J. Bot. 56:2434–2440.
Hooper-Reid, N.M. & Robinson, G.G.C., 1978b. Seasonal dynamics of epiphytic algae growth in a marsh pond: composition, metabolism and nutrient availability. Can. J. Bot. 56:2441–2448.
Hudon, C. & Bourget, E., 1981. Initial colonization of artificial substrate: community development and structure studied by scanning electron microscopy. Can. J. Fish. Aquat. Sci. 38:1371–1384.
Hunding, C., 1971. Production of epibenthic micro-algae in the littoral zone of a eutrophic lake. Oikos 22:389–397.
Jones, J.G., 1974. A method for observation and enumeration of epilithic algae directly on the surface of stones. Oecologia 16:1–8.
Jones, J.G., 1979. Spatial variation in epilithic algae in a stony stream, Wilfin Beck, U.K., with particular reference to Cocconeis placentula. Freshwat. Biol. 8:539–546.
Jones, R.C. & Adams, M.S., 1982. Seasonal variations in photosynthetic response of algae epiphytic on Myriophyllum spicatum L. Aquatic Bot. 13:317–330.
Loeb, S.L., 1981. An 'in situ' method for measuring the primary productivity and standing crop of the epilithic periphyton community in benthic systems. Limnol. Oceanogr. 26: 394–399.
Margalef, R., 1949. A new limnological method for the investigation of thin layered epilithic communities. Hydrobiologia 1:215–216.
Mason, C.F. & Bryant, R.J., 1975. Periphyton production and chironomid grazing in Alderfen Broad, Norwich, England. Freshwat. Biol. 5:271–277.
McRoy, C.R. & Goering, J.J., 1974. Nutrient transfer between seagrass Zostera marina and its epiphytes. Nature (London) 248:173–174.
Moss, B., 1977 Adaptations of epipelic and epipsammic freshwater algae. Oecologia 28:103–108.
Moss, B., 1981. The composition and ecology of periphyton communities in freshwaters. 2: Interrelationships between water chemistry, phytoplankton populations and periphyton populations in a shallow lake and associated experimental reservoir's Lund tubes. Br. Phycol. J. 16:59–76.
Pip, E. & Robinson, G.G.C., 1982. A study of the seasonal dynamics of three phycoperiphytic communities, using nuclear track autoradiography. I. Inorganic carbon uptake. Arch. Hydrobiol. 94:341–371.
Pip, E. & Robinson, G.G.C., 1982. A study of the seasonal dynamics of three phycoperiophytic communities, using nuclear track autoradiography. II. Organic carbon uptake. Arch. Hydrobiol. 96:47–64.
Prowse, G.A., 1959. Relationship between epiphytic algae and their macrophyte roots. Nature (London) 183:1204–1205.
Robinson, G.G.C. & Putt, M., 1982. Evidence for light adaptation of vertically distributed periphyton (unpublished).
Rosemarin, A.S. & Gelin, C., 1978. Epilithic algal presence and pigment composition on naturally occurring and artificial substrates in Lakes Trumen and Fiolen, Sweden. Verh. Internat. Verein. Limnol. 20:808–813.
Schindler, D.W., Frost, V.E. & Schmidt, R.V., 1973. Production of epilithiphyton in two lakes of the Experimental Lakes Area, Northwestern Ontario. J. Fish. Res. Bd. Canada 30:1511–1524.
Shamess, J.J. & Robinson, G.G.C., 1982. The structure and comparison of periphytic and planktonic algal communities in two eutrophic prairie lakes. (Unpublished.)
Sicko-Goad, L., Stoermer, E.F. & Ladewski, B.G., 1977. A morphometric method for correcting phytoplankton cell volume estimates. Protoplasma 93:147–163.
Stanley, D.W., 1976. Productivity of epipelic algae in tundra ponds and a lake near Barrow, Alaska. Ecology 57: 1015–1024.
Swanson, C.D. & Bachmann, R.W., 1976. A model of algal exports in some Iowa streams. Ecology 57:1076–1080
Tai, Y.C. & Hodgkiss, I.J., 1975. Studies on Plover Cove Reservoir, Hong Kong. Freshwat. Biol. 5:85–103.
Twilley, R.R., Brinson, M.M. & Davis, G.J., 1977. Phosphorus absorbtion, translocation and secretion in Nuphar lutea. Limnol. Oceanogr. 22:1022–1032.
VanRaalte, C.D., Valiela, I. & Teal, J.M., 1976. Production of epibenthic salt marsh algae: light and nutrient limitation. Limnol. Oceanogr. 21:862–872.
Wetzel, R.G., 1975. Limnology. W.B. Saunders Co. Philadelphia. 743 pp.
Wetzel, R.G., 1964. A comparative study of the primary productivity of higher aquatic plants, periphyton, and phytoplankton in a large, shallow lake. Int. Rev. ges. Hydrobiol. 49:1–61.
Woelkerling, W.J., 1976. Wisconsin Desmids. I. Aufwuchs and plankton communities of selected acid bogs, alkaline bogs, and closed bogs. Hydrobiologia 48:209–232.

Author's address:
G. Robinson
Department of Botany
University of Manitoba
Winnipeg, Canada R3T 2N2

CHAPTER 31

A light and dark trough methodology for measuring rates of lotic periphyton settlement and net growth

An evaluation through intersite comparison

MAX L. BOTHWELL and STEVE JASPER

Abstract. Continuous-flow light and dark troughs were used to measure and compare passive settlement and net growth rates of periphyton communities in the North Thompson, South Thompson and lower Thompson rivers, British Columbia. At all sites chlorophyll *a* concentration in the dark increased linearly for up to 2–3 weeks, while light-trough values increased exponentially. Use of dark-trough data to correct light-trough accumulation for passive settlement resulted in net growth rate estimates which were as low as one-third of unadjusted estimates. This correction was most important when algal growth rates were low relative to settlement. Algae in the North Thompson River always had higher substratum affinity coeffients and more rapid shifts in the chlorophyll acid ratio in the light than communities in the South Thompson and lower Thompson rivers. We hypothesize that these differences are due to the greater tychoplankton:phytoplankton ratio in the North Thompson. This river has little upstream lake influence compared to the South and lower Thompson rivers, both of which drain from large lakes.

Introduction

The biomass accrual procedure for estimating the primary productivity of periphyton communities on *in situ* artificial substrata has been justifiably criticized (Wetzel 1964, 1965). The high magnitude of both predatory and non-predatory losses of autochthonously fixed carbon under uncontrolled *in situ* conditions severely limits the interpretation of biomass accumulation data in terms of specific cellular growth rate, the most crucial parameter in studies of eutrophication of lotic systems. However, the use of rigorous experimental control with artificial troughs greatly improves the reproducibility and reliability of artificial substrata biomass accrual data. With critical experimental control, the procedure can be used to identify nutrient factors limiting primary production in lotic environments (Stockner & Shortreed 1978; Dickman 1973) and to document changes in the rate of exponential increase of algal biomass with increments in the concentration of limiting nutrient (Bothwell & Stockner 1980). While algal cell division rates calculated from increases in chlorophyll in short-term experiments may still underestimate actual division rates, the fact that rapid, exponential increases are observed suggests that losses are minimal.

Another uncertainty about biomass accrual assessments of primary productivity, which is unabated even with close experimental control, is the recognition that accumulated biomass results from both passive settlement and active growth (Newcombe 1949, 1950; Sládeček & Sládečková 1964). The problem of separating these two simultaneous processes was acknowledged by early workers. Kevern & Ball (1965) suggested that the time course of accumulation be divided into two phases, a 10–15 day colonization-lag phase, during which no

Wetzel, R.G. (ed.), Periphyton of Freshwater Ecosystems.
 ISBN 90 6193 768 X.

growth occurred, followed by an arithmetic growth phase during which no colonization occurred. King & Ball (1966) attempted to indirectly calculate the rates of organic accumulation due to settlement and growth. Both of these approaches suffered from untested and questionable assumptions. The early suggestion by Newcombe (1950) of mathematically separating these components by using the finance principle of compound interest at a constant deposit rate was not pursued because it re-

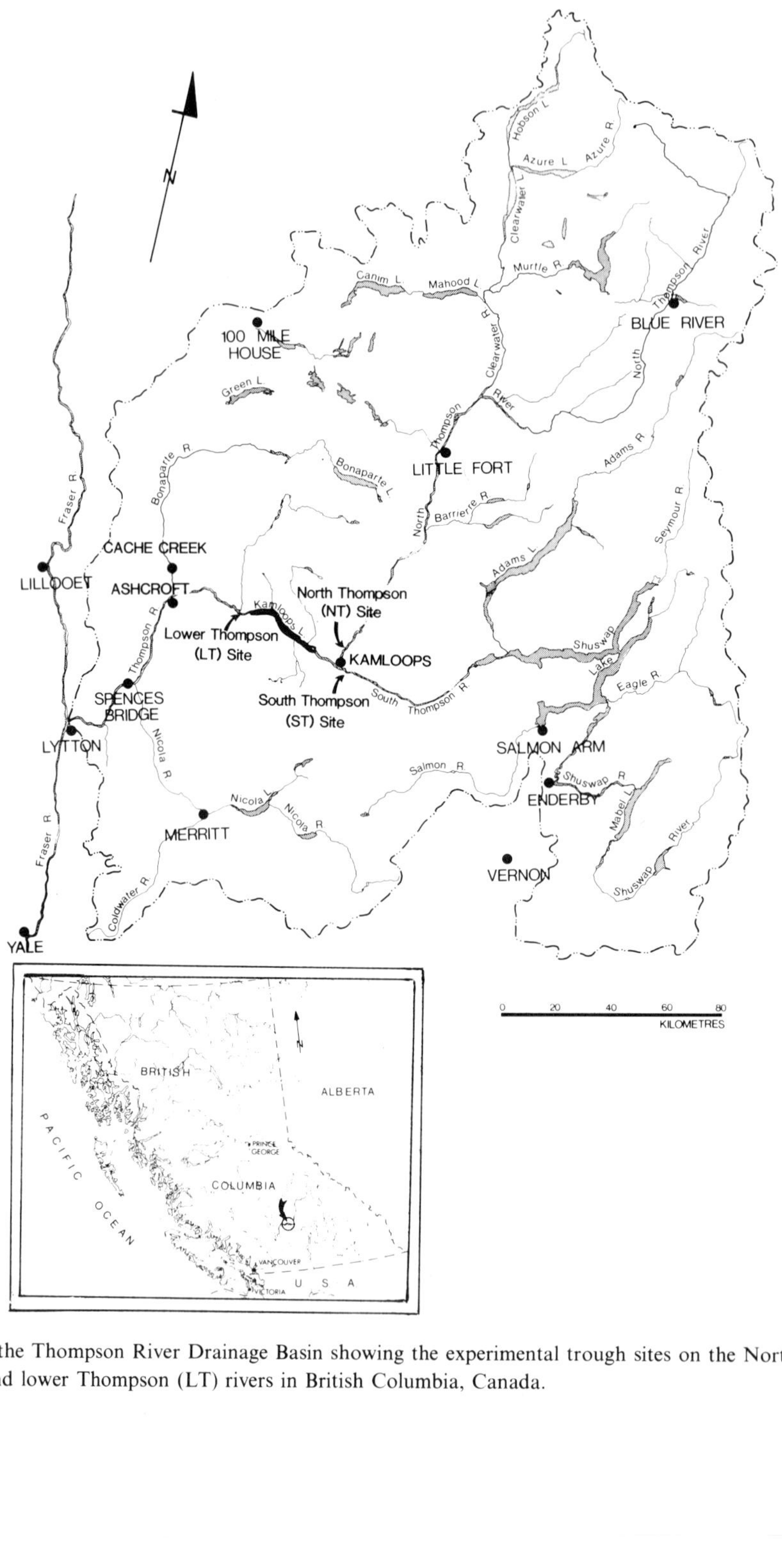

Fig. 1. Location map of the Thompson River Drainage Basin showing the experimental trough sites on the North Thompson (NT), South Thompson (ST) and lower Thompson (LT) rivers in British Columbia, Canada.

quired both a knowledge of the deposit rate (settlement) and the demonstration that it is fairly constant during the period of growth. Hence, the ability to accurately correct algal biomass accrual rates for passive settlement has been hampered by the lack of a procedure which enables direct and reproducible measurements of the rate of algal deposition on substrata in the absence of phototrophic growth.

This paper describes a series of periphyton biomass accrual experiments conducted during 1980–1982 on three rivers in the Thompson River Basin, British Columbia. Chlorophyll *a* accumulation rates in dark troughs served as a direct measure of the rate of passive algal settlement on substrata and was used to correct accrual rates in identical parallel light troughs. Specific growth rates of algal periphyton determined in this manner were compared to division rates computed without this correction. The simultaneous conduct of these experiments on three different rivers in the basin allowed intersite comparisons of the effectiveness of dark-troughs in determining and correcting for rates of algal settlement.

Site description

The Thompson River Basin is located in south central British Columbia and drains approximately 55 000 km^2 of the province's Interior Plateau (Fig. 1). The two major tributary systems to the Thompson River are the North and South Thompson rivers with mean annual flows of about 430 $m^3 \cdot s^{-1}$ and 280 $m^3 \cdot s^{-1}$, respectively. The North Thompson River, originating in the Cariboo Mountains, flows for 300 km to a confluence with the South Thompson near the City of Kamloops. With the exception of flow contributed by the Clearwater River system, much of the water in the North Thompson is unmodified by lake impoundment. On the other hand, nearly all of the water carried in the South Thompson River passes through at least one large lake. Arising at the outlet of Shuswap Lake, the South Thompson flows westward 64 km before merging with the North Thompson to form the Thompson River. The main river continues its westward flow through Kamloops Lake. The Thompson River emerging from the western end of Kamloops Lake is referred to in this study as the lower Thompson River.

Methods

Experimental troughs

Experimental continuous-flow trough facilities were located on the banks of the North Thompson (NT) and South Thompson (ST) rivers, near their confluence, and on the lower Thompson (LT) River (Fig. 1). The details of construction and operation of these trough facilities are given elsewhere (Bothwell 1983). Briefly, each facility contained two identical parallel Plexiglas troughs. One was enclosed in complete darkness by a black plywood cover. The other trough was exposed to ambient sunlight, filtered of UV-light by plate glass. Water was continuously pumped directly from midstream at each site to a rapidly-flushing constant headtank (<60 sec hydraulic residence time). A steady flow to each trough of 47 $L \cdot min^{-1}$ provided a uniform water velocity of 50 $cm \cdot s^{-1}$. Styrofoam lining the bottom of the troughs served as a colonization and growth substratum for natural periphyton from each river.

Chlorophyll accumulation studies

Four chlorophyll accumulation experiments under light and dark conditions were conducted during 1980–1981 on the North Thompson, South Thompson and lower Thompson rivers. These experiments were done under low flow, pre-freshet conditions in late winter – early spring (February-April). Each trial began by placing fresh styrofoam sheets in the troughs and commencing the flow of water. For chlorophyll analyses, quadruplicate cores (4.75 cm^2) of the styrofoam were removed with a corkborer every day for the first 4–5 days and thereafter at 2 or 3 day intervals for a total period of 17–24 days. The cores were frozen at −20° C in the dark until analysis, usually within 1–2 weeks. Algal pigment was extracted in 90% acetone by homogenizing each complete core with a Polytron grinder. Determination of chlorophyll *a*,

Table 1. Mean chlorophyll *a* concentrations in river water [$mg \cdot m^{-3} \pm$ coeff. var. (n)] in grab samples taken between 1000–1400 h from the trough headtank system at each site during three experiments.

Date	River		
	North Thompson	South Thompson	Lower Thompson
April, 1981	0.55 ± 15% (8)	1.01 ± 12% (8)	0.33 ± 12% (8)
February, 1982	0.24 ± 17% (10)	0.53 ± 13% (10)	0.32 ± 25% (10)
March, 1982	0.23 ± 26% (10)	0.71 ± 24% (10)	0.23 ± 17% (10)

corrected for *a* phaeopigments, was made fluorometrically before and after acidification using the method of Holm-Hansen *et al* (1965). A Turner Designs fluorometer was equipped with a far-red sensitive photomultiplier (R 446), a Corning CS-5-60 excitation filter and the combination emission filter system, Corning CS-2-64/Wratten-70, suggested by Loftus & Carpenter (1971). The change in fluorescence on acidification (acid ratio) was used to express the relative amount of active chlorophyll *a* and *a* phaeopigments in samples. During calibration, an acid ratio of 2.10 was found for pure chlorophyll *a*.

Periphyton samples, removed from the troughs during each experiment, were preserved with Lugol's solution for algal taxonomic examination.

Two additional dark-trough chlorophyll accumulation experiments were run in February and March of 1982. During these trials and during the April 1981 light and dark-trough experiment, chlorophyll concentrations of the river water were measured at each site. One-liter water samples collected from the headtanks on each sampling day were filtered through Whatman GF/F filters, homogenized in 90% acetone and analyzed for chlorophyll *a* using the same procedure described above for the styrofoam cores.

During February and March of 1982 the possibility of diurnal fluctuations in the chlorophyll concentration of the North and South Thompson rivers was examined. A Manning Autosampler was programmed to collect 450-ml samples from the headtanks at each site at hourly intervals for 24 h. The following day these samples were filtered in the lab, extracted in 90% acetone and analyzed for chlorophyll *a* as above. Manual grab samples taken simultaneously with Autosampler collected water, but filtered and analyzed immediately, were used to check the results obtained with the Autosampler.

Growth rate calculations

Net growth rates of the algal periphyton communities were calculated from the chlorophyll *a* accumulation data using two different methods. The first approach ignored information provided by the dark-trough. Chlorophyll *a* data from the light-troughs were fitted by least-squares regression to the simple exponential growth equation:

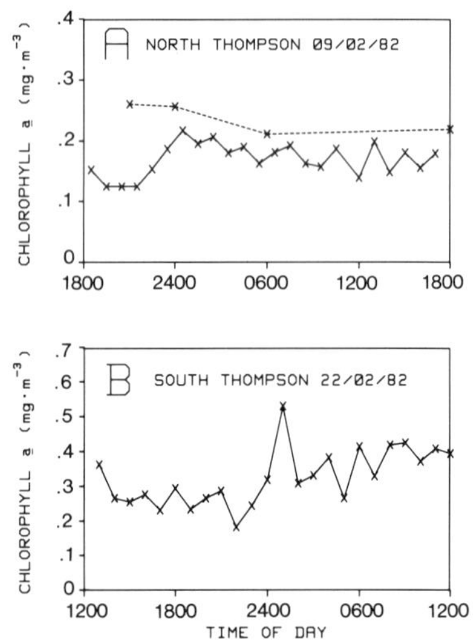

Fig. 2. Chlorophyll *a* concentrations in river water ($mg \cdot m^{-3}$) during a 24-h interval from the North Thompson and South Thompson rivers. Samples were taken from the headtank system by an Autosampler. A: North Thompson experiment from 1800 hours, 9 February 1982 to 1800 hours, 10 February 1982. × - - - ×, manual grab samples. B: South Thompson experiment from 1200 hours, 22 February 1982 to 1200 hours, 23 February 1982.

$$y = a\, e^{kt} \tag{1}$$

where y is the chlorophyll *a* concentration (mg · m^{-2}) at day, t; 'a' is the initial chlorophyll *a* concentration; and k is the specific net growth rate (ln · d^{-1}). Values of k, divided by 0.693, result in growth expressed as divisions · d^{-1}. In an attempt to limit the influence of rapid initial settlement on the calculation, chlorophyll data from the first 3–5 days of each experiment were omitted.

The second approach, originally proposed by Newcombe (1950), was the use of an equation in which increasing biomass through time is due to both active growth of existing periphyton and passive settlement of new algal cells. Two assumptions are explicit in the application of this principle of compound fixed interest at a constant deposit rate. First, the rate of passive settlement of algal biomass on substrates is constant throughout an experiment. Second, no growth occurs in the settling populations on the day of settlement. With these assumptions the algal periphyton biomass, y (mg Chl*a* · m^{-2}), after t days of growth is given by:

$$y = s \sum_{j=0}^{t-1} (1+k)^j \tag{2}$$

where s is the daily rate of algal settlement (mg Chl*a* · m^{-2} · d^{-1}), and k is the daily fractional growth of the adhering population. The rate of chlorophyll accumulation in the dark troughs provides a direct measure of s. Hence:

$$y/s = \sum_{j=0}^{t-1} K^j \tag{3}$$

where K = (1 + k). This polynomial:

$$y/s = 1 + K + K^2 + \ldots + K^{t-1} \tag{4}$$

can be numerically solved for K by iteration. Values for k, thus calculated, directly translate into divisions · d^{-1}. To ascertain the effectiveness of dark-trough controls in correcting light-trough data for passive settlement, calculations of k starting on Day 1 were compared to values computed after allowing for a few days of rapid colonization.

Results

River chlorophyll concentrations

During the course of accumulation trials in April, 1981 and February and March, 1982 the chlorophyll *a* content of the rivers remained reasonably stable (Table 1). Coefficients of variation were between 12–26%, provided that the experiments were of short duration and did not coincide with periods of rapid fluctuation in river discharge.

Chlorophyll *a* concentrations were always highest in the South Thompson River, usually 2–3 times greater than samples taken from the lower or North Thompson rivers at the same time (Table 1).

A check for diurnal variation in chlorophyll concentration of the rivers was made to see if mid-day grab samples were biased. While some variation was found there was no evidence of a diurnal trend at either the NT or ST site during February 1982 (Fig. 2). Repeated trials in March produced similar results. Occasional manual grab samples taken to

Table 2. Dark-trough chlorophyll *a* accumulation rates (mg · m^{-2} · d^{-1} ± S.E.) calculated by least-squares linear regression of the initial phase of the arithmetic increase.

Date	River		
	North Thompson	South Thompson	Lower Thompson
February, 1980	0.18 ± .018	0.082 ± .006	0.031 ± .002
March, 1980	0.21 ± .009	0.11 ± .004	0.075 ± .005
April, 1980	0.19 ± .004	0.045 ± .002	0.048 ± .006
April, 1981	0.32 ± .027	0.053 ± .004	0.070 ± .021
February, 1982	0.13 ± .005	0.18 ± .006	0.059 ± .002
March, 1982	0.11 ± .005	0.08 ± .006	0.071 ± .007

verify the operation of the Autosampler, usually produced higher chlorophyll recoveries with less variability, suggesting pigment loss in samples stored for more than a few hours. Continuous-flow fluorometry for *in vivo* chlorophyll concentration at the LT site during February and March, 1981 also failed to detect any major diel change (J. Vande Castle, unpublished). Hence estimates of river chlorophyll *a* from headtank grab samples taken at mid-day appear to be unbiased.

Dark-trough accumulation

Under conditions of complete darkness, chlorophyll *a* was found to increase at a linear rate at all three sites for a period of up to 2–3 weeks (Fig. 3). In some instances there was evidence of a plateau in chlorophyll values reached after the initial uniform rate of increase had ceased. Dark accumulation rates of chlorophyll *a* were determined from the slope of a least-squares linear regression using data from the period of uniform increase. With the exception of February, 1982, the North Thompson (NT) always had the highest dark slope (Table 2). Often, in 1980 and 1981, the dark rate at NT was 5–6 times greater than values observed at ST or LT. Not only were the dark slopes at NT usually higher, they were also more consistent. In the three experiments in 1980 no significant ($P<0.05$) difference in NT slopes existed between the trials. On the other hand, dark chlorophyll *a* accumulation rates at ST and LT were usually much lower and highly variable. Significant differences ($P<0.05$) between experiments run in successive months were found at both these sites.

Because of the identical physical and hydrodynamic characteristics of the troughs at each site, the rate of accumulation of chlorophyll *a* in darkness was largely determined by the concentration of colonizing organisms in the river water and the degree of affinity of those organisms for substrata attachment. To assess the relative substratum affinity of the algae suspended in the rivers at each site, the dark accumulation rates (mg Chl*a* $\cdot$ $m^{-2} \cdot d^{-1}$) were divided by the mean chlorophyll concentration (mg Chl*a* $\cdot$ m^{-3}) in the rivers during that increase. This yields a coefficient ($m \cdot d^{-1}$) which we interpret as a relative measure of the substratum affinity of the algal cells suspended in the river. It is recognized, and emphasized here, that these numbers have no significance in absolute terms. However, the rigorous physical control afforded by the experimental troughs makes these coefficients an excellent means of quantitatively comparing the substratum affinity of the algal com-

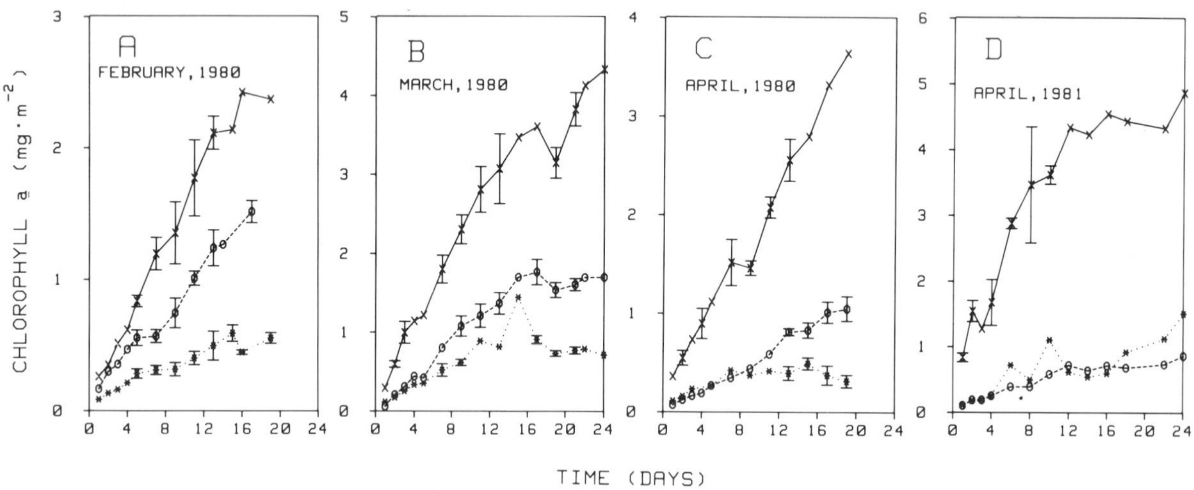

Fig. 3. Chlorophyll *a* ($mg \cdot m^{-2}$) concentration in the dark-troughs during four accumulation experiments in 1980 and 1981. North Thompson ×—×—×; South Thompson O- O- O; lower Thompson *···*···*. A: Experiment 1, 6–25 February 1980; B: Experiment 2, 3–27 March 1980; C: Experiment 3, 29 March 1980–16 April 1980; D: Experiment 4, 31 March 1981–23 April 1981. Values shown are the means of quadruplicate analyses with 95% error bars. Some bars omitted for clarity.

Table 3. Relative substratum affinity coefficients (m · d^{-1}) for river chlorophyll *a* during April 1981, February 1982 and March 1982 in the North Thompson, South Thompson and lower Thompson experimental troughs. Values were calculated by dividing the dark-trough chlorophyll *a* accumulation rate by the mean chlorophyll *a* concentration in the headtank system at each site.

Date	River		
	North Thompson	South Thompson	Lower Thompson
April, 1981	0.58	0.053	0.21
February, 1982	0.54	0.34	0.18
March, 1982	0.48	0.11	0.31

munities in the three rivers investigated here. The general patterns shown by these relative substratum affinity coefficients were similar to those seen in the dark-trough slopes. The highest affinity values always occurred at NT (Table 3). Although in February, 1982, dark accumulation rate was highest at ST, normalization for river chlorophyll *a* resulted in a higher substratum affinity at NT. Affinities at NT were also very constant, changing by less than 20% in experiments replicated in different years. Algal affinity in the South Thompson and lower Thompson rivers was always lower (Table 3). In April, 1981 substratum affinity at ST was an order of magnitude less than at NT. Substratum affinities at ST and LT also varied considerably between experiments.

The higher substratum affinities of algae in the North Thompson River were also indicated by the much higher chlorophyll *a* values reached in the dark, compared to plateau values observed at ST and LT (Fig. 3).

Light-trough accumulation

The influence of differences in the rate of passive settlement and the rate of cell division is apparent in semi-log plots of light trough chlorophyll concentration vs. time. This type of plot, often used to graphically estimate cell division rate in batch algal cultures, will show downward curvature near the origin if passive settlement is occurring. The degree of curvature is determined by the rate of cell division. This point is illustrated in Fig. 4. Assuming a constant growth rate of 0.10 divisions · d^{-1}, different rates of settlement serve only to shift the position of the curves on the ordinant scale (Fig. 4A) i.e., the size of the inoculum is changed. However, assuming the same settlement rate (0.10 mg Chl*a* · m^{-2} · d^{-1}), higher growth rates result in less curvature (Fig. 4B). Both of these effects are seen in light-trough data from NT, ST and LT (Fig. 5). The higher dark settlement rates at NT are reflected in the much higher ordinate scale position of the North Thompson curves. The lower Thompson plots show the least curvature and in three of the four experiments had the highest calculated growth rates.

One outcome of settlement resulting in this curvature is that at low growth rates, chlorophyll *a* accumulation rates on light exposed substrata may require a week or longer to approach the actual growth rate constant. Because of this, growth rates computed by a least-squares fit to an exponential model, even after allowing 3–5 days for settlement, were higher than those estimated by equation (2)

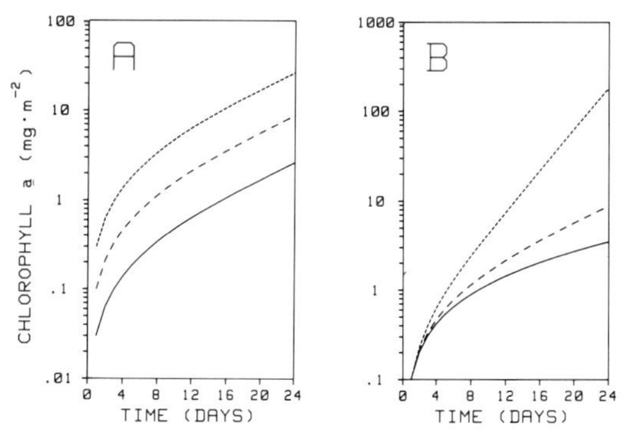

Fig. 4. Calculated theoretical accumulation curves for chlorophyll. A: Growth rate constant at 0.1 divisions · d^{-1} for various settlement rates – 0.3 (····); 0.1 (- - -); 0.03 (——) mg · m^{-2} · d^{-1}; B: Settlement rate constant at 0.1 mg · m^{-2} · d^{-1} for various growth rates – 0.3 (····); 0.1 (- - -); 0.03 (——) divisions · d^{-1}.

(Table 4). The long and variable time interval needed for the effects of settlement to become insignificant makes the possibility of dark-trough correction attractive.

If the chlorophyll accumulation rate in the dark corrects for passive algal settlement in the light, then growth rates computed using equation (2) should not be affected by the time interval chosen for the calculation. This assumption was found to be true in the South Thompson and lower Thompson rivers. In seven out of eight comparisons at ST and LT the standard errors of growth constants calculated from Day 1 overlapped those computed after allowing for 3–5 days of settlement (Table 4). In these two rivers the dark-trough appears to adequately correct for non-growth chlorophyll accrual.

However, this was not the case in the North Thompson River. In three of the four experiments at NT, higher growth constants were obtained when calculations were started from Day 1, than after allowing for a period of rapid initial settlement (Table 4). At this site, initial chlorophyll increases in the light were usually more rapid than could be explained by the dark-trough. A comparison of chlorophyll *a* levels in the light and dark-troughs during the first two days of each experiment confirms this conclusion. With the exception of the March, 1980 experiment, highly significant differences between light and dark appeared within 24 h (Table 5). In one case, the chlorophyll *a* level in the NT light trough was over 400% higher than in the dark-trough after one day. This never occurred at ST and was observed only once at LT (April 1981).

The rapid increase in periphyton chlorophyll in the light-trough at NT was paralleled by the rapidity with which the light and dark-troughs at NT differed in their chlorophyll acid ratio. North Thompson River periphyton communities only needed 1–2 days for this ratio, expressed as percent chlorophyll *a*, to be consistently different in the light and dark (Fig. 6). In contrast, the attached algal communities at LT and ST required 3–5 days and 5–10 days, respectively, for divergence of pigment acid ratios in the light and dark. An example

Table 4. Periphyton growth constants (divisions · d^{-1} ± S.E.) using two growth models and varying experimental intervals at the North Thompson, South Thompson and Lower Thompson sites during February 1980, March 1980, April 1980 and April 1981.

Experiment	Model		
	Equation (1)	Equation (2)	
	interval[a]	interval[a]	interval[b]
North Thompson			
February 1980	0.26 ± .01	0.17 ± .02	0.22 ± .02
March 1980	0.25 ± .02	0.17 ± .01	0.17 ± .01
April 1980	0.17 ± .02	0.11 ± .01	0.25 ± .01
April 1981	0.04 ± .001	0.006 ± .002	0.06 ± .02
South Thompson			
February 1980	0.14 ± .02	0.05 ± .02	0.06 ± .02
March 1980	0.19 ± .001	0.10 ± .01	0.08 ± .01
April 1980	0.30 ± .01	0.19 ± .02	0.17 ± .02
April 1981	0.17 ± .02	0.13 ± .01	0.17 ± .01
Lower Thompson			
February 1980	0.29 ± .01	0.21 ± .02	0.24 ± .02
March 1980	0.33 ± .01	0.22 ± .01	0.22 ± .01
April 1980	0.30 ± .02	0.18 ± .02	0.19 ± .02
April 1981	0.31 ± .03	0.29 ± .03	0.26 ± .02

[a] Data for the initial 3–5 day period omitted in the computation.
[b] Data from Day 1 and the date of experiment termination used in computation.

Table 5. Chlorophyll *a* concentration (mg·m^{-2} ± 95% confidence limits) in the light and dark troughs on the North Thompson River, during the initial two days of experiments in February 1980, March 1980, April 1980 and April 1981.

Experiment	Day	Light	Dark
February 1980	1	0.47 ± .031	0.21 ± .034
	2	0.94 ± .10	0.35 ± .070
March, 1980	1	0.30 ± .038	0.29 ± .066
	2	0.76 ± .14	0.60 ± .070
April, 1980	1	1.9 ± .34	0.36 ± .079
	2	4.6 ± .80	0.55 ± .130
April, 1981	1	1.3 ± .05	0.84 ± .048
	2	3.3 ± .10	1.5 ± .12

of this is shown in Fig. 6. In most cases the acid ratio remained fairly stable in the dark during each experiment. Consequently, divergence between light and dark was due to increases in the percent chlorophyll *a* in the light, as would be expected from active growth.

Although the measurements of chlorophyll *a* and the chlorophyll acid ratio are obviously related, the rapid increases in chlorophyll at NT were much too large to have been caused by the observed shifts in the acid ratio *per se*. Also, on one occasion, a large difference in chlorophyll concentration between light and dark at NT preceeded any difference in the acid ratios. The opposite was sometimes true at ST.

Discussion

The use of chlorophyll *a* accumulation in dark-troughs to correct for algal settlement in the light involves three basic assumptions. The first premise is that increases in chlorophyll in the dark are due solely to settlement. While most algae are able to synthesize chlorophyll *a* in darkness (Meeks 1974) and facultative heterotrophic growth is well known in pennate diatoms and particularly prevalent in benthic forms (Droop 1974), it is generally believed that in most natural, open-water environments growth rates supported in this manner are slow compared to photoautotrophy (Hellebust & Lewin 1977). This, coupled with the evidence of linear increases in pigment, which in most cases could be back-extrapolated to near zero on Day O, strongly suggest that chlorophyll *a* accrual in the dark was largely the result of passive settlement.

It should be noted incidently that the dark-trough periphyton communities were exposed to light for 10–15 min each sampling day. Although this brief exposure is inadequate to support photoautotrophic cell division, it may have increased the stability of photosynthetic pigments in cells settled in the dark. Such brief exposures to light are known to increase algal survival in prolonged darkness (Smayda & Mitchell-Innes 1974; Umebayashi 1972).

The second assumption implicit in the use of dark-trough controls is that any chlorophyll increases unrelated to growth are the same in the light as in the dark. This assumption is questionable for two reasons. First, high variability in cellular chlorophyll content has been thoroughly documented for algal cultures in response both to changes in nutrient status and to degree of illumination (Strickland 1960; Meeks 1974). Hence, while biomass settlement might be the same in light and dark, alterations in the Chl*a*:living carbon ratio would complicate the interpretation of biomass settlement based on pigment increases. Secondly, settlement rates in the light might not be the same as in the dark. The possibility of some light-mediated mechanism influencing substratum settlement, even though generally thought to be a random process, cannot be dismissed.

In spite of these reservations, evidence from the ST and LT sites indicated that chlorophyll *a* accumulation rates were similar in light and dark. With one exception, no significant difference was found between chlorophyll *a* levels in the light and dark after two days in experiments on these rivers.

On the other hand, the extremely rapid increase in chlorophyll in the light at the North Thompson site indicated that at least one and perhaps both of the aforementioned processes were occurring in our experiments. Chlorophyll *a* levels in the light, 100–400% greater than in the dark after just 24 h at the NT site, would be most simply explained by an increase in the cellular content of this pigment in response to high light exposure in the trough. However, this explanation seems unlikely. Although the cellular chlorophyll content of diatoms, dominant in all experiments reported here, is known to turnover rapidly (Riper *et al* 1979) and change within a generation time in response to altered light intensity (Hitchcock 1977; Marra 1980), photosynthetic pigment cell quotas normally vary inversely with the degree of light exposure (Perry *et al* 1981; Beardall & Morris 1976; Falkowski 1980), i.e. initially, chlorophyll values would be expected to be higher in the dark than in the light, not lower. Furthermore, the rates of intracellular pigment adjustment to new light levels is temperature dependent (Hitchcock 1980). The rapid chlorophyll shifts reported here have been observed at temperatures of less than 1° C. Another explanation appears necessary. A detailed, 48 h light-dark accumulation trial at NT showed rapid ATP increases of 100% accompanying chlorophyll increases of 300% during the first 9 h light period (Bothwell & Jasper, unpublished). Although such rapid ATP fluctuations must be interpreted with caution (Bottomley & Stewart 1976), their non-transient nature suggests that dramatic algal biomass increases occur in the light as well as pigment increases. While further study will be needed to clarify the mechanisms of these light-induced changes, it is clear that dark-trough chlorophyll accumulation at the NT site does not account for all non-growth pigment increases in the light.

The third assumption in the application of dark-trough controls is that settlement not only be the same in the light and dark, but that it also be constant during the experimental period. While the data from these trials do not permit a direct test of the validity of this assumption in the light, the dark-trough results certainly confirm the constancy of settlement in the absence of light.

This analysis shows that at the sites investigated here on the lower and South Thompson rivers all three of the assumptions above were reasonably

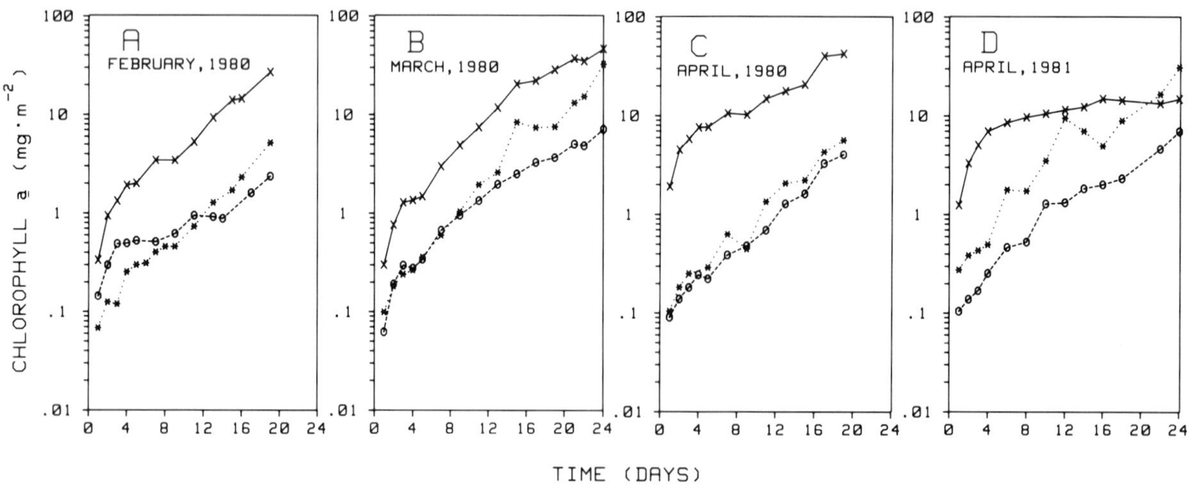

Fig. 5. Chlorophyll *a* (mg · m^{-2}) concentration in the light-troughs during four growth experiments in 1980 and 1981. North Thompson ×—×—×; South Thompson O- O- O; lower Thompson *···*···*. A: Experiment 1, 6–25 February 1980; B: Experiment 2, 3–27 March 1980; C: Experiment 3, 29 March 1980–16 April 1980; D: Experiment 4, 31 March 1981–23 April 1981. Values shown are the means of quadruplicate analyses.

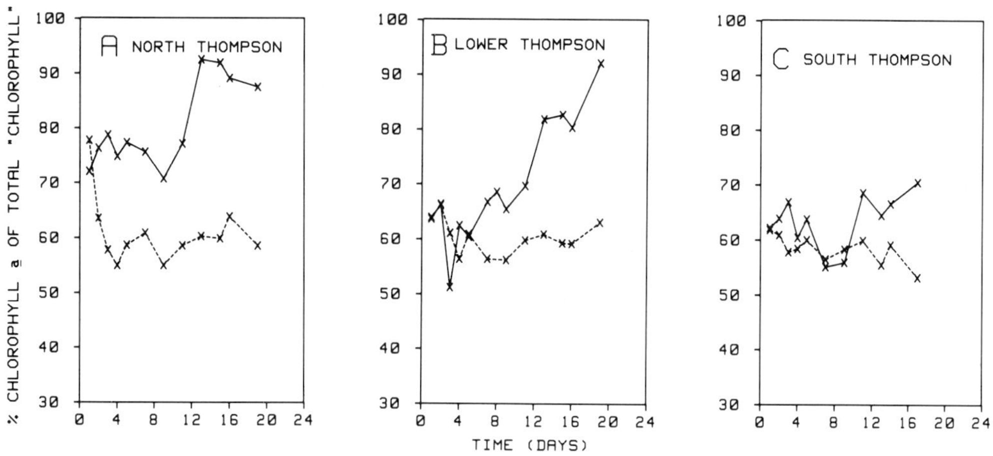

Fig. 6. Chlorophyll *a* as a percent of total 'chlorophyll' (chlorophyll *a* and *a* phaeopigments) in light and dark-troughs during the February experiment in 1980. The values were determined from the acid ratios of the 90% acetone pigment extracts i.e. a ratio of 2.1 = 100% Chl *a* and a ratio of 1.0 = 0% Chl *a*. Solid line, light-troughs; dashed line, dark-troughs.

true. Hence, dark-trough controls provided an accurate correction for passive algal settlement in the light. In such cases, the use of dark-troughs has two major advantages. First, when algal growth rates are slow relative to passive settlement, light exposed substrata alone can potentially overestimate actual net production. For example, in the South Thompson River during February, 1980, when water temperatures were near 0°C and periphyton were showing physiological signs of extreme phosphorus deficiency (Bothwell & Daley 1981) growth rate corrected for settlement was near zero, 0.05 divisions · d^{-1}. Without this correction, estimates of cell division rate were almost three times higher, 0.14 divisions · d^{-1}. Second, by correcting for settlement, dark troughs theoretically permit accumulation experiments to be completed in a very short period of time; less than one week is possible, depending on growth rate. This period is advantageous because it minimizes error resulting from algal losses due to sloughing.

In instances like that encountered on the North Thompson River where the assumption of equal settlement in light and dark is invalid, the correction for dark settlement is conservative. Since non-growth chlorophyll increases in the light are higher than in the dark, growth rates computed from the first day, even with the correction, are still likely to be overestimates. In such cases growth rate calculations made after allowing for a few days of settlement will usually be lower and probably more accurate.

The anomalous behaviour of chlorophyll accumulation in the North Thompson compared to the South and lower Thompson rivers deserves further comment. Not only did the suspended algae in the former river have a higher substratum affinity, they also responded more rapidly to the favorable substrata and light conditions provided by the troughs. This response was reflected both in the rapid divergence of chlorophyll acid ratios in the light and dark and in the sharp increase in chlorophyll *a* in the light within 24–48 h. The only difference between the North Thompson and the other two rivers which might be responsible for all these observations is the degree of upstream lake influence. On an annual basis at least 40% of the water in the North Thompson River at the point of confluence with the South Thompson has never been in a lake while much of the remaining 60% comes from oligotrophic lakes in the Clearwater River drainage and flows 150–200 km before reaching Kamloops. Long stretches of the North Thompson River bed are composed of stable rock substrata and well suited for attached algal growth. Thus, while no direct cell counts were made from the river, it seems reasonable to expect that the majority of live algal cells in suspension in this river are actually periphytic in origin i.e., tychoplankton. The situation at the LT and ST sites is quite different in this regard. Although the lower Thompson River supports very high benthic algal biomass, the siting of the trough facility only 500 m from the outlet of Kamploops Lake assured that lake phytoplankton were well represented in water supplied to the troughs. Similarily, nearly all of the water in the South Thompson is derived from lakes. Though the river travels 64 km from the outlet of Shuswap Lake to the experimental trough site, the intervening reach of river bed is composed entirely of silts and sands which are constantly reworked and thus unsuitable for attached algal accumulation. Therefore, the percent of algal chlorophyll due to phytoplankton is likely much higher at the LT and ST locations. Since some phytoplankton species would be unsuitable for substratum attachment and growth, it follows that the communities in these two rivers would have lower substratum affinity relative to those in the North Thompson River. This difference was corroborated by taxonomic analysis of the diatom communities in the troughs. The most abundant taxa at NT were *Hannea arcus, Diatoma tenue, Achnanthes minutissima* and *Synedra ulna,* all frequently regarded as periphytic. Although the trough communities at LT and ST eventually contained many periphytic forms of *Fragilaria, Achnanthes* and *Diatoma,* there was evidence that some of the early colonizing diatoms at these sites were planktonic forms not seen at NT. We thus hypothesize that both the high substratum affinity and rapid pigment response seen at the NT location are related to the high tychoplankton:phytoplankton ratio of the North Thompson River. Whether these responses represent some strategy characteristic of certain lotic tychoplankton deserves further evaluation.

The large difference in algal settlement rates observed in the three rivers further highlights the futility in attempting to estimate average daily primary productivity of periphyton from single time exposures of artificial substrata. Uncontrolled biomass losses are normally cited as the major source of error. However, the results here document that varying settlement rates can also lead to erroneous conclusions. After one week chlorophyll levels at NT were often an order of magnitude or more greater than at LT or ST and yet growth rates at NT were never as high as at LT.

Acknowledgements

Mr. R. Wiegand provided the mathematical expertise and computer program for calculating growth rates at constant settlement. The effort of R. Mitchell who conducted the majority of the field sampling is gratefully acknowledged. K. Suzuki and J. Vande Castle performed many of the laboratory analyses and all of the data manipulation. Algal enumeration was done by M. Bolin. Dr. R.J. Daley provided comments on the first draft of this manuscript which improved the clarity of presentation. This project was supported in part by Weyerhaeuser Canada Ltd.

References

Beardall, J. & Morris, I., 1976. The concept of light intensity adaptation in marine phytoplankton: Some experiments with Phaeodactylum tricornutum. Mar. Biol. 37:377–387.

Bothwell, M.L., 1983. All-weather troughs for periphyton studies. Water Res. In press.

Bothwell, M.L. & Daley, R.J., 1981. Determination of phosphorus sufficiency and growth rates of benthic algae in the Thompson River B.C. using experimental flowing troughs. Regional IWD Management Report, Vancouver, B.C., 32 pp.

Bothwell, M.L. & Stockner, J.G., 1980. Influence of secondarily treated kraft mill effluent on the accumulation rate of attached algae in experimental continuous-flow troughs. Can. J. Fish. Aquat. Sci. 37:248–254.

Bottomley, P.J. & Stewart, W.D.P., 1976. The measurement and significance of ATP pools in filamentous blue-green algae. Br. Phycol. J. 11:61–62.

Dickman, M., 1973. Changes in periphytic algae following bicarbonate additions to a small stream. J. Fish. Res. Bd. Can. 30:1882–1884.

Droop, M.R., 1974. Heterotrophy of carbon. In: Stewart, W.D.P. (ed.) Algal Physiology and Biochemistry. Univ. Calif. Press, Blackwell Sci. Publications pp. 530–560.

Falkowski, P.G., 1980. Light-shade adapatation in marine phytoplankton. In: Falkowski, P.G. (ed.) Primary Productivity in the Sea. Brookhaven Symposia in biology. Plenum Press pp. 99–119.

Hellebust, J.A. & Lewin, J., 1977. Heterotrophic nutrition. In:

Werner, D. (ed.) The Biology of Diatoms, Botanical Monographs, Vol. 13. Univ. Calif. Press, Blackwell Sci. Publications pp. 169–197.

Hitchcock, G.L., 1977. The Time Course of Photosynthetic Adaptation, the Growth Response and Variation in Pigment, Carbohydrate, and Protein of Skeletonema costatum and Detonula confervacea to Changes in Light Intensity. Ph.D. thesis, Univ. of Rhode Island, Narragansett, Rhode Island.

Hitchcock, G.L., 1980. Diel variation in chlorophyll a, carbohydrate and protein content of the marine diatom Skeletonema costatum. Mar. Biol. 57:271–278.

Holm-Hansen, O., Lorenzen, C.J., Holmes, R.W. & Strickland, J.D.H., 1965. Fluorometric determination of chlorophyll. J. Cons. Perm. Int. Explor. Mer. 30:3–15.

Kevern, N.R. & Ball, R.C., 1965. Primary productivity and energy relationships in artificial streams. Limnol. Oceanogr. 10:74–87.

King, D.L. & Ball, R.C., 1966. A qualitative and quantitative measure of aufwuchs production. Trans. Amer. Micros. Soc. 85:232–240.

Loftus, M.E. & Carpenter, J.H., 1971. A fluorometric method for determining chlorophylls a, b, and c. J. Mar. Res. 29: 319–338.

Marra, J., 1980. Time course of light intensity adaptation in a marine diatom. Mar. Biol. Lett. 1:175–183.

Meeks, J.C., 1974. Chlorophylls. In: Stewart, W.D.P. (ed.) Algal Physiology and Biochemistry, Univ. Calf. Press, Blackwell Sci. Publications. pp. 161–176.

Newcombe, C.L., 1949. Attachment materials in relation to water productivity. Trans. Amer. Micros. Soc. 68:355–361.

Newcombe, C.L., 1950. A quantitative study of attachment materials in Sodon Lake, Michigan. Ecology 31.204–215.

Perry, M.J., Talbot, M.C. & Alberte, R.S., 1981. Photoadaptation in marine phytoplankton: response of the photosynthetic unit. Mar. Biol. 62:91–101.

Riper, D.M., Owens, T.G. & Falkowski, P.G., 1979. Chlorophyll turnover in Skeletonema costatum, a marine plankton diatom. Pl. Physiol., 64:49–54.

Sládeček, V. & Sládečková, A., 1964. Determination of periphyton production by means of the glass slide method. Hydrobiologia 23:123–158.

Smayda, T.J. & Mitchell-Innes, B., 1974. Dark survival of autotrophic, planktonic marine diatoms. Mar. Biol. 25:195–202.

Stockner, J.G. & Shortreed, K.R.S., 1978. Enhancement of autotrophic production by nutrient addition in a coastal rainforest stream on Vancouver Island. J. Fish. Res. Board Can. 35:28–34.

Strickland, J.D.H., 1960. Measuring the production of marine phytoplankton. Fish. Res. Board Canada. Bull. No. 122. 172 pp.

Umedayashi, O., 1972. Preservation of some cultured diatoms. Bull. Tokai Reg. Fish. Res. Lab. No. 69:55–62.

Wetzel, R.G., 1964. A comparative study of the primary productivity of higher aquatic plants, periphyton, and phytoplankton in a large, shallow lake. Int. Revue. ges. Hydrobiol. 49:1–61.

Wetzel, R.G., 1965. Techniques and problems of primary productivity measurements in higher aquatic plants and periphyton. Mem. Ist. Ital. Idrobiol. 18 (Suppl):249–267.

Authors' address:
M.L. Bothwell/S. Jasper
National Water Research Institute
Pacific and Yukon Region
Inland Waters Directorate
4160 Marine Drive
West Vancouver, B.C., Canada V7V 1N6

CHAPTER 32

The application of a nuclear track autoradiographic technique to the study of periphyton photosynthesis

G.G.C. ROBINSON and E. PIP

Abstract. The application of nuclear track autoradiography to the determination of inorganic carbon assimilation rates of fifteen periphytic algal species is described. Vertical profiles of carbon assimilation for each species following three periods of substrate colonization are presented, along with the contribution made by each species to total algal productivity. It is demonstrated that cellular carbon assimilation rates are a function of cell surface area and that, as a consequence, small periphytic algae have shorter generation times than large algae.

Introduction

Most current methodology appears to have restricted investigation of the metabolism of periphyton to the whole attached community. It is, however, possible that the application of micro-autoradiography might permit investigation of some aspects of the metabolism of individual periphytic species and combinations of the same. Such techniques have been successful in the detection of metabolically active nannoplankters (Paerl 1977) and have proved effective in distinguishing active from inactive phytoplankton (Paerl 1978). When used carefully, both grain density and nuclear-track techniques are applicable. Knoechel & Kalff (1976a, 1979) maintain that grain density procedures are not without problems, whereas Paerl & Stull (1979) find a high degree of correlation between both techniques, while also listing potential problems of both techniques. Knoechel & Kalff (1976b) have successfully quantified the productivity of individual species of phytoplankton by the use of nuclear-track autoradiography and the contribution of individual species to phytoplankton productivity has been estimated by Maguire & Neill (1971); Watt (1971) and Stull *et al* (1973) by the application of grain-density techniques. Pip & Robinson (1982a, 1982b) have applied nuclear-track autoradiography in the investigation of inorganic and organic carbon assimilation by individual taxa of epiphytic periphyton. The purpose of this study has been to investigate some general aspects of the usefulness of this technique to studies of periphytic algae.

Materials and methods

The study site was a bed of *Typha latifolia* L. situated at the periphery of a shallow (30–40 cm) blind channel in the Delta Marsh at the southern end of Lake Manitoba, Canada. Physical and some chemical parameters of standing waters in the Delta Marsh are described elsewhere (Hooper & Robinson 1976). Cellulose acetate substrata were vertically positioned for colonization by periphyton amongst the *Typha* (Hooper & Robinson 1976) on 10 June 1977 and these substrata were used for

Wetzel, R.G. (ed.), Periphyton of Freshwater Ecosystems.
 ISBN 90 6193 768 X.

Table 1. Surface areas of major periphytic species.

Species	Surface area (μm^2)	Standard error
Oedogonium idioandrosporum (Nordt. & Wittr.) Tiffany	7166.9	284.3
Stigeoclonium nanum Kuetzing	1028.2	32.3
Chlorococcum humincola (Naeg.) Rabenhorst	330.9	14.3
Ankistrodesmus falcatus (Corda) Ralfs	338.8	13.2
Scenedesmus quadricauda (Turp.) de Brébisson	253.8	7.7
Coleochaete soluta (de Bréb.) Pringsheim	452.6	14.5
Navicula radiosa Kuetzing	165.7	8.9
Fragilaria vaucheriae Kuetzing	446.6	16.9
Diatoma tenue var. *elongatum* Lyngb.	1260.1	34.5
Synedra acus Kuetzing	730.5	19.1
Nitzschia filiformis (W. Smith) Hust.	479.4	23.9[a]
	191.3	12.1[b]
Nitzschia acicularis A.Cl.	447.1	26.4
Anabaena circinalis Rabenhorst	171.0	3.8
Oscillatoria limosa (Roth.) C.A. Agardh	87.3	1.4
Oscillatoria limnetica Lemmermann	50.6	1.9

[a] Colonization period 'A'.
[b] Colonization periods 'B' and 'C'.

experimentation following 8 and 15 days of colonization. Substrata were also positioned on 18 June and removed for experimentation following 7 days of colonization. At approximately midday on each sampling date, 3 substrata were removed and 5 cm^2 portions cut from each strip at 10-cm intervals. Substrata portions were placed individually into 30 mL light bottles containing filtered (Whatman GF/C) marsh water, inoculated with 0.5 ml of standardized $NaH^{14}CO_3$ solution (approximately 1.0 μCi), stoppered, and placed on a suspension apparatus which was then placed in the water for a 1-h incubation period, each substratum thus being incubated at its depth of origin. The ^{14}C inoculum and incubation period were such that cells would be marginally labelled. Following incubation, samples were fixed with Lugol's iodide solution and returned to the laboratory. The inorganic carbon content of marsh water was determined from alkalinity and dissolved CO_2 estimates (A.P.H.A. 1971) and during each incubation period illumination throughout the water column was determined with a Li-Cor (Li-155A) quantum meter equipped with a submersible quantum sensor.

Fixed periphyton was subsequently scraped from the acetate substrata and resuspended in the 30-mL water samples in which it had been incubated. The number of cells per unit of substratum surface area of each major algal species was determined by microscopic examination of cell suspensions in Palmer counting cells. Although undoubtedly not all species were accounted for, cell counts were made until no unaccounted for species were observed.

For the preparation of micro-autoradiographs, subsamples of the fixed periphyton were sedimented onto microscopic slides according to the method of Knoechel & Kalff (1976b). Slides were washed thoroughly and coated in the dark with Kodak NTB3 nuclear track emulsion. Emulsion-coated slides were air dried and exposed in total darkness for 5 days at 4° C. Following exposure, slides were developed and dried and the algal material mounted under cover clips in 30% glycerine (Knoechel & Kalff 1976b).

Nuclear tracks, as defined by Knoechel & Kalff (1976b), resulting from the ^{14}C disintegrations during the exposure period, that eminated from the viable cells of each periphytic algal species were counted using phase-contrast microscopy. Tracks per cell fitted a Poisson distribution and sufficient numbers of cells of each species were scored where possible to provide mean values of tracks per cell which had standard errors of less than 10%. Track data were then corrected for emulsion geometry, by the proportion of beta particles with insufficient energy to create a track and self-absorbtion (Knoechel & Kalff 1976b), and expressed as dpm/cell. We did, however, feel that dpm/cell data were inadequate for the comparison of species since the cells of each species were of considerably different sizes. Accordingly, the surface area of the cells of each algal species was estimated by determining the length of the major axes of each organism and estimating surface area from the formula for the most similar geometric form. Sufficient measurements were made so that calculated mean surface areas had a standard error of less than 10% (Table 1). This determination permitted track data for each species from each sample to be expressed as dpm/1000 μm^2. Both these data transformations and dpm/cell data could then be expressed in terms of μg carbon assimilated.

On each occasion that the above procedures were employed, a ^{14}C-labelled culture of *Cyclotella*, which had been preserved with Lugol's iodine, was subjected to identical steps. The similarity of tracks per cell of *Cyclotella* from one occasion to the next served as an indication that the method was internally consistent and that any leakage of label from preserved cells was minimal. According to Silver & Davoll (1978), the use of Lugol's solution entails minimal loss of label from cells.

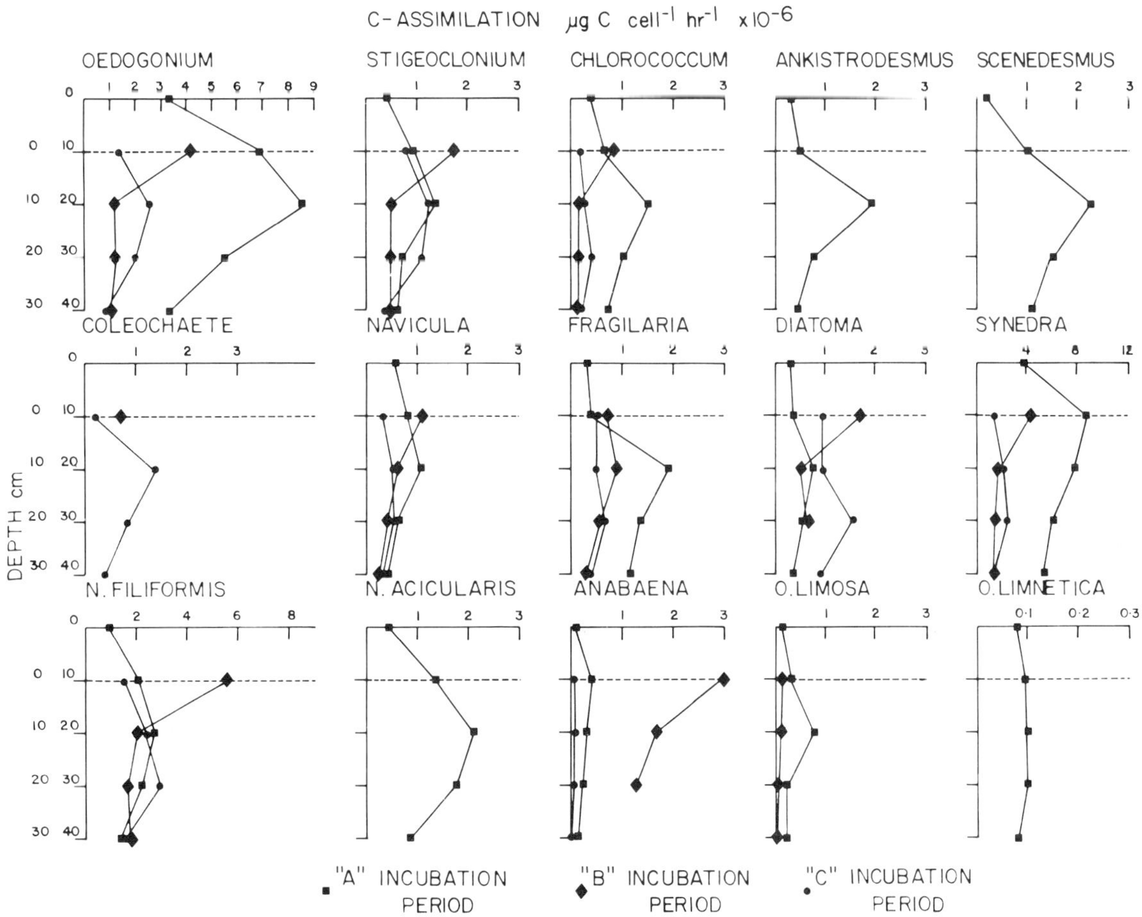

Fig. 1. Vertical profiles of carbon assimilation rates of 15 major periphytic species following three periods of substrata colonization.

Table 2. Ranges of carbon-assimilation rates for 15 major periphytic species.

Species	Carbon-assimilation (μg C cell^{-1} h^{-1} × 10^{-6})
Oedogenium idioandrosporum	0.84 ± 0.16–8.48 ± 1.70
Stigeoclonium nanum	0.40 ± 0.07–1.77 ± 0.25
Chlorococcum humicola	0.14 ± 0.03–1.52 ± 0.23
Ankistrodesmus falcatus	0.32 ± 0.06–1.96 ± 0.21
Scenedesmus quadricauda	0.18 ± 0.06–2.27 ± 0.56
Coleochaete soluta	0.21 ± 0.04–1.36 ± 0.21
Navicula radiosa	0.32 ± 0.06–1.11 ± 0.13
Fragilaria vaucheriae	0.28 ± 0.07–1.85 ± 0.32
Diatoma tenue var. *elongatum*	0.31 ± 0.06–1.54 ± 0.11
Synedra acus	1.26 ± 0.12–8.50 ± 0.60
Nitzschia filiformis	0.96 ± 0.17–5.53 ± 0.65
Nitzschia acicularis	0.38 ± 0.09–2.11 ± 0.26
Anabaena cirnicalis	0.02 ± 0.01–2.77 ± 0.52
Oscillatoria circinalis	0.08 ± 0.02–0.74 ± 0.08
Oscillatoria limnetica	0.08 ± 0.02–0.10 ± 0.01

Results and discussion

Vertical profiles of carbon assimilation rates for the 15 major organisms considered (Fig. 1) demonstrate that for the incubation period 10–18 June (A) production of all organisms present showed depression of the rate of assimilation at the water surface and maximum rates at either 10 or 20 cm below the surface. Following the longer, 10–25 June (B), incubation period, during which time water levels had dropped approximately 10 cm, uptake/cell generally decreased at most depths and surface inhibition was lacking. Following the incubation period 18–25 June (C) individual uptake values were again generally lower than had been apparent following the 'A' period, but once again maximum uptake values were found at 10 or 20 cm depth. Since light conditions were very similar during each experiment (approximately 550 μE m^{-2} s^{-1} at 0 cm; 440 at 10 cm; 320 at 20 cm; 240 at 30 cm and 40 at 40 cm), the absence of surface inhibition following the 'B' period may reflect the presence of self-shading on the more densely colonized substrata. Self-shading may also have contributed to the generally lower carbon assimilation rates following the 'B' period, although some other factor must have had some bearing, since rates following the shorter 'C' period were also generally lower than those following 'A' (Fig. 1). Although it cannot be said what actual light conditions prevail for cells in a progressively self-shaded community, it does appear that maximum uptake rates occurred between 320 and 550 μE m^{-2}s^{-1}, and as such are consistent with the ranges of values found by Harris & Lott (1973) for members of the Bacillariophyceae and Chlorophyceae in Lake Ontario. Ranges of carbon fixation rates of individual species (Table 2) expressed on a per cell basis indicate *Oedogonium idioandrosporum* to have the highest rate amongst the Chlorophyceae, *Synedra acus* to have the highest amongst the diatoms, and *Anabaena circinalis* to have the highest amongst the Cyanophyceae. Although comparative data from other periphyton studies are not available, Knoechel & Kalff (1976b) have reported similar absolute values for planktonic *Tabellaria* and *Anabaena*..

Since the amount of each species (as μm^2 cell surface area/unit of substratum area) and carbon uptake rates (as C uptake/1000 μm^2 cell surface area) for all species were determined, vertical profiles of total productivity could be derived (Fig. 2) and the contribution of each species to that total estimated (Fig. 3). Although total productivity values derived from the summation of individual species' productivity may well be underestimates by virtue of the exclusion of some species, such values are categorically values of primary productivity and not community productivity, which is normally the form that data has taken in the past. In this particular case the total values are in fact of the

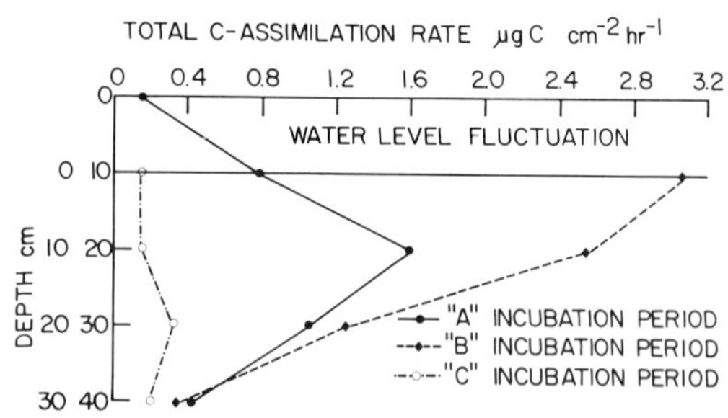

Fig. 2. Vertical profiles of total carbon assimilation rates for periphyton, following three periods of substrata colonization.

same order as those of Hooper-Reid & Robinson (1978) which were obtained from conventional ^{14}C methodology for a number of epiphyte communities in the Delta Marsh. From Fig. 2 it can be seen that the maximum total productivity value following colonization period 'A' was 1.60 μgC cm^{-2} h^{-1}; following period 'B' was 3.06 μgC cm^{-2} h^{-1} and following period 'C' was 0.33 μgC cm^{-2} h^{-1}. Vertical profiles demonstrate surface inhibition following the 'A' and 'C' periods, as would be expected from the carbon assimilation rates of individual cells (Fig. 1).

The contributions of individual species to total productivity (Fig. 3) are such that *O. idioandrosporum*, *Stigeoclonium nanum*, *S. acus* and *Nitzschia filiformis* were the major contributors following the 'A' period; *Fragilaria vaucheriae* and *A. circinalis* were major contributors following the 'B' period, and *Coleochaete soluta* and *Diatoma tenue* var. *elongatum* following the 'C' period.

Although it seems that three major contributors, *O. idioandrosporum, S. acus,* and *N. filiformis* are also relatively productive on a per cell basis, they are not necessarily so on the basis of computed carbon assimilation per unit of cell surface area. A logarithmic relationships exists between the maximum carbon assimilation rate per unit of cell surface area and total cell surface area (Fig. 4). This relationship was most distinct for the Chlorophyceae examined but less clear, although still present,

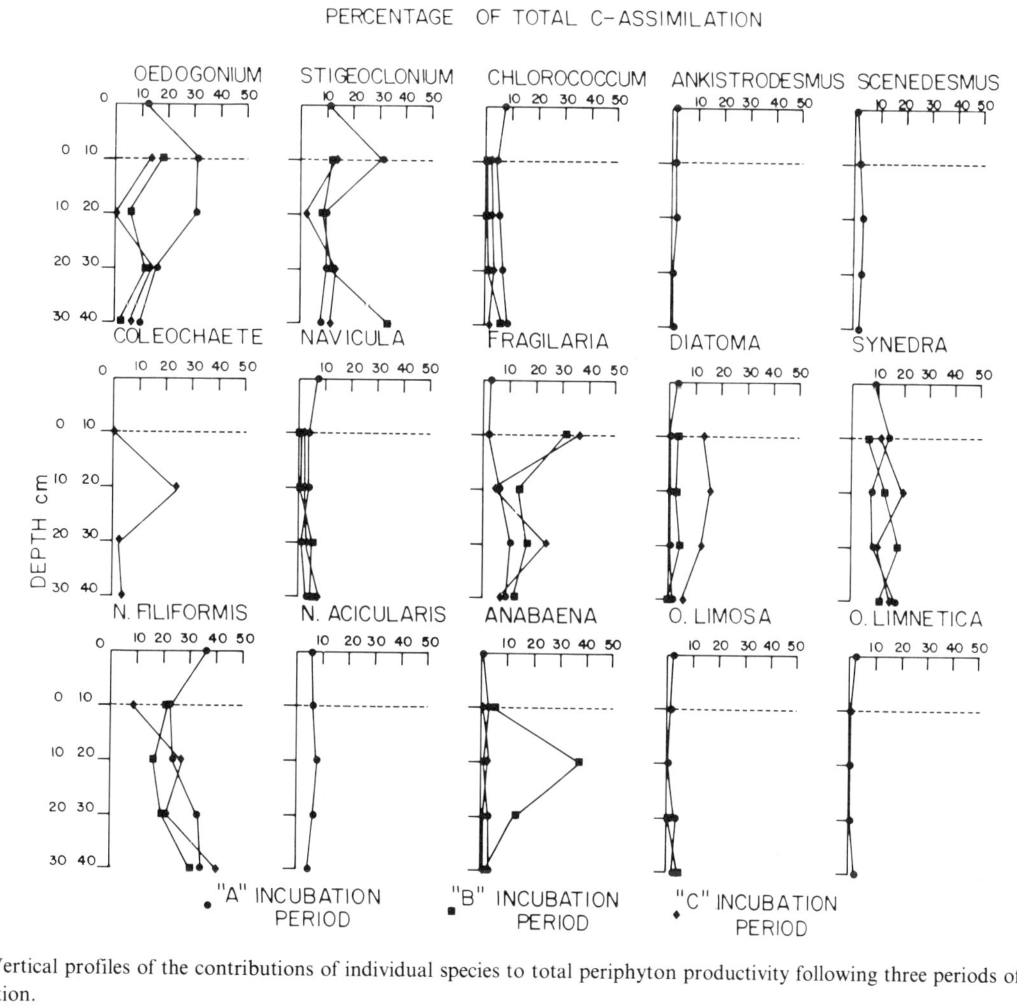

Fig. 3. Vertical profiles of the contributions of individual species to total periphyton productivity following three periods of substrata colonization.

Table 3. The relationship between cell surface area and the ratio of cell surface area per unit of substratum area to carbon assimilation (index of generation time) 'r' values.

	Depth (cm)			
	0	10	20	30
Chlorophyceae	0.639 (n=11)[a]	0.640 (n=10)[a]	0.680 (n=11)[a]	0.743 (n=11)[a]
Bacillariophyceae	0.536 (n=16)[a]	0.655 (n=16)[a]	0.603 (n=16)[a]	0.289 (n=14)
Cyanophycae	0.469 (n=6)	0.357 (n=6)	0.063 (n=5)	0.208 (n=6)

[a] Significant.

for the Bacillariophyceae and Cyanophyceae considered. From the combined data on all three classes, a minimum carbon uptake rate per 1000 μm^2 cell surface was achieved by organisms with a cell surface area of approximately 1400 μm^2. It has similarly been shown that nannoplankters may often exhibit primary productivity rates greater than their biomass proportion would suggest (Watt 1971).

If small cells are more productive per unit of cell surface area than large cells, one would expect generation times of small cells to be shorter than those of large cells. To obtain a relative measure of growth potential of each organism, a ratio of the cell surface area of each species per unit of substratum surface area to the carbon assimilation rate of each species was computed to provide an abstract index of division or turnover time. As might

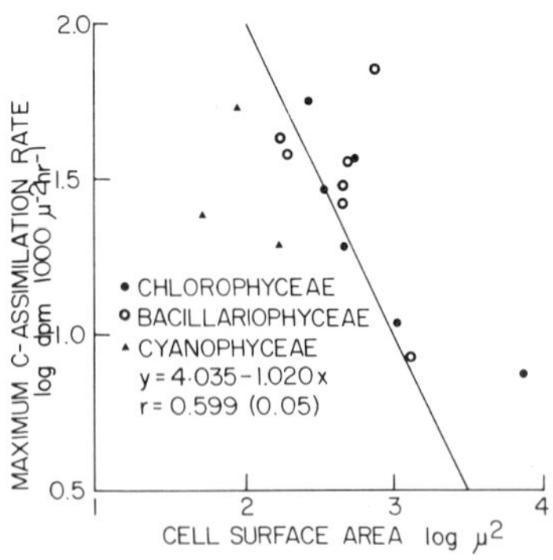

Fig. 4. The relationship between the maximum rates of carbon assimilation and the cell surface areas of major periphytic algae.

be expected, for the Chlorophyceae at each depth and the Bacillariophyceae at all but one depth there existed a significant linear relationship between this index of generation time and cell size (Table 3).

In summary, we hope to have demonstrated the applicability of nuclear-track autoradiography for the provision of information on the *in situ* rates of carbon assimilation of individual periphytic algal species and the contribution of individual species to the true primary productivity of a complex community.

Acknowledgements

An N.S.E.R.C. operating grant to the first author and the provision of a research facility at the University of Manitoba Field Station, Delta Marsh are gratefully acknowledged. R. Knoechel is also acknowledged for his guidance in the use of nuclear track autoradiography.

Contribution 8, University of Manitoba Field Station, Delta Mars, Manitoba, Canada.

References

American Public Health Association, 1971. Standard methods for the examination of water and wastewater. 13th ed. Washington, D.C.

Harris, G.P. & Lott, J.N.A., 1973. Light intensity and photosynthetic rates in phytoplankton. J. Fish. Res. Bd. Can. 30:1771–1778.

Hooper, N.M. & Robinson, G.G.C., 1976. Primary production of epiphytic algae in a marsh pond. Can. J. Bot. 54:2810–2815.

Hooper-Reid, N.M. & Robinson, G.G.C., 1978. Seasonal dynamics of epiphytic algal growth in a marsh pond: productivity, standing crop, and community composition. Can. J. Bot. 56:2434–2440.

Knoechel, R. & Kalff, J., 1976a. The applicability of grain density autoradiography to the quantitative determination of algal species production: a critique. Limnol. Oceanogr. 21:583–590.

Knoechel, R. & Kalff, J., 1976b. Track autoradiography: a method for the determination of phytoplankton species productivity. Limnol. Oceanogr. 21:590–596.

Knoechel, R. & Kalff, J., 1979. The advantages and disadvantages of grain density and track autoradiography. Limnol.

Oceanogr. 24:1170–1171.

Maquire, B. & Neill, W.E., 1971. Species and individual productivity in phytoplankton communities. Ecology 52:903–907.

Paerl, H.W., 1977. Ultraphytoplankton biomass and production in some New Zealand lakes. N.Z.J. Mar. Freshwat. Res. 11:297–305.

Paerl, H.W., 1978. Effectiveness of various counting methods in detecting viable phytoplankton. N.Z.J. Mar. Freshwat. Res. 12:67–72.

Paerl, H.W. & Stull, E.A., 1979. In defense of grain density autoradiography. Limnol. Oceanogr. 24:1166–1169.

Pip, E. & Robinson, G.G.C., 1982a. A study of the seasonal dynamics of three phycoperiphytic communities, using nuclear track autoradiography. I. Inorganic carbon uptake. Arch. Hydrobiol. 94:341–371.

Pip, E. & Robinson, G.G.C., 1982b. A study of the seasonal dynamics of three phycoperiphytic communities, using nuclear track autoradiography. II. Organic carbon uptake. Arch. Hydrobiol. 96:47–64.

Silver, M.W. & Davoll, P.J., 1978. Loss of ^{14}C activity after chemical fixation of phytoplankton: error source for autoradiography and other productivity measurements. Limnol. Oceanogr. 23:362–368.

Stull, E.A., de Amezaga E. & Goldman, C.R., 1973. The contribution of individual species of algae to primary productivity of Castlelake, California. Verh. Int. Verein. Limnol. 18:1776–1783.

Watt, W.D., 1971. Measuring the primary production rates of individual phytoplankton species in natural mixed populations. Deep-Sea Res. 18:329–339.

Authors' address:
G.G.C. Robinson/E. Pip
Department of Botany
University of Manitoba
Winnipeg, Manitoba, Canada

CHAPTER 33

A method for measuring the dehydrogenase activity of river periphyton

B. BOYANOVSKY

Abstract. The use of tetrazolium salts to measure dehydrogenase activity (DA) of epilithic communities was more appropriate and sensitive than the oxygen change methods. While modifying the DA assay for estimation aerobic decomposition by periphyton, the effects of a defined medium on nonadapted periphyton was found to be insignificant. For DA analyses of periphyton we suggest using: triphenyl tetrazolium, an incubation temperature 28° C, a modified termination of enzyme reactions, and methods for avoiding the interference of photosynthetic pigments. The technique suggested gave reproductable results obtained from various samplings at the same place of a moderately polluted river during a fortnight within the range of variability of subsamples (10%). The DA of periphyton (per unit stone surface) from an area with a current velocity of 1.5 m s^{-1} was tenfold lower than that of the same water with a current velocity of 0.15 m s^{-1}. The activity of the autumnal periphyton was twice greater than that during summer in areas shadded by forested banks. The DA was greater (up to 68 × 10^{-3} μmoles H_2 h^{-1} cm^{-2}) near organic pollution sources than a few kilometers downstream (min. 4.7 × 10^{-3} μmoles H_2 h^{-1} cm^{-2}).

Introduction

The dehydrogenase assay using tetrazolim salts and its advantages over the other methods for estimating the rate of biological oxidation was of much interest among scientists working on problems of water purification (Hubena 1975; Simon & Dore 1977).

In this paper we propose a modification of the dehydrogenase assay for studying the catabolic activity of river periphyton. While working on the assay, several points of principle and many methodological problems became apparent. These problems were more or less satisfactorily solved by mechanical modifications of this originally biochemical method (Kun & Abood 1949) for studying the efficiency of oxidation of activated sludge. Most of the main problems are analyzed and possible solutions are suggested. An attempt is made to statistically evaluate the precision and variability of results.

Preliminary results from a study of periphyton taken from organically polluted rivers (Sazava and Vltava near Prague, Czechoslovakia) and a heavily polluted brook near Sofia, Bulgaria are presented here. Samples were taken from places with different current velocity, different distances from the source of pollution and during different seasons.

The term 'periphyton' is used for the total mass attached to the stones and it includes bacteria, algae, small animals and non-living matter.

Wetzel, R.G. (ed.), Periphyton of Freshwater Ecosystems.
 ISBN 90 6193 768 X.

Main properties of TZS and formazans and considerations for their use in measuring dehydrogenase activity

Tetrazolium salts (TZS) are important indicators of energy absorption and release in chemical and biochemical reactions. The advantages of tetrazoles over the other redox indicators (such as methylene blue) are associated with their properties (Jambor 1960):

a) They are readily soluble in water giving colourless solutions.

b) The biochemical reduction of TZS leads to the formation of intensively coloured water insoluble formazans, which remain at the place of reduction, i.e. in the cell, provided it is intact.

c) Formazans dissolve well in organic solvents.

d) TZS have low toxicity.

The choice of TZS is principally based on the redox potential of the system salt/formazan. TZS with low E_0 values are more suitable; they may be used under aerobic conditions and they do not compete with O_2 for the electrons (H_2 resp.) of the substrate. TZS with higher E_0 values have the principal advantage that with the addition of phenazine methosulphate (PMS) to the sample one can distinguish quantitatively the hydrogen that is directed to the respiratory chain from the hydrogen which enters biosynthetic pathways. The disadvantage of tetrazoles with higher E_0 values is that they require anaerobic conditions of the samples.

The reduction of TZS in the animal cells takes place mainly in the mitochondria, in the chloroplasts in plants, and in the cytoplasmic membrane in bacteria. Questions can be raised about the rate of tetrazole penetration through the corresponding membrane as an eventual rate limiting factor of the enzymic reactions. The larger di-TZS molecules do not always freely penetrate mitochondrial membranes. Their permeability is enhanced by the presence of phosphate ions and thus the yield of formazans is increased. Therefore the use of phosphate buffers is recommended. Monotetrazoles penetrate membranes better than ditetrazoles (Altman 1972).

For the work with the periphyton we chose 2, 3, 5-triphenyltetrazolium chloride (TTC), a monotetrazole with the highest redox potential (E_0 490 mV). Its reduction leads to the formation of triphenyl formazan (TF).

Methodological problems

The main problems associated with the dehydrogenase assay are: the presence of dissolved O_2 in the sample and the reagents, the achievement of controlled anaerobic conditions and optimum tetrazole concentration, pH and temperature, as well as the addition or not of defined nutrition substrate, the termination of the enzyme reaction, and the interference of assimilation pigments co-extracted with the formazan.

The various assays with tetrazoles used in studies mainly on activated sludge can be divided into two groups: (a) those that simulate natural conditions, and (b) assays which are optimized from a biochemical point of view with the purpose of maximum formazan yield and increased sensitivity of the assay.

Our experiments showed the possibility of a compromise between the two groups for part of the conditions in order to shorten the duration of the assay and to preserve the sample integrity.

Conditions of incubation

Effect of dissolved oxygen

Oxygen competes with TZS with similar, highly positive E_0 values as an acceptor of hydrogen and electrons from the respiratory pathway and the cytochrome P 450 system. PMS also react predominantly with oxygen. The sensitivity of TZS reduction to O_2 is due to several interrelated factors. (Altman 1972). Several modifications of the dehydrogenase assay ignore the effect of dissolved O_2 on the reduction of TTC from activated sludge (Ford *et al* 1966), resulting from the experiments of Kun & Abood (1949). Differences in the enzymic reduction of TTC under aerobic and anaerobic (N_2) conditions were observed in the course of one hour. After that, by the end of the reaction, the yield of formazan was equal under both conditions. The material studied (rat liver homogenate), however, was significantly more active than either periphyton or activated sludge.

There are several ways of reducing the effects of oxygen: evacuation of the samples, addition of substances that bind the dissolved O_2, or purging with inert gas, usually oxygen-free N_2.

The best results were obtained by the latter method, which is however labour consuming. For routine work it seems quite sufficient to add the necessary, carefully dosaged, quantity of Na_2S and $CoCl_2$ (Klapwijk *et al* 1974). It is desirable to compare these results with those obtained by the use of an inert gas.

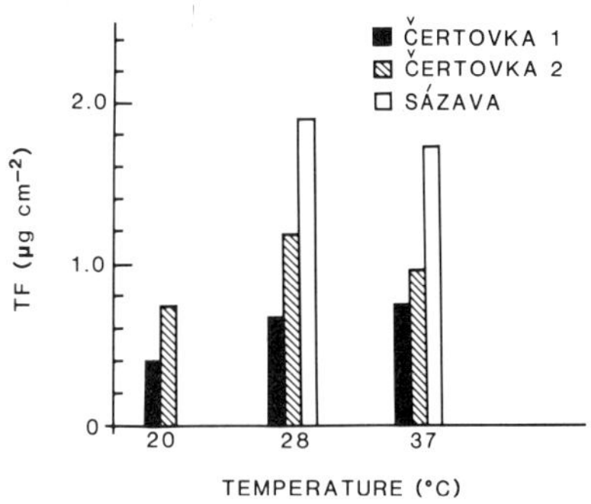

Fig. 1. Dehydrogenase activity at different incubation temperatures.

Photosensitivity

Aqueous solutions of tetrazolium salts are photosensitive. Incubation should be carried out in the dark.

Toxicity and concentration of TZS

Concentrations of TZS in the range of hundredths of a percent do not have inhibitory effects on almost all studied organisms or on entire aquatic communities. Reduced concentrations are associated with reduced yield of formazan, but this reduction is not due to exhaustion of TZS in the solution. TZS are generally less toxic to plants than they are to animals but even some higher freshwater invertebrates can survive several days in dilute TTC solution (Jambor 1960).

Table 1. Effect of defined nutrition medium on yield of TF (μg h^{-1}) in mixed bacterial culture (per ml) and river periphyton (per cm^2).

	With glucose	Without glucose
Bact. culture 1	0.33	0.71
Bact. culture 2	0.56	1.64
Periphyton:		
River Vltava	6.57	6.00
Canal Certovka	0.69	0.83
River Sazava		
Station 1	2.07	2.50
Station 2	4.9	7.13
Station 3	19.54	26.16
Station 4	4.75	6.55
Station 5	5.87	24.26
Station 6	6.17	11.84

Effect of defined nutrition medium

The addition of substrate to the sample should allow measurement of the maximum dehydrogenase activity. On the other hand, organisms which are not adapted to a given substrate react to its addition in a different way. Addition of the same substrate in which bacteria had been cultivated rapidly increases the yield of formazan. Samples of periphyton with varying activity are influenced in a different way by the addition of 0.167 mg l^{-1} glucose; in some highly active samples the addition of substrate even causes reduction of the yield of TF (Table 1). These results and similar studies on activated sludge (Klapwijk *et al* 1979; Simon & Dore, 1977) imply the necessity of careful use of defined substrates for study of respiration rates.

Effect of pH

In order to obtain comparable results the samples should be buffered in a well-defined pH medium. The pH of the aqueous environment which is optimal for the water organisms as a whole is not identical to the pH optimum of the enzyme reduction of TZS. Reactions at a higher pH, up to 8.0, give a greater yield of TF from activated sludge (Hubena 1975). Increasing pH above 8.5 may lead to non-enzymatic reduction of TZS or reduction of formazan itself. Maintenance of a high pH (above 8.5) is also not ecologically justified.

Effect of the temperature

At 28° C the activity of samples stored three days at 5° C (Sazava) and of fresh autumn samples (Certovka 2) was higher than at 37° C, and for fresh summer samples the difference in activity was insignificant (Fig. 1). At 20° C the yield of formazan was lower. These data show that aquatic communities adapted to lower temperature did not exhibit higher dehydrogenase activity at 37° C, as was indicated by recent studies on activated sludges (Effenberger & Zahradka 1966; Lenhard 1968; Ford *et al* 1966; Klapwijk *et al* 1974; Simon & Dore 1977).

Duration of incubation

With the very heterogeneous material represented in our samples, the problem for defining the time of incubation is associated with the problem of possible changes in the sample as a result of bacterial activity and alterations in the structure of the biocenosis rather than with the exhaustion of the substrate and accumulation of inhibitory reaction products. With periphyton and activated sludge, the prevailing quantity of TF is formed in the first 2–3 h. of incubation, after which the rate of formation decreases (Boyanovsky 1979). The duration of the incubation is the first condition of the experiments that has to be defined and strictly observed.

Termination of the enzyme reaction

Until now this termination has been achieved by the addition of acid, heating, or addition of the extracting agent itself. Small amounts of acid or organic solvents, however, have uncertain effects while the larger concentrations of acid and heating destroy the cellular membranes and thus increase the variability of the results.

Formaldehyde appears to be the most appropriate for our purposes. It blocks the enzymic transfer of hydrogen prior to its interactions with TZS (Altman 1972) and has no effect on the blank samples (Boyanovsky 1979).

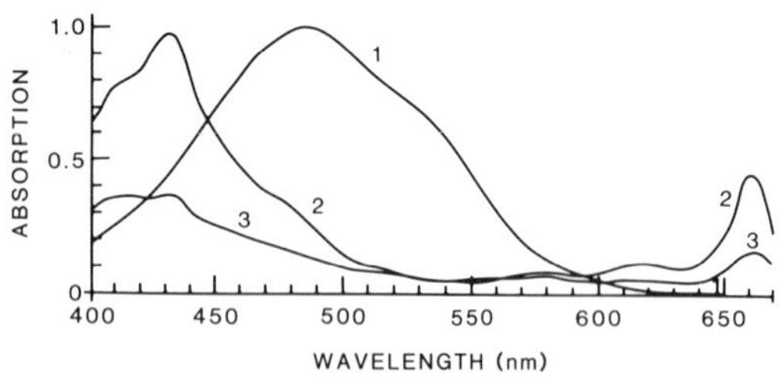

Fig. 2. Absorbance curves: *1*, Ethanol solution of pure triphenyl formazan, and *2*, *3*, Ethanol extract of periphyton from different sampling points of the River Sazava.

Interference by assimilation pigments

Extraction of formazan from the cellular mass is also accompanied by inevitable co-extraction of the pigments and products of their decomposition. The absorbance maximum of the ethanol extract of periphyton from different rivers was near 430 nm (Fig. 2), and was due mainly to the carotenoids and probably chlorophyll *b*. These plant pigments interfere considerably with the measurement of the extracted TF at its maximum absorbance of 485 nm. The blank sample may even have greater light absorbance than the increase due to TF. Two ways have been proposed for avoiding this interference (Dockal 1976): (a) chromatographic separation of the formazan from the chlorophylls, and (b) use of a relationship between the absorbance of the extract at 663 nm and 472 nm, which is supposed to be linear. We find both corrections unacceptable. The absorbance should instead be measured above the maximum, by which the interference is practically avoided. At 520 nm we measured only a few percent (5.9 to 17%) of the maximum absorbance due to assimilation pigments, although the absorbance of TF was 75% of the maximum (Fig. 2).

Presentation of results

For comparing the results of studies within the ecosystem it is sufficient that they be presented as quantity of formazan per unit time. For comparison with results obtained from different ecosystems or by different methods, we recommend that the quantity of formazan is expressed in molar concentrations or in terms of hydrogen which is related to oxygen uptake. In a number of cases more information is derived by relating the results of the assay to the corresponding biomass.

In our experiments we could not find a relationship between the dehydrogenase activity and the parameters which characterize the biomass (Kjeldahl-N), autotrophic biomass (chlorophyll a), or the total number of bacteria.[1]

Thus, for certain purposes, for analyzing the role of periphyton in a given river stretch we can use data related to a unit stone surface (e.g. 1 cm^2). Other methodological problems such as separation of the microorganisms from the suspension and extraction of TF by lower alcohols are reviewed by Hubena (1975) and Simon & Dore (1977).

Proposed method of measuring dehydrogenase activity

The periphyton suspension was obtained by the method of Puncochar (1977). Samples were taken in triplicate with two blanks. The reaction was carried out at 28° C in the dark, for 1 h after the addition of TTC and stopped with formaldehyde. The samples were sparged with N_2 throughout the incubation.

The periphyton suspension, phosphate buffer and TTC were used in a ratio 7:2:1. An aliquot of the suspension (7 to 14 ml depending on its density) was taken with a pipette with a cut-off tip and placed in a pear-shaped flask with a rubber stopper and two glass tubes passing through (Fig. 3) and buffered with K-Na phosphate (Sorensen), pH 7.73, to a final concentration of 0.01 M. After 15 min of sparging with N_2[2], a 0.5% solution of TTC was added to yield a final concentration of 0.05% and the flasks were transferred to darkness. One hour later the N_2 stream was terminated and the mixture transferred to a centrifuge tube containing 1 to 2 ml filtered concentrated formaldehyde solution (final concentration 4%). The sample was centrifugated for 10 min at 4000 rpm, the clear supernatant syphoned off, the pellet resuspended in 3 ml

[1] Unpublished data P. Punčochář, Hydrobiol. Lab.; Czechosl. Acad. of Sci., Prague.

[2] For routine analyses Na_2S may be used for deoxygenation in concentration up to 0.1% in the sample, which is, however, a less cautious way than the proposed one. Moreover, the sample is also stirred by N_2 sparging.

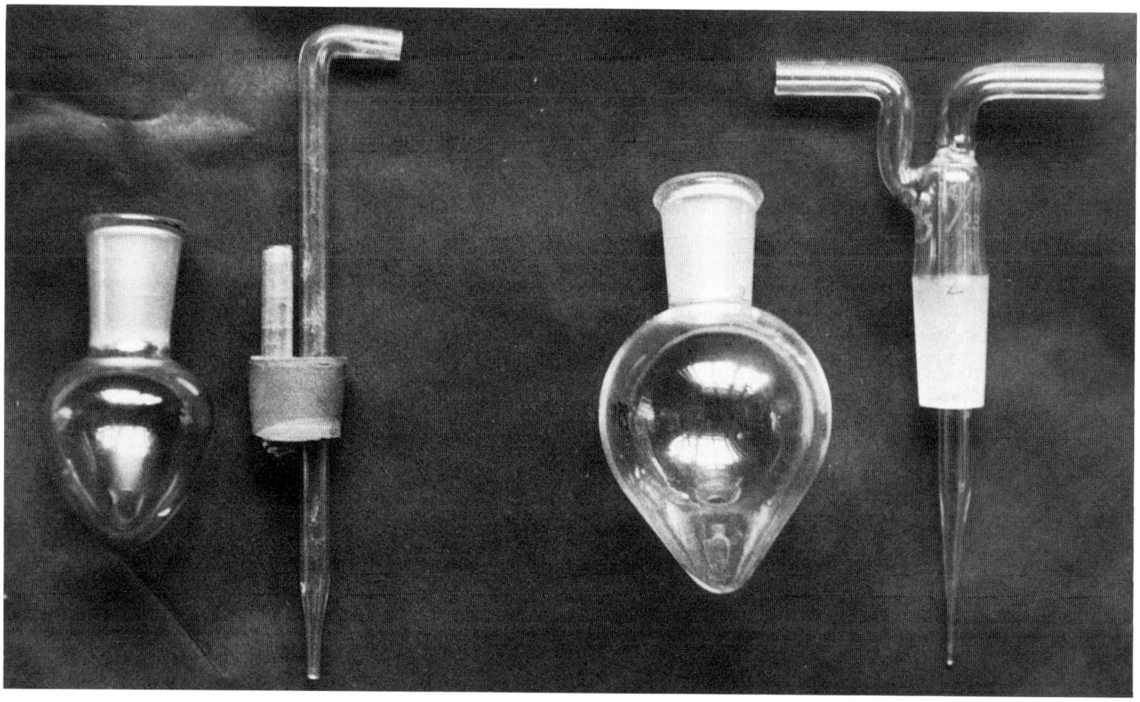

Fig. 3. Incubation flasks used in the analyses.

ethanol, and the tube sealed. The TF formed was extracted for 1 h at room temperature in darkness with occasional shaking. If after 5–10 min of extraction the colour was more intense than that of weak tea, more alcohol was added. After centrifugation for 10 min at 4000 rpm, absorbance of the extract was measured at 420 nm against a blank without TTC, and the results were expressed per unit stone surface using a standard curve plotted for pure TF at 520 nm.

The stone surface was measured by the method of Puncochar (*ibid.*) by carefully wrapping the substrata in a layer of aluminium foil and comparing the weight of the foil piece with foil of known surface areas.

Results

The proposed method was used for testing the activity of samples taken from rivers with various organic loads. Values varied within the range of 0.5 to 20.4 μg TF h^{-1} cm^{-2} or 1.6×10^{-3} to 68×10^{-3} μ moles H_2 h^{-1} cm^{-2} if expressed in H_2 equivalent (H_2 which accepted TTC from oxidized substrates). The periphyton from a brook, polluted with sewage of a linen factory near Sofia, was composed of lower heterotrophic organisms and had a summer mean of activity 31 μg TF h^{-1} cm^{-2}.

For statistical evaluations of variability of results obtained by the described technique, five experiments were carried out with 28 periphyton subsamples from moderate and low activity. Variability of the subsamples was higher (variation coefficient up to 10.3%) at the lowest measured activity, probably because of higher blanks. A statistical valuation of sampling technique is given by Puncochar (*ibid.*).

From four observations in Certovka, a canalized branch of the river Vltava in Prague, during a fortnight we measured nearly the same activities of 0.63 to 0.66 μg TF h^{-1} cm^{-2}. In the same river profile and at the same time, the other branch of the river with current velocities at the site of sampling, 0.15 m s^{-1} (one order of magnitude lower than the first site), the periphyton was much thicker, not so compact, and had a 10 times greater dehydrogenase activity (6.6 μg TF h^{-1} cm^{-2}).

In the Certovka reaches which had forested banks, the activity of the autumnal periphyton was twice as large as that during the summer. This difference could be explained by the effect of shading on the periphytic community and by the greater stability of periphyton towards climatic conditions in comparison with river plankton. Similar comparisons were made using the data for the standing crop (chlorophyll *a* and Kjeldahl N) of the periphyton in the eutrophic lower part of the river Sazava (Boyanovsky 1979).

The summer periphyton activity was much greater near the sources of organic pollution (small towns) than it was several kilometers downstream (Table 2). The differences among the sampling stations according to Student's t-test were significant ($p = 0.99$).

Conclusion

The method is sufficiently sensitive and is informative of periphyton catabolic activity, especially below the limit of detection by oxygen uptake techniques. The method gives reliable results within a large range of values. Moreover the method appears to be simple and quite applicable to routine work. Further development of the procedure is needed in order to distinguish between aerobic, anaerobic and biosynthetic processes in periphyton.

Table 2. Summer periphyton dehydrogenase activity (DA) above and below pollution inputs in the River Sazava.

Station	Sampling in relation to the pollution	DA (μg TF h^{-1} cm^{-2})
1	below	20.36
2	above	2.56
3	below	5.95
4	no poll.	1.66
5	above	1.70
5′	below	4.89

References

Altman, F.P., 1972. Quantitative dehydrogenase histochemistry with special reference to the pentose shunt dehydrogenases. Progr. Histochem. Cytochem. 4:223–273.

Boyanovsky, B.B., 1979. Characteristics of Attached and Suspended Biomass in the Lower Part of the River Sazava and Measurement of Dehydrogenase Activity of River Periphyton. Ph. D. Thesis, Hydrobiol. Lab. Czechoslov. Acad. of Sci., Prague, (in Czech), 116 p.

Dockal, P., 1976. Slimes in Cooling Systems and Their Suppression. Water Research Institute, Prague, Czechoslovakia, 213 p.

Effenberger, M. & Zahradka, V., 1966. Vyzkum method stanoveni oxidacni richlosti smesi a enzymaticke aktivity kalu. Zaverecna sprava VUV, Praha, 69 p.

Ford, D.L., Yang, J.T., Eckenfelder, W.W., 1966. Dehydrogenase enzyme as a parameter of activated sludge activities. Proc. 21st Ing. Waste Conf. Purdue Univ., pp. 534–543.

Jambor, B., 1960. Tetrazolium Salze in der Bioligie. Gustav Fischer Verl. Jena, 151 p.

Hubena, J., 1975. Activity of the Activated Sludge . . ., Ph.D. Thesis, VSCHT, Prague, (in Czech), 139 pp.

Klapwijk, A., Drent, J., Steenworden, J.H.A.M., 1974. A modificated procedure for the TTC dehydrogenase test in activated sludge. Wat. Res., 8:121–125.

Kun, E. & Abood, L.G., 1949. Colorimetric estimation of succinic dehydrogenase by triphenyltetrazolium chloride. Science 109:144.

Lenhard, G., 1968. A standardized procedure for the determination of dehydrogenase activity in samples from anaerobic treatment systems. Wat. Res., 2:161–167.

Puncochar, P., 1977. Bestimung von Bakterien im Aufwuchs (Periphyton) Proc. IInd. Int. Hydromicrobiol. Symp. Smolenice, 1975, ed. I. Daubner, Veda Vrl., Bratislava, pp. 73–78.

Simon, P. & Dore, M., 1977. Etude des paramètres du dosage des déhydrogenasés par le TTC applique aus boues activees. CEBEDEAU-BECEWA, 403:252–265.

Author's address:
B. Boyanovsky
Department of Hydrobiology and Ichthyology
Sofia University
8 Dragan Tzankov Str.
1421 Sofia, Bulgaria

CHAPTER 34

Artificial substrata for reducing periphytic variability on replicated samples

PETER G. MEIER, DAVID O'CONNOR and DAVID DILKS

Abstract. Periphyton accrual (chlorophyll *a*) was evaluated utilizing two different geometrically shaped artificial glass substrata. The first consisted of small coverslips attached to a glass slide (Dilks & Meier 1981), while the other method employed both horizontally and vertically oriented small culture tubes. Sufficient numbers of each type were placed in Fleming Creek (Ann Arbor, Michigan), a second order temperate stream, to permit four collections of three replicates for each type at 5-day intervals. With the exclusion of the first collection (after 5 days), the variability among the replicates was less than 8 percent. Analysis of the chlorophyll *a* data by ANOVA and simple linear regression showed least amount of variation in horizontally positioned test tubes, followed by those of vertical orientation, and then by the coverslips. The highest concentration of periphyton biomass measured in chlorophyll *a* was, however, observed in the vertically placed test tubes. The coverslips and horizontally incubated tubes collected comparable concentrations of chlorophyll *a* over the exposure period.

Introduction

The concept of 'periphyton biomass' as water quality monitors is limited because of the problem of reproducibility between replicated samples (Weitzel 1979). Hence, pollutional perturbations that would either stimulate or reduce this periphyton community by 15 to 25 percent cannot presently be isolated, since variability between replicated substrata samples exceeds the upper value. The components responsible for this 'lack of agreement' between replicates are attributed to the irregularity of periphytic growth on substrata and the errors associated with sample collection and analysis.

The patchiness of periphytic growth can be overcome by sub-sampling a larger substratum area. Dilks & Meier (1981) showed that variability in chlorophyll *a* concentration on weight sub-areas of a glass slide averaged 18 percent. However, comparisons of sub-area means between three glass slides resulted in a reduction of variability to less than 4 percent.

It also was pointed out in this study that a considerable amount of time was saved. The coverslips were removed in the field, immediately preserved (Standard Methods 1981), and stored at −20° C for 24 h prior to chlorophyll *a* analysis. Hence the task of scraping off the periphyton in the laboratory and the associated potential error were circumvented.

Since the method of sub-area sampling via coverslips and immediate initiation of pigment extraction in the field appeared promising, a study was carried out that would incorporate this concept. In addition, the effect of the shape and orientation of the substratum on periphyton accrual was investigated.

Wetzel, R.G. (ed.), Periphyton of Freshwater Ecosystems.
 ISBN 90 6193 768 X.

Methods

Six two-dimensional substrata in the form of round glass coverslips (203 mm^2) were attached with stop-cock grease on a 50×75 mm glass slide. Three slides were retained in a converted plastic microscope slide tray secured with wire to a stake driven into the stream bed (Dilks & Meier 1981). The slides were oriented vertically to the water flow and 13 cm above the stream bottom. Four trays each containing 3 slides were placed at the same time in Fleming Creek (Ann Arbor, Michigan) on 8August, 1981. All stations were similar in physical and chemical parameters. One tray and its slides were retrieved at random every 5 days over the 20-day period. One coverslip from each glass slide was transferred to a separate plastic centrifuge tube containing 10 ml of 90% buffered acetone and immediately preserved on ice. The remaining coverslips were removed for microscopic examination of the attached community and for organic ash-free weight analysis.

The three-dimensional artificial substrata consisted of glass culture tubes that were 6 mm in diameter and 50 mm long and had a exposed surface area of 970 mm^2. These tubes were secured by slipping the open end of the test tube over the protruding part of the tygon tubing, which in turn was attached to a rectangular PVC frame anchored to a masonry brick (Fig. 1). Depending upon the orientation of the substrata, the frames were constructed to either accommodate six horizontal or vertical tubes. Four of each type were also deposited in the stream. The placement was such that the horizontal tubes were exposed to those conditions as those existing at the mid-depth of the vertical tubes. At 5-day intervals a brick containing each type of test tube was retrieved. Removal of test tubes was accomplished simply by pulling the tygon tubing through the PVC support. In this manner the glass test tube could be either collected in the plastic centrifuge tube containing the acetone or in a plastic petri dish, without actually touching the exposed surface. This method was repeated for each collection period.

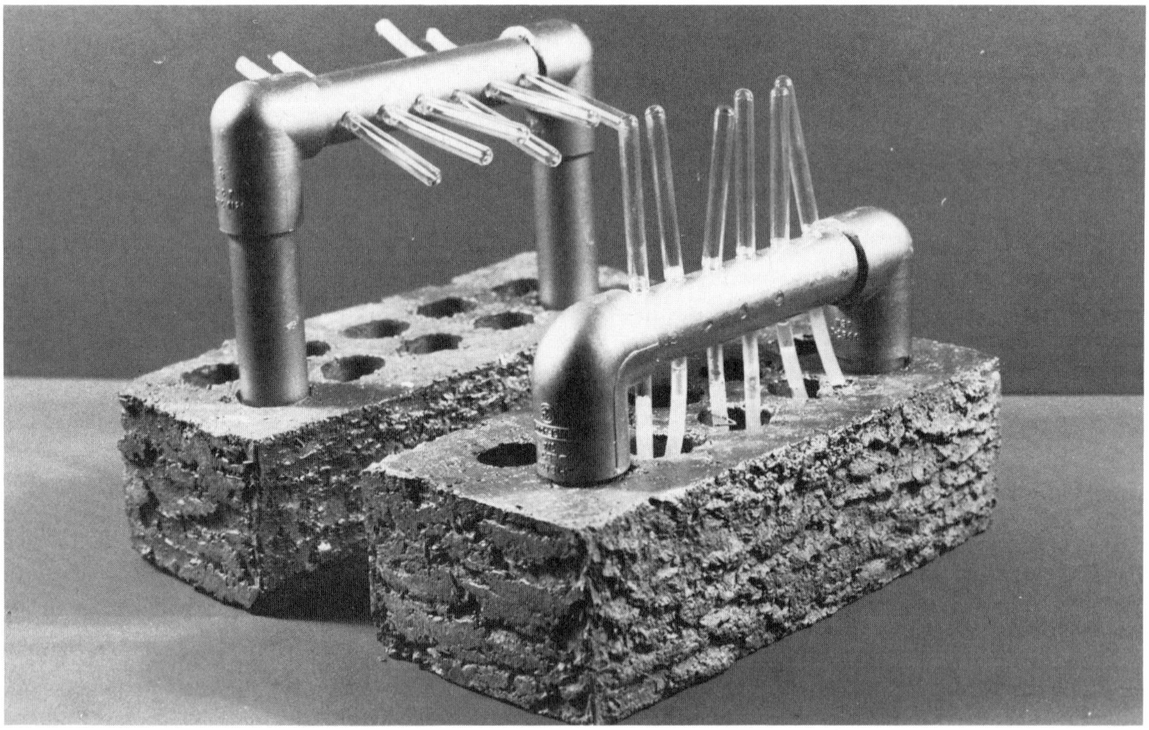

Fig. 1. Simple apparatus for holding tubular periphyton substrata.

Current velocities were made at each incubation site with a pigmy current meter.

All samples collected were treated in the same fashion for analysis of chlorophyll a. Prior to storage at −20° C, tubes were exposed to ultrasonification for 1 min. The plastic tubes were removed from the freezer and agitated for one minute on a vortex mixer then centrifuged for 10 min at 3000 rpm. The supernatant was then fluorometrically analyzed on the 10× scale for chlorophyll *a* and phaeophytin concentration (Yentsch & Menzel 1963; Standards Methods 1981).

Results and discussion

Water samples collected at 5-day intervals over the study period showed little variability in terms of chlorophyll *a* concentration, pH, turbidity, and temperature. Even velocity changes were minimal and averaged 12 cm sec^{-1} at the depth of the substrata (13 cm above bottom). The water depth at the study site was only 35 cm and due to the clarity, light penetration was to the bottom with little attenuation with depth. Even though other analyses such as ash-free weight and species composition were performed, only chlorophyll *a* concentration as a function of periphyton biomass will be discussed.

Table 1. Mean periphytic chlorophyll *a* concentration and standard deviation calculated from three replicates of each type of substratum for designated time intervals.

Substrata	Days exposed	Chlorophyll *a* mg m^{-2} (±S.D.)	Variation percent
Vertical tubes	5	7.45 (±0.85)	11.4
Horizontal tubes	5	5.78 (±1.48)	25.6
Coverslips	5	2.66 (±0.88)	33.1
Vertical tubes	10	16.08 (±0.34)	2.1
Horizontal tubes	10	6.28 (±0.45)	7.2
Coverslips	10	4.92 (±0.37)	7.7
Vertical tubes	15	13.53 (±0.59)	4.4
Horizontal tubes	15	7.06 (±0.29)	4.1
Coverslips	15	7.49 (±0.43)	5.7
Vertical tubes	20	22.75 (±1.36)	6.0
Horizontal tubes	20	14.75 (±0.08)	0.5
Coverslips	20	12.64 (±1.68)	7.2

Results of mean chlorophyll *a* concentration for 3 replicates for each type of substratum are presented in Table 1. Large variance is observed with all substrata at 5 days. This variance is mostly attributed to the patchiness of colonization and the relatively low chlorophyll *a* concentration on each exposed site. However, with that exception, the reproducibility for each substratum is good. The somewhat greater variability of the two-dimensional surfaces (coverslips) to the glass tubes was associated with the fact that only one coverglass from each glass slide was used for the analysis. If two or three separate analyses for each glass slide would have been carried out, this reported variability might have been reduced to that of the glass tubes.

Reproducibility of the replicates for the test tubes was very good and were comparable to those values reported by Dilks & Meier (1981). However, the two methods of orientation showed that the vertical tubes accrued significantly higher biomass than that observed with the horizontally placed glass tubes and glass coverslips, because the vertical tubes had a significant amount of surface area shielded from the full force of the current, in contrast to the horizontal tubes and coverslips which were completely exposed. The shielded area may offer microhabitats suitable for higher periphyton densities, possibly due to less sloughing. Visual inspection of the vertical tubes supported this opinion; periphyton densities were noticably higher on the portion of the tubes not directly exposed to the current.

The average velocity near the substrata was 12 cm sec^{-1}. As current velocity decreases, periphyton densities on the vertical tubes would be expected to become more similar to the other methods. Castenholz (1961) found that horizontally placed slides supported significantly more biomass than vertically placed slides in lakes, although this was likely because organic material settled onto the horizontal slides. The orientation of substrata that best reproduce natural conditions was not within the scope of this work and warrants further investigation.

In conclusion it can be stated that artificial substrata are useful for water quality monitoring. Data employing two methods of artificial substrata gave comparable results in that the variability between replicates can be reduced to eight percent or less. Employing ANOVA and simple regression, the horizontally oriented culture tube method showed less variability than that observed with the coverslips and vertically placed tubes. However, the vertically positioned substrata accrued significantly more biomass.

References

Castenholz, R.W., 1961. An evaluation of a submerged glass method of estimating production of attached algae. Verh. Int. Ver. Limnol. 14:155–159.

Dilks, D.W. & Meier, P.G., 1981. The use of coverslips in estimating periphyton accrual. J. Freshwat. Ecol. 1:321–326.

Standard Methods for the Examination of Water and Wastewater. 1981. 15th ed. American Public Health Association, Washington, D.C. 1134 pp.

Weitzel, R.L., 1979. Periphyton measurements and applications. In: Weitzel, R.L. (ed.) Methods and Measurements of Periphyton Communities: A Review. ASTM STP. 690. American Society for Testing and Materials. 33 pp.

Yentsch, C.S. & Menzel, D.W., 1963. A method for determination of phytoplankton chlorophyll and phaeophytin by fluorescence. Deep-Sea Res. 10:221–226.

Authors' address:
P.G. Meier/D. O'Connor/D. Dilks
Department of Environmental and Industrial Health
School of Public Health
The University of Michigan
Ann Arbor, MI 48109, U.S.A.

CHAPTER 35

A rapid extraction technique for photosynthetic pigments and ATP from periphyton communities

ROBERTO ANTONIETTI

Abstract. A brief treatment with liquid nitrogen of periphyton samples is more effective than overnight exposure to solvent for extraction of photosynthetic pigments, but it is not as effective as grinding for the total extraction. ATP concentrations measured after freezing with liquid nitrogen are greater than values obtained with boiling Tris-buffer extraction.

Introduction

During recent years, interest in cellular ATP content has increased, even though data application is still discussed. Cellular ATP content is often considered to be a good estimate of microbial biomass (Holm-Hansen 1973; MacKinnon *et al* 1977; Clark *et al* 1978; Oades & Jenkinson 1979; Nelson &Lawrence 1980). Other authors (Ausmus 1973; Herman & Temple 1979; Kennicut 1980; Wissmar 1981), however, consider ATP content to be variably related to the physiological state of the organisms.

ATP determination by boiling Tris-buffer extraction is not readily applicable to periphyton communities for two reasons (Clark *et al* 1978). (a) If the substrate is immersed directly into the boiling extraction solution, the temperature will rapidly decline below 95° C and the ATP will be partially destroyed. (b) If the periphyton community is scraped from natural substrata, the intercellular ATP pool will change rapidly during this treatment.

The procedure for the simultaneous extraction of ATP and photosynthetic pigments in cold acetone (Clark *et al* 1978) avoids these problems. Unfortunately this method requires overnight extraction that can potentially introduce additional errors by losses. The method described here exposes periphyton samples to liquid nitrogen with the objective of reducing the time and increasing efficiency of extraction.

Methods

Periphyton communities attached to an artificial plastic tank were collected with soft scraping and were suspended by means of magnetic stirrer for subsampling.

The methods proposed by Clark *et al* (1978) were used both for pigment and ATP extraction, as well as for acetone distillation. The Tris-buffer method of Holm-Hansen (1973) was also applied for ATP extraction. Five-ml volumes of periphyton suspensions were added to 20 ml of boiling Tris-buffer (0.02M, pH 7.75) in 2-ml fractions to avoid cooling of the extraction solution.

Treatment with liquid nitrogen was realized as follows: 5 ml of sample were put into an aluminium tube and then immersed for 3 min in liquid nitrogen. After holding for 5 min at room temperature, 20 ml of acetone (100%, buffered with sodium bi-

Wetzel, R.G. (ed.), Periphyton of Freshwater Ecosystems.
 ISBN 90 6193 768 X.

Table 1. Statistical parameters of 665 nm absorbance data obtained by different methods of extraction.

	Overnight extraction	Nitrogen treatment	Grinding and overnight extraction	Nitrogen treatment
N samples	5	5	5	5
Average values	0.163	0.199	0.300	0.200
Standard dev.	0.015	0.024	0.033	0.035
CV %	9	12	11	17
	$F(1, 8) = 7.93$ $p<0.025$		$F(1,8) = 22.03$ $p<0.01$	

carbonate) were added. For pigment extraction, another 25-ml quantity of acetone was added; acetone distillation for ATP was made without this addition. Immediately after collection, ATP samples were assayed at room temperature by the firefly luminescence reaction method of a LKB Wallac luminometer.

Results

The extraction efficiency of the photosynthetic pigments after treatment with liquid nitrogen was greater than with overnight extraction at −20° C (Table 1). However, extraction following exposure to liquid nitrogen was less efficient than extraction by grinding. The absorption spectra of the pigments did not differ (Fig. 1); these results indicate an absence of denaturation processes on pigments during the nitrogen treatment.

The efficiency of ATP extraction after freezing with liquid nitrogen was significantly greater than with Tris-buffer extraction (Table 2). These preliminary data are in accordance with the results of other investigators (Clark *et al* 1978). The data in Table 2 are corrected for the firefly reaction efficiency by means of internal standards of ATP (LKB Wallac), and they are expressed in mV because the relation ATP = f (mV) is linear in the range measured. The reduction of ATP levels in samples ground for 1 min before Tris-buffer extraction of ATP indicates possible degradative alternations of the ATP pool during mechanical treatment.

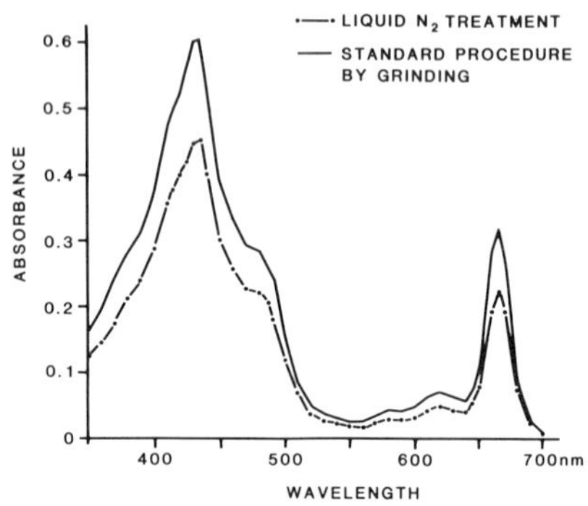

Fig. 1. Absorption spectra for photosynthetic pigments extracted in acetone by grinding and by liquid nitrogen treatments.

Table 2. Statistical parameters of relative ATP values obtained by different methods of extraction.

	Tris-buffer extraction	N_2 treatment, acetone extraction	Tris-buffer extr. after grinding
N samples	3	3	3
Average values	60.5	84.0	33.5
Standard dev.	12.5	9.27	3.33
	$F(1.4) = 6.93$ $0.05<p<0.10$		

References

Ausmus, B.S., 1973. The use of ATP assay in terrestrial decomposition studies. Bull. Ecol. Res. Comm. (Stockholm) 17: 223–234.

Clark, R.J., Messenger, D.I., Dickson, K.L. & Cairns, J., Jr, 1978. Extraction of ATP from Aufwuchs communities. Limnol. Oceanog. 23:1055–1059.

Herman, L.E. & Temple, K.L., 1979. Comparison of ATP, phosphatase, pectinolyase and respiration as indicators of microbial activity in reclaimed coal strip mine spoils. Soil Sci. 127:70–73.

Holm-Hansen, O., 1973. The use of ATP determinations in ecological studies. Bull. Ecol. Res. Comm. (Stockholm) 17:215–222.

Kennicut, M.C., 1980. ATP as indicators of toxicity. Water Res. 14:325–328.

MacKinnon, M.L., Chassaing, B. & Olive, Ph., 1977. First data on ATP in Lake Leman. Verh. Internat. Verein. Limnol. 20:914–920.

Nelson, P.O. & Lawrence, A.W., 1980. Microbial viability measurements and activated sludge kinetics. Water Res. 14: 217–255.

Oades, J.M. & Jenkinson, D.S., 1979. Adenosine triphosphate content of the soil microbial biomass. Soil Biol. Biochem. 11:201–204.

Wissmar, R.C., 1981. Chemical composition and adenylate energy charge ratios of Scenedesmus abundans (KIRCHNER) CHOD. as a function of nutrient limitation. Verh. Internat. Verein. Limnol., 21:1449–1453.

Author's address:
R. Antonietti
Institute of Ecology
University of Parma
Via Primo, Groppi 16
Parma, 43100, Italy

CHAPTER 36

Periphyton pigment analyses using paper chromatography

P.V. ELORANTA

Abstract. Paper chromatography was used to study the community structure, the proportion of degraded chlorophylls to the total pigment content, and the chlorophyll and carotenoid content of the epilithic periphyton algae from Lake Tahoe, California-Nevada. Analyses were made using both samples directly from the lake and cultured periphytic algae. The extraction was made with boiling 100% methanol; two dimensional chromatography was used to separate the pigments. The relative contribution of different pigment fractions to total light absorption by the complete periphyton extract was estimated by planimetric integration of absorption curves. From the pigment structure, estimates were made of the proportions of different algal groups in the samples. Further, the results gave information on the proportions of active and degradated forms of chlorophyll *a*, as well as the chlorophyll *a* content per unit biomass. The results showed the rarity of phaeophytin *a* in the epilithic algal communities in Lake Tahoe, but often some proportion of chlorophyll *a* occurred as other forms of pigment degradation products. The percentage of chlorophyll *a* of the complete pigment extract varied according to environmental conditions, and the productivity and vitality of the algae.

Introduction

Evaluation of the biomass and the community structure of periphyton usually encounters several well-known difficulties due to the varying proportions of inorganic sediments, heterotrophic organisms and detritus to the total mass of the periphyton from natural substrates. Studies concerning the structure of algal communities in periphyton commonly account only for diatom frustules; the proportion of other algal groups to the total algal mass is often disregarded because of counting problems. The diatoms, however, are often attached on filamentous algae, which provide a secondary substrate for attaching diatoms (e.g., *Synedra, Achnanthes*) and protection for moving diatoms (*Rhopalodia, Navicula*, etc.). The proportion of organic matter of periphyton samples can be measured by combustion and estimation of carbon content, but it is difficult to evaluate organic matter content of algae, animals, and organic detritus. Algal biomass has often been estimated by measuring the chlorophyll *a* content spectrophotometrically using various extracting solvents and equations for calculations. However, spectrophotometric determinations of chlorophylls contain several sources of error and even in the best case the information obtained is limited (Rai 1980; Marker *et al* 1980).

Many results are reported for freshwater plankton and periphyton communities concerning the proportion of pheopigment *a* (and pheophytin *a*) to total chlorophyll *a*, but rather few analyses by chromatography exist (Hallegraeff 1976a, b, 1977;

Wetzel, R.G. (ed.), Periphyton of Freshwater Ecosystems.
 ISBN 90 6193 768 X.

Rieman 1978a, 1980, 1982). Already ten years ago Rai (1973) assumed that for a complete analysis of all chlorophylls and their degradation products there is probably no alternative to chromatographic methods. The cited works with chromatography have demonstrated the rarity of pheophytin *a* in freshwater plankton communities and the commonness of other degradation products of chlorophyll *a* (especially chlorophyllides). The occurrence of pheophytin *a* in freshwaters have been shown in connection with high densities of zooplankton grazing on algae (Currie 1962; Rieman 1978a, 1982).

The main objectives of the present study were 1) to study, using paper chromatography, the proportion of chlorophyll *a* to the complete periphyton pigment extract, 2) to estimate the main community structure by pigment composition, and 3) to study the changes in pigment structure and pigment concentration during bioassays with cultured and natural periphyton algae.

Material

Samples from natural substrates were collected for pigment analyses by Scuba-diving (Dr. Stan L. Loeb and Dr. John E. Reuter, Tahoe Research Group, University of California, Davis) from the eu- and sublittoral zones of Lake Tahoe (California-Nevada). The periphytic algal communities in the lake consist mainly of epilithic algae extending from the shore line down to the depth of 150–200 m (Loeb 1980).

Filamentous green algae dominated in the eulittoral splash-zone of Lake Tahoe, mixed in some areas with diatoms. Typical green algal genera in this zone were *Ulothrix, Spirogyra, Mougeotia* and *Zygnema.* In the deeper zones, diatoms and filamentous blue-green algae dominated, growing sometimes as mixed, sometimes as almost pure, communities. Blue-green algae grew near the surface of the substrata (rocks), whereas diatoms during optimum growth almost covered the blue-green

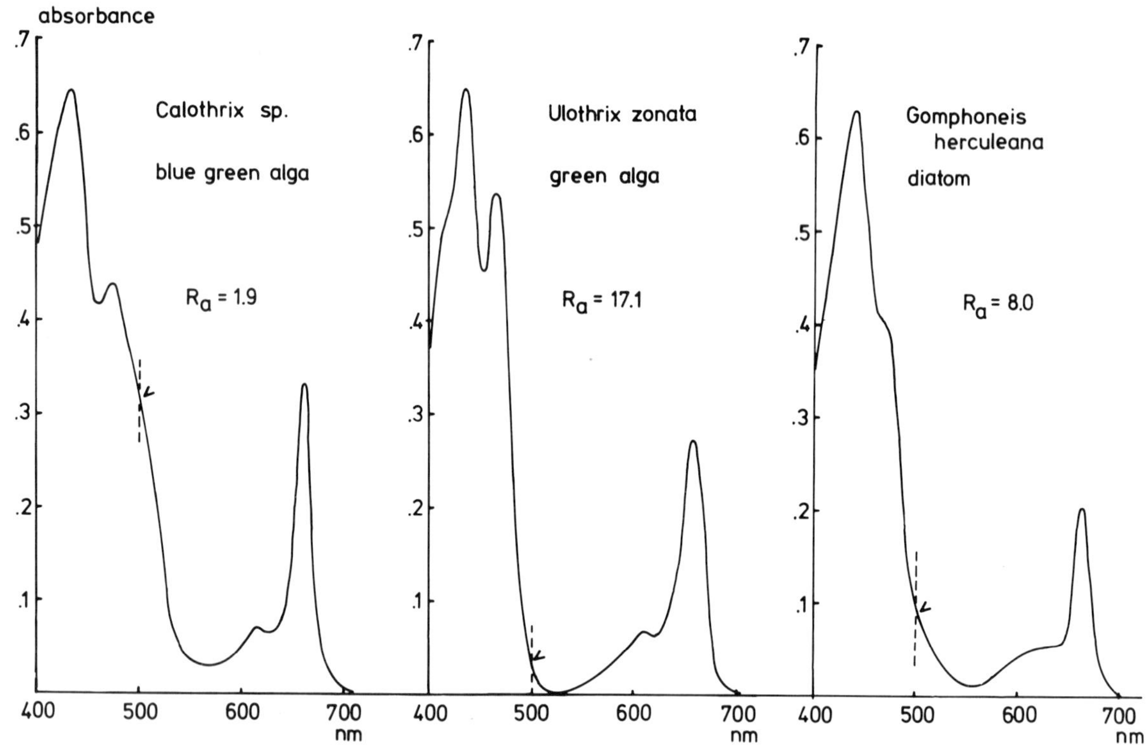

Fig. 1. Light absorption curves between 400 and 700 nm of the complete 100% methanol extract of three epilithic algae isolated from Lake Tahoe, California-Nevada. For R_a see explanation in text; the arrows idicate the absorbance at the wavelength of 500 nm.

algae by growing on the blue-green colonies and/or making long, branched stalked colonies (up to 10 cm long) and shading the other algae. Dominant diatoms in epilithic communities in Lake Tahoe were *Gomphoneis herculeana, Synedra ulna* var. *spathulifera, Rhopalodia gibba, Fragilaria* spp., and *Epithemia* spp. *Calothrix, Tolypothrix, Dichotrix* and *Nostoc* were common blue-green algal genera in the epilithon.

Samples were collected by brush-syringe method (see more detailed description of the method in Loeb 1980) from the depths of 0.5, 2, 8, and 16 m at different localities during the winter and spring 1981–1982. Some species of each major algal group were isolated and cultured in laboratory by Dr. Varpu Eloranta for the further experimental work and for comparative pigment analyses.

Methods of chromatographic pigment analyses

Whole quantitative samples collected from the rocks in the lake, or a sufficient quantity of cultured algae, were centrifuged and the water was removed as completely as possible. Some samples were extracted as fresh material immediately after sampling; others were stored in a freezer (−15° C) for a few days. According to Marker (1980) this pre-freezing may increase extraction from green algae, especially when grinding is impractical. The samples were extracted in 2.0 ml of 100% methanol boiling in a water bath for 2 3 min with simultaneous mixing and grinding with a glass pestle. After the second centrifugation, 1.0 ml of the extract was taken for the starting spot of the chromatogram and 0.5 ml was diluted with 100% methanol for the absorption spectrum of the complete extract.

The initial spot was placed in dim light on Whatman No. 3 paper and the methanol extract was evaporated with flowing nitrogen gas. Two dimensional chromatograms were developed using an eluent of benzine (b.p. 100–140° C) (10): petrol ether (b.p. 40–60° C) (2.5): acetone (2) (Bauer 1952; Hallegraeff 1976a) as the first eluent and the mixture of benzine (10): petrol ether (2.5): acetone (1): methanol (0.25) as the second eluent (Hallegraeff 1976a). After separation of pigments, a facsimile was drawn and the spots were cut from the paper and eluted again with 4.0 ml of hot 100% methanol. Light absorption curves of all separated pigments were made spectrophotometrically (400–700 nm). The pigments were identified by their absorption spectra and their relative position on the chromatogram. No attempt was made to estimate the water-soluble phycobilins and the different forms of transformed carotenoids. The relative contribution of different pigment fractions to total light absorption (400–700 nm) of the complete extract was calculated by planimetric integration of absorption curves. The absolute concentrations of chlorophyll *a* in the extracts were calculated using the equation of Iwamura, *et al* (1970) and measuring directly from the absorption spectra using the specific adsorption coefficient of 77.0 (cf. Riemann 1978b).

Results

Estimation of community structure by chromatography

The epilithic communities in Lake Tahoe were rather simple in structure, consisting mainly of green algae, blue-green algae, and diatoms. The proportions of those major groups varied both spatially and seasonally. The green algae and diatoms showed wide seasonal variation in growth and occurrence, whereas blue-green algae grew more slowly and remained in place throughout the year.

The first attempt to estimate the proportions of main algal groups with the pigment analyses was based on the light absorption spectra of the complete pigment extract. Because of differences in the carotenoid composition in different algal groups, the blue portions of the absorption spectra (400–550 nm) differed clearly between the algal groups (Fig. 1). In the pigment extract from the blue-green alga, the absorbance at 500 nm was almost 50% of the maximum absorbance at the blue part of the absorption curve, whereas the same percentage for the diatom was only *ca.* 15% and less than 10% for green alga (Fig. 1). The R_a-ratio of absorbances (average of absorbances at

Table 1. The ranges of the percentages of some characteristic pigments of the total light absorption of the methanol extract for some cultured epilithic blue-green algae, green algae and diatoms from Lake Tahoe.

Pigment	Blue-greens	Diatoms	Greens
Myxoxanthophyll	34–48	—	—
Ketocarotenoids (2)	5–14	—	—
Chlorophyll *c*	—	13–19	—
Fucoxanthin	—	25–39	—
Diadinoxanthin	—	5–10	—
Chlorophyll *b*	—	—	17–32
Lutein	—	—	9–32

430 and 450 nm/average of absorbances at 500 and 520 nm) was calculated for cultured periphytic algae and for samples from natural communities (Fig. 2). R_a-values for pure cultures ranged 1.8–2.4 for blue-green algae, 6.1–10.7 for diatoms and 12.9–51.5 for green algae. Plotting of results for natural communities from different sampling depths in the lake showed a decreasing tendency of R_a-values with increasing depth. This result indicates the increasing proportion of blue-green algae in the deeper waters. The communities from the shallow water were dominated by green algae with high R_a-values and, in some places, by diatoms (*Gomphoneis-Synedra*), whereas at some other stations at the same depth, mainly blue-green algae grew (Fig. 2).

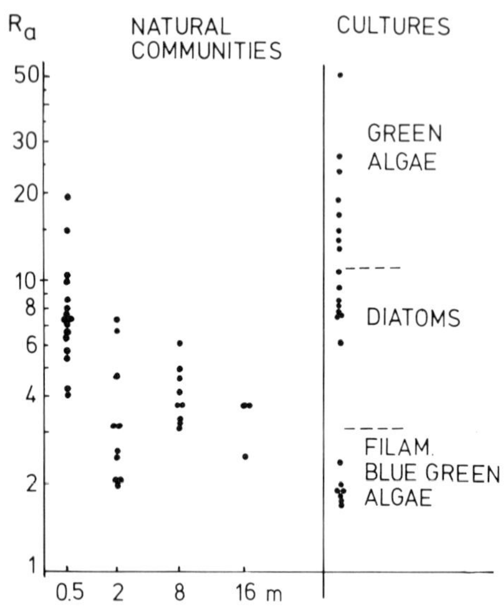

Fig. 2. R_a-values for samples collected from different depths of Lake Tahoe littoral and for algae isolated from epilithic communities of Lake Tahoe and cultured in laboratory (R_a = aver. absorbance at 430 and 450 nm/average absorbance at 500 and 520 nm).

The pigment structure of the samples studied by chromatography also provided information on the main algal groups occurring in the communities. Each major algal group contains some characteristic pigments (e.g., chlorophyll *b* and lutein in green algae, but which are missing in blue-greens and diatoms). Characteristic pigments for blue-green algae are myxoxanthophyll, and some ketocarotenoids; chlorophyll *c*, fucoxanthin, and diadinoxanthin occur in diatoms. The presence or absence of those pigments indicate the relative abundance of main algal groups. By knowing the wide-range percentage of each pigment in growing cells of dominating species/groups as measured from unispecies cultures (Table 1), it was possible to estimate the approximate contribution of each group to the total community.

Chlorophylls and chlorophyll a derivatives

Chlorophyll *a* is the main photosynthetic pigment in most algae, but its content in the cells is known to vary widely depending on the algal species and on several environmental factors (light, nutrients, temperature, etc.) and on the physiological stage of the cells. Thus, it was found also in Lake Tahoe that the same mass of periphyton from the same place contained different amounts of chlorophyll *a* on different sampling dates. The proportion chlorophyll *a* of the light absorption of the complete pigment extract varied similarly with time and com-

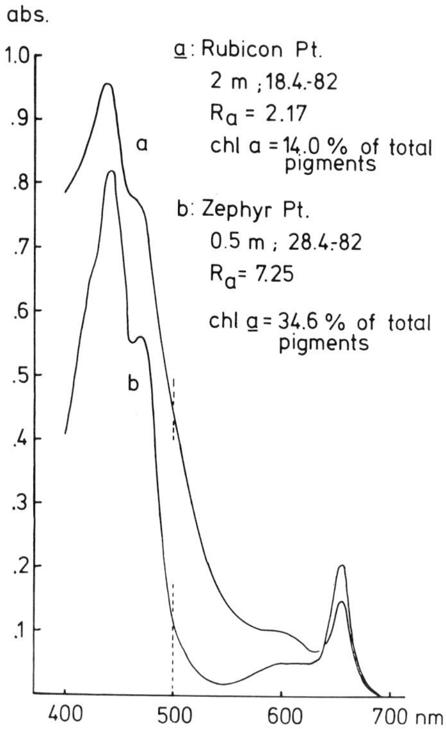

Fig. 3. The light absorption curves of the complete pigment extracts in methanol of two epilithic epiphyton samples from Lake Tahoe.

munity type (Fig. 3).

As measured from the light absorption curves of the chromatographically pure chlorophyll *a* from different algal communities, the structure of chlorophyll *a* differed somewhat between major algal groups. The ratio of integrated areas of the absorption curve between blue (400–550 nm) and red parts (550–700 nm) was $1.47 \pm 0.19 : 1$ (n = 13) for blue-greens, $1.67 \pm 0.23 : 1$ (n = 10) for green algae, and 1.80 ± 0.08 : (n = 6) for diatoms.

The ratio between chlorophyll *a* and chlorophyll *b* content in periphytic green algae was $1.78 \pm 0.52 : 1$. The corresponding ratio for chlorophyll *a* and chlorophyll *c* in periphytic diatoms was $2.25 \pm 0.35 : 1$.

Mg-free degradation products of chlorophyll *a* were not found on the chromatograms of the samples from the lake. In contrast, other degraded products of chlorophyll *a* were found frequently (0–20% of the total chlorophyll *a*). In the presence of myxoxanthophyll, the major carotenoid in blue-green algae in Lake Tahoe and chlorophyll *c* of diatoms (which like the chlorophyllides) did not migrate from the center of chromatogram. The estimation of the proportion of these pigments to the total extract was difficult and proximate.

Chlorophyll a *in algal assays*

During the bioassays made with cultured periphytic blue-green species (*Calothrix* sp.) and with diatoms collected as fresh from the lake (*Gomphoneis herculeana* mixed with some other diatoms) also the proportion of chlorophyll *a* of the complete pigment extract, as well as the content of chlorophyll *a* per unit mass, was also measured to demonstrate growth activity under different conditions. It was assumed that under similar environmental conditions, increased chlorophyll *a* content indicated increased growth rate. Simultaneous measurements of ^{14}C-fixation were made to measure the algal production, but those results are not yet available for comparison.

Growth of *Calothrix* was fastest at the warmest temperature (+ 25° C), and simultaneously had the greatest proportion of chlorophyll *a* to total pigment extract and the highest concentration of chlorophyll *a* after 4 weeks (Fig. 4). Corresponding results obtained in the example at + 5° C with low chlorophyll *a* percentage (10.6% of the total pigments) was seen in the natural blue-green communities at Rubicon Point (2 m; 18.4.82) growing in cold water (Fig. 3).

In bioassays with diatoms, the growth stimulated by added nutrients was indicated by an increased chlorophyll *a* percentage of the total pigments (Fig. 5).

Discussion

Analysis of the concentrations of the photosynthetic pigments is almost the only way to estimate the share of algae to the total periphytic matter containing also variable proportions of inorganic material, detritus, and animals. The quantitites of chlorophyll *a* have been commonly used also in periphyton studies as a measure of algal biomass. The analyses have normally made using acetone, meth-

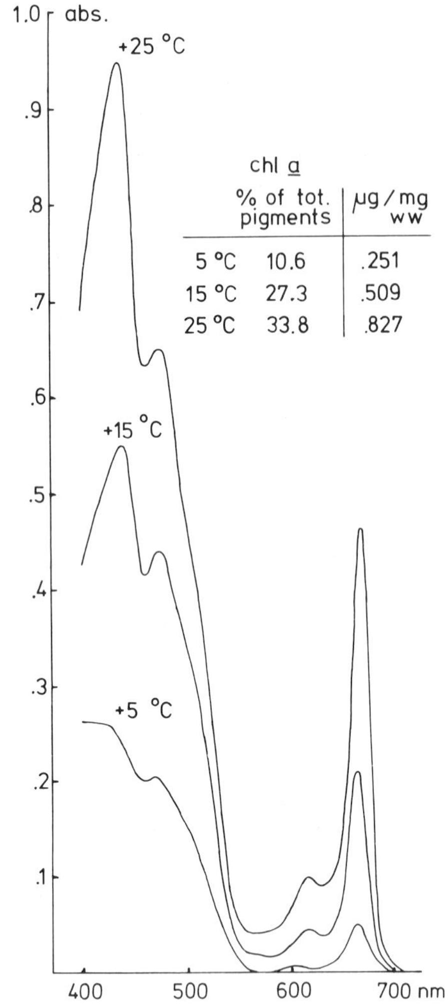

Fig. 4. The light absorption curves of the complete pigment extracts in methanol of *Calothrix* sp. cultures, cultured 4 weeks in N-free media at different temperatures.

anol and ethanol as extraction solvents and readings have been taken at the various wavelengths depending on equations developed from specific extinction coefficients. The degradation products, often called 'pheophytins', have been estimated after acidification and rereading of the absorbances.

During the last ten years many doubts are cast on the spectrophotometric methods of chlorophyll determination. In those methods, e.g. for chlorophyll *a* determination, other chlorophylls, especially chlorophyll *b* (Reimann 1978b), carotenoids (Reimann 1978a) and by the degradation products of chlorophyll *a* (e.g., Rai 1973) interfere with the specific analyses. Further, trichomatic equations give positive concentrations for chlorophyll *b* and *c* also for samples which undoubtedly do not contain those pigments (Rai 1973, 1980).

The results reported on the degradated products of chlorophyll *a* in freshwater plankton communities (Hallegraeff 1976a, b, 1977; Riemann 1978a, 1982) demonstrate how rarely pheophytin *a* occurs in the algal communities, although the degradation products analyzed with the spectrophotometric acidification procedure are often called pheophytin.

The results from the periphyton of Lake Tahoe on the degradation products of chlorophyll *a* coincided with those of Hallegraeff and Riemann, because I did not find pheophytin *a* on any chromatogram of epilithic periphyton algae. The occurrence of pheophytin is usually connected to the dense zooplankton population grazing on algae, but in Lake Tahoe animals in periphyton communities were sparse; small chironomid larvae were found in samples only occasionally. Instead of pheophytin, other degradation products (chlorophyllides) were found in the periphyton samples and those are undetectable without chromatography (cf. Hallegraeff 1976b, 1977; Riemann 1978a).

Most of the quantitative community analyses on periphytic algae contain results only on diatom

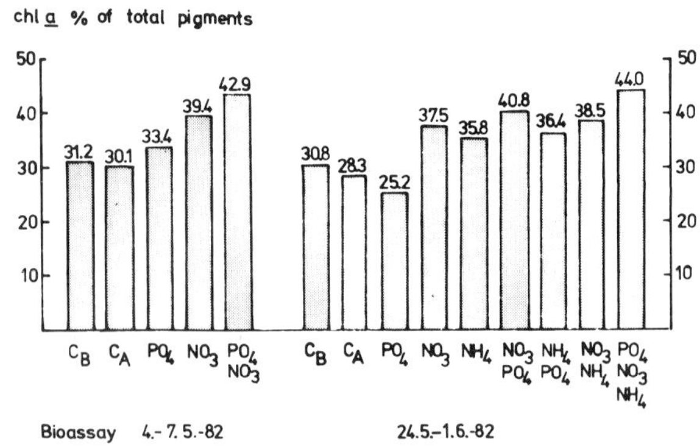

Fig. 5. The response of *Gomphoneis herculeana* communities to nutrient additions indicated by the percentage of chlorophyll *a* of the total pigment extract (C_B = control with lake water at the beginning of the incubation, C_A = control after the incubation).

numbers. Diatom cell counts are possible and the results can be transformed to fresh weight biomass from the average cell volume of each species. The counting of filamentous or colonial periphytic algae by cells or by volume in mixed periphyton samples is very difficult. Thus, the estimation and description of the proportions of major algal groups by microscopical techniques is difficult. Chromatography provides a tool for more precise, although still rather approximate, estimation of the proportion of main algal groups within the total community (Hallegraeff 1977). However, rather few results are reported on freshwater algal communities using chromatographic methods (Hallegraeff 1976, 1977; Riemann 1978a, 1982; Rott 1980). All of these reports analyzed pigments in phytoplankton communities.

Hallegraeff (1977) mentioned that chlorophyll *c* was found only in small quantities and that it was more important in plankton samples dominated by dinoflagellates than in samples dominated by diatoms. In the epilithic periphyton samples from Lake Tahoe, chlorophyll *c* had an important contribution to the total pigment extract whenever diatoms were present in significant quantities. Sometimes, however, difficulty was encountered with separation of chlorophyll *c* and chlorophyllides, both of which remained in the center of the chromatogram.

Myxoxanthophyll of the blue-green algae was also recorded by Hallegraeff (1977) in plankton samples, but only in small quantities, in spite of the blue-green algal bloom. In the epilithic filamentous blue-green found in Lake Tahoe, myxoxanthophyll was the major carotenoid; the percentage ranged from 34–48% of total light adsorption of the complete pigment extract in methanol. These examples demonstrate the need of controlled analyses done with the isolated algal strains from the waterbody and community studied before estimations of the contribution of algal groups to total community by pigment extracts are possible.

Chlorophyll *a* determination has been used as an indirect biomass measure also in bioassays and it is shown by several authors that the chlorophyll maximum is reached before the maximum of particle volume or cell density, and indicates decreasing chlorophyll *a* content in aging cells with nutrient depletion (e.g., V. Eloranta 1976; Hallegraeff 1976b). Halldal & French (1958) found that the maximal growth temperature for blue-green *Anacystis* resulted in maximal chlorophyll concentrations. According to the results of Castenholz (1969), suboptimal growth temperatures of thermophilic algae result in lower chlorophyll concentrations than those obtained at optimal temperatures. In the experiments with the filamentous blue-green species (*Calothrix* sp.) isolated from Lake Tahoe periphyton and cultured in constant illumination, growth was slowest at the low temperature (+ 5° C) and fastest at the highest temperature (+ 25° C). The low chlorophyll *a* content per unit mass and also the low percentage of chlorophyll *a* of the total pigment extraction may explain the slow growth of attached blue-green algae in Lake Tahoe (cf. Loeb 1980), where the water temperature is higher than + 15° C for only short periods of the year.

The growth of diatom community of *Gomphoneis herculeana* was found to be limited in Lake Tahoe by nitrogen (Loeb *et al,* unpubl. results). Stimulation of growth by nitrogen was seen both as an increased percentage of chlorophyll *a* to total pigments and also as increased chlorophyll *a* content per unit mass. The results demonstrate the usefulness of chromatographic pigment analyses in the algal experiments. Chromatographical analyses give much more information on algal response to the changes in environmental factors or to toxicity of water or other compounds.

Acknowledgements

I wish to thank Dr. Stan Loeb and Dr. John Reuter, Tahoe Research Group, University of California, Davis, for diving the samples from the cold lake and both Prof. Dr. Charles R. Goldman and Dr. Stan Loeb (UCD) for supporting all needs in the analyses. My warm thanks are also due to Dr. Varpu Eloranta, who isolated and cultured several periphytic species from Lake Tahoe for pigment analyses and algal assays. This work was supported by State Waters Resources Control grant No. 440341-20742.

References

Bauer, L., 1952. Trennung der Karotenoide und Chlorophylle mit Hilfe der Papier-Chromatographie. Naturwissenschaften 39:88.

Castenholz, R.W., 1969. Thermophilic blue-green algae and the thermal environment. Bact. Rev. 33:476–504.

Currie, R.I., 1962. Pigments in zooplankton faeces. Nature 193:956–957.

Eloranta, V., 1976. Effects of different process wastes and main sewer effluents from pulp mills on the growth and production of Ankistrodesmus falcatus var. acicularis (Chlorophyta). Biol. Res. Rep. Univ. Jyväskylä 2:3–33.

Halldal, P. & French, C.S., 1968. Algal growth in crossed gradients of light intensity and temperatures. Pl. Physiol. 33: 249–252.

Hallegraeff, G.M., 1976a. Pigment diversity in freshwater phytoplankton. 1. A comparison of spectrophotometric and paper chromatographic methods. Int. Revue ges. Hydrobiol. 61:149–168.

Hallegraeff, G.M., 1976b. Pigment Diversity, Biomass and Species Diversity of Phytoplankton of Three Dutch Lakes. Ph.D. Thesis, Limnol. Lab., Inst. Anim. Physiol., Univ., Amsterdam, 177 pp.

Hallegraeff, G.M., 1977. Pigment diversity in freshwater phytoplankton. 2. Summer-succession in three Dutch lakes with different trophic characteristics. Int. Revue ges. Hydrobiol. 62:19–39.

Iwamura, T., Nagai, H. & Ichimura, S., 1970. Improved methods for determining contents of chlorophyll, protein, ribonucleic acid, and desoxyribonucleic acid in plankton populations. Int. Revue ges. Hydrobiol. 55:131–147.

Loeb, S.L., 1980. The Production of the Epilithic Periphyton Community of Lake Tahoe, California-Nevada. Ph.D. Thesis, Univ. Calif., Davis., 165 pp.

Marker, A.F.H., 1980. A note on the extraction of chlorophyll from benthic algae using methanol. Arch. Hydrobiol, Beih. 14:88–90.

Marker, A.F.H., Nusch, E.A. & Rai, H., 1980. The measurement of photosynthetic pigments in freshwaters and standardization of methods: Conclusions and recommendations. Arch. Hydrobiol, Beih. 14:91–106.

Rai, H., 1973. Methods involving the determination of photosynthetic pigments using spectrophotometry. Verh. int. Ver. Limnol. 18:1864–1875.

Rai, H., 1980. Some problems in determination of photosynthetic planktonic pigments and their decomposition products. Arch. Hydrobiol., Beih. 14:3–13.

Riemann, B., 1978a. Quantitative and qualitative determinations of chlorophylls and phaeopigments in Lake Mossø. Verh. int. Ver. Limnol. 20:674–677.

Riemann, B., 1978b. Absorption coefficients for chlorophylls a and b in methanol and a comment on interference of chlorophyll b in determination of chlorophyll a. Vatten 34:187–194.

Riemann, B., 1980. A note on the use of methanol as an extraction solvent for chlorophyll a determination. Arch. Hydrobiol., Beih. 14:70–78.

Riemann, B., 1982. Measurements of chlorophyll a and its degradation products: A comparison of methods. Arch. Hydrobiol., Beih. 16:19–24.

Rott, E., 1980. Spectrophotometric and chromatographic chlorophyll analysis: comparison of results and discussion of the trichromatic method. Arch. Hydrobiol., Beih. 14:37–45.

Author's address:
P.V. Eloranta
Department of Biology
University of Jyväskylä
SF-40100 Jyväskylä 10, Finland

CHAPTER 37

The utilization of periphyton in waterworks pre-treatment for nutrient removal from enriched influents

A. SLÁDEČKOVÁ, P. MARVAN and J. VYMAZAL

Abstract. The utilization of periphyton activity for the removal of surplus nutrients from enriched influents in drinking water reservoirs has been examined as a possible method of waterworks pre-treatment. By means of batch laboratory, as well as continuous-flow *in situ* model experiments, a marked uptake of nitrate and phosphate by the periphyton biofilm has been demonstrated. The results of model experiments are being tested in a pilot scale 'bioeliminator' with respect to the demands of engineering waterworks practice.

Introduction

The ability of algae to assimilate and accumulate nutrients, especially nitrogen and phosphorus compounds, from the culture media or from enriched waters is well known and often discussed in the literature (e.g. Likens 1972). By means of bioassay techniques using primarily chlorococcal algae as test organisms, the physiological basis as well as kinetics of the nutrient uptake have been studied in detail. The results of these studies have been utilized in tertiary waste treatment processes (Symposium 1978). Much less attention has been given to nutrient uptake by attached periphyton algae, mainly because of their difficult cultivation under laboratory conditions.

In 1961 a group of biologists from Moscow University published a remarkable book summarizing the results of laboratory as well as outdoor experiments with the periphyton community functioning as the so-called 'bioeliminator' (Skadovski 1961; Sládečková 1966). They showed a relatively high uptake of nitrate by attached filamentous green algae and proposed to utilize the process in waterworks pre-treatment. Unfortunately, their projects have not been accepted yet by the engineering practice. Therefore, a similar research program has started in Czechoslovakia aimed at the critical evaluation of former results and at their final engineering application.

Methods

Two types of model have been constructed. The batch laboratory model (Fig. 1) was 1 m long, 0.3 m wide, 0.15 m high and had a volume of 24 l. The circulation of water in the model was provided by a pump motor with a propeller. The flow rate in the central section was 1.45 cm s^{-1}, and in both lateral sections 0.8 cm s^{-1}. The model was illuminated by fluorescent tubes from above. As artificial substrata for the periphytic growth forming the so-called biofilm, a fine mesh plastic screen was used. The model was filled with a brook water enriched with K_2HPO_4 and KNO_3 (for initial concentrations of NO_3-N and PO_4-P, see Figs. 3 & 4). Evaporation losses were compensated for with distilled water. The operation period was during 19 October–2 December 1981.

Wetzel, R.G. (ed.), Periphyton of Freshwater Ecosystems.
 ISBN 90 6193 768 X.

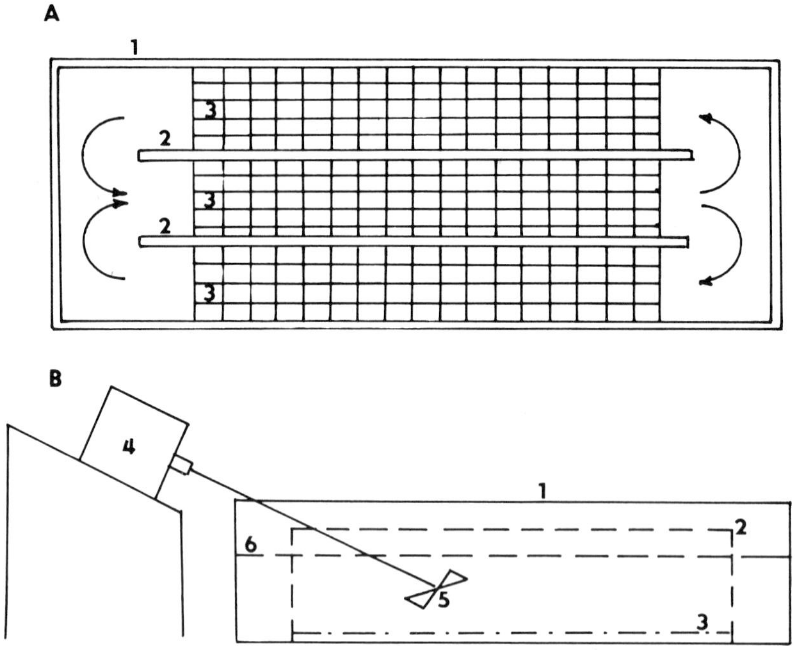

Fig. 1. Batch laboratory model. A. Top view. B. Side view. 1. plastic trough; 2. plastic partition walls; 3. fine mesh plastic screen placed near the model bottom, hole diameter 0.1–0.2 mm; 4. pump motor; 5. propeller; 6. water level.

The continuous-flow *in situ* model (Fig. 2) had several characteristics of an artificial stream (Clark *et al* 1980), but it was placed directly in the outflow from the biological sewage treatment plant. The model was 1 m long, 0.2 m wide, and 0.1 m high with plastic screens in vertical positions, perpendicular to the direction of flow. The operation period was 5–9 April 1982.

The nutrient uptake by the periphyton biofilm was followed by means of control chemical analyses, outside of the model, by an algal assay (AGP), and by documenting the periphyton composition. Ammonia-nitrogen was determined by direct Nesslerization, nitrite by coupling of diazotied sulphanilic acid with N-(1-naphthyl)-ethylenediamine dihydrochloride, nitrate by a photometric procedure with sodium salicylate, and phosphates by the molybdenum blue method (Hofmann *et al* 1965). Total phosphorus was determined according to Popovský (1970).

The determination of algal growth potential (AGP) by a laboratory algal assay was carried out according to Marvan *et al* (1981) with the test alga *Scenedesmus quadricauda*, strain Greifswald/15.

The periphyton species composition was evaluated by the abundance estimation using the five-parted estimation scale (Sládečková 1962).

Results of the batch laboratory model experiment

The uptake of nitrate by the periphyton biofilm in the course of the batch experiment is demonstrated in Fig. 3, and the uptake of phosphorus in Fig. 4. Algal growth potential determined several times in the course of the model experiment showed a similar pattern (Fig. 5). The periphyton biofilm formation was spontaneous, without any inoculation. The development stages, spores, etc. came mostly from the orignial brook water and to a small extent also from the air. In the course of the experiment a conspicuous succession of dominant periphyton constituents could be observed: at the beginning the biofilm was formed predominantly by filamentous blue-green algae belonging to the genus *Plectonema*; later on they were replaced by fila-

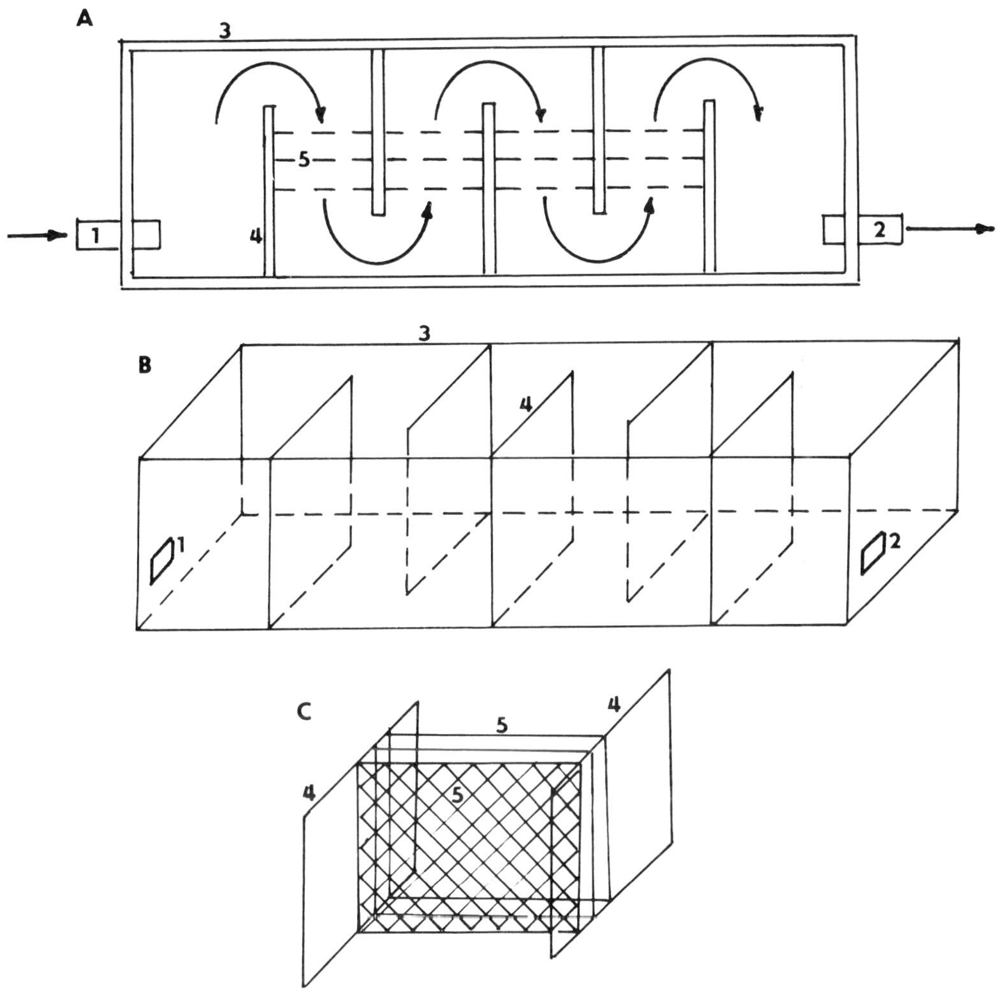

Fig. 2. Continuous-flow *in situ* model. A. Top view. B. Side view. C. Screens, detail. 1. inflow; 2. outflow; 3. plastic trough; 4. plastic partition walls; 5. medium mesh plastic screens; hole diameter 2–3 mm.

mentous green algae *Ulothrix zonata* and *Rhizoclonium hieroglyphicum*. At the end, when the concentrations of nutrients became extremely low, the biofilm consisted of filamentous iron bacteria (*Leptothrix ochracea*). The diatom *Achnanthes minutissima* showed a permanently high abundance during the whole experiment. Other diatoms were much less abundant: *Cocconeis pediculus, C. placentula, Diatoma vulgare, Gomphonema olivaceum, G. parvulum, Melosira varians, Navicula cryptocephala, N. gracilis, N. gregaria, Nitzschia dissipata, Rhoicosphenia curvata, Synedra tenera, S. ulna* and *S. vaucheriae*.

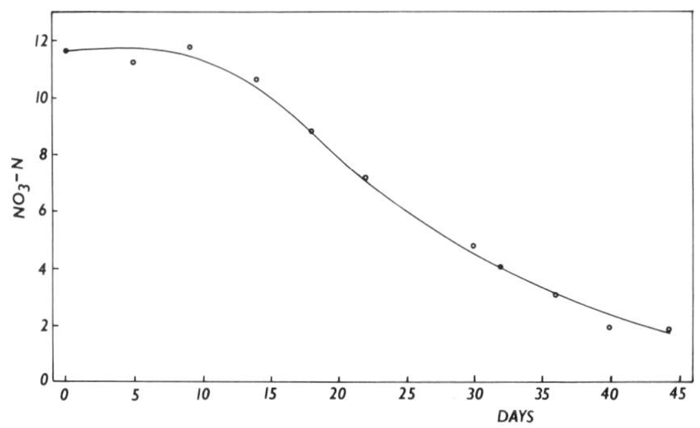

Fig. 3. Uptake of nitrate during the batch experiment. Concentrations of NO_3-N in mg l^{-1}.

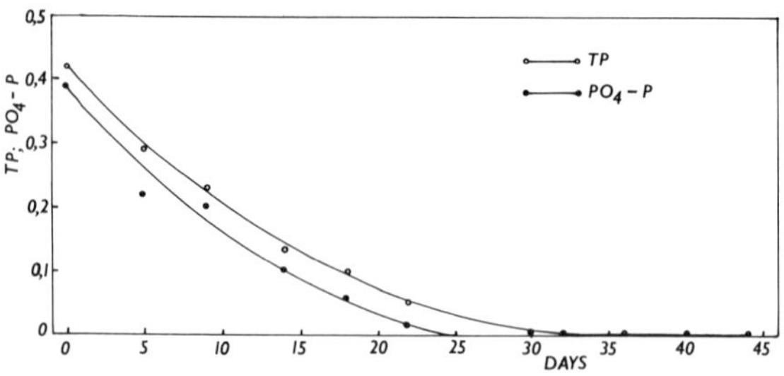

Fig. 4. Uptake of phosphorus during the batch experiment. Concentrations of total phosphorus (TP) and PO_4-P in mg l^{-1}.

Results of the continuous-flow *in situ* model experiment

The nutrient uptake measured during this short-term experiment is demonstrated in Fig. 6. Samples for chemical analyses were taken simultaneously from the inflow and outflow of the model. Colonization of screens and periphyton formation were very rapid, only a few days before the beginning of the experiment, because particles of slimes and algal mats were permanently washed out from the walls and edges of the secondary settling tank in the sewage treatment plant and transported by the current into the model. The dominant periphyton constituents were in this case filamentous green algae *Stigeoclonium tenue* and *Oedogonium* sp. steril. Less abundant were: *Microspora quadrata, Ulothrix zonata, Microthamnion kuetzingianum,* blue-green algae *Phormidium autumnale, Lyngbya*

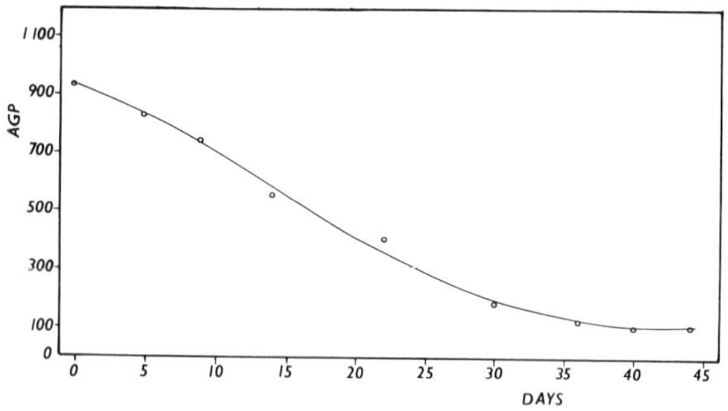

Fig. 5. Nutrient uptake during the batch experiment, demonstrated by the changes of algal growth potential (AGP) during the batch experiment. AGP values = dry weight of the test alga biomass reached in the stationary phase of growth curve (mg l^{-1}).

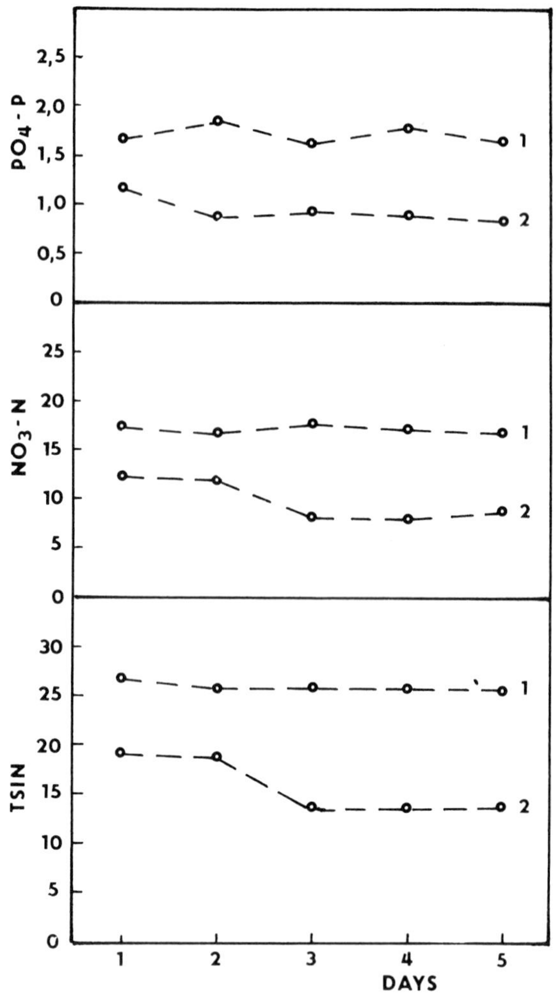

Fig. 6. Nutrient uptake during the continuous-flow experiment. 1. Inflow; 2. Outflow. Concentrations of PO_4-P, NO_3-N, and total sum of inorganic nitrogen (TSIN) in mg l^{-1}.

kuetzingii, and diatoms *Achnanthes minutissima. A. lanceolata, Amphora ovalis, Gomphonema parvulum, Navicula cryptocephala* var. *veneta, Nitzschia fonticola, N. gregaria, N. palea* and *N. thermalis.*

Conclusions

In both model experiments the nitrate uptake by the periphyton biofilm was quite in accordance with the previous results of Skadovski (1961). Moreover, the uptake of phosphorus compounds was also shown, and was confirmed by the determination of algal growth potential. The nutrient uptake was most efficient during the first period of the batch experiment and was observable even from the very beginning of the short-term continuous-flow experiment. In practical applications these results indicate that it would be best to allow the biofilm formation only during a limited period of the exponential growth phase and then replace the colonized screens by new ones. The periphyton biomass may be washed off the screens and utilized as fodder or fertilizer.

At present these results are being tested in a pilot scale model of bioeliminator: a 10 m long channel with the possibility of variable insertion of plastic screens. Such a device should be placed into small enriched influents transporting surplus nutrients from agricultural land into drinking water reservoirs and function as the biological method of waterworks pre-treatment. In the temperate climate zone it could be in operation nearly all the year through, with minimum cost and energy requirements. Moreover, the periphyton biomass may be harvested and utilized as a valuable nutrient-rich material itself.

References

Clark, J.R., Rogers, J.H., Dickson, K.L. & Cairns, J. Jr., 1980. Using artificial streams to evaluate perturbation effects on Aufwuchs structure and function. Wat. Res. Bull. 16: 100–104.

Hofmann, P. *et al*, 1965. Standard Methods of Chemical Analysis of Water. SNTL Praha, 449 pp. (in Czech).

Likens, G.E., 1972. Nutrients and Eutrophication: The limiting-nutrient Controversy. Am. Soc. Limnol. Oceanogr., Spec. Sym. Vol. 1, 328 pp.

Marvan, P., Přibil, S., Sládečková, A & Źáková, Z., 1981. Proposal of a standard method for the determination of the trophic potential of water. Wat. Mgmt B 31:5–8 (in Czech).

Popovský, J., 1970. Determination of total phosphorus in fresh waters. Int. Revue ges. Hydrobiol. 55:435–443.

Skadovski, S.N., 1961. Periphyton communities in the function of bioeliminator. Moscow University Press, 363 pp (in Russ.).

Sládečková, A., 1962. Limnological investigation methods for the periphyton ('Aufwuchs') community. Bot. Rev. 28: 286–350.

Sládečková, A., 1966. The significance of the periphyton in reservoirs for theoretical and applied limnology. Verh. int. Ver. Limnol. 16:753–758.

Slebodník, L., 1982. Elimination of Nutrients by Means of Periphyton. Thesis, Inst. Chem. Technol., Praha, 90 pp. (in Slovakian).

Symposium, 1978. Experimental use of algal cultures in limnology. Mitt. int. Ver. Limnol. 21, 607 pp.

Authors' address:
A. Sládečková/J. Vymazal
Department of Water Technology
and Environmental Engineering
Institue of Chemical Technology
Praha, Czechoslovakia

P. Marvan
Department of Hydrobotany
Botanical Institute
Czechoslovak Academy of Sciences
Brno, Czechoslovakia

CHAPTER 38

Nutrient removal by epiphytic microorganisms of *Phragmites communis*

A. KURATA

Abstract. Many lagoons occur around Lake Biwa and some of these lagoons are extensively eutrophicated because of influx of the domestic waste water and cultivated field runoff. *Phragmites communis* grows abundantly and it occupies vast areas in these lagoons. The denitrifying activity of epiphytic microorganisms colonized on *P. communis* and artificial substrata was examined using the Giltmy medium. For 20 days' incubation, nitrite and nitrate were actively reduced by epiphytes. Most of the epiphytes colonized on different kinds of macrophytes or collected from various lagoons showed active denitrification. The denitrifying activity of epiphytes can be very effective in nitrogen removal in the eutrophicated lagoons.

Introduction

Lake Biwa is the largest freshwater lake in Japan and is located in the center of the main island. There are many lagoons and reservoirs around Lake Biwa. Recently, many lagoons have been eutrophicated extensively by the influx of the domestic waste water and cultivated field runoff. It is reported that very high community densities of epiphytes grow on substrate in eutrophicated rivers (Aizaki 1978). It is assumed that the high concentrations of nutrients contained in the waste water are purified in the lagoons by the very high population densities of microorganisms, phytoplankton and epiphytes on *Phragmites communis,* which cover a vast area of lagoons. In particular, it is believed that the epiphytes on *P. communis* are very important in the function of self-purification and nutrient removal in many lagoons. However, the ecological role and function of epiphytes on the emergent plants in lagoons and reservoirs have not yet been made clear, even qualitatively.

In this paper, an attempt is made to clarify the denitrifying activity of epiphytic microorganisms on *P. communis* and artificial substrata colonized in lagoons.

Materials and methods

Two stations were established in Nishinoko Lagoon, which is one of the largest lagoons, though shallow (maximum 2.0 m depth), of Lake Biwa. In this lagoon, *P. communis* grows well with very high densities and it colonizes about 167 ha. Locations of the sampling stations and the distribution of *P. communis* are shown in Fig. 1. Very high nutrient concentrations of influent flow into Station 1 and water from the lagoon flows out from Station 4.

For the enumeration of heterotrophic bacteria among epiphytic microorganisms, three kinds of media were employed. The composition of these media are shown Tables 1, 2 & 3, respectively. To ascertain the denitrifying activity of heterotrophic bacteria, two kinds of media were employed for the nitrite reducer and nitrate reducer according to the

Wetzel, R.G. (ed.), Periphyton of Freshwater Ecosystems.
 ISBN 90 6193 768 X.

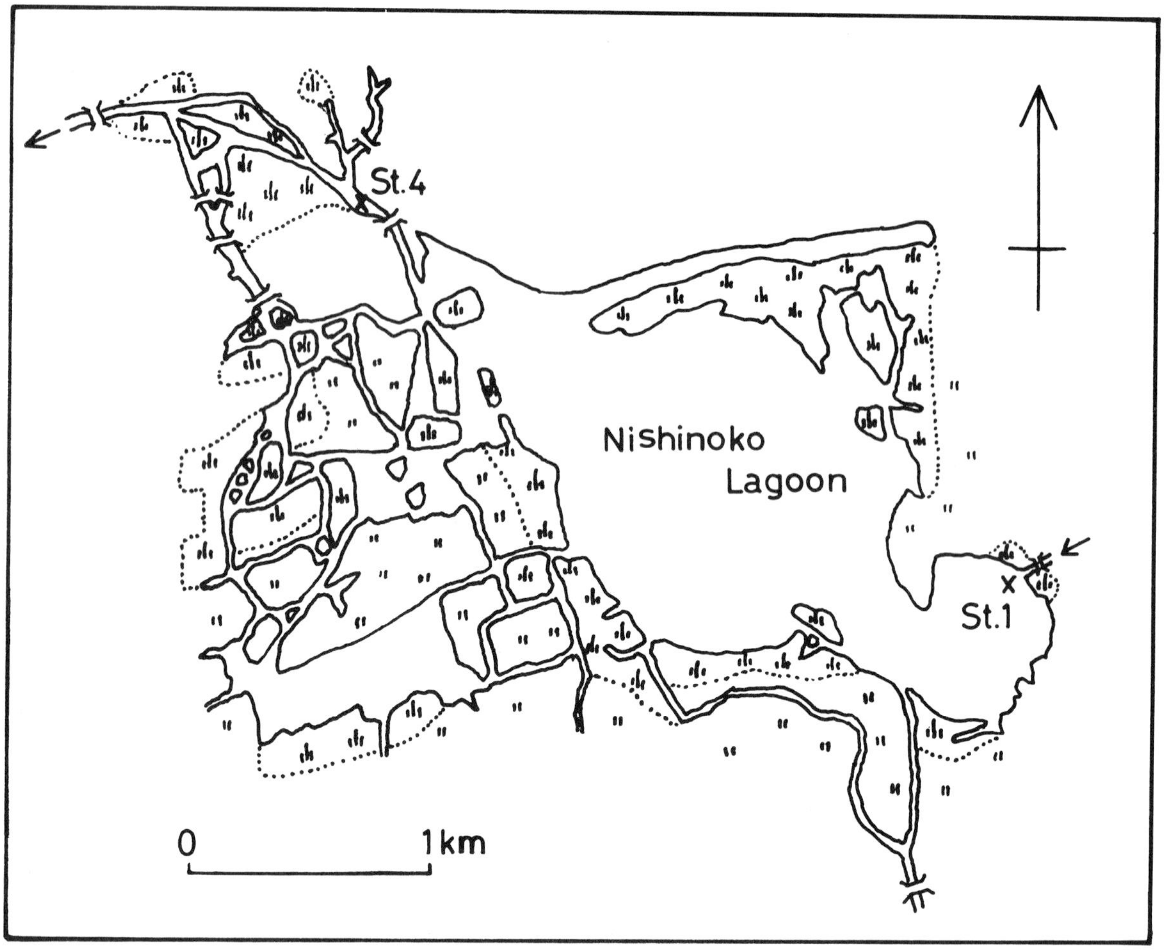

Fig. 1. Location of the sampling stations set up in Nishinoko Lagoon. Triple markings = *Phragmites communis* zone; double markings = paddy fields.

Table 1. The broth agar medium for the enumeration of epiphytic heterotrophs.

Component	Amount
Beef extract (g)	10.0
Polypeptone	10.0
NaCl	5.0
Agar	12.0
Tap water (l)	1.0
pH 7.2	

Table 2. The YPA medium for the enumeration of epiphytic heterotrophs.

Component	Amount
Polypeptone (g)	5.0
Yeast extract	3.0
Agar	12.0
Tap water (l)	1.0
pH 7.2	

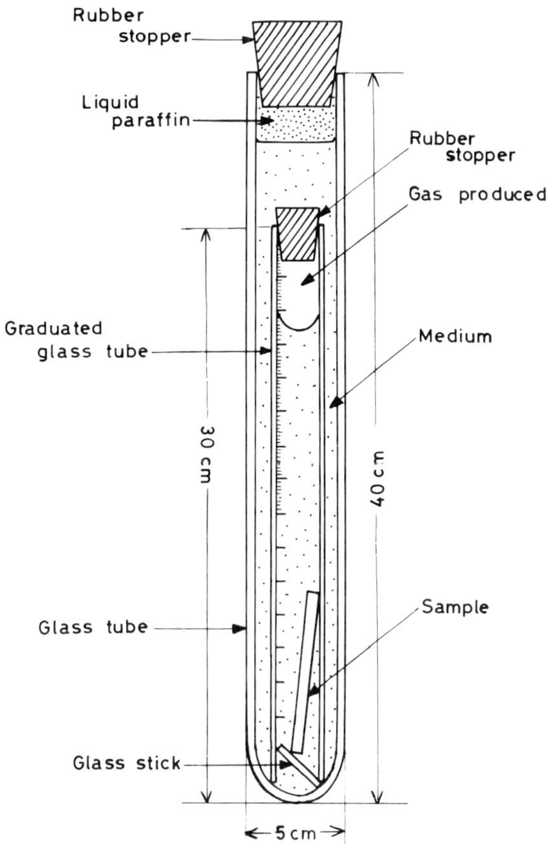

Fig. 2. The epiphyte incubation tube for the measurement of nitrite and nitrate reduction activities.

method of Giltmy (composition of these media are shown in Tables 4 & 5). For the enumeration and isolation of actinomycetes and fungi, antibiotics amended media were employed for each organism. Colonies of both organisms grown on petri-dishes were observed and identified microscopically. As necessary, isolation of colonies was made from

Table 3. The ZoBell 2216 modified F medium for the enumeration of epiphytic heterotrophs.

Component	Amount
Polypeptone (g)	5.0
KH_2PO_4	0.05
$FeSO_4 \cdot 7H_2O$	0.05
Tap water (l)	1
pH 7.2	

Table 4. The composition of medium for denitrifying bacteria (nitrite reducer).

	Component		Amount
(A)	$NaNO_2$ (g)		0.3
	Asparagine		1.0
	1% BTB alcohol (ml)		1.0
	Distilled water		100
(B)	Sodium citrate (g)		8.5
	$MgSO_4 \cdot 7H_2O$		1.0
	$FeSO_4 \cdot 6H_2O$		0.05
	KH_2PO_4		1.0
	$CaCl_2 \cdot 2H_2O$		0.2
	Distilled water (ml)		500
(A) + (B)		pH 7.0–7.2	

Table 5. The composition of medium for denitrifying bacteria (nitrate reducer).

	Component		Amount
(A)	KNO_3 (g)		1.0
	Asparagine		1.0
	1% BTB alcohol (ml)		1.0
	Distilled water		100
(B)	Sodium citrate (g)		8.5
	$MgSO_4 \cdot 7H_2O$		1.0
	$FeCl_3 \cdot 6H_2O$		0.05
	KH_2PO_4		1.0
	$CaCl_2 \cdot 2H_2O$		0.2
	Distilled water (ml)		500
(A) + (B)		pH 7.0–7.2	

Table 6. The medium for the enumeration and the isolation of actinomycetes.

Component		Amount
Glycerol (g)		12.5
Arginine		1.0
K_2HPO_4		1.0
NaCl		1.0
$MgSO_4 \cdot 7H_2O$		1.0
Agar		15.0
Tap water (l)		1.0
	pH 6.9–7.1	

Streptmycin was added (40 $\mu g\ ml^{-1}$) aseptically after autoclaving.

Table 7. The medium for the enumeration and the isolation of fungi.

Component	Amount
Glucose (g)	1.0
Yeast extract	0.1
K_2HPO_4	1.0
$MgSO_4 \cdot 7H_2O$	0.5
Agar	15.0
Distilled water (l)	1.0
pH 7.0	

Streptmycin was added (40 μg ml^{-1}) aseptically after autoclaving.

petri-dishes to agar slants for further observation. The composition of these media is shown in Tables 6 & 7.

For the measurement of nitrite and nitrate reducting activity by epiphytic microorganisms on the submerged stems of *P. communis* and colonized glass rods in the lagoon, the graduated incubation tubes shown in Fig. 2 were used. Many glass rods, about 7.0 cm length and 1.0 cm diameter, were incubated in the lagoon for two weeks, during which various kinds of epiphytes colonized abundantly on them. Each of the epiphytes colonized on the glass rods was inoculated into the incubation tubes in an aseptic bench immediately after the sampling and incubated in the dark at 25° C for 20 days. Gas production was easily measured by reading the minute scale. After the incubation, the concentration of nitrite or nitrate in the medium was always determined. All experiments were done in duplicate.

Table 8. Comparison of the quality of water at Station 1 and Station 4.

Item	Station 1	Station 4
Temp. (° C)	24.8	28.5
pH	6.84	7.18
D O (O_2 mg l^{-1})	9.17	7.19
Saturation degree of DO (%)	105.4	88.2
Optical density (640 mμ)	0.040	0.002
C O D (O_2 mg l^{-1})	6.53	4.24
B O D (O_2 mg l^{-1})	7.07	4.51

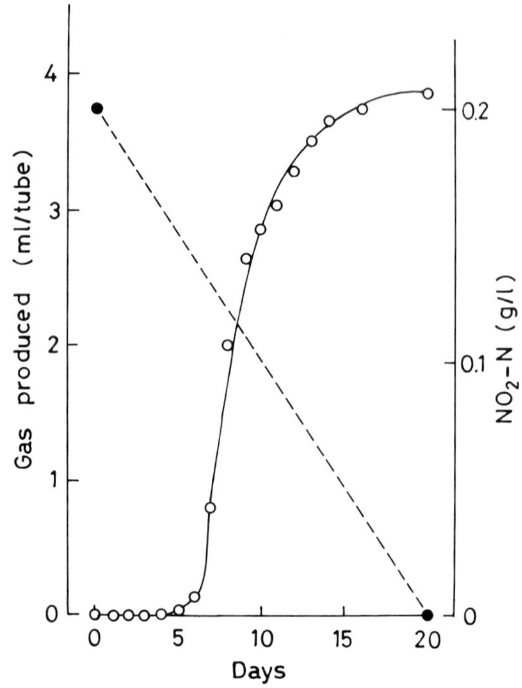

Fig. 3. The activity of nitrite reduction by epiphytes on the colonized glass rods. Gas: (○), NO_2: (●).

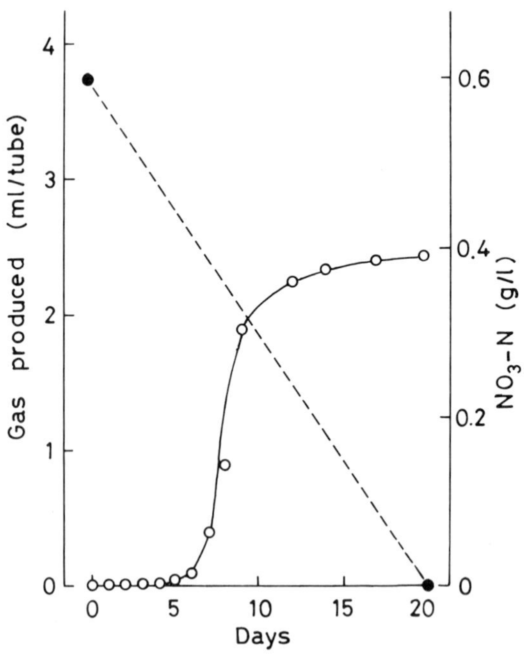

Fig. 4. The activity of nitrate reduction by epiphytes on the colonized glass rods. Gas: (○), NO_3: (●).

Table 9. The population density of epiphytic microorganisms on the soaking glass sticks in the lagoon.

Microorganism (cell cm^{-2})	Station 1	Station 4
Heterotrophic bacteria	8.3×10^7	2.1×10^6
Actinomycetes	8.9×10^4	1.7×10^3
Fungi	9.5×10^4	1.5×10^3
Yeast	3.0×10^2	2.0×10^1

Results and discussion

The quality of water collected at Stations 1 and 4 is shown in Table 8, in which very high concentrations of nutrients contained in water of Station 1 are compared with those of Station 4. As indicated by this Table, the influent that flows into Station 1 is self-purified considerably in the process of passing through the lagoon by various kinds of organisms. However, the mechanism of self-purification in the ecosystem of lagoons has not been clarified. In shallow lagoons with abundant *P. communis,* it is generally thought that much of the bio- and abioseston contained in the influent is precipitated primarily by the stagnation of the water current, and that very high nutrient concentrations are utilized secondarily by high population densities of phytoplankton and the emergent macrophytes. Besides the nutrient removal of phosphorus and nitrogen from the discharged influent by both processes, biological nitrogen removal, such as by denitrification, must be occurring in these eutrophicated aquatic environments.

Denitrifying bacteria are widely distributed and they have a very high denitrifying activity in the sediments of eutrophicated lakes and reservoirs (Gamble *et al* 1977). It is also reported (Nakajima 1981) that sessile microbes isolated from the water and river beds of polluted rivers had extensive activity of N_2O production, and the activity was not inhibited even at 20% partial pressure of oxygen. However, it seems that denitrification of epiphytic microorganisms has not been examined extensively.

Glass rods of a diameter similar to that of the stems of *P. communis* were incubated in water (50-cm deep layer) at Stations 1 and 4 in the lagoon for two weeks. After incubation, an abundant colonization of epiphytes was always observed. In each incubation tube inoculated with a colonized glass rod, nitrite and nitrate were very actively reduced and a visible amount of gas was produced. In the final period of incubation, sulfide production was observed. After the incubation, the concentrations of nitrite and nitrate were always checked, but not detected. The results are shown in Figs. 3 & 4. It was assumed that N_2 and N_2O must be contained in the gas produced in the tubes. The composition of the gas and stoichiometric analysis will be reported in the next paper.

As reported in a previous paper (Kurata 1981), the epiphytic microorganisms on the submerged macrophytes in Lake Biwa were composed of heterotrophic bacteria, actinomycetes, fungi, and yeasts. The composition of bacteria and fungi colonized on *P. communis* was also described (Mason 1976; Taligoola 1972; Apinis 1972). In this paper, therefore, the composition of epiphytic microbes colonized on glass rods was examined using different kinds of media (Table 9). The population densities of epiphytic microorganisms colonizing glass rods was always higher at polluted Station 1 than at Station 4 at the outlet of the lagoon. However, denitrification must depend only on heterotrophic bacteria among various kinds of epiphytic microorganisms. Accordingly, the population densities of epiphytic heterotrophs were examined using different kinds of media (Table 10). Almost the same population densities of heterotrophs were found with each medium. No great differences occurred in the bacterial numbers isolated from the colonized glass rod samples and the submerged stems of *P. communis* (Table 9). Therefore, it is

Table 10. The population density of epiphytic heterotrophs on *Phragmites communis*.

Medium	Epiphytic heterotroph (cells cm^{-2})
Broth agar	1.2×10^7
Y P A medium	1.3×10^7
ZoBell 2216 F medium	1.8×10^7

Table 11. Nitrite and nitrate reduction by epiphytic microorganisms on *Phragmites communis* collected from different lagoons.

Lagoon	Nitrite reduction	Nitrate reduction
Matsunoki L.	++	+++
Otomegaike L.	–	–
Komatsunuma L.	+++	+++
Nishinoko L.	+++	+++
Hirako L.	++	+++

Table 12. Nitrite and nitrate reduction by epiphytic microorganisms on various emergent plants.

Emergent plant	Nitrite reduction	Nitrate reduction
Phragmites communis	+++	+++
Phalaris arundinacea	++	+++
Zizania latifolia	+++	++
Trapa bispinosa	+++	+++
Carex amplifolia	++	+++
Nuphar japonicum	–	++

suggested that the population density of epiphytic microorganisms colonized on the submerged materials is strongly affected by the water quality and trophic conditions of surrounding water in the *P. communis* zone of this lagoon.

There are many lagoons around Lake Biwa and each lagoon differs in nutrient concentrations because of polluted or nonpolluted water influx. Denitrifying activities by epiphytic microorganisms on *P. communis* collected from several lagoons were examined (Table 11). Except for one lagoon, denitrifying activity was observed in most of the epiphyte samples collected.

Epiphyte samples colonized on various other emergent macrophytes were also collected and analyzed for denitrifying activity in the same way (Table 12). *Nuphar japonicum* was obtained only in Otomegaike Lagoon. Denitrification occurred by epiphytes colonizing the submersed portions of all plants. It is likely that denitrification is commonly caused by epiphytes colonized on various types of submersed and emergent macrophytes in eutrophicated lagoons around the lake. This process must be effective in nitrogen removal in the lagoons if one considers their wide and abundant distribution and high activities.

References

Aizaki, M., 1978. Seasonal changes in standing crop and production of periphyton in the Tamagawa River. Jap. J. Ecol. 28:123–134.

Apinis, A.E., Chesters, C.G. & Taligoola, H.K., 1972. Microfungi colonizing submerged standing culms of Phragmites communis Trin. Nova Hedwigia 23:473–480.

Gamble, T.N., Betlach, M.R. & Tiedje, J.M., 1977. Numerically dominant denitrifying bacteria from world soils. Appl. Envir. Microbiol. 33:926–939.

Kurata, A., 1981. The production of B group vitamins by epiphytic microorganisms on macrophytes in Lake Biwa. Verh. int. Ver. Limnol. 21:596–599.

Mason, C.F., 1976. Relative importance of fungi and bacteria in the decomposition of Phragmites leaves. Hydrobiologia 51:65–69.

Nakajima, T., 1981. Denitrification by sessile microbes in a polluted river. Verh. int. Ver. Limnol. 21:1400–1405.

Taligoola, T.K., Apinis A.E. & Chesters, C.G.C., 1972. Microfungi colonizing collapsed aerial parts of Phragmites communis Trin. in water. Nova Hedwigia 23:465–472.

Author's address:
A. Kurata
Lake Biwa Research Institute
Uchide-hama
Otsu-city
Shiga Prefecture, 520, Japan

Effects of agricultural wastewater on benthic algae in ditches in the Netherlands

S.P. KLAPWIJK, T.F. DE BOER and M.J. RIJS

Abstract. A modified artificial substrate method was used to study the effects of agricultural wastewater on benthic algae in ditches in a polder in the western part of the Netherlands. In the ditches chemical gradients were found particularly in orthophosphate and ammonium nitrogen, showing decreasing values with increasing distances (up to 1 km) from the discharging farmhouses. The effects of the wastewater discharges on the benthic algae were studied by measuring several parameters, e.g. biomass, species composition, diversity (number of species mm^{-2}, Margalef index, Shannon-Wiener index) and Sørensen similarity indices. Some benthic algae, e.g. *Chaetopeltis orbicularis* and *Stigeoclonium tenue* showed respectively a negative and a positive correlation with the concentration of eutrophicating nutrients. Saprobity indices revealed no great differences between sampling stations in the gradients, indicating the need to modify this method for use in small stagnant bodies of water in the Netherlands. Periphyton productivity proved to be a good indicator of pollution, showing a decrease of biomass growth (in two weeks) with increasing distance from the source of pollution. However, an increase of floating algae and aquatic angiosperms during summer can severely limit the use of artificial substrates due to light interception.

Introduction

Although the policy of the water authorities in the Netherlands is aimed at diverting direct discharges of wastewater from surface waters, especially small ones to sewage treatment plants, nevertheless discharges of wastewater into ditches still occur due to the economic costs which are needed to bridge the sometimes large distance to the nearest sewer system. This situation holds especially for the discharges from farms and houses situated in rural polder areas far from villages and cities. Calculations have shown that 8% of the total number of dwellings at present in existence will probably never be connected to a main sewerage system in view of the high cost involved (Ministry for Transport and Public Works, 1975). This can result locally in a severe deterioration of the water quality, especially if the discharges occur into small polder ditches.

The wastewater from farms and country-houses is to a small extent of domestic nature and to a large extent agricultural: rinsewater from stables and milktanks, drainwater from dung-hills and silos. Though illegal, the liquid manure of dairy cattle and of pig- and calf-batteries is sometimes also discharged into ditches.

In this investigation the effects of agricultural wastewater are studied in a normal Dutch grassland polder. This study is part of a research program which was carried out in 1975 and 1976 on the effects of different kinds of wastewater on the benthic algae in ditches in the western part of the Netherlands (Klapwijk 1978, 1980; Bergh *et al* 1978).

Wetzel, R.G. (ed.), Periphyton of Freshwater Ecosystems.
 ISBN 90 6193 768 X.

Study-area, materials and methods

The study was carried out in the Meije polder, in the community of Bodegraven, in the western part of the Netherlands (Fig. 1). In this grassland polder all farms and houses are situated along the northwest edge, discharging their household and agricultural wastewater into many of the southeast trending ditches.

The influence of the wastewater may be expected to gradually decrease in proportion to increasing distance from the farms and houses through dilution with rainwater and/or by self-recovery.

The ditches are intersected by the main watercourse of the polder at approximately 1 km distance from the farms. This main watercourse is connected directly with 'De Vliet', from which water can be pumped up to the 'boezem' and into which, in times of drought, water can be directed out of the boezem.

In the polder 108 sampling stations were chosen for investigation of the water chemistry, and 45 sampling stations for investigation of the benthic algal flora (Fig. 1). The sampling stations were arranged in several zones at the following distances from the farms: zone A at 100 m, zone B at 300 m, zone C at 1300 m and zone D at 2300 m. Some sampling stations were also located in the main watercourse.

The investigation period from April to July, 1976 can be characterized as extremely dry and sunny. In this period the following hydrochemical parameters were measured 4 times at each station: PO_4^{3-}, NH_4^+, NO_2^-, NO_3^-, Cl^- and specific conductivity (K 25). The first 4 parameters were analyzed in the laboratory with a Technicon AA II autoanalyzer and the last 2 with a Philips Ion Activity meter (type PW 9413) and a W.T.W. conductivity meter (type LF 54), respectively.

Benthic algae were sampled with glass microscope slides as artificial substrates, fixed in Plexi-

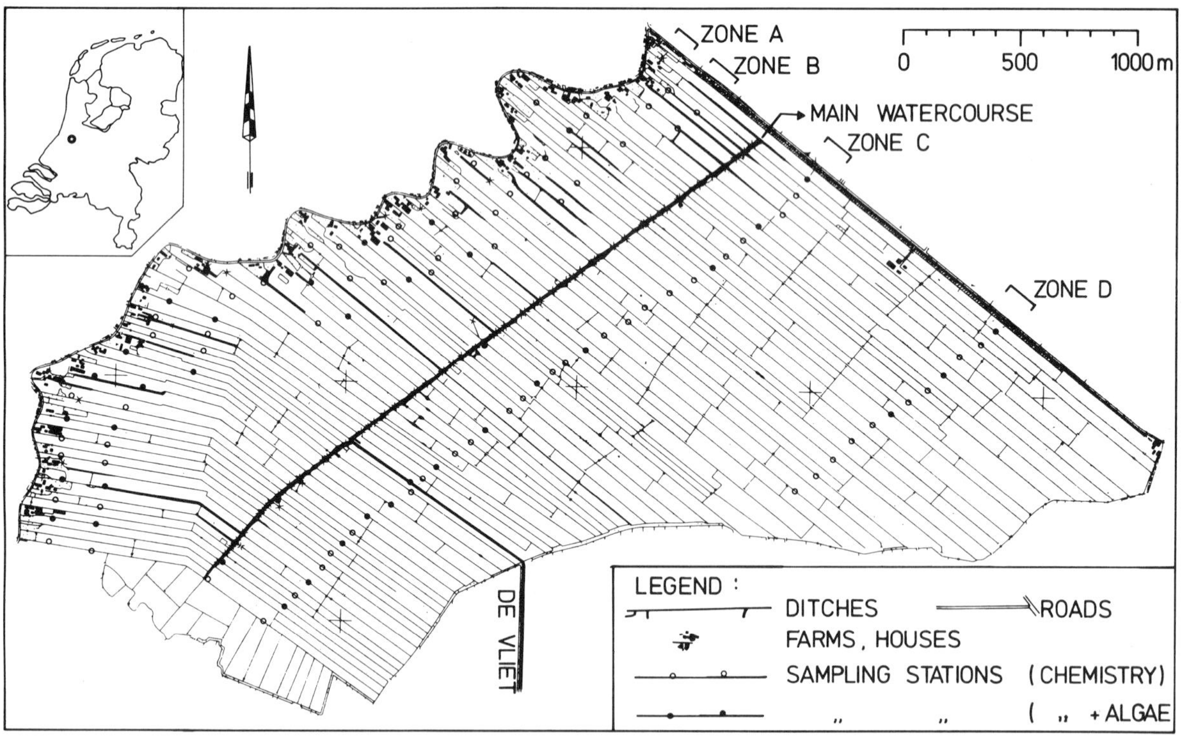

Fig. 1. The Meije-polder with regular pattern of ditches, location of sampling stations and farms, and its location in the Netherlands (inset).

Table 1. Means (±s.d.) of measured chemical and physical parameters in different ditch zones on four sampling dates and test statistics of the regression test over zones A, B and C (n = number of sampled ditches).

Parameter		Zone A (n = 30)	Zone B (n = 30)	Main watercourse (n = 8)	Zone C (n = 30)	Zone D (n = 10)	Test statistic of regression test (n = 30)
PO_4^{3-}	($mgPl^{-1}$)	2.15 ± 1.63	1.16 ± 0.84	0.35 ± 0.08	0.23 ± 0.12	0.13 ± 0.13	−7.03****
NH_4^+	($mgPl^{-1}$)	1.47 ± 2.14	0.32 ± 0.64	0.09 ± 0.04	0.06 ± 0.04	0.09 ± 0.08	−5.03****
NO_2^-	($mgNl^{-1}$)	0.04 ± 0.07	0.02 ± 0.03	0.04 ± 0.03	0.01 ± 0.01	0.01 ± 0.00	−4.20****
NO_3^-	($mgNl^{-1}$)	0.11 ± 0.20	0.15 ± 0.46	0.33 ± 0.48	0.04 ± 0.14	0.00 ± 0.01	−3.87****
Cl^-	(mgl^{-1})	301 ± 61	309 ± 72	281 ± 21	294 ± 28	310 ± 32	−0.13
K 25	(μScm^{-2})	1053 ± 176	1005 ± 105	893 ± 41	884 ± 156	916 ± 68	−4.78****

The least significant value ($\alpha = 0.05$; two-tailed test) is ± 1.96; ****$p < 0.0001$.

glas racks, which were suspended from floats in the water, as described by Klapwijk (1980). The slides were exposed in the ditches for a period of 2 weeks all in the same direction (southwest). The slides with attached benthic algae were sampled twice, viz. in April and June, in order to get some impressions of the benthic flora in different seasons (cf. Hillebrand 1977).

From a rack of 12 slides 3 were mounted in Aquamount (Edward Gurr Ltd., England) for microscopical examination. Algal counts were made of the number of individuals on randomly selected viewing fields with the aid of cell finder culture slides (model Leiden, Microlab Holland, 1974) until a total of at least 1 mm^2 was searched. The remaining slides per rack (7–9 replicates) were used for determinations of biomass by weighing oven-dried (60°C, 48 h) glass slides before and after exposure in the ditches.

To evaluate the algal results the same computational methods were applied, as in the other published studies (cf. Klapwijk 1978, 1980): Sørensen similarity indices, diversity by means of Margalef's formula and the Shannon-Wiener function and saprobity using the Zelinka & Marvan (1961) procedure with the number of individuals mm^{-2} expressing the abundance of a given species. In order to obtain also one figure per sampling station, a kind of weighted mean (SI) of the saprobic valencies was calculated using the following formula:

$$SI = \frac{(0\times x) + (1\times o) + (2\times \beta) + (3\times \alpha) + (4\times p)}{10}$$

where: x = xenosaprobic valency
o = oligosaprobic valency
β = β-mesosaprobic valency
α = α-mesosaprobic valency
p = polysaprobic valency

This value of SI is almost comparable to Pantle & Buck's (1955) saprobic index.

Table 2. Mean values (± s.d.) of Sørensen similarity indices (in %) between sampling stations in different ditches within the same zone in April and June 1976. The number of comparisons on which the means are based, are given in parentheses.

Month	Zone A	Zone B	Main watercourse	Zone C	Zone D
April	59 ± 14	61 ± 13	70 ± 12	64 ± 12	60 ± 10
	(55)	(66)	(10)	(66)	(3)
June	54 ± 11	50 ± 12	51 ± 10	54 ± 10	43 ± 10
	(66)	(66)	(10)	(66)	(3)

Table 3. Mean values of Sørensen similarity indices (in %) between sampling stations in different zones within the same ditch in April and June 1976. The number of comparisons (= ditches) on which the means are based, are given in parentheses.

Month		Zone A	Zone B	Main watercourse	Zone C	Zone D
April	Zone A	100 (11)	68 (11)	53 (3)	60 (11)	53 (2)
	Zone B		100 (12)	63 (4)	68 (12)	61 (3)
	Main watercourse			100 (4)	69 (4)	60 (2)
	Zone C				100 (12)	68 (3)
	Zone D					100 (3)
June	Zone A	100 (12)	60 (12)	47 (4)	52 (12)	46 (3)
	Zone B		100 (12)	45 (4)	55 (12)	44 (3)
	Main watercourse			100 (4)	61 (4)	38 (2)
	Zone C				100 (12)	52 (3)
	Zone D					100 (3)

Results

The hydrochemical analyses of the ditches and the main watercourse are summarized in Table 1. With the exception of chloride, all parameters, especially orthophosphate and ammonium nitrogen tended to decrease from zone A to zone D. The results of a non-parametric regression test for k-related samples (de Jonge 1963) in 30 ditches, running through A, B and C, showed that all courses were significant at the 5% probability level ($P<0.0001$) except for Cl^-. The effects of the discharges of agricultural wastewater were most clearly shown in the high concentrations of PO_4^{3-} and NH_4^+ in zone A, so that this water can be characterized as hypertrophic (Leentvaar 1980).

The species composition of the benthic algal communities cannot be given in detail. In total, 105 species of algae were found on the artificial substrates, belonging to the following taxonomical groups: Cyanophyta (13%), Euglenophyta (5%), filamentous Chlorophyta (24%), non-filamentous Chlorophyta (38%), Chrysophyta (21%) and algae from other groups (1%). Some frequently occurring species are presented in Table 6. Because of insufficient taxonomic knowledge of benthic algae, only 48% of the species could be identified with certainty.

Sørensen similarity indices were used to compare the species composition of the algal communities at the sampling stations. In Table 2 the average values and standard deviations were calculated of Sørensen indices between sampling stations in different ditches within the same zone. In April the mutual resemblance of the benthic algal flora in zone A was slightly smaller than in zone B and C as could be expected as a consequence of the discharge of different kinds of wastewater in some of the ditches near zone A. In June this tendency was not at all apparent. A possible explanation of this high similarity may be the development of large masses of floating algae and duckweed (*Lemna* spp.) in more than 50% of the ditches in June (see discussion).

In Table 3 the means of Sørensen similarity indices between sampling stations in different zones within the same ditch are presented in correlation diagrams. In a gradient situation the indices in such a diagram would be expected to decrease from left to right and from bottom to top. This pattern indeed emerged, both in April and June, especially when the main watercourse was not included. Comparing the indices between the zones A, B and C, it can be seen from Table 3 that in April the mean similarity between zone A and B was equal to the similarity between zone B and C (68%), while in June the similarity between zone A and B (60%) was greater than that between B and C (55%) in accordance with the geographical situation of the zones. Similarity indices have also been computed (but not shown here) between the algal vegetation at the same sampling stations in the two months of sampling (April and June). These appeared to be rather low (average similarity = 39%), which can

Table 4. Biomass, numbers of individual algae and of algal species and diversity indices (means ± s.d.) of different ditch zones in April and June 1976 with statistics of the regression test over zones, A, B and C (n = number of sampled ditches).

Month	Parameter	Zone A	Zone B	Main watercourse	Zone C	Zone D	Test statistic of regression test
April		(n = 11)	(n = 12)	(n = 5)	(n = 13)	(n = 3)	(n = 11)
	Biomass ($mgdm^{-2}$)	58.3 ± 36.2	34.1 ± 26.5	44.2 ± 46.4	26.4 ± 20.5	24.2 ± 17.5	−3.77***
	No of individuals mm^{-2}	603 ± 445	702 ± 268	742 ± 289	621 ± 176	697 ± 59	+1.50
	No of species mm^{-2}	24 ± 7	24 ± 6	29 ± 1	27 ± 5	34 ± 7	+1.70
	Margalef index (D)	8.26 ± 2.50	8.18 ± 1.85	9.88 ± 0.59	9.40 ± 1.69	11.50 ± 2.37	+1.28
	Shannon-Wiener index (H)	3.30 ± 0.80	3.29 ± 0.49	3.54 ± 0.43	3.65 ± 0.42	3.71 ± 0.12	+0.85
	Shannon-Wiener index (E)	0.73 ± 0.14	0.72 ± 0.07	0.73 ± 0.09	0.77 ± 0.06	0.73 ± 0.03	+0.21
June		(n = 12)	(n = 12)	(n = 5)	(n = 13)	(n = 3)	(n = 12)
	Biomass ($mgdm^{-2}$)	61.1 ± 60.2	21.6 ± 31.0	56.4 ± 74.7	13.8 ± 18.0	12.8 ± 18.7	−0.50
	No of individuals mm^{-2}	1205 ± 1068	821 ± 697	1127 ± 960	608 ± 447	631 ± 1029	−2.65**
	No of species mm^{-2}	25 ± 7	24 ± 10	23 ± 6	21 ± 5	17 ± 8	−0.92
	Margalef index (D)	8.15 ± 1.96	8.39 ± 2.70	7.91 ± 2.18	7.80 ± 1.71	7.76 ± 2.03	−0.61
	Shannon-Wiener index (H)	2.91 ± 0.48	2.75 ± 0.93	2.76 ± 0.90	2.84 ± 0.83	3.07 ± 0.59	−0.72
	Shannon-Wiener index (E)	0.63 ± 0.09	0.61 ± 0.20	0.61 ± 0.17	0.64 ± 0.18	0.79 ± 0.21	+0.20

The least significant value ($\alpha = 0.05$; two-tailed test) is ±1.96; ** $p<0.01$; *** $p<0.001$.

be interpreted as an indication of algal periodicity.

The results of the measured algological parameters are summarized in Table 4. Biomass tended to decrease from zone A to zone D in both months. In April the number of individuals mm^{-2}, the number of species mm^{-2} and the Margalef (D) and Shannon-Wiener (H) diversity indices tended to increase from zone A to zone D, while in June the opposite tendency was perceptible. A possible reason for this was again the already mentioned covering of the water surface with floating masses of filamentous algae and duckweed in June. The equitability component of the Shannon-Wiener index (E) was remarkably stable over the zones in both months, proving that the distribution of individuals among the species was more or less constant at all sampling stations. The results of a non-parametric regression test for k-related samples (de Jonge 1963), running through zones A, B and C, showed that most courses were not significant at the 5% probability level except for biomass in April and number of individuals mm^{-2} in June.

Table 5. Average saprobic valencies (x–p), their weighted means ± s.d. (SI), average number ± s.d of species used in the computations (n_i) and number of sampled ditches (n) in different zones in April and June 1976.

Month	Zone	x	o	β	α	p	SI	n_i	n
April	A	0.04	0.45	3.82	5.46	0.25	2.54 ± 0.21	7.3 ± 1.7	11
	B	0.06	0.75	3.68	5.24	0.27	2.49 ± 0.26	6.4 ± 2.4	12
	Main watercourse	0.15	1.43	3.27	4.74	0.41	2.38 ± 0.30	7.2 ± 2.4	5
	C	0.16	1.67	3.03	4.03	1.07	2.41 ± 0.63	5.5 ± 2.1	13
	D	0.32	2.06	3.28	3.76	0.54	2.22 ± 0.73	7.0 ± 1.7	3
June	A	0.09	0.13	3.13	5.77	0.87	2.73 ± 0.22	7.5 ± 2.5	12
	B	0.11	0.39	3.69	5.34	0.48	2.57 ± 0.31	5.7 ± 3.2	12
	Main watercourse	0.03	0.04	2.67	6.38	0.88	2.80 ± 0.21	6.0 ± 2.6	5
	C	0.05	0.27	2.97	5.28	1.44	2.78 ± 0.22	4.9 ± 2.2	13
	D	0.00	0.00	0.74	6.96	2.29	3.16 ± 0.27	3.0 ± 2.7	3

Table 5 shows the average saprobic valencies according to Zelinka and Marvan for the different zones in the two months of exposure. In the Table also the weighted means and standard deviations of the saprobic valencies (SI), the average number of species used in the computations (n_i), and the number of sampled ditches (n) are given. The results suggest that the saprobic valency in both months lie between β- and α-mesosaprobity with no clear differences between the zones. Testing the results for regression over the zones A, B and C, then, did not reveal significance at the 5% probability level. Comparing the average number of species used in the saprobity computations (Table 5) with the average number of species mm^{-2} found in the different zones (Table 4), it is obvious that only a small proportion of the observed species were used in the computations. The reason for this is that not all observed species could be exactly identified and not all of them had known indicator values.

Spearman's rank-correlation coefficients with correction for equal values (de Jonge 1963) were computed between a number of chemical parameters (PO_4^{3-}, NH_4^+, Cl^-, K 25) and several algological characteristics: number of individuals mm^{-2}, number of species mm^{-2}, Margalef's and Shannon-Wiener's diversity indices (D, H, E), biomass, Zelinka and Marvan's weighted saprobity value (SI), and the number of individuals of some frequently occurring species (Table 6). From this it could be concluded that the number of individuals mm^{-2} and of species mm^{-2} were positively correlated with NH_4^+ and biomass in June but not in April. The different diversity characteristics (individuals mm^{-2}, D, H) showed in April a negative correlation with K 25 and in June a positive correlation with biomass. Moreover, H was negatively correlated with PO_4^{3-}, but only in April, while E showed no significant correlation at all. Zelinka & Marvan's weighted saprobity index (SI) showed a weak correlation with PO_4^{3-}, NH_4^+ and Cl^- in April, but not in June.

Of all the algological characteristics biomass was most positively correlated with the parameters NH_4^+ and PO_4^{3-} which were affected by agricultural wastewater. Both in April and June biomass was correlated very significantly with NH_4^+, but also to a lesser extent with PO_4^{3-}. In June biomass was also positively correlated with the number of individuals mm^{-2}, the number of species mm^{-2}, and D, showing that in that month a high biomass was also exhibited by a high diversity.

The calculation of correlation coefficients between the number of individuals of particular species and some chemical parameters and biomass gave varying results, as may be expected. While the two Cyanophyta did not show any significant correlation at all, some of the filamentous Chlorophyta did reveal a clear positive or negative correlation with the chemical parameters. This rotation was most notable for *Chaetopeltis orbicularis,* showing a negative correlation with PO_4^{3-} and NH_4^+ in April and June, but also with Cl^- and K 25 in April. This held more or less also for *Microspora* spec. 1 in April. *Oedogonium* spec. 2 and *Stigeoclonium farctum* were both negatively correlated with NH_4^+ in April and positively with biomass in June. *Stigeoclonium tenue* was in April as well as in June positively correlated with PO_4^{3-} and NH_4^+ and also with K 25 in April and with biomass in June. The non-filamentous Chlorophyta did not show many significant correlations, except for a negative correlation of *Tetraspora gelatinosa* with K 25 and Cl^- and a positive correlation of *Ectogeron elodeae* with PO_4^{3-} and K 25, both in April. The diatom *Nitzschia hungarica* showed a positive correlation with NH_4^+ and biomass, both in April and June, while *Nitzschia palea* showed these significant relations only in June. Therefore, both species seemed relatively resistant to high NH_4^+ concentrations and contributed substantially to the amount of biomass. The same applied to *Stigeoclonium tenue* in June.

Discussion

The impact of agricultural wastewater was well reflected in the concentration of eutrophicating nutrients. As could be expected, Cl^- was not influenced by wastewater discharges and consequently more or less constant over the zones. In fact, the Cl^- content in the polder showed a strong increase over time (mean values: April: 199 mg l^{-1}; May

Table 6. Spearman's rank correlation coefficients between some chemical parameters (PO_4^{3-}, NH_4^+, Cl^-, K 25) and several algological characteristics (no. of individuals mm^{-2}, no of species mm^{-2}, Margalef and Shannon-Wiener diversity indices, Zelinka & Marvan's weighted saprobity values, biomass and no. of individuals mm^{-2} of some frequently occurring species) in April and June 1976.

	April					June				
	PO_4^{3-}	NH_4^+	Cl^-	K 25	Biomass	PO_4^{3-}	NH_4^+	Cl^-	K 25	Biomass
No. of individuals mm^{-2}	+0.15	+0.02	0.00	−0.08	−0.08	+0.26	+0.43**	−0.35*	−0.01	+0.67***
No. of species mm^{-2}	−0.19	+0.07	−0.01	−0.53***	+0.12	+0.20	+0.33*	−0.23	+0.17	+0.62***
Margalef index (D)	−0.15	−0.02	−0.05	−0.39**	+0.09	+0.04	+0.08	−0.02	+0.21	+0.34*
Shannon-Wiener index (H)	−0.33*	−0.03	−0.22	−0.40**	+0.12	−0.03	0.00	+0.10	+0.23	+0.10
Shannon-Wiener index (E)	−0.23	−0.12	−0.14	−0.12	+0.09	−0.08	−0.20	+0.27	+0.18	−0.27
Zelinka & Marvan (SI)	+0.29*	+0.32*	+0.31*	+0.27	+0.16	−0.12	+0.05	−0.10	+0.04	+0.03
Biomass	+0.30	+0.53***	+0.31*	+0.09		+0.40**	+0.69***	−0.22	+0.21	
Cyanophyta										
Phormidium fragile	−0.24	−0.09	−0.20	−0.18	−0.17	−0.06	−0.19	−0.04	−0.01	+0.02
Pseudanabaena catenata						−0.04	+0.05	−0.17	−0.09	+0.08
Chlorophyta (filamentous)										
Chaetopeltis orbicularis	−0.56***	−0.36*	−0.39**	−0.44**	+0.28	−0.33*	−0.35*	+0.06	−0.15	−0.07
Enteromorpha intestinalis	+0.28	+0.08	−0.36*	+0.19	+0.25					
Microspora spec. 1	−0.53***	−0.32*	−0.19	−0.38**	−0.31*					
Oedogonium spec. 2	−0.26	−0.41**	−0.28	−0.33*	+0.15	−0.04	+0.07	−0.03	−0.02	+0.33*
Stigeoclonium farctum	−0.08	−0.37**	+0.01	+0.02	−0.50***	−0.01	+0.25	−0.22	−0.10	+0.36**
Stigeoclonium tenue	+0.37**	+0.31*	+0.04	+0.42**	+0.30	+0.40**	+0.60***	−0.11	+0.24	+0.76***
Chlorophyta (non-filamentous)										
Ectogeron elodeae	+0.46***	−0.06	−0.08	+0.29	−0.04	+0.14	+0.29	−0.23	+0.07	+0.29
Sphaerobotrys fluviatilis	−0.16	+0.08	−0.17	−0.08	−0.27	−0.18	−0.23	−0.17	−0.10	−0.12
Tetraspora gelatinosa	−0.29	−0.18	−0.32*	−0.39**	0.00					
Bacillariophyta										
Nitzschia hungarica	+0.22	+0.55***	+0.08	+0.25	+0.42**	+0.17	+0.37**	−0.18	+0.28	+0.50***
Nitzschia palea	+0.15	+0.26	+0.23	−0.01	+0.22	+0.13	+0.35*	+0.03	+0.28	+0.48***
Other algae										
Diploeca flava	−0.19	−0.24	−0.20	−0.41**	−0.20					
Trachelemonas volvocina	−0.05	−0.07	−0.19	−0.09	+0.17					

* $p<0.05$; ** $p<0.01$; *** $p<0.001$; n = 44 in April and 45 in June.

355 mg l^{-1}; June 309 mg l^{-1}; July: 340 mg l^{-1}), due to evaporation and the inlet of relatively brackish boezemwater to maintain the water level in the ditches.

The advantages of using artificial substrates for collecting algae in saprobiological research have already been stressed before (Sládečková 1962; Klapwijk 1978, 1980). The present investigation, however, also brought some drawbacks of the method to light, particularly the influence on the results of floating algal masses and duckweed, which interfere with the benthic algal flora in two possible ways: (1) by light interception and (2) by utilization of the nutrients in the water. A covering of the water surface with duckweed and floating algae was associated in this study with low benthic algal biomass and low NH_4^+ concentrations. This last association has also been recorded by Pott (1980) in a series of *Lemna*-rich waters in West Germany, which shows again the influence of macrophyte and algal periodicity (cf. Hillebrand 1977) as well in research on the effects of pollution (see also Klapwijk 1978, 1980).

Benthic algal biomass proved to be a good indicator of pollution in this study, but this may be dependent on the nature of the waste-water (household and agricultural) and on the degree of pollution (cf. Klapwijk 1980).

The need to revise and develop the saprobity system for stagnant ditches and other stagnant waters in the Netherlands is obvious, as has been stated before (Klapwijk 1978, 1980). In this study the insufficiency of taxonomic knowledge and literature of benthic algae, especially of juvenile forms, became apparent, as well as the lack of saprobity indicating values for the majority of the benthic algal species. The results of this study, particularly the correlation coefficients between chemical parameters and several algal species (Table 6), offer a first indication of the saprobic valency of a species. Based on this study, *Chaetopeltis orbicularis* and *Stigeoclonium tenue*, for example, appear to be rather good indicators for, respectively, clean and polluted waters. In the case of *Stigeoclonium tenue* this was well known (cf. Sládeček 1973; McLean & Benson-Evans 1974; Rosemarin 1980, 1982), but for *Chaetopeltis orbicularis* and several other species this was not, as far as we know, recorded before (cf. Mauch 1976). However, such field observations should be followed by experimental work in order to verify them and to determine the tolerance of particular species to selected chemical parameters (cf. Francke & Hillebrand 1980; Rosemarin 1980, 1982; Francke & den Oude, 1983).

Conclusions

The influence of agricultural wastewater discharges is clearly reflected both in some hydrochemical parameters, especially orthophosphate and ammonium nitrogen, as well as in the biomass and species composition of the benthic algal community on artificial substrates. However, through the interaction of floating algae and aquatic angiosperms with nutrients in the water, not only can the value of chemical analyses be reduced, but the effects of wastewater on benthic algae can also obscured, because of other factors, such as underwater light, which can be more limiting for algal growth than nutrients. These findings, therefore, demonstrate the need for an integrated survey of the total waterbody.

Acknowledgements

This investigation was supported by a grant (R 489) from the Dutch Ministry of Public Healt and Environmental Hygiene, which also enabled the first author to participate in the First SIL Workshop on Periphyton in Freshwater Ecosystems, in Växjö, Sweden, 14–17 September 1982. The authors are greatly indebted to Prof. Dr. W.H.O. Ernst and Prof. Dr. M. Vroman for their help in preparing the manuscript, to Paul de Vocht for the chemical analyses, to Mr. P. van Doorn for drawing Fig. 1, to Paul Gutteridge for correcting and to Miss C.A.L.M. van Dijk for typing the English text.

References

Bergh, M.L.E., van Doorn, J. & Bos, P., 1978. Effect of manure runoff on phosphate absorption characteristics of polderditch sediments. Verkenningen I.V.M. - VU, Ser. C 9.

de Jonge, H., 1963. Inleiding tot de medische statistiek. 2 d ed. Verh. Ned. Inst. Praev. Geneesk. 41, Leiden (in Dutch).

Francke, J.A. & Hillebrand, H., 1980. Effects of copper on some filamentous Chlorophyta. Aquat. Bot. 8:285–289.

Francke, J.A. & Den Oude, P.J., 1983. Growth of Stigeoclonium and Oedogonium species in artificial ammonium-N and phosphate-P gradients. Aquat. Bot. 15:375–380.

Hillebrand, H., 1977. Periodicity and distribution of multicellular green algae in two lake areas in the Netherlands. Ph. D. Thesis, Free Univ. Amst.

Klapwijk, S.P., 1978. Saprobiological evaluation of Dutch ditches using benthic algae on artificial substrates. Verh. int. Ver. Limnol. 20:1811–1815.

Klapwijk, S.P., 1980. Effects of laundry wastewater on benthic algae in ditches in the Netherlands. Hydrobiol. Bull. 14: 142–151.

Leentvaar, P., 1980. Comparison of hypertrophy an a seasonal scale in Dutch inland waters. In: Barica, J. & Mur, L.R. (eds.), Hypertrophic Ecosystems. Developments in Hydrobiology 2: pp. 45–55.

Mauch, E., 1976. Leitformen der Saprobität für die biologische Gewässeranalyse. Cour. Forsch. Inst. Senckenberg 21, Teil 1–5, Frankfurt am Main.

McLean, R.O. & Benson-Evans, K., 1974. The distribution of Stigeoclonium tenue Kutz in South Wales in relation to its use as an indicator of organic pollution. Br. Phycol. J. 9:83–89.

Ministry for Transport and Public Works, 1975. The combat against surface water pollution in the Netherlands. Prospectivie multiannual Program 1975–1979.

Pantle, E. & Buck, H., 1955. Die biologische Überwachung der Gewässer und die Darstellung der Ergebnisse. Gas Wasserfach 96:604.

Pott, R., 1980. Die Wasser- und Sumpfvegetation eutropher Gewässer in der Westfälichen Bucht. Pflanzensoziologische und hydrochemische Untersuchungen. Abh. Landesmus. Naturk. Münster Westf. 42(2):1–156.

Rosemarin, A.S., 1980. Aspects of biological competition between Stigeoclonium tenue and Cladophora glomerata, two filamentous green algae, characteristic of eutrophic waters. In: Barica, J. & Mur, L.R. (eds.), Hypertrophic Ecosystems. Developments in Hydrobiology 2: p. 347.

Rosemarin, A.S., 1982. Phosphorus nutrition of two potentially compating filamentous algae, Cladophora glomerata (L.) Kütz. and Stigeoclonium tenue (Agardh) Kütz. from Lake Ontario. J. Great Lakes Res. 8:66–72.

Sládeček, V., 1973. System of water quality from the biological point of view. Arch. Hydrobiol. Beih. 7:1–218.

Sládečková, A., 1962. Limnological investigation methods for the periphyton (Aufwuchs) community. Bot. Rev. 28: 286–350.

Zelinka, M. & Marvan, P., 1961. Zur Präzisierung der biologischen Klassifikation der Reinheit fleissender Gewässer. Arch. Hydrobiol. 57:389–407.

Authors' address:
S.P. Klapwijk
Waterboard of Rijnland
Breestraat 59
2311 CJ Leiden, The Netherlands

T.F. de Boer/M.J. Rijs
Department of Biology and
Institute for Environmental Studies
Free University
Amsterdam, The Netherlands

CHAPTER 40

Attached and suspended bacterial populations in shallow running waters under different levels of pollution

P. PUNČOCHÁŘ

Abstract. Numbers of heterotrophic bacteria (plate counts on beef-peptone agar), direct counts and the biovolume of bacteria were investigated. In addition, chlorophyll *a* and kjeldahl nitrogen concentrations were determined in epilithon.

Bacteria attached in epilithon predominated in all localities, forming in very shallow (0.1–0.3 m) or polluted localities more than 95% of the bacterial population.

Differences in bacterial densities in epilithon indicated the input of organic pollution more distinctly than suspended bacteria since the former are dependent on the available organic substrates. This relation was especially true regarding the numbers of epilithic heterotrophic bacteria which ranged from 10^2 cm^{-2} in unpolluted mountain streams to 10^9 cm^{-2} in a heavily polluted river. Suspended heterotrophic bacteria in these localities ranged from 10 to 10^5 ml^{-1} indicating level bacterial pollution. The total cell counts in epilithon of the same localities ranged between 10^8–10^{10} cm^{-2} and in water between 10^5–10^7 ml^{-1}.

Introduction

Microbial populations developing on the bottom (mainly stones) in running waters form the attached component of the community, causing the accumulation of microbial biomass along the stream.

The importance of attached biomass for self-purification in running waters and its preponderance over the suspended biomass was described in experimental channels (Traaen *et al* 1972; Wuhrmann 1972, 1974; Wuhrmann & Eichenberger 1975). Our earlier data (Punčochář 1977, in press; Punčochář & Straškrabová 1980; Straškrabová *et al* 1978) demonstrated that attached bacteria were major components of the bacterial population in polluted river sites and, in addition, that sharp changes in densities of bacteria in the epilithon occurred in response to pollution inputs. High concentrations of sessile bacteria compared with planktonic bacteria have been found in unpolluted mountain streams (Geesey *et al* 1978).

The increasing amount of slime biomass and appearance (outbreak) of sewage fungus covering the bottom of running waters below sources of organic pollution demonstrate a close relation between the attached component and pollution level (Curtis & Harrington 1971; Costerton *et al* 1978; Aizaki 1979). Use of attached component as an index of water quality has been recognized for many years (e.g. Fjerdingstadt 1965). However, difficulties in quantitative sampling from natural substrata have limited routine use of these investigations (Lind 1974). The methodological difficulties are probably the main cause of few reports on attached bacterial populations in running waters in spite their importance for the microbial community (Lock 1981).

The aim of this paper is to quantify the attached

Wetzel, R.G. (ed.), Periphyton of Freshwater Ecosystems.
 ISBN 90 6193 768 X.

component of bacterial populations in epilithon of various running waters having a different level of organic pollution. The suspended component was estimated at the same localities, and the relation of both components to pollution level and input was examined. In addition, an attempt was made to compare the share of both components in bacterial standing crop, based on the estimation of bottom surface to water volume ratio at the respective sites.

Methods and materials

Epilithon samples

Sampling procedures developed from the methodology previously described (Punčochář 1977, 1981, in press; Straškrabová *et al* 1978) were used. Three to five stones were collected from the bottom at each sampling site, placed in polyethylene bags and transported to the laboratory. The average diameter of the stones did not exceed 10–15 cm, and total surface area of stones usually ranged between 400–800 cm^2. After 6–8 h the epilithon was carefully removed by using a wire brush and washed with water from the same locality. Then the suspension was mixed in a food mixer for 1 min. Two subsamples were then taken for chlorophyll *a* analysis. Another two subsamples were taken for bacteriological analyses. Each of these was further homogenized in two piston homogenizers to destroy flocs in the epilithon suspension. In the first homogenizer (30–40-ml volume), a slow piston movement was applied for 30–40 s. In a second homogenizer (10–20-ml volume) the piston rotation reached about 2000 rpm for 1 min. After this, the appropriate dilutions were made and the same methods as for water samples were used for bacteriological analyses. Kjeldahl nitrogen was estimated in the same two subsamples as bacteria.

Sterile conditions during sampling and preparation of the material were hopefully attained by using sterile glassware and sterilization with 96% ethyl alcohol (for hand washing, brushes, blender etc.).

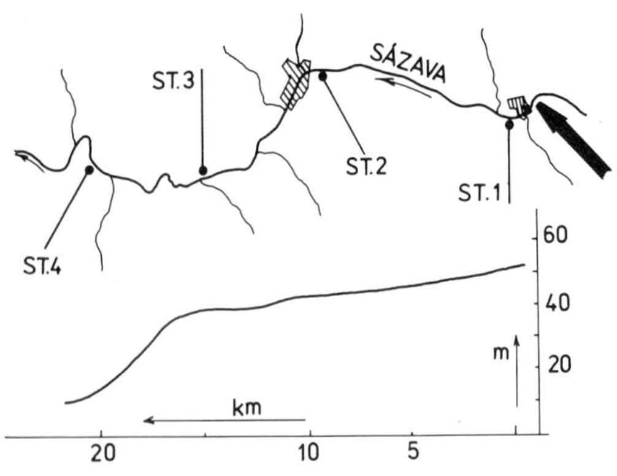

Fig. 1. Scheme of the stretch of the Sázava River with the marked source of pollution in winter, sampling stations (1–4) and river bed slope.

Water samples

Water samples for bacteriological analyses of suspended bacteria and for chemical analyses were taken in sterile glass bottles (500–1000 ml). Water for chlorophyll *a* determination and washing of periphyton was taken in alcohol-cleaned polyethylene bottles.

Determination of bacteria

Numbers of heterotrophic bacteria (BPA) were determined as colony counts on standard beef-peptone agar (No. 1, Lachema, Czechoslovakia) after 7 days of incubation at 20° C. Two-three dilutions in duplicate or triplicate were used for each epilithon subsample and water sample analyses.

Total numbers of bacteria (DC) were determined by the direct microscopic count technique (see e.g. Sorokin & Kadota 1972; Straškrabová 1973 or Straškrabová & Komárková 1979) on membrane filters (cellulose acetate, Synpor 8, Czechoslovakia, porosity 0.23 μm, diameter 35 mm). Water and diluted epilithon samples were fixed with formalin (3%), then filtered, dried, stained with erythrosin and counted at a magnification of 1200×. Two different dilutions (with two parallel filters each) for epilithon samples and two parallel filters for water samples were made. A constant volume of 7 ml was filtered in all cases, using bacteria-free tap water for dilutions.

Bacterial biomass (BB), was determined as the cell volume biomass (see Straškrabová & Komárková 1979). One hundred cells were measured for each sample. No correction was made for shrinkage.

For chlorophyll *a* determination (chl. *a*) 3–10 ml of epilithon suspension or 500–1000 ml of water was filtered through Whatman (GF/C) filters, dried and stored frozen until analyzed according to Lorenzen (1967). Kjeldahl nitrogen (Kj–N), NH_4–N and COD were determined according to Procházková *et al* (1970). For Kj–N, 5–10-ml volumes of each epilithon subsample were digested.

Variability of the epilithon characteristics
95% confidence limits (i.e. $\bar{x} \pm 2\sigma$ or $\bar{y} \div^{\times} 2\sigma$) were calculated for two subsamples from one epilithon sample procedure. The limits estimated for BPA = $\bar{y} \div^{\times} 1.97$, for DC = $\bar{y} \div^{\times} 1.59$ and for chl. *a* = $\bar{x} \pm 12.4\%$. More details on the evaluation of variability of epilithon samples and analyses will be published elsewhere.

Determination of stone surface
Sampled stones were dried after the removal of epilithon and carefully wrapped in aluminium foil. The total stone surface was measured by weighing the foil. Similarly, the upper parts of stone surfaces (i.e. continously washed with water current and illuminated) were measured in cases when it was possible to distinguish them. Generally, the stone surface in contact with the stream bottom was about 40% of total stone surface area.

Evaluation of epilithon characteristics
Bacteria and Kj–N were usually related to the unit area of the total stone surface whereas chl. *a* concentrations were related per unit area of the illuminated stone surface. However, if heavy accumulation of attached biomass appeared (visible thickness more than 1–2 mm), the data were related to the illuminated area of the stone surface only assuming the colonization of lower part of the stones to be negligible.

Localities
Epilithon samples were taken from various shallow flowing waters having the flow rates of 0.03–8.5 $m^3 s^{-1}$. The basic characteristics of the particular localities and profiles are included in the results section. All the localities were located in the crystallinicum regions; therefore the stones were similar (mainly granite).

Various streams in the High Tatra and Krkonoše Mountains, and upper parts of rivers were sampled, most of them above the forest border at 1200–2000 m a.s.l., in October, 1981 and June and July, 1982.

Žebrakovský Brook is located in drainage area of Sázava River at about 400 m a.s.l. and it is surrounded by agricultural fields, grasslands, and small forests. Average flow rate is about 0.03 $m^3 s^{-1}$. The brook has no apparent source of pollution, being a local source of drinking water.

Pstružný Brook is also located in the drainage area of Sazava River; its distance from Žebrakovský Brook is about 20 km. The investigated site is polluted with municipal sewage from the nearby town Humpolec. Average flow rate is about 0.1 $m^3 s^{-1}$.

The Malše River is located in South Bohemia, and the average flow rate is about 2.02 $m^3 s^{-1}$.

Two sampling sites were investigated, i.e. above and below effluent of partially purified municipal sewage from the town Kaplice.

River Sázava is in the drainage area of Vltava River, and the investigated stretch is located in the upper part of its flow, with an average flow rate of about 8.5 $m^3 s^{-1}$. The stretch (Fig. 1) is heavily polluted seasonally from starch factories (October–February). The main pollution sources are located above St. 1. A stony river bed occurs in some places with about a 4–7% slope. No other serious pollution source influenced the stretch during winter.

The investigated sites were chosen with the respect to different levels of organic pollution, i.e. at various distances from pollution sources (mainly municipal sewage and wastes of food factories).

Some of the localities were sampled through the season and the number of samples (n) used for evaluation of means and ranges is given in the tables.

The depths of the sampling sites ranged between 0.1–0.7 m, mostly within 0.2–0.4 m. The stones were taken from the areas with current velocities of 0.1–0.5 m s^{-1} (measured at a 5-cm distance above the stone surfaces by a hydrometrical wing).

Table 1. Characteristics of bacterial populations at two sites of the River Malše. Site A above and Site B below pollution input. Ranges of 11 samples taken throughout one year. Concentration of epilithic bacteria related to the total stone surface area. Percentage of attached BPA given as mean ranges.

Site	A	B
Mean depth (m)	0.20–0.70	0.20–0.70
COD (mg O_2 l^{-1})	4.1–15.8	9.4–39.6
BPA in water (10^3 ml^{-1})	1.3–23	6.5–192
BPA in epilithon (10^6 cm^{-2})	0.09–1.9	3.0–15.0
DC in water (10^6 ml^{-1})	1.8–9.3	5.8–60.2
DC in epilithon (10^9 cm^{-2})	0.26–1.3	0.7–2.4
SVR (cm^2 ml^{-1})	0.027–0.164	0.027–0.164
Attached bacteria (%)	65–86	73–89

The profiles investigated within the same streams or in streams of similar flow rates were chosen with the respect to the similarity of the stream-bed shape and slope, mean depth, occurrence of vegetation, distribution and sizes of stones on the bottom, and minimum shading (if possible) from the banks, etc.

Bottom surface to water volume ratio (SVR = cm^2 ml^{-1})

The ratio of bottom surface area to water volume below the unit area of water surface was estimated to evaluate the percentage of attached (or suspended) component in the standing crop at a given site. Once SVR was estimated, the percentage of epilithon was calculated under the assumption that characteristics determined in the epilithon sample would hold for the site bottom. Mean water volume below a unit area of the stream surface was estimated by measuring the site mean depth. The bottom surface was estimated by measuring the surface of stones taken from the bottom below a determined stream surface area (measurements done usually below 0.5–1.0 m^2). Several measurements in various localities indicated that bottom surface area (based on total stone surface measurements) can be estimated to be about 2–6 times larger than the area of stream surface. When only the illuminated area of stone surfaces is taken into account, the bottom surface can be assumed to be equal to or up to three times larger than the area of the stream surface. Because of this, the percentage of attached component was calculated according to what surface areas were used for calculations of concentration.

Percentages of the attached component of the population standing crop of respective sites are presented in the above ranges to cover the variability in the area of bottom surface.

Results

Influence of point pollution source on bacterial populations in a river is evident in Table 1. The BPA concentrations of both suspended and attached bacteria increased below pollution sources in agreement with the increase in oxidizable organics (COD). BPA more remarkably increased in epilithon, despite a very wide range in the data from Site A. DC varied in epilithon to a substantially lower extent both within and between sites.

Data indicating the pollution level in the two brooks are summarized in Table 2. In Žebrakovský Brook the bottom surface area was 1–3 times greater than a unit area of brook surface. The microbial characteristics were in good agreement with those observed in the unpolluted site of the above locality.

Pstružný Brook is polluted with municipal sewage and a visible slime occurred on the bottom. The illuminated area of the bottom was the same or twice as large as the area of brook surface. Conspicuous fluctuations in water quality occurred in the site as indicated by a wide range in the data. As compared with Žebrakovský Brook, BPA concentrations in both water and epilithon were two orders of magnitude higher. DC and BB differed accordingly by approximately one order of magnitude. Similarly, chl. *a* and Kj–N concentrations in epilithon were about 7 times higher in the polluted brook. In both localities, concentrations of BPA, DC, and BB were about one to three orders of magnitude higher in epilithon than in water and the predominance of the attached bacterial component in standing crops was also evident.

Unpolluted localities mentioned above are, however, located in the agricultural-forest regions and, consequently, an input of nutrients and allochthonous organic matter might not be negligible. In checking localities without any apparent input of nutrients, samples were taken from the streams in mountains. In these streams (Table 3) considerably lower concentrations of BPA in epilithon and water were found as compared with the localities considered above. The chl. *a* concentrations, however, reached the same level. It should be noted that stones overgrown with mosses were not taken for epilithon analyses in these mountain streams.

The largest concentrations of bacteria found in epilithon were obtained from heavily polluted sites of the River Sázava (see also Table 4). The ranges in Table 3 indicated an increase of about 7 orders of magnitude for BPA in epilithon from polluted waters, whereas direct counts increased in the range by about two orders. Hence, the percentage of BPA in total bacterial flora of epilithon changed,

Table 2. Mean values and ranges of bacterial epilithon in Žebrakovský and Pstružný brooks. In addition, characteristics of water samples are given (means). Samplings from Aug. 1979 to Dec. 1980 (n = 12). Epilithon data related to the total stone surface at the former and to illuminated area at the latter locality.

Locality		Žebrakovský Brook	Pstružný Brook
COD (mg O_2 l^{-1})	mean	17.6	80.6
NH_4 N (mg l^{-1})	mean	0.093	2.26
Organic N (mg l^{-1})	mean	0.49	4.29
Site mean depth (m)	range	0.05–0.20	0.09–0.36
BPA in water (10^3 ml^{-1})	mean	10.4	2200
	range	0.88–41	31–8000
BPA in epilithon (10^6 cm^{-2})	mean	0.3	170
	range	0.05–0.9	2.6–600
DC in water (10^6 ml^{-1})	mean	3.6	52.1
	range	0.4–7.3	2.9–223
DC in epilithon (10^9 cm^{-2})	mean	1.1	8.0
	range	0.53–2.0	0.65–34.5
BB in water (10^6 μm^3 ml^{-1})	mean	0.6	13.7
	range	0.09–1.19	0.6–42.5
BB in epilithon (10^9 μm^3 cm^{-2})	mean	0.17	4.1
	range	0.04–0.27	0.3–25.2
Chl. a in epilithon (μg cm^{-2})	mean	2.5	17.1
	range	0.7–6.2	2.0–97.4
Kj N in epilithon (μg cm^{-2})	mean	47.0	363
	range	20.2–93.0	37.8–1170
SVR (cm^2 ml^{-1})	range	0.05–0.6	0.027–0.22
Attached BPA (%)	mean range	77–88	77–84
Attached BB (%)	mean range	97	88–93

Table 3. Ranges of minimum and maximum data on attached and suspended components of microbial populations found in the running waters having very different pollution levels. Epilithon data of unpolluted streams are related to total surface of stones whereas in polluted river they are related to illuminated surface of stones.

Locality	Unpolluted mountain streams		Heavily polluted river	
	n	range	n	range
BPA in water (numbers ml^{-1})	18	$2.2 \times 10 - 6.6 \times 10^2$	13	$8.3 \times 10^4 - 8.6 \times 10^5$
BPA in epilithon (numbers cm^{-2})	22	$9.4 \times 10^2 - 4.2 \times 10^4$	13	$5.6 \times 10^7 - 1.1 \times 10^9$
DC in water (numbers ml^{-1})	7	$1.9 \times 10^5 - 1.8 \times 10^6$	10	$3.0 \times 10^6 - 8.3 \times 10^7$
DC in epilithon (numbers cm^{-2})	7	$3.9 \times 10^8 - 1.3 \times 10^9$	10	$1.3 \times 10^{10} - 8.2 \times 10^{10}$
Chl. *a* in epilithon ($\mu g\ cm^{-2}$)	7	0.88–3.52	10	17.0–200

Table 4. Data on bacterial populations in several sites of the River Sázava sampled in Feb. 1978. Epilithon characteristics (means and 95% limits) were related to the illuminated area of stone surfaces.

	Site 1	Site 2	Site 3	Site 4
Mean depth (m)	0.3	0.5	0.3	0.2
BPA in water ($10^5\ ml^{-1}$)	8.6	8.0	1.7	1.5
BPA in epilithon ($10^8\ cm^{-2}$)	10.0	7.0	1.7	0.9
	5.3–20.0	3.4–13.0	0.9–3.3	0.5–1.8
DC in water ($10^6\ ml^{-1}$)	38.2	19.8	19.3	24.0
DC in epilithon ($10^{10}\ cm^{-2}$)	8.2	2.5	1.1	0.9
	5.2–13.2	1.6–4.0	0.7–1.7	0.6–1.4
BB in water ($10^6\ \mu m^3\ ml^{-1}$)	3.4	2.9	5.6	10.8
BB in epilithon ($10^9\ \mu m^3\ cm^{-2}$)	24.5	5.4	2.7	2.5
SVR ($cm^{-2}\ ml^{-1}$)	0.03–0.13	0.02–0.08	0.03–0.13	0.05–0.2
Attached BPA (%)	97–99	94–99	97–99	97–99
Attached DC (%)	99	97–99	98–99	96–99
Attached BB (%)	>99	97–99	94–99	92–98

Table 5. Changes of attached and suspended components of bacteria at St. 4 of Sázava River (see Table 4 & Fig. 1) during the period of heavy pollution (winter) and outside this period (summer). Epilithon data of winter period were related to illuminated stone surface and to total stone surface in summer samples.

Period		Summer (n = 4)	Winter (n = 7)
BPA counts in water ($10^3\ ml^{-1}$)	mean	67	110
	range	5.0–180	31–224
BPA counts in epilithon ($10^6\ cm^{-2}$)	mean	0.9	19
	range	0.12–1.4	1.2–100
DC in water ($10^6\ ml^{-1}$)	mean	12.6	26.6
	range	10.7–15.7	6.3–44.7
DC in epilithon ($10^9\ cm^{-2}$)	mean	2.4	4.7
	range	0.5–5.6	2.7–7.5
Kj–N concentration ($\mu g\ cm^{-2}$)	mean	88.1	376
	range	46.0–153	170–1100
Chl. *a* concentration ($\mu g\ cm^{-2}$)	mean	12.1	14.0
	range	2.50–31.3	7.1–25.0

thus exhibiting a positive relation to pollution level. The ranges of bacterial concentrations in water from these very different localities exhibited similar relationships, though to a lower extent (about 4 orders of magnitude).

The relationship between bacterial components and pollution was not surprising and can generally be expected. This general trend proved to be significant especially for BPA, and a regression was performed on the logarithmical transformation of the data (Fig. 2). However, in spite of the significance of the correlation coefficient, the relationship between correlated variables was not very close. Thus, for a particular value of one variable, a rather wide range of the other might be found.

This situation was shown, for example, by the data obtained within a stretch of Sázava River which is heavily polluted in winter. Sewage fungus (see Curtis & Harrington 1971) composed of *Sphaerotilus* and *Leptomitus* filaments occurred on the bottom. Very high densities of bacteria in epilithon strikingly decreased (by one order) along the stretch, whereas changes of suspended component were remarkably lower. Nevertheless, the attached bacterial component formed in all profiles about 96% of the total bacterial population.

Similar disagreement between changes of suspended and attached BPA is represented in Table 5, containing the data on the influence of seasonal pollution from starch factories on the microbial community at Station 4 of the Sázava River. BPA

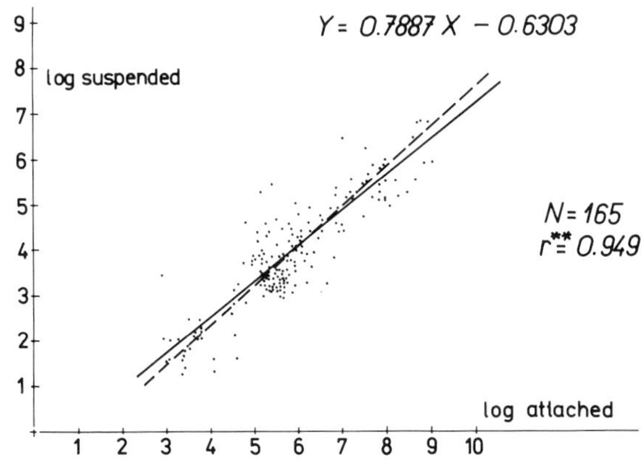

Fig. 2. Correlation between log BPA in epilithon and water in running waters with different levels of pollution.

Table 6. Characteristics of microbial components at two sites of the mountain stream Zelený Brook (High Tatra Mts., Oct. 1981), located above the forest border. A: above a mountain hotel, B: below.

	Site A	Site B
BPA in water (ml^{-1})	20	2.5×10^3
BPA in epilithon (10^2 cm^{-2})	23.9	8.7
DC in water (10^6 ml^{-1})	0.3	0.2
DC in epilithon (10^9 cm^{-2})	1.1	1.1
Chl. *a* concentration (μg cm^{-2})	3.50	3.30

in epilithon sharply increased in winter whereas a negligible change was noted in the water. DC accordingly was twice as high in winter. However, the epilithon biomass, being of a heterotrophic nature (see chl. *a*), increased considerably.

The opposite situation occurred in a mountain stream (Table 6). BPA increased about 100 times in the water at the site below a mountain hotel, whereas BPA in the epilithon was only slightly decreased. DC in both water and epilithon, as well as chl. *a*, did not change.

Discussion

These data indicate very high concentration of bacteria in epilithon in both unpolluted and polluted running waters, with a range 10^8–10^{10} cells cm^{-2} for total cell counts. Bacterial biomass in epilithon of these various polluted localities had a wider range, i.e. 10^7–10^{10} μm^3 cm^{-2} (based on a substantially lower number of samples). Concentrations of suspended bacteria were mostly 1000 times lower at these localities.

The same relationship between concentrations of both components was found by Geesey *et al* (1978) in Canadian mountain streams. With the exception of maximum values they found concentrations of both bacterial components about one order of magnitude lower as compared with our lowest findings. However, our data from unpolluted localities (i.e. 10^8 cells cm^{-2}) are in good agreement with bacterial concentrations found in subarctic rivers (Lock & Wallace: in Lock 1981) despite of differences in methodology (esp. epifluorescent microscope technique). For polluted or eutrophic waters no other data on epilithic DC have been published except ours (Punčochář, in press; Punčochář & Straškrabová 1980; Straškrabová *et al* 1978). An even wider range in concentrations were found for epilithic BPA (10^2–10^9 cells cm^{-2}) indicating that BPA counts comprised an increasing percentage of total bacteria with increasing pollution. These data extend the validity of our earlier observations of the relationship between BPA in epilithon and pollution changes in river sites due to both selfpurification and pollution increase (Punčochář, in press; Punčochář & Straškrabová 1980). The changes in suspended BPA were less remarkable in particular downstream changes as selfpurification in the river progressed.

In the literature, the only notice on epilithic BPA has been by Bott (1975), who found in a small American stream a concentration range 10^5–10^7 cells cm^{-2} in epilithon and about 1000 times lower concentrations in water. These results are in good accordance with our data despite differences in methodology.

The dependency of epilithic bacteria on changes in the level of pollution is not too surprising since they develop under specific conditions of locality and form an autochthonous component. In contrast, the suspended organisms of stream communities are continuously washed downstream and represent an allochthonous origin for individual sites (see Costerton *et al* 1978). This suspended component is composed of both bacteria allochthonous to the stream and stream bacteria which develop along the flow (Baker & Farr 1977). The growth rates of bacteria in running waters (Baker 1981) are, however, much longer than the retention time of water in very short stream transects. Consequently, the relation of suspended components to substrate inputs is separated in time and space, being influenced by flow rate fluctuations as well as water temperatures and other parameters. In addition, fluctuations might occur due to irregularities in allochthonous input (e.g. effluent flow rates through the day, etc.). Thus, the characteristics of the suspended component are like chemical characteristics of water samples.

In spite this, we found significant a correlation between suspended and attached BPA (Fig. 2). The main cause was evidently the simultaneous input of bacteria and organic matter in municipal sewage or organic-rich wastewater. In addition, a part of attached bacteria continuously emigrate (detach) to water (Sanders 1967; Hossel & Baker 1979), thus influencing the suspended component.

Clearly, simultaneous analysis of attached and suspended components enable information on substrate enrichment by the former and on bacterial input to the site by the latter.

The attached component can be sensitive to very low inputs of biodegradable substrate as indicated by data of Curtis & Harrington (1971) who found sewage fungus in some localities with no noticable changes in water quality. Our data from St. 4 of Sázava River (Table 5) showed a similar effect because BPA concentrations increased in epilithon in winter although mostly negligible changes of BOD and COD were found between summer and winter data (Punčochář *et al* 1975; Nejedlý 1974) in this locality.

Very low concentrations of epilithic BPA in mountain streams indicate that input of allochthonous organics can be a factor of primary importance for their development.

The very large difference between standing crops of suspended and attached bacteria in running waters is primarily the result of the attached component being additionally supported by an increasing bottom surface area to water volume ratio in shallow localities.

In our estimations the percentage of attached bacteria predominated at all localities studied (mean depth up to 0.7 m). However, our procedure for evaluation of the percentage of epilithic component in the standing crop is open for discussion.

Concerning bottom surface area, the data by our simple method reached exactly the same average values as Kairesalo (1976) found by another methodology in a stony lake littoral (i.e. bottom surface 4 times larger than the area of water surface).

Assessment of the suspended component is made without any serious difficulties of samplings due to rather homogeneous distribution of free-floating bacteria in the water column of running waters. However, fluctuations might occur as already mentioned. Obviously, increasing allochthonous input can substantially change the percentage of the suspended component in a river system. At the same time, an increase in the percentage of stream microorganisms might occur along a stream as a result of their progressive growth (e.g. see higher percentage of algal suspended component in lower parts of flow, Straškrabová *et al* 1978).

Flow rate fluctuations (in particular sudden changes) can change concentrations of both plankton and epilithon concentrations and alter bottom surface area-water volume ratios. In spite of this importance, little information is still available on the influence of flow rates on epilithon (Liaw & MacCrimmon 1978; Punčochář, in press).

More difficulties are involved in the estimation of the epilithic component. Thus, the weakest point in our determinations of the percentage attached component is the assumption that a sample of 3–5 stones is sufficiently representative for the whole site. Due to this problem, much attention has been paid to the choice of comparable stream sites and places for stone collections.

To analyze the variability of both bacteria and chl. *a* in epilithon five samples (five stones each) were taken from a river bottom area of *ca.* 2 m^2 (water current 0.2–0.4 m s^{-1} at 5 cm above it; riffle type). We found higher variability within subsamples than between samples (preliminary report in Punčochář 1981). Clumping and fast sedimentation during collection of subsamples were presumably responsible. Since increasing the number of subsamples is impracticable for obvious reasons, 95% limits based on the variability of subsamples were used for evaluation of data.

Nevertheless, we sometimes observed large differences in chlorophyll *a* between samples from adjacant places, some of them with positive relationships to water current velocity. But light intensity variations within the bottom microenvironment are usually considered more important (see McConnel & Sigler 1959; Saunders & Eaton 1977). Surprisingly, most of these chl. *a* fluctuations were not accompanied by significant differences in epilithic bacteria densities (Punčochář, unpubl. data). Geesey *et al* (1978) also observed an inverse relationship between chlorophyll *a* and bacteria in one mountain stream. A simultaneous increase in chl. *a* and bacteria concentrations is, however, evident in profiles with enriched sewage pollution (in accordance with the experimental observations of Wuhrmann & Eichenberger 1975). Evidently, more data dealing with epilithon variability and controlling factors are needed to increase our knowledge of the microbial communities in running waters.

Summary

Concentrations of bacteria in epilithon of both unpolluted and heavily polluted running waters (0.03–8.5 m^3 s^{-1}) ranged between 10^8–10^{10} cells cm^{-2} for total cell counts and between 10^2–10^9 cells cm^{-2} for heterotrophic bacteria.

The concentrations of suspended bacteria were about 100–1000 times lower in the same localities. A predominance of the attached component in the instantaneous bacterial population (standing crop) was found in all studied sites, based on the estimation of bottom surface-water volume ratios.

Input of organic pollution (mainly municipal sewage) to streams was accompanied by an increase in the concentration of epilithic, especially heterotrophic, bacteria, which increased in percentage in relation to the total bacterial numbers. The suspended bacterial component changed similarly, but to a lower extent, depending on the allochtonous input to the particular site.

Acknowledgements

The author would like to thank Dr. B. Desortová for the chlorophyll *a* determinations and also Mrs. L. Mikeštíková for assistance in bacteriological analyses of the material.

Stimulating discussions with and the continuous interest of Dr. J. Hrbáček during the investigation of epilithon are appreciated very much.

Many thanks are due to Mr. R.T. Bell (Limnological Institute, University of Uppsala) for the improvement of the English and the style of this manuscript.

References

Aizaki, M., 1979. Growth rates of microorganisms in a periphyton community. Jap. J. Limnol. 40:10–19.

Baker, J.H., 1981. Growth rates of bacteria in rivers. Rep. Freshwat. Biol. Ass. 49:57–62.

Baker, J.H. & Farr, I.S., 1977. Origins, characterizations and dynamics of suspended bacteria in two chalk streams. Arch. Hydrobiol. 80:308–326.

Bott, T.L., 1975. Bacterial growth rates and temperature optima in a stream with fluctuating thermal regime. Limnol. Oceanogr. 20:191–197.

Costerton, J.W., Geesey, G.G. & Cheng, K.J., 1978. How bacteria stick. Scient. Am. 238:86–95.

Curtis, E.J.C. & Harrington, D.W., 1971. The occurrence of sewage fungus in rivers in the United Kingdom. Wat. Res. 5:281–290.

Fjerdingstadt, E., 1964. Pollution of streams estimated by benthal phytomicroorganisms. 1. A saprobic system based on communities of organisms and ecological factors. Int. Revue ges. Hydrobiol. 49:63–131.

Geesey, G.G., Mutch, R., Costerton, J.W. & Green, R.B., 1978. Sessile bacteria: An important component of the microbial population in small mountain streams. Limnol. Oceanogr. 23:1214–23.

Hossel, J.C. & Baker, J.H., 1979. Estimation of the growth rates of epiphytic bacteria and Lemna minor in a river. Freshwat. Biol. 9:319–327.

Kairesalo, T., 1976. Measurements of producton of epilithon and littoral plankton in Lake Pääjärvi, Southern Finland. Ann. Bot. Fenn. 13:114–118.

Liaw, W.K. & MacCrimmon, H.R., 1978. Assessing changes in biomass of riverbed periphyton. Int. Rev. ges. Hydrobiol. 63:155–171.

Lind, O.T., 1974. Handbook of Common Methods in Limnology. C.V. Mosby Co., Saint Louis, 154 pp.

Lock, M.A., 1981. River epilithon- a light and organic energy transducer. In M.A. Lock & D.D. Williams (eds), Perspectives in Running Water Ecology. Plenum Press, N.Y.:3–40.

Lorenzen, C.J., 1967. Determination of chlorophyll and phaeopigments: spectrophotometric equations. Limnol. Oceanogr. 12:343–346.

McConnel, W.J. & Sigler, W.F., 1959. Chlorophyll and productivity in a mountain river. Limnol. Oceanogr. 4:335–351.

Nejedlý, A., 1974. Studie prostorového a časového průběhu deoxigenačního procesu v řece Sázavě v úseku Havlíčkův Brod-Vilémovice. Rep. Hydrol. Inst. Prague, Czech., 44 p.

Procházková, L., Blažka, P. & Králova, M., 1970. Chemical changes involving nitrogen metabolism in water and particulate matter during primary production experiment. Limnol. Oceanogr. 15:797–807.

Punčochář, P., 1977. Bestimmung von Bakterien in Aufwuchs. Proc. 2nd int. Hydromikrobiol. Symp. Smolenice, Veda Vlg., Bratislava: 73–78.

Punčochář, P., 1981. Kvantitativní vzorky přisedlé složky bakteriálních populací v mělkych tocích, odběry a hodnocení. In J. Krajča (ed.), Sb. 2. konf. Metody technika vzorkovaní vod. Brno: 91–100 p.

Punčochář, P., in press. Attached component of bacterial populations in the river community. Proc. 3rd int. Hydromikrobiol. Symp. Smolenice, June, 1980.

Punčochář, P. & Straškrabová, V., 1980. Bakteriální složka nárostů v řece Sázavě (Bacterial component of growths in the River Sazava). Vodní hospodářství, Ser. B 11:301–304. In Czech with English summ.

Punčochář, P., Zajíček, V. & Justýn, J., 1975. Problémy hydrologie a čistoty vody Sázavy (Problems of hydrology and water quality of the River Sazava). Acta sci. nat. Mus. Bohem. merid. (Česke Budějovice) 15:67–84. (in Czech with English summ.).

Sanders, W.M., 1967. The growth and development of attached stream bacteria. Wat. Resour. Res. 3:81–87.

Saunders, M.J. & Eaton, J.W., 1976. A method for estimating the standing crop and nutrient content of the phytobenthos of stony streams. Arch. Hydrobiol. 78:86–101.

Sorokin, Y.I. & Kadota, H., 1972. Techniques for the Assessment of Microbial Production and Decomposition in Fresh Waters. IBP Handbook 23, Blackwell Scientifie Publication, Oxford, 112 p.

Straškrabová, V., 1973. Methods for counting water bacteria-comparison and significance. Acta hydrochim. hydrobiol. 1:433–454.

Straškrabová, V. & Komárková, J., 1979. Seasonal changes of bacterioplankton in a reservoir related to algae. 1. Numbers and biomass. Int. Revue ges. Hydrobiol. 64:285–302.

Straškrabová, V., Punčochář, P., Bojanovski, B.B. & Fuksa, J., 1978. Suspended and attached microorganisms in a polluted river. Verh. int. Ver. Limnol. 20:2278–2283.

Strickland, J.D. & Parsons, T.R., 1965. A Practical Handbook for Seawater Analysis. Bull. Fish. Res. Bd Can. 167.

Traaen, T.C., Ormerod, K.S. & Efraimsen, H., 1972. Heterotrophic growth in indoor research channels; selfpurification and production. Verh. int. Ver. Limnol. 18:891–895.

Wuhrmann, K., 1972. Stream purification. In R. Mitchell (ed.), Water Pollution Microbiology. Willey, N.Y.:119–151.

Wuhrmann, K., 1974. Some problems and perspectives in applied limnology. Mitt. int. Ver. Limnol. 20:324–402.

Wuhrmann, K. & Eichenberger, E., 1975. Experiments on the effects of inorganic enrichment of rivers on periphyton primary production. Verh. int. Ver. Limnol. 19:2028–2034.

Author's address:
P. Punčochář
Hydrobiological Laboratory
Institute of Landscape Ecology
Czechoslovak Academy of Sciences
Vltavská 17
151 05 Praha 5, Czechoslovakia

CHAPTER 41

A study of the periphyton of Lake Saimaa, polluted by waste waters of the pulp industry

A method for water pollution control analysis

I. KETTUNEN

Introduction

The effects of effluents of the pulp and paper industry on recipient waters have been studied in Finland since the 1930's (Hagman 1936). The most important of the adverse effects of waste water has generally been held to be organic solids and their direct and indirect oxygen consumption.

Investigations carried out in recent years have taken particular account of the increasingly varied

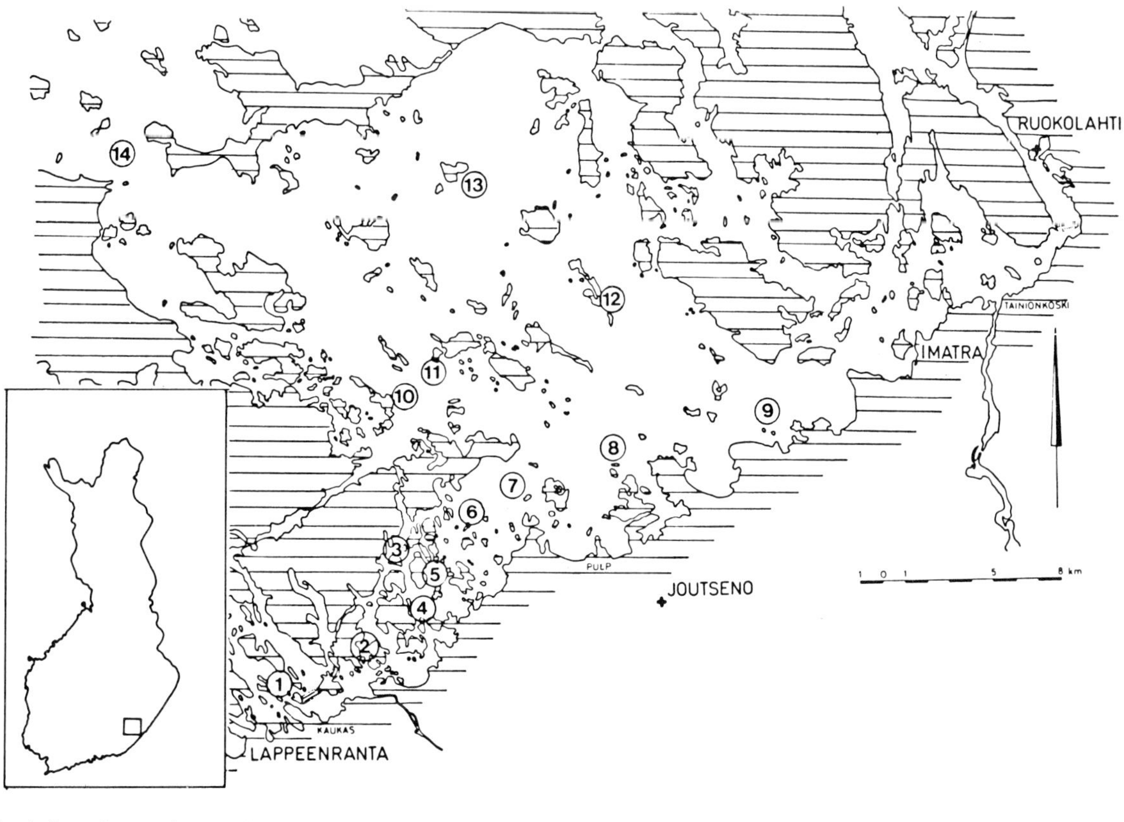

Fig. 1. Sampling stations, Lake Saimaa.

Wetzel, R.G. (ed.), Periphyton of Freshwater Ecosystems.
© 1983, Dr W. Junk Publishers, The Hague. ISBN 90 6193 768 X.

effects of waste waters, above all as a result of the growing prevalence of cellulose bleaching. In the area of southern Lake Saimaa influenced by industrial wastes, inhibiting effects can be seen in the macroscopic vegetation as well as in primary production (Heinonen 1972, 1980; Lehmusluoto & Heinonen 1970). Koli & Turkia (1964) found that various *Valvata*- and *Lymnea*-mollusk species were good indicators of the overall effects of waste waters. Falck (1969) reported that the distribution and species composition of fish populations were dependent on the level of pollution.

The aim of this investigation was to evaluate the state of pollution of southern Lake Saimaa by periphyton analyses.

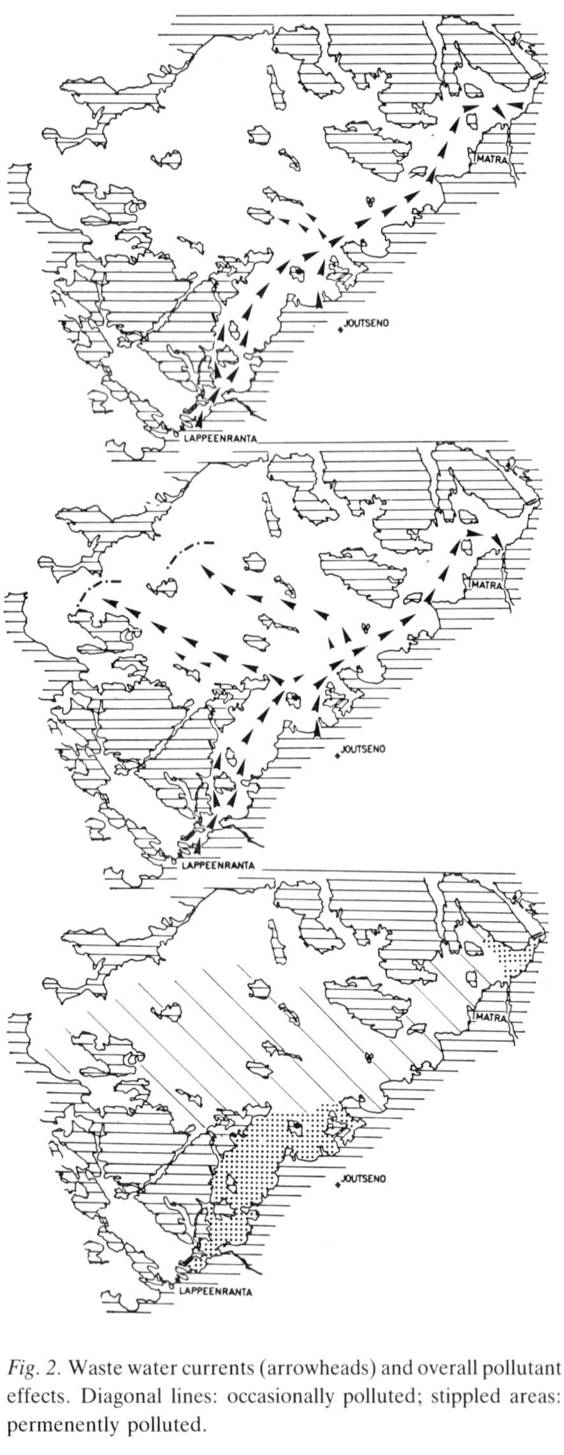

Fig. 2. Waste water currents (arrowheads) and overall pollutant effects. Diagonal lines: occasionally polluted; stippled areas: permenently polluted.

Area investigated

The area investigated (Fig. 1) is a part of Lake Saimaa, the largest lake in Finland. Permanent effects of effluent pollution extend over an area of about 70–80 km^2 (Fig. 2). The area investigated contains a very large number of islands and the shore line is very complex. In the southern regions the lake is shallow, with an average depth of only about 5 m. Further north the deepest parts of the open lake are 40–68 m deep.

Water currents affect the direction of flow of effluents. The prevailing directions of water currents and the spread of pollutants in the area investigated in summer and winter are presented in Fig. 2. During the non-frozen season effluents have been shown to spread mainly in the surface water layers, presumably due to the higher temperature of the waste water. Mixing of the pollutants with the lake water depends therefore mainly on the prevailing winds, but is in any case much more

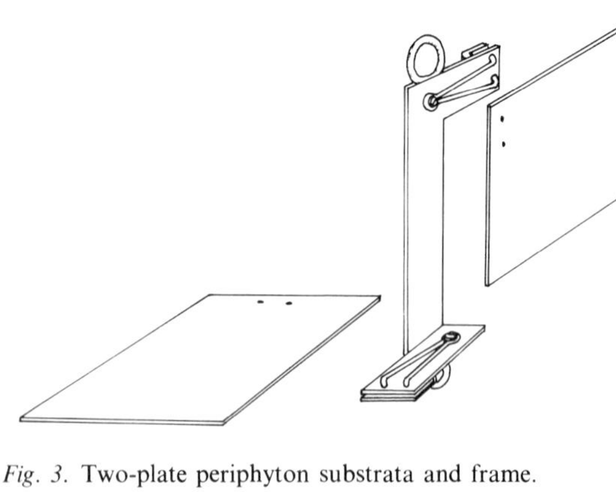

Fig. 3. Two-plate periphyton substrata and frame.

efficient than during the frozen season, when the effluents dispense regularly into the bottom layers. During the winter, bottom currents may travel for considerable distances (up to 20 km) against the main direction of flow in the watercourse (Heinonen 1972, 1973).

Methods

In this investigation the artificial substrata method was used. The plates were 10 × 15 cm in size and were prepared from transparent hard plastic sheets. The incubation frame was at the beginning of the investigation a stainless steel support holding two of the test plates (Fig. 3). One of the plates was in the vertical and one in the horizontal plane. In order to obtain duplicate results, and in less productive waters in order to obtain a greater surface area of incubation, the original design of the framework was modified to provide a frame containing eight vertically positioned plates in a circle (Fig. 4).

Spring mounting of the plates enabled plate removal and replacement with the frame submersed in water. The method was developed by Eloranta & Kunnas (1976), Heinonen (1981), and in the final form by the author. Similar apparati have been

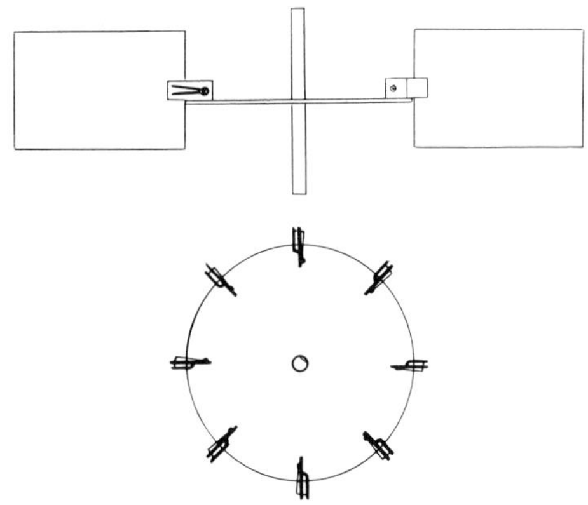

Fig. 4. Eight-plate periphyton substrata sampler and supporting frame.

used in other investigations in Finland (Heinonen 1981).

The equipment as presented in Fig. 4 was anchored in offshore waters about 1.5–2 m deep so that the plates (10 × 15 cm) were at a depth of 0.5 m. The plates were incubated *in situ* for three weeks. Immediately after lifting, the plates were fastened into individual water-tight plastic bags. Drying of the plates during transport to the laboratory was prevented by addition of 5–10 ml of distilled water to the bags. Laboratory examination of

Table 1. Chlorophyll *a* and temperatures.

Station	Chlorophyll *a* mg m^{-2}			Temperature during incubation °C		
	June	August	October	June	August	October
1	3.1		2.1	13.1–15.1	20.2–17.0	9.6–4.7
2	9.6		0.9	12.3–15.3	20.0–17.1	9.9–5.2
3	26.0	26.0	2.5	11.8–15.4	19.7–16.7	9.6–5.0
4	48.0	22.0	1.9	12.8–15.6	19.6–16.6	9.6–4.2
5	28.0	34.0	0.5	11.1–15.2	18.8–16.7	9.7–4.5
6	30.0	18.0	0.9	10.8–14.4	18.5–16.3	9.4–4.3
7	29.0			9.2–16.2	18.7–16.3	9.6–4.5
8	2.4		1.0	11.2–14.6	18.1–16.2	9.9–5.0
9	3.5			9.2–14.0	17.9–16.2	
10	0.4	2.7	0.7	8.3–12.7	18.6–16.2	10.1–5.3
11	1.5	6.7		8.7–14.7	18.7–15.7	10.0–5.3
12	2.1			8.3–14.7	18.2–16.5	10.0–5.6
13	1.0	3.7	1.6	8.7–13.6	18.7–16.3	10.0–6.2
14	0.2	2.4		7.7–11.5	18.4–16.0	10.2–7.1

the plates was carried out on the same day.

Periphyton adhering to the plates was removed by scraping and simultaneous rinsing with distilled water. The biomass obtained was diluted in a known volume of water, in most cases 500 ml. The sample was shaken thoroughly and 3–5 duplicate samples were taken from each sampling site and sampling occasion.

The periphyton samples obtained as described above were analyzed for evaporation residue, ignition residue, ignition loss, and chlorophyll *a*. The results were calculated and reported per unit area. The area of each plate was 300 cm^2.

Results

Chlorophyll *a* was a good indicator of the effects of pulping effluents as stimulators of primary production. Chlorophyll *a* values obtained in the unpolluted waters were 0.5–2.0 mg m^{-2} per 3 weeks, whereas the corresponding values in the polluted areas were 20–40 mg m^{-2} per 3 weeks (Fig. 5; Table 1). The solid contents expressed in Fig. 5 show that during the warm water season the values obtained varied randomly and independently of the results of the chlorophyll *a* analyses. During the cold water period of early summer the solids adhering to the test plates described the extent of distribution of clay and kaoline.

During the winter, plate incubations carried out under the ice layer gave no indication of chlorophyll *a* formation except during the final winter period in the unpolluted control site, where very low levels of chlorophyll formation were recorded.

The chemical quality of the lake water varied throughout the period of investigation. This variation was greatest in the areas with the highest levels of production (Fig. 6). The totally unpolluted and also the most polluted waters had a considerably more stable chemical quality.

On the basis of the chemical analyses water quality changed from pure to polluted within a very short period of time. It is therefore apparent that random variation may result in major errors in water quality estimation.

In water quality monitoring research it is of considerable importance to know the overall effect and the degree of permanence of waste water pollution. The periphyton incubation carried out in this

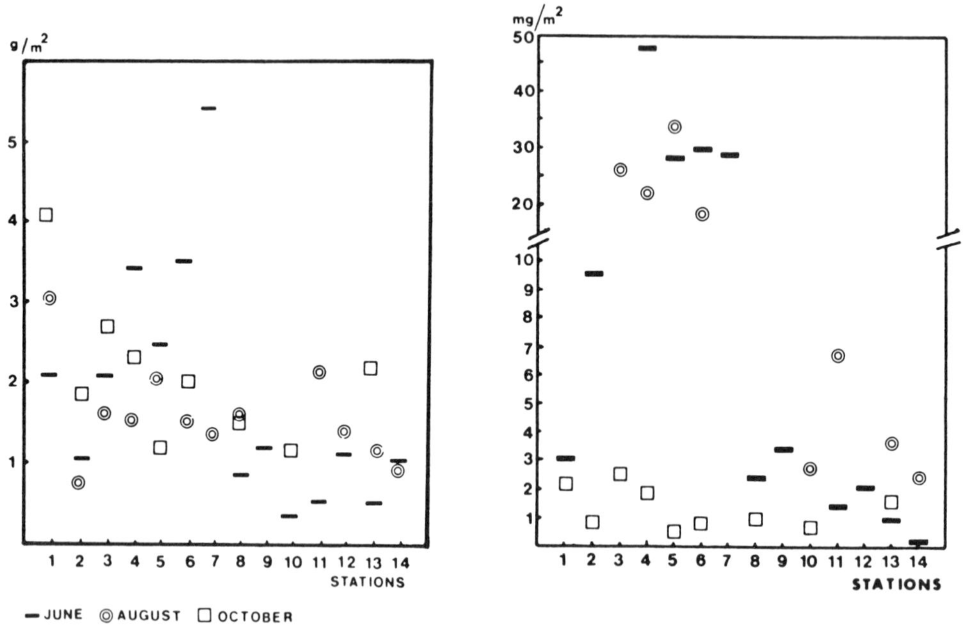

Fig. 5. Seasonal differences in solid substances (left) and chlorophyll *a* (right) on substrata at different stations in Lake Saimaa.

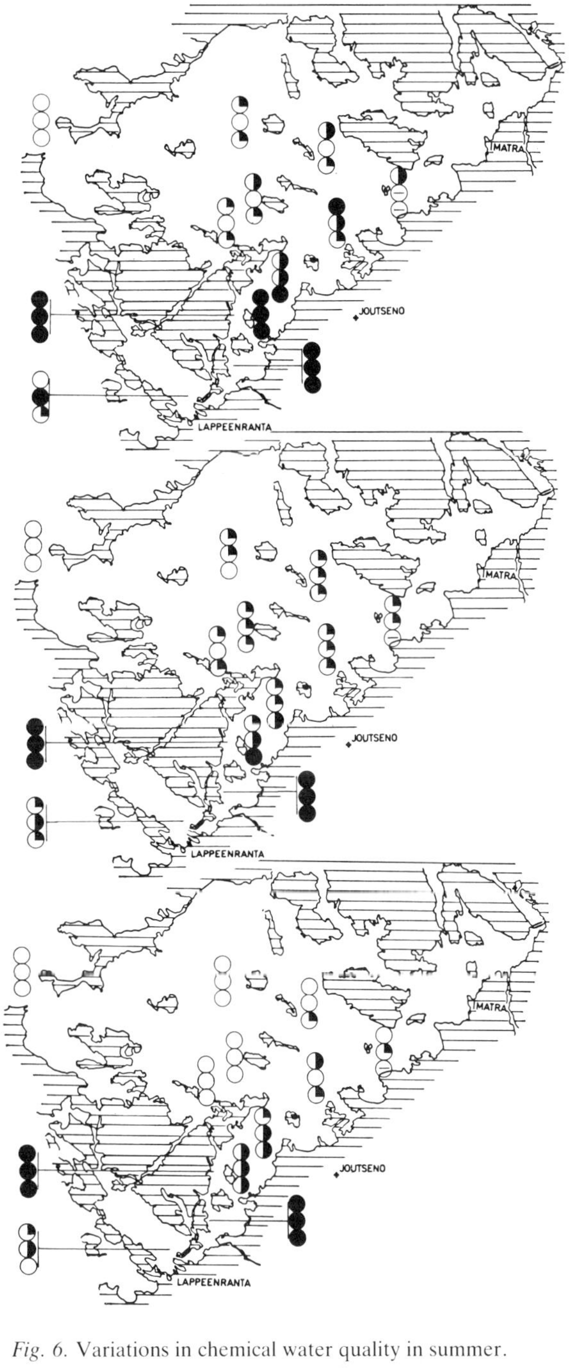

Fig. 6. Variations in chemical water quality in summer. ○: unpolluted ◔: slightly polluted ◑: polluted ●: heavily polluted.

Na mg l^{-1}	cond. mS m^{-1}	COD mg l^{-1} O_2
○ <2.2	○ <5.0	○ <6.1
◔ 2.3–3.5	◔ 5.1–7.0	◔ 6.2–7.0
◑ 3.6–7.0	◑ 7.1–10.0	◑ 7.1–10.0
● >7.0	● >10.0	● >10.0

investigation indicated that the method could give a clearer picture of the pulping wastes during the productive season than could rapidly changing results of chemical analyses. Periphyton alone is not sufficient to indicate the polluting effects of pulping effluents. However, it provides and excellent refinement of the general picture obtained by chemical analysis.

References

Eloranta, P. & Kunnas, S., 1976. A comparison of littoral periphyton in some lakes of Central Finland. Biol. Res. Rep. Univ. Jyväskylä 2:34–50.

Falck, P., 1969. Kalaston koostumus eräillä Vuoksen vesistön likaantuneilla ja puhtailla alueilla vuosina 1966–1967. Saimaan vesiensuojeluyhdistyksen tiedonantoja 13:1–40.

Hagman, N., 1936. Vattenundersökning betr. Kymmeneälv. Mimeographed report.

Heinonen, P., 1972. Jätevesien vaikutus järvien rehevöittäjänä. Vesientutkumuslaitoksen julkaisuja 5:1–87.

Heinonen, P., 1973. Jätevesien kulkeutuminen talvella vastavirtaan. Limnologisymphosion 1970, 67–72.

Heinonen, P., 1980. Quantity and composition of phytoplankton in Finnish inland waters. Vesientutkimuslaitoksen julkaisuja 37:1–91.

Heinonen, P., 1981. Pohjakasvustotutkimukset (Perifyton) rehevöitymisen arvioinnissa. Vesihallituksen tiedotuksia 212:21–42.

Koli, L. & Turkia, E., 1964. Über die Wassermollusken im Südteil des Saimaa-Sees in süd-Finnland. Ann. zool. fenn. 1:81–88.

Lehmusluoto, P. & Heinonen, P., 1970. Eräiden jätevesien vaikutus Saimaan perustuotantoon. Vesi 4:1–8.

Author's address:
I. Kettunen
Sutelantie 6
45130 Kouvola, Finland

CHAPTER 42

Concluding remarks

ROBERT G. WETZEL

As we conclude an intensive week of exhange, I am encouraged by the dedication with which you have addressed the tasks of evaluating the complexities of freshwater periphytic communities. Although I still feel that reseach on periphyton has somewhat stagnated, I sense an infusion of enthusiasm, even optimism by the collective expertise of this work shop. We all recognize that optimism does not emerge easily. We have sufficient experience to appreciate the enormous problems facing us in the extreme heterogeneity and complexity of periphytic communities.

For years we have been subtly, some of us not so subtly, encouraging intensified investigation of the littoral zone and wetland producers as the driving forces and these zones as major controlling metabolic regions of most freshwater ecosystems. Seeing your reports and discussion only reaffirm the opinions that not only are growth and productivity of the sessile communities of major importance in most freshwater ecosystems, but that important regulatory functions exist in periphytic communities that can radically alter rates and pathways of ecosystem biogeochemical cycling.

Expressions have been voiced that we are simply evaluating periphytic communities in greater detail than in the past. I agree, we are encouraging exactly that, but in very different ways than what was done previously. We are finally recognizing the highly structured, stratified community components. We emphasized the importance of characterization of the periphyton at the species level, both in architecture and functionally. Importantly, we recognize that the periphytic community cannot be adequately understood without simultaneously understanding the metabolism, population dynamics, and competitive interactions of the primary integrating components of the bacteria and other microflora in close juxtaposition to the algae, as well as the metabolically dynamic substrata upon which the periphyton rests.

Certainly a major future direction in periphyton research focuses upon the rate processes and primary metabolic pathways. Examples were many. The pathways and rates of inorganic carbon utilization as it is distributed among dynamic sources and competing carbon sinks by bacteria, algal, sediment, and macrophytic sites of utilization and production; rates of phosphorus utilization and release by algae and macrophytes in relation to competitive sinks by release to the water, bacteria, and inorganic adsorption; simultaneous rates of utilization and generation of inorganic and organic nitrogen compounds by competing and interrelated sources, processes, and sinks within the periphytic components, to name only a few. Despite the emphasis on physiological processes and their rate regulation, we emphasized that these analyses compliment, not supplant, analyses of biomass and productivity at the community level.

Emphasis was also given to the necessity of applying modern methodology at the spatial and temporal levels of the microorganisms constituting periphyton. Much of the microtechnique needed for

Wetzel, R.G. (ed.), Periphyton of Freshwater Ecosystems.
 ISBN 90 6193 768 X.

an understanding of the metabolism and control mechanisms already exists. Although methodology is not a panacea, application of modern techniques can permit much greater understanding than was possible previously. As detailed knowledge of the functioning of the periphytic components increases, the complexity increases as we approach closer to the true, integrated community. Collaboration among interactive disciplines becomes increasingly important.

As we progress further and further into the operational detail of periphytic communities, we should not renounce the search for structural and functional unity. Although such commonality exists at both levels, metabolic similarities are most likely and hopefully will emerge rather soon.

Our recommendations for future research only scratch the surface. Although they will surely be looked upon as a fledgling struggle in the future, they represent a beginning, an analysis of where we are and should fruitfully direct attention into productive research areas in the immediate future. Hopefully we have challenged the young, sparked their interest, and demonstrated that the fascination of periphyton resides in that elusive complexity of its integration.

Author's address:
Robert G. Wetzel
W.K. Kellogg Biological Station
Michigan State University
Hickery Corners, MI 49060, U.S.A.

CHAPTER 43

Recommendations for future research on periphyton

ROBERT G. WETZEL

Several days of the First International Workshop on Periphyton in Freshwater Ecosystems were devoted to intensive discussions on evaluations of the present understanding of periphytic communities and future needs and directions of research on periphyton. Participants divided into four topical subjects; their recommendations follow.

Structure and Dynamics of Periphytic Populations and Communities (K. Sand Jensen & R. Bohr, Chairpersons).

Structure

1. Periphyton is generally defined as the community of all organisms living upon the surfaces of submersed objects in water.
2. In order to avoid misunderstanding, the spatial arrangement of the elements of periphytic communities should be referred to as its architecture.
3. In describing the architecture of periphyton, no particular methods can yet be recommended. Attention should be directed to development of reliable methods for the analyses of spatial distribution of periphyton components.
4. Particular attention should be directed toward the roles of bacteria in the development of the assemblages.
5. Development of standard methods, if possible, for sampling of periphytic communities on different substrata should be attempted. Particular attention should be directed to the efficiencies of different sampling methods.
6. Knowledge of the species taxonomy of the attached organisms is of the utmost importance, and requires appropriately trained taxonomists.
7. The use of artificial substrata should be restricted only to relative comparative analyses of periphyton, e.g. along a pollutional gradient.
8. Knowledge of the structure and architecture of the community is the foundation for interpretation of its functional characteristics.
9. Quantification of the community structure should be done by evaluations of biomass and enumeration of live individuals.
10. For generalizations to emerge, researchers must work towards unification in analyses and presentation of results.
11. The word periphyton should be used as an important keyword in information retrieval systems.

Dynamics

1. Periphyton is a physical link between the terrestrial and open water systems. On a microscale, periphyton acts as an interface between the substratum and the surrounding water. Research needs include:
   a. The roles of periphyton in the regulation of whole ecosystem nutrient dynamics and the

Wetzel, R.G. (ed.), Periphyton of Freshwater Ecosystems.
© 1983, Dr W. Junk Publishers, The Hague. ISBN 90 6193 768 X.

ways in which this community modifies inputs of nutrients from external sources.
   b. Evaluation of the ability of periphyton to function as site-specific indicators of pollution problems (bio-stimulatory or inhibitory).
2. The periphyton community is functionally a microcosm which includes internal autotrophic and heterotrophic processes occurring simultaneously within a boundary layer. Research needs include:
   a. Application of micro-techniques to determine pools and rates of processes with fine spatial and temporal resolutions (e.g. microelectrodes for changes in oxygen, pH, etc.; chemical analyses of inorganic nutrients, isotopic tracers; enzymatic analyses).
   b. Exploration of the feasibility of examining ecological processes at the community versus the species levels.
3. Periphyton dynamics should be measured using time scales which are appropriate to the rates of change of the structure (species composition and biomass) and the rates of those processes being investigated.
4. Investigation of periphyton dynamics in specific environments is often confounded by the interaction of many regulating parameters. The regulatory roles of specific parameters can be determined by analyzing:
   a. Alteration of a specific parameter at a single site in the natural environment.
   b. Select sites for periphyton analyses which differ only with respect to a single parameter.
   c. Manipulate specific parameters under semi-natural or controlled conditions.
   d. Integrated mathematical models of biochemical processes, as well as models of population dynamics, could be used to preliminarily evaluate periphyton dynamics. Hypotheses concerning the complex interaction between the regulating parameters can be formally stated and synthesized by this means, and provide focus, direction, and a common background for the research advocated in parts a, b, and c above.
5. Structure, quantity (biomass), and processes should be measured in the community and in its developmental stages (rapidly growing, stationary, and secondary-declining phases). Specific investigations can address:
   a. Growth form (size, longevity, mobility) of species.
   b. Calculations of the elements of the growth equation: $B = (G + C) - (Gr + D + Dec)$ for single species of groups of organisms (bacteria, algae, animals). [B = biomass; G = growth; C = colonization; Gr = grazing; D = detachment; Dec = decay].
   c. Interstitial physicochemical parameters (e.g. light, temperature, oxygen, pH, inorganic nutrients, etc.)
   d. Processes of photoautotrophs (photosynthesis, nutrient kinetics-internal pools) and of heterotrophs (decomposition, oxidation-reduction of inorganic substrates).
   e. The rate of internal processes in part d in proportion to the rate of exchange with the external water phase.
   f. These investigations can be beneficially used in a comparison of the same type of periphytic community under differing environmental conditions (e.g. climate), trophic status, and in comparisons among periphytic communities attached to different substrata.
6. Periphyton is proposed as model communities ideally suited for experimental treatment of several problems in theoretical ecology (e.g. diversity-stability, succession, competition for space in sessile communities, grazing by 'keystone' species). The communities exhibit rapid dynamics, are amenable to experimental manipulations, and allow for adequate replication.

Periphyton–Substrata Interrelationships (R.G. Wetzel, Chairperson)

Structural aspects of the community

1. An understanding of the functional couplings between the periphyton community and the substrata to which they are attached requires a detailed understanding of the composition and

dynamics of periphyton components at micro-levels within the communities.

2. Despite apparent heterogeneity among periphyton community structure, generalized patterns of colonization and succession spatially and temporally are probable and should be sought. Spatial and temporal sequences in composition and architecture should be analyzed in relation to:
   a. Bacterial and fungal components; particularly in the early stages of colonization;
   b. Unicellular/colonial algae;
   c. Filamentous algae and bacteria (including their epiphytes);
   d. Inorganic detritus, particularly $CaCO_3$;
   e. Organic detritus from living and senesced cells, the substratum and imported to the community from outside sources.
3. In both structural and functional analyses, it is important to be able to distinguish between living and dead cells. Continued development of analytical techniques is urgently needed to evaluate differences in live and dead species distributions both in relation to successional development of the community as well as in directly coupled associations on or within the supporting substrata. Several of the following methods were discussed and encouraged for continued development.
   a. Algae, bacteria, and fungi: Analyses of species of algae, pigment fractionation and quantitative chromatographic separation of pigment degradation products; and microscopy at many levels: light with selective staining and fluorescence analysis; scanning electron microscopy; transmission electron microscopy particularly in relation to micro-organismal intrusions into the substratum; specific organic compounds, including selective enzyme activities.
   b. Community: Elemental analyses (Si, $CaCO_3$, etc.) of inorganic components and of organic fractions (total organic matter, C, N, proteins, carbohydrates, lipids, specific end product organic compounds).
   c. Animals: Functional analyses (e.g. on the basis of feeding types) are more illuminating than only species-biomass analyses of many periphytic animals (e.g. protozoans, rotifers, gastropods, oligochaetes, nematodes, chironomids, etc.). Food selectivity on the basis of size and food quality are of particular need of quantitative evaluation.

Time of sampling

Time of sampling of periphytic populations and communities and their functional analysis are in need of careful study. In past investigations, sampling intervals have not been in accord with rapid changes in reproductive and metabolic rates. The following time scales were recommended:

1. Algae: For distribution and motility studies, 0.5 day or less; 1 to 2-day sampling should be adequate for population changes.
2. Bacteria: one day or less.
3. Detritus: 3 to 5-day intervals.
4. Animals: Sessile populations: 3 to 5 days; grazers: variable with questions being addressed.

Functional aspects of the community

1. It was concluded that the complex and tightly coupled metabolism among the living, senescing, and dead components of periphyton and the relationships to their substratum must be analyzed in much more rigorous ways than has previously been the usual case. Conclusions drawn from large-scale community analyses of the surrounding medium could be misleading in understanding the operation of the periphyton species within the community and their couplings with nutrient sources from and metabolism within the substrata.
2. In order to effectively evaluate the interrelationships among the microorganisms and to the metabolism within the substrata (both macrophyte tissue and within sediments), several experimental and *in situ* approaches were suggested, which singly and in combination would assist in understanding the processes involved. In addition to assays of changes in population and community structure, direct assays of growth rates and nutrient kinetics are important.

a. It will likely be necessary to compartmentalize the community in order to minimize heterogeneity and to concentrate analyses on key mechanisms.
b. Some problems can be effectively evaluated by use of simplified experimental systems, e.g. analyses of epipelon in the laboratory in which manipulations of external organic matter loading would alter redox conditions and nutrient release rates at the sediment-water interface; use of rapidly growing macrophytes on which epiphytic communities would be kept simple by rapid colonization of the developing tissue.
c. Use of artificial substrata in metabolic analyses can be only applied in a relative way (e.g. responses to a toxicant); their use in simulation of natural substrata does not permit accurated comparison to living systems and is not recommended.
d. Most substrata are living, such as viable macrophyte tissue or a steeply stratified benthic community of microorganisms that are metabolically coupled to the inorganic and organic constituents of the substrata. The couplings and potential synchrony of the substratum metabolism must be understood in sufficient detail to permit quantitative determinations of flux rates to the periphyton and from the periphyton to the substratum.
e. A combination of parameters were recommended to analyze both community and species specific metabolism and cycling of nutrients and organic compounds. Present and predicted technology, such as microelectrodes, microautoradiography, enzymatic activities, etc. permit direct analyses of metabolism and nutrient fluxes within the community and in some cases at the species level. It was argued that analyses must be made at this level and that only then can accurate statements be made of uptake from and release rates to the overlying water.

3. The metabolic cycling of several important elements were considered of major importance for analysis under a variety of differing natural (oligotrophic, eutrophic, acidified, etc.) and manipulated conditions.
a. Inorganic-organic carbon cycling.
   i. The sources (inputs) of CO_2–HCO_3 include the water (overlying, interstitial), inorganic (e.g. $CaCO_3$), and organic (respiratory, photorespiratory, bacterial-fungal mineralization of organic substrates released extracellularly from living and senesing cells or plant tissue) or adsorption of DOM to inorganic particulates (e.g. $CaCO_3$).
   ii. The sites and rates of utilization (photosynthetic fixation by sessile algae and macrophytic substrata) and losses (to water; to inorganic compounds such as nucleating $CaCO_3$) of inorganic and organic carbon.
   iii. Parameters influencing these metabolically-mediated rates of carbon production and losses need to be evaluated, including:
      (a) Competition by differential species uptake capabilities (K_m values).
      (b) Physical variables, in particular light intensity and quality, temperature, and turbulence.
      (c) Other biotic variables, e.g. release of organic compounds by the substrata, which may be stimulatory or inhibitory to the sessile microflora.
b. Phosphorus cycling.
   i. The separate rates of algal, bacterial, and substratum utilization and release of phosphorus in inorganic and organic forms are confused and inadequately understood, particularly in relation to interactions of photosynthesis, senescence, and adsorption (e.g. to nucleating $CaCO_3$).
   ii. Marked differences in phosphorus cycling are anticipated between epiphytic and epipelic communities. It is anticipated that phosphorus is tightly conserved and that little is released to the water under most conditions.
   iii. Modern techniques, such as partitioned labelling with ^{32}P and ^{33}P, hold promise to

elucidate these pathways and rapid cycling rates.

c. Nitrogen cycling.
   i. The quantitative pathways of the many forms of inorganic and organic nitrogen within periphytic communities and across the substratum interface are very pooly understood. Some evidence indicates that periphyton can be exceedingly important in the regulation of inorganic nitrogen flux rates into and from sediment and macrophyte substrata, in nitrogen fixation of ecosystems, dissolved inorganic and organic nitrogen fluxes from groundwater, and other processes.
   ii. Modern methodology (e.g. enzyme activities, isotopic tracers, metabolic end product analyses, etc.) permits detailed analyses of the functions of periphyton in nitrogen cycling and should be applied intensively.

d. The quantitative exchange fluxes of silica between periphyton and substrata are similarly very poorly understood and in need of detailed evaluation. Substrata sources of silica from both inorganic particulates and certain macrophytes are likely of major significance to succession and development of periphyton communities. The application of heavy Si and analogues of Si are particularly encouraged in future analyses.

e. The importance of organic micronutrients to periphytic communities is probably of major interactive significance, in particular to algae on a selective competitive basis. Some evidence indicates that periphyton may synthesize excessive quantities of organic micronutrients, e.g. B vitamins, and release significant quantities to the littoral and pelagial regions of lakes. It is probable that organic substrates released extracellulary from algae, macrophytes, and other sources are utilized by bacteria which synthesize essential organic micronutrients for algae in a passive reciprocal or symbiotic exchange. Availability and use of these products by periphytic invertebrates are unknown but are likely of functional significance.

Methodology
(G.G.C. Robinson, Chairperson)

Whole periphyton community analyses

1. The use of several artificial substrata was described. These were: (a) Mini glass substrata (both cover-circles and small glass tubes). Their use permitted reduction in variance of chlorophyll determinations and in so doing increased the sensitivity of a chlorophyll bioassay of stream conditions. (b) Sheets of polystyrene in light and dark troughs positioned in riverine situations for the estimation of, and distinction between, algal settlement and periphyton growth. (c) Smooth plates of granite colonized *in situ* by epilithic communities and used in a laboratory incubator for the development of a relationship between irradiance and primary productivity. (d) Small granite pieces, which conveniently permitted SEM examination of community development. (e) Artificial roughened plastic plants which served to mimic natural conditions of shading and movement of natural macrophyte substrata, and which could withstand rigorous treatment. These substrata are recommended for the purposes for which they were used, and it was clear that the choice of artificial substratum should depend upon the purpose for which it will be used.

2. From discussion of chlorophyll technology, the following recommendations are made: (a) The choice of solvent is important. It is recommended that this choice depend upon the community that is to be sampled, and that this might best be accomplished by comparison of acetone, methanol and ethanol. (b) Extraction aids are recommended. Grinding and homogenization are certainly effective and ultrasonification may assist. Although fast freezing in liquid nitrogen was seen to improve ATP extraction, it did not markedly influence chlorophyll extraction. (c) It is recommended that total extraction should not be assumed under any circumstance. Microscopic examination of material following extraction is recommended, as is verification that residual chlorophyll on substratum has not been neglected. (d) It is recommended that one not assume that traditional trichromatic formulae, which are based upon assumptions of rather spe-

cific community structures (and some of these marine), will be appropriate. (e) Assessment of degradation products is important, but it is recommended that one not assume phaeophytin to be the only important degradation product. Detection of others may require chromatographic separation.

3. Methodology for determination of periphytic bacterial biomass was considered, and particularly as an indicator of stream conditions. Kjeldahl nitrogen and cell counts (both direct and from plate counts) were used. It is recommended that in the use of bacterial biomass as an assay of stream conditions, biomass per unit area of substratum as well as per unit volume of overlying water be taken into account.

4. Where algal periphyton is to be used for assessment of riverine conditions, it was demonstrated that failure to distinguish between passive algal settlement and active growth may create a 2–3 fold error. Consequently, it is recommended that this distinction be incorporated. Flow through light and dark troughs may provide a means by which this can be accomplished.

5. In community metabolism studies of low carbonate waters where metabolically induced changes in dissolved inorganic carbon are significant, periphyton net and gross productivity rates may be assessed by measurement of changes in DIC within enclosed incubation systems. The effectiveness of this approach has been demonstrated and a predictive relationship between irradiance and gross production developed. For assessment of decomposer metabolism, it is recommended that dehydrogenase activity may be a more suitable indicator of respiratory activity than traditional O_2 methodology. There is the potential here for distinction between anaerobic and aerobic processes in periphyton.

6. The necessity for careful monitoring of environmental parameters of periphytic communities was stressed. This necessity became apparent in a clear demonstration of the influence of stream velocity upon periphyton community structure.

7. In response to the recent application of Gleasonian terrestrial ecological methods and the use of indices of dissimilarity, for purposes of community comparison, it is recommended that the choice of the most appropriate index to use should be made carefully, for the abilities of available indices to assess community information in the form of both proportionality of components and density of components differ.

8. The following additional recommendations relate to community studies. (a) In recognition of the differences between conditions governing lotic and lentic periphyton, the development of accurate micro-scale systems for measuring water velocity is recommended. (b) It is further recommended that effort be expended in the use and further development of miniaturized instrumentation for the monitoring of the micro-environments of periphyton as exhaustively as possible. (c) It is recommended that the devlopment of effective littoral enclosures or 'corrals' for purposes of manipulative experimentation be pursued. (d) The need for technology for assessment of lake macro-communities, and perhaps the more extensive exploration of remote sensing methodology, was identified.

Within periphyton community analyses

1. The details of a nuclear track microautoradiographic (NTA) method which permits the location, quantification and estimation of assimilation rates of certain nutrient elements were explored. The use of such a method to estimate inorganic and organic carbon assimilation by individual periphytic populations was demonstrated. It is suggested that application of NTA and grain-density autoradiographic techniques can provide needed information on the kinetics of supply and assimilation of nutrients within periphytic communities. It is recommended that such methods be applied to sectioned, but otherwise spatially intact, periphyton communities. Adaptation of procedures for transmission electron microscopy might further permit the tracking of nutrients between substratum and periphyton, between periphytic organisms and within periphyton cells. Electron image analysis, which may provide automated video radioautographic grain counting, would greatly enhance such methodology.

2. It is suggested that needed information on functional relationships among components of pe-

riphytic communities may be obtainable from experimental exclusion of components. For example, the competitive heterotrophic role of algal periphyton might be examined from experiments in the presence and absence of bacteria. The alternative approach of building synthetic communities of increasing complexity may be a profitable avenue for future exploration.

3. It is suggested that much is to be gained in the understanding of periphyton populations from extensive and intensive autecological studies of periphytic organisms under batch and steady state culture conditions.

4. In view of some progress being made with chemostats and turbidostats on phytoplankton populations, it may well be profitable to investigate the construction of equivalent systems for periphyton investigations. Although technical difficulties may be large, the concept of a 'periphytotron' was suggested, in which the nutrient and metabolic fluxes can be manipulated and evaluated experimentally above and below a simulated substratum interface of known permeability.

Periphyton and Water Quality
(A. Sládečková & J. Barica, Chairpersons)

Status

1. The taxonomic knowledge of periphytic organisms is inadequate, particularly among juvenile forms. Good identification keys with instructive illustrations are needed.

2. Periphyton research has lagged behind that in other areas of limnology. The roles of periphyton in aquatic ecosystems, particularly in running waters, has not been sufficiently defined. The advantages of analyzing ecosystems biologically are still underestimated. No trophic state guidelines exist for periphyton in relation to nutrient dynamics, biomass, and productivity.

3. Periphyton analyses from both natural and artificial substrata are useful in the assessment of water quality. Application of indicator organisms in water quality evaluation, however, is still rather subjective and lacks generality. Cause and effect relationships are not adequately defined, particularly among bacteria, aquatic mosses, and invertebrates.

4. A communication gap exists between researchers in periphyton an in water management, hygiene, and environmental projects. Students are often insufficiently trained in both theoretical and applied limnological research.

5. The ability to use periphyton research in numerous applied areas of water quality management is underestimated.

6. Serious discrepancies in periphyton data presentation and interpretation result from differences in methodology and can result in contradictory conclusions. Applied periphyton research tends to be too individualistic and insufficient in direction and integration.

Future applied needs and directions

1. A need exists for a periphyton manual, i.e. an illustrated methodological guide for routine applied research users. It is recommended that a working group be established under the auspices of the Societas Internationalis Limnologiae to undertake this task at the 22nd Congress of SIL in France. The manual should be completed within two years. Dr. G. Cronberg was recommended as chairperson of the periphyton manual working group.

2. We encourage increased research, particularly emphasizing the ecosystem concept as well as experimental and modelling approaches that allow predictions of periphyton communities and related water quality conditions. Since certain physical and chemical parameters vary in predictable ways as drainage basin area increases, ecological studies should be interpreted within this natural framework. Increased research is also needed on the effects of losses caused by current-related turbulence and grazing, as well as plant-plant and plant-animal interactions. Research is also required to evaluate the ontogeny of periphytic organisms and specific environmental conditions within which they exist.

3. In the field of education, we suggest that students receive, in addition to their specialized disci-

plines, a sufficiently broad base in related disciplines (e.g. hydrology, hydraulics, statistics, modelling, water quality management, etc.), as well as training in cooperative team research. We further suggest that consideration be given to continuing, up-grading education in limnological sciences, particularly for professionals in management and decision-making positions.

4. Experimental studies should be encouraged to evaluate the effectiveness of periphyton in treatment of drinking and wastewater supplies, particularly for removal of inorganic and organic contaminants, taste and odor causing substances, bacteria and viruses, etc.

5. A unification of methods should be encouraged, and intercalibration assays should be conducted among leading laboratories. These activities and their results should be incorporated into the preparation of the applied periphyton manual.

6. We encourage a more coherent, coordinated, and integrated approach to both basic and applied periphyton research and stress the urgent need for support of these projects by funding agencies.